Low-Dimensional Applications of Quantum Field Theory

NATO ASI Series

Advanced Science Institutes Series

A series presenting the results of activities sponsored by the NATO Science Committee, which aims at the dissemination of advanced scientific and technological knowledge, with a view to strengthening links between scientific communities.

The series is published by an international board of publishers in conjunction with the NATO Scientific Affairs Division

A	**Life Sciences**	Plenum Publishing Corporation
B	**Physics**	New York and London
C	**Mathematical and Physical Sciences**	Kluwer Academic Publishers
D	**Behavioral and Social Sciences**	Dordrecht, Boston, and London
E	**Applied Sciences**	
F	**Computer and Systems Sciences**	Springer-Verlag
G	**Ecological Sciences**	Berlin, Heidelberg, New York, London,
H	**Cell Biology**	Paris, Tokyo, Hong Kong, and Barcelona
I	**Global Environmental Change**	

PARTNERSHIP SUB-SERIES

1. Disarmament Technologies	Kluwer Academic Publishers
2. Environment	Springer-Verlag
3. High Technology	Kluwer Academic Publishers
4. Science and Technology Policy	Kluwer Academic Publishers
5. Computer Networking	Kluwer Academic Publishers

The Partnership Sub-Series incorporates activities undertaken in collaboration with NATO's Cooperation Partners, the countries of the CIS and Central and Eastern Europe, in Priority Areas of concern to those countries.

Recent Volumes in this Series:

Volume 359—Correlations and Clustering Phenomena in Subatomic Physics
edited by M. N. Harakeh, J. H. Koch, and O. Scholten

Volume 360—Surface Diffusion: Atomistic and Collective Processes
edited by H. C. Tringides

Volume 361—Functional Integration: Basics and Applications
edited by Cecile DeWitt-Morette, Pierre Cartier, and Antoine Folacci

Volume 362—Low-Dimensional Applications of Quantum Field Theory
edited by Laurent Baulieu, Vladimir Kazakov, Marco Picco, and Paul Windey

Series B: Physics

Low-Dimensional Applications of Quantum Field Theory

Edited by

Laurent Baulieu

CNRS and Université Pierre et Marie Curie (Paris VI)
Paris, France

Vladimir Kazakov

Université Pierre et Marie Curie (Paris VI) and
École Normale Superieure
Paris, France

Marco Picco

CNRS and Université Pierre et Marie Curie (Paris VI)
Paris, France

and

Paul Windey

Université Pierre et Marie Curie (Paris VI)
Paris, France

Plenum Press
New York and London
Published in cooperation with NATO Scientific Affairs Division

Proceedings of a NATO Advanced Study Institute on
Low-Dimensional Applications of Quantum Field Theory,
held July 11 – 29, 1995,
in Cargèse, France

Library of Congress Cataloging-in-Publication Data

Low-dimensional applications of Quantum field theory / edited by
Laurent Baulieu ... [et al.].
p. cm. -- (NATO ASI series. B, Physics ; v. 362)
"Published in cooperation with NATO Scientific Affairs Division."
"Proceedings of a NATO Advanced Study Institute on Low-Dimensional
Applications of Quantum Field Theory, held July 11-29,1995, in
Cargèse, France"--T.p. verso.
Includes bibliographical references and index.
ISBN 0-306-45686-9
1. Quantum field theory--Congresses. 2. Integrable equations-
-Congresses. 3. Mathematical physics--Congresses.
4. Itzykson,Claude. I. Baulieu, Laurent. II. North Atlantic
Treaty Organization. Scientific Affairs Division. III. NATO
Advanced Study Institute on Low-Dimensional Applications of Quantum
Field Theory (1995 : Cargèse, France) IV. Series.
QC174.45.A1L65 1997
530.14'3--DC21 97-17158
CIP

ISBN 0-306-45686-9

A Division of Plenum Publishing Corporation
233 Spring Street, New York, N. Y. 10013

http://www.plenum.com

10 9 8 7 6 5 4 3 2 1

Printed in the United States of America

PREFACE

The Cargèse Summer School "Low Dimensional Applications of Quantum Field Theory" was held in July 1995.

The School was dedicated to the memory of Claude Itzykson.

This session focused on the recent progress in quantum field theory in two dimensions with a particular emphasis on integrable models and applications of quantum field theory to condensed matter physics. A large fraction of the school was also devoted to a detailed review of the exciting developments in four dimensional supersymmetric Yang-Mills theory.

The diversity of the topics presented constitute, in our opinion, one of the most attractive features of these proceedings. Some contributions constitute a very thorough introduction to their subject matter and should be helpful to advanced students in the field while others present entirely new research, not previously published, and should be of considerable interest to the specialist.

There were in depth introductory lectures on the application of conformal field theory techniques to disordered systems, on the quantum Hall effect, on quantum integrable systems, on the thermodynamic Bethe Ansatz and on the new developments in supersymmetric gauges theories. The computation of the three point function of the Liouville model using conformal bootstrap methods was presented in detail. Many topics involved interesting developments in matrix model techniques, two dimensional quantum gravity, Calogero-Sutherland model, the Ising model in a magnetic field, random manifolds as well new methods to tackle some of the fundamental problems in turbulence theory. Most of the contributions are included in this volume.

We would like to thank all the people who contributed to the successful organization of the conference and in particular the staff of the Cargèse Institute. We are greatly indebted to the NATO Division for Scientific Affairs for funding and for their constant attention to our meeting, to the *Bureau de la formation permanente du Centre national de la recherche scientifique* for an important grant and to the *Ministère des affaires étrangr̀es.*

We are grateful to Ms. J. Lawrence, from Plenum, for her gentle reminders and her infinite patience as well as for her help with the publication.

Finally, we would like to thank all the participants for creating an excellent working atmosphere and especially the contributors of this volume who took the time to write up their contributions.

This school was the first Cargèse session which was held after Marie France Hanseler left her position as a General Secretary of the Cargèse Institute. Marie France was one of the pioneers who founded this beautiful institution. She became known and appreciated among an ever growing circle of scientists from all around the world for her dedication to the Institute and for her professional skill. She helped organize every Cargèse school and workshop for the last 25 years. She had a special talent to help the organizers of each session offer to all participants the best work and cultural atmosphere, while relieving them from the burden of administrative duties. She became a respected friend of many of us. We thank her warmly for all her achievements.

Laurent Baulieu
Vladimir Kazakov
Marco Picco
Paul Windey

In Memoriam CLAUDE ITZYKSON 1938-1995

Claude Itzykson died in May 1995 in Paris. The theoretical physics community has been deeply saddened by this loss. He was one of its most distinguished members and with his keen taste for mathematics he played a unique rôle in fertilizing many areas of physics with new and elegant approaches.

When Claude died we were putting the last touches on the preparation of the School. The simple homage of dedicating the School to his memory appeared to the four of us, independently, as the most natural thing in the world. It was probably the first sign of incredulity at the idea of not seeing him in Cargèse anymore. Since then, many conferences and workshops have been held in his memory in France and abroad and many tributes to his contributions to theoretical physics have appeared in the international scientific press. They all demonstrate the depth of Claude's scientific influence and they also show that his personal warmth had made him a large circle of friends among physicists. Writing these lines more than a year after his death, we would just like to evoke the dear memory of Claude in Cargèse.

Claude came to Cargèse from the very beginning of the *Institut de Physique* which at the time was entirely devoted to particle and theoretical physics. He came back many times for more than twenty years either as a lecturer or as an organizer of some of its finest sessions. He greatly enjoyed the style of Cargèse with its mixture of very hard and intense work and its moments of lazy Mediterranean life. It was the perfect setting to do what he seemed to love the most: learn, teach, and invent new physics.

For the people who never had chance of being in Cargèse when Claude was there it is probably hard to imagine how deeply he influenced its scientific atmosphere by his relentless curiosity and his complete availability. Given his devouring passion for science, he rarely missed a lecture, always sitting in the back of the hall, always asking questions. He also made a point of attending every single talk given by young students. It was partly his gentle way of encouraging them but mostly a sign of his joy at listening to new ideas.

When the lectures were not in session he was never very far from the main building, pacing in front of the library with a colleague or a student, learning and sharing his formidable knowledge. He was always trying to go a step further but he insisted on doing science on firm ground. Often, when a topic was all the rage of the day, he would, pretending to be impressed by the new arcane developments of the field, ask a few very simple questions over and over again as a reminder to whomever was wise enough to listen that there was foundation work yet to be done.

Claude was a wonderful person. He was above all a man and a scientist of great elegance.

We will miss him deeply and in Cargèse, for many years to come, we will look for him smoking his cigarette during the breaks, just outside the lecture hall.

Laurent Baulieu
Vladimir Kazakov
Marco Picco
Paul Windey

CONTENTS

CLASSICAL GEOMETRY AND TARGET SPACE DUALITY

ORLANDO ALVAREZ

DEPARTMENT OF PHYSICS
UNIVERSITY OF MIAMI
P.O. BOX 248046
CORAL GABLES, FL 33124 USA

1 INTRODUCTION

Target space duality is a remarkable phenomenon where different 2-dimensional non-linear sigma models are physically equivalent. A key reason for interest in this subject is that duality often turns a strong coupling problem into an equivalent weak coupling one thus transforming an intractable problem into a manageable one. There are two questions which immediately come to mind. The difficult one is: given a sigma model does there exist a dual model? A more accessible one is: when are two sigma models dual to each other? In these lectures we will attempt to address the latter question within the framework of classical hamiltonian mechanics.

Basically, two sigma models are target space duals of each other if there exists a canonical transformation between the phase spaces which preserves the respective hamiltonians [1, 2, 3, 4]. The existence of such a map is a difficult question to determine because the phase spaces are infinite dimensional. To make progress, we look for guidance in explicit examples. The case of toroidal target spaces suggests a promising approach. The duality transformation between the infinite dimensional phase spaces, in the case of toroidal target spaces, may be viewed as being induced by a special map between some finite dimensional bundles over the target spaces. In these lectures we address whether a similar phenomenon can arise between more general targets. We will see that it is possible to look for a special type of "restricted" target space duality which leads to an intriguing problem in classical differential geometry. There is a known duality transformation which arises when one of the target spaces is a simple Lie group. The explicit form of the generating function suggests a generalization to more general manifolds. This plays a pivotal role in our formulation.

Low Dimensional Applications of Quantum Field Theory
Edited by L. Baulieu *et al.*, Plenum Press, New York, 1997

Table 1.1. The key analogy

Euclidean geometry	*is to*	Riemannian geometry
	as	
toroidal target space duality	*is to*	"restricted" target space duality

There is a lot known about target space duality when the target is a torus. The excellent review article of Giveon, Porrati and Rabinovici [5] discusses the physics arising from toroidal targets in great detail. Additionally, there are roughly 300 references to the literature in this review which allow the reader to explore the historical development of the subject. We will use [5] as our unique reference on toroidal target spaces. The more recent phenomenon of "non-abelian" duality goes back the work presented in [6] but can actually be traced back to a much older paper [7]. The notion that duality can be formulated as canonical transformation goes back to [1] even though in the 1970's much work was done in statistical mechanics on the study of abelian duality in lattice systems from the partition function viewpoint. The explicit construction of a canonical transformation including the generating function for the target space $SU(2)$ is due to [2].

The approach taken in these lectures is different from the traditional approaches to duality presented in the literature (see for example the discussions in [6, 8, 2, 9, 10, 11]). In all these approaches one has explicitly symmetries in the target space which play a central role in the discussion of duality. The duality transformation in these theories with symmetry is some type of generalized Fourier transform. I was looking for a formulation which did not depend on the existence of symmetries. I wanted something which might be applicable to mirror symmetry [12]. The key analogy to keep in mind while reading these lectures is presented in Table 1.1. Mostly I will present ideas and concepts rather than detailed mathematical formulas. The derivation of explicit formulas requires a discussion of the theory of G-structures, a discussion of the differential forms version of the Froebenius theorem, and a presentation of Cartan's equivalence method [13]. These topics are outside the scope of these lectures and will be presented in Part I of [14]. The ideas presented here can be generalized by weakening the requirement that the canonical transformation be induced by a finite dimensional map. This requires the full machinery of Cartan-Kahler theory [15] and will be presented in Part II of [14]. These ideas can be extended to complex manifolds as discussed in [14].

2 Preliminaries

The classical nonlinear sigma model is defined by a map x from a Lorentzian world sheet Σ to a target manifold M and some additional geometric data which specifies the lagrangian. In this article we will take the world sheet Σ to be either $\mathbf{R} \times \mathbf{R}$ or $\mathbf{R} \times S^1$. The first factor is time and the second factor is space. Local coordinates on Σ will be denoted by (τ, σ). The target space M is endowed with a metric tensor $ds^2 = g_{\mu\nu} dx^\mu \otimes dx^\nu$ and a 2-form $B = \frac{1}{2} B_{\mu\nu} dx^\mu \wedge dx^\nu$. The lagrangian density for this model is

$$\mathcal{L} = \frac{1}{2}\eta^{ab}g_{\mu\nu}(x)\partial_a x^\mu \partial_b x^\nu - \frac{1}{2}\epsilon^{ab}B_{\mu\nu}(x)\partial_a x^\mu \partial_b x^\nu \,, \tag{2.1}$$

where η is the two dimensional Lorentzian metric on the world sheet. A good way of denoting the sigma model is to use the notation (M, ds^2, B) which incorporates all the relevant geometrical data. The canonically conjugate momenta are given by

$$\pi_\mu = \frac{\partial \mathcal{L}}{\partial \dot{x}^\mu} = g_{\mu\nu}(x)\dot{x}^\nu + B_{\mu\nu}(x)\frac{\partial x^\nu}{\partial \sigma} \,. \tag{2.2}$$

Local coordinates in phase space may be taken to be $(x(\sigma), \pi(\sigma))$. If the spatial part of Σ is a circle then $(x(\sigma), \pi(\sigma))$ is a loop in T^*M, the cotangent bundle of M. If the spatial part of Σ is $\mathbf{R}$ then $(x(\sigma), \pi(\sigma))$ is a path in T^*M. The symplectic form on the phase space is given by

$$\int \delta\pi_\mu(\sigma) \wedge \delta x^\mu(\sigma)\, d\sigma \,, \tag{2.3}$$

where δ is the differential on the phase space. With this symplectic structure we see that the basic Poisson bracket is given by

$$\{x^\mu(\sigma), \pi_\nu(\sigma')\}_{\mathrm{PB}} = \delta^\mu{}_\nu\, \delta(\sigma - \sigma') \,. \tag{2.4}$$

The hamiltonian density and the worldsheet momentum density are respectively given by

$$\mathcal{H} = \frac{1}{2}g^{\mu\nu}(x)\left(\pi_\mu - B_{\mu\kappa}\frac{dx^\kappa}{d\sigma}\right)\left(\pi_\nu - B_{\nu\lambda}\frac{dx^\lambda}{d\sigma}\right) + \frac{1}{2}g_{\mu\nu}(x)\frac{dx^\mu}{d\sigma}\frac{dx^\nu}{d\sigma} \,, \tag{2.5}$$

$$\mathcal{P} = \pi_\mu \frac{dx^\mu}{d\sigma} \,. \tag{2.6}$$

Target space duality is the phenomenon that a nonlinear sigma model (M, ds^2, B) is *equivalent* to $(\widetilde{M}, \widetilde{ds}^2, \widetilde{B})$, a different nonlinear sigma model. We need to define *equivalent*. For the moment we will make a preliminary definition which will be modified later.

Definition 2.7 (Preliminary)

Two sigma models (M, ds^2, B) and $(\widetilde{M}, \widetilde{ds}^2, \widetilde{B})$ are said to be target space dual *to each other if there exists a canonical transformation from the phase space of (M, ds^2, B) to the phase space of $(\widetilde{M}, \widetilde{ds}^2, \widetilde{B})$ which maps the hamiltonian $\mathcal{H}$ of the first model to the hamiltonian $\widetilde{\mathcal{H}}$ of the second model.*

Later we will see that there are some important domain and range issues which must be addressed to have a good definition of target space duality.

3 Examples

3.1 Circular target space

As an example of the above we consider the case where the target space is a circle of radius R (for a more detailed discussion look in [5]). The coordinate x on the circle has period $2\pi R$. The hamiltonian density is given by

$$\mathcal{H} = \frac{1}{2}\pi^2 + \frac{1}{2}\left(\frac{dx}{d\sigma}\right)^2 . \tag{3.1}$$

We can define a *formal* canonical transformation by the ordinary differential equations

$$\frac{d\tilde{x}}{d\sigma} = \pi(\sigma) , \tag{3.2}$$

$$\tilde{\pi} = \frac{dx}{d\sigma} . \tag{3.3}$$

It is clear that the hamiltonian is preserved by this map. The new hamiltonian tells us that the new target space is either $\mathbf{R}$ or S^1.

The transformation above is formal because of certain domain and range issues. First we note that there is an S^1 action on our circle given by translating x. Noether's theorem leads to the conserved target space momentum $P = \int \pi(\sigma)d\sigma$ associated with translations in the target space. For the moment let us assume the world sheet is $\Sigma = \mathbf{R} \times S^1$. In this case, the difference $x(2\pi) - x(0)$ is quantized in units of $2\pi Rw$ where the winding number w is an integer. Consequently, phase space divides into sectors labeled by $[w, p_T]$ where w is the winding number and p_T is the total target space momentum. Since both these quantities are conserved, the subspace labeled by $[w, p_T]$ will be invariant under the hamiltonian flow. This observation has important consequences when we examine the duality transformation in more detail. Integrating equations (3.2) and (3.3) we see that:

$$\tilde{x}(2\pi) - \tilde{x}(0) = \int_{S^1} \pi(\sigma)d\sigma , \tag{3.4}$$

$$\int_{S^1} \tilde{\pi}(\sigma)d\sigma = x(2\pi) - x(0) . \tag{3.5}$$

The above indicates that the dual target manifold should be a circle of radius $\widetilde{R}$ and we should have relations $2\pi\widetilde{R}\tilde{w} = p_T$ and $2\pi Rw = \tilde{p}_T$. Since the winding number $\tilde{w}$ must be an integer we see that p_T must be "classically quantized" and likewise $\tilde{p}_T$. At the classical level we have the following: given two radii R and $\widetilde{R}$, the sigma model with radius R on the reduced phase space characterized by $[w, 2\pi\widetilde{R}\tilde{w}]$ is dual to the sigma model on a circle of radius $\widetilde{R}$ on the reduced phase space characterized by $[\tilde{w}, 2\pi Rw]$. Note that some type of "pre-quantization" has taken place in trying to define duality. There is no "good" map from the full phase space of the nonlinear sigma model on a circle of radius R to the nonlinear sigma model on a circle of radius $\widetilde{R}$. Under quantization we observe that the momentum P must be quantized in units of $1/R$, likewise, the momentum $\widetilde{P}$ must be quantized in units of $1/\widetilde{R}$. Incorporating this we see that there is now a relation $2\pi R\widetilde{R} = 1$ between R and $\widetilde{R}$. Note that if we take $\Sigma = \mathbf{R} \times \mathbf{R}$ then the winding number is not defined and the domain and range issues do not appear. The lesson learned in this example is that there are delicate domain and range issues which must be understood if one wants to be mathematically precise.

Preliminary definition (2.7) must be expanded to include domain and range information. A better definition of classical duality would be

Definition 3.6 *Two sigma models* (M, ds^2, B) *and* $(\widetilde{M}, \widetilde{ds}^2, \widetilde{B})$ *are said to be* target space dual *to each other if there exists a canonical transformation from a reduced phase space of* (M, ds^2, B) *to a reduced phase space of* $(\widetilde{M}, \widetilde{ds}^2, \widetilde{B})$ *which maps the hamiltonian* $\mathcal{H}$ *of the first model to the hamiltonian* $\widetilde{\mathcal{H}}$ *of the second model. The reduced phase spaces must be invariant under the respective hamiltonian flow.*

3.2 Toroidal target spaces

In this example we choose the target space to be an n-dimensional torus $\mathbf{T}^n$ (for a more detailed discussion look in [5]). The metric ds^2 and the 2-form B are chosen to be constant. The basic Poisson brackets for this model are given by equation (2.4). By differentiating they may be written as

$$\left\{ \frac{dx^\mu}{d\sigma}(\sigma), \pi_\nu(\sigma') \right\}_{\mathrm{PB}} = \delta^\mu{}_\nu \delta'(\sigma - \sigma') . \tag{3.7}$$

If we now put $dx/d\sigma$ and π into a $2n$-vector

$$z(\sigma) = \begin{pmatrix} dx/d\sigma \\ \pi \end{pmatrix} \tag{3.8}$$

then the Poisson brackets may be written as

$$\left\{ z^A(\sigma), z^B(\sigma') \right\}_{\mathrm{PB}} = Q^{AB} \delta'(\sigma - \sigma') , \tag{3.9}$$

where

$$Q = \begin{pmatrix} 0 & I_n \\ I_n & 0 \end{pmatrix} . \tag{3.10}$$

The indices A and B range over $\{1, \ldots, 2n\}$. We see that a constant linear transformation $T : z \mapsto Tz$ will preserve the symplectic structure if T is in the pseudo-orthogonal group $O_Q(2n)$ consisting of linear transformations which preserve the quadratic form Q. This group is isomorphic to $O(n, n)$. Since the hamiltonian is a quadratic form in the z's with constant coefficients, we see that a $T \in O_Q(2n)$ leads to a new sigma model hamiltonian with constant coefficients. A similar phenomenon will be studied in detail in a more general setting later. For the moment we make a few observations. Since the hamiltonian a positive definite quadratic form in $\{dx/d\sigma, \pi\}$, the linear transformations in $O_Q(2n)$ that preserve the quadratic form belong to a certain maximal compact subgroup K which is isomorphic to $O(n) \times O(n)$. The dimension of the coset space $O_Q(2n)/K$ is n^2 which is precisely the total number of independent components in the metric ds^2 and the 2-form B. In fact the coset space $O_Q(2n)/K$ parameterizes the space of nonlinear sigma model hamiltonians with constant coefficients. Actually one has to be careful in the quantum theory. One can show that transformations in the subgroup $O(n, n; \mathbf{Z})$ lead to equivalent hamiltonians.

In this report we are interested in local conditions which necessarily guarantee the existence of a restricted type of target space duality. Because we are only interested in local issues we will generally take $\Sigma = \mathbf{R} \times \mathbf{R}$.

3.3 Toroidal target spaces revisited

We revisit toroidal target spaces and adopt a different viewpoint which will generalize to generic targets. There is a very interesting structure which arises in the models we have been studying. Introduce a new space isomorphic to $\mathbf{R}^n$ with coordinates p. Define $p(\sigma)$ by

$$\frac{dp}{d\sigma} = \pi(\sigma) . \tag{3.11}$$

Note that there is an ambiguity in the definition of p due to the constant of integration. Instead of working in phase space $(x(\sigma), \pi(\sigma))$ we can work in a space with coordinates $(x(\sigma), p(\sigma))$. In terms of these new variables the symplectic form may be written as

$$\int \delta x^\mu(\sigma) \wedge \frac{d}{d\sigma}\delta p_\mu(\sigma)\, d\sigma = \frac{1}{2}\int d\sigma \begin{pmatrix} \delta x & \delta p \end{pmatrix} \wedge \frac{d}{d\sigma}\begin{pmatrix} 0 & I_n \\ I_n & 0 \end{pmatrix}\begin{pmatrix} \delta x \\ \delta p \end{pmatrix} . \tag{3.12}$$

This symplectic form is degenerate. We have also discarded a surface term. We will ignore these technical issues. Note that the matrix Q enters into this formulation of the symplectic form. Also, a constant coefficient linear transformation $T \in O_Q(2n)$ acting on (x,p) space will preserve this symplectic form. Using the variables (x,p) we can integrate equations (3.2), (3.3), and obtain

$$\tilde{x}(\sigma) = p(\sigma) + a , \tag{3.13}$$
$$\tilde{p}(\sigma) = x(\sigma) + b , \tag{3.14}$$

where a and b are constants. In passing we mention that if $\Sigma = \mathbf{R} \times S^1$ and if we naively quantize then there will be appropriate periodicity requirements on (x,p) and $(\tilde{x},\tilde{p})$. The map $(x,p) \mapsto (\tilde{x} = p + a, \tilde{p} = x + b)$ between the toroidal target spaces $\mathbf{T}^{2n}$ induces transformations (3.13) and (3.14) on the space of paths.

The question which will occupy our attention throughout this report is whether it is possible to formulate target space duality as an induced map between the infinite dimensional phase spaces arising from an ordinary map from a $2n$-manifold to another $2n$-manifold.

The variables p have an important physical significance. They are the variables which describe the "solitons" in toroidal models. It is best to take a semi-classical viewpoint. By the canonical commutation relations we have that

$$\pi_\mu(\sigma) = -i\frac{\delta}{\delta x^\mu(\sigma)} . \tag{3.15}$$

Therefore we have

$$\exp[i\alpha^\mu p_\mu(\sigma)] = \exp\left(\alpha^\mu \int_{-\infty}^{\sigma} d\sigma' \frac{\delta}{\delta x^\mu(\sigma')}\right) . \tag{3.16}$$

This equation tells us that $\exp[i\alpha \cdot p(\sigma)]$ is the operator which creates a "kink" with jump of size α. To see this let us assume the target space is a circle of radius R and consider a state with the property that $\langle x(\sigma)\rangle = x_{\text{cl}}(\sigma)$. The action of the operator

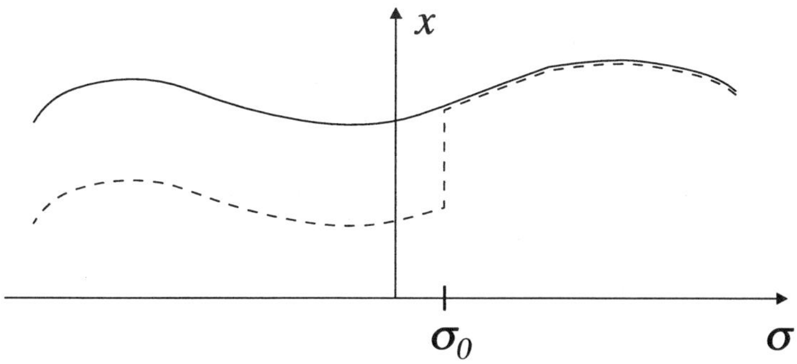

Figure 1. The action of $\exp[2\pi i R p(\sigma_0)]$ takes a state with expectation value $x_{\rm cl}(\sigma)$ (the solid curve) to a state with expectation value $x_{\rm cl,\,new}(\sigma)$ (the dashed curve).

$\exp[2\pi i R p(\sigma_0)]$ on the state leads to a new state with expectation value $x_{\rm cl,\,new}(\sigma)$ which contains a $2\pi R$ kink at σ_0 as in Figure 1.

In the toroidal models we see that the space with coordinates (x, p) is a space which describes both the "particles", *i.e.*, the x's, and the "solitons", *i.e.*, the p's. The torus $\mathbf{T}^{2n}$ is determined by a lattice in $\mathbf{R}^{2n}$. The group $O_Q(2n)$ naturally acts on this $\mathbf{R}^{2n}$. The lattice will be invariant under $O(n,n;\mathbf{Z}) \subset O_Q(2n)$. This action preserves the symplectic form (3.12).

4 Generating functions

The term *generating function* is used in a variety of different contexts in mechanics. Assume we have a hamiltonian system. The equations of motion may be derived as the extremals of the variational principle defined by the action $I = \int (pdq - H(q,p)dt)$. Assume we have a second system with canonical coordinates $(\tilde{q}, \tilde{p})$, time independent hamiltonian $\widetilde{H}(\tilde{q}, \tilde{p})$ and action $\tilde{I}$. The variational principle for I is equivalent to the variational principle for $\tilde{I}$ if $pdq - Hdt$ differs from $\tilde{p}d\tilde{q} - \widetilde{H}dt$ by a total differential:

$$\tilde{p}d\tilde{q} - \widetilde{H}dt = pdq - Hdt + dG \, . \tag{4.1}$$

If the function G is time independent then

$$\tilde{p} = \frac{\partial G}{\partial \tilde{q}} \, , \tag{4.2}$$

$$p = -\frac{\partial G}{\partial q} \, , \tag{4.3}$$

and $H = \widetilde{H}$. The function $G(q, \tilde{q})$ is called the *generating function.* A geometrical discussion from the viewpoint of symplectic geometry may be found in [16, 17].

For example, in the simple harmonic oscillator with $H = \frac{1}{2}(p^2 + q^2)$, the generating function $G(q, \tilde{q}) = q\tilde{q}$ leads to the "duality transformation" $\tilde{p} = q$ and $p = -\tilde{q}$.

In a field theoretic context we have that for a circular target space the duality transformation is generated by the functional

$$G[x,\tilde{x}] = \int \tilde{x}(\sigma)\,\frac{dx}{d\sigma}\,d\sigma\ . \tag{4.4}$$

A simple computation shows that

$$\begin{aligned}\tilde{\pi} &= \frac{\delta G}{\delta \tilde{x}} = \frac{dx}{d\sigma}\ ,\\ \pi &= -\frac{\delta G}{\delta x} = \frac{d\tilde{x}}{d\sigma}\ .\end{aligned}$$

Note that the integrand is the pullback of a 1-form on the product space $S^1 \times S^1$ of the two target circles, and that G is reparametrization invariant.

5 Generic target space

5.1 Background

We now address whether it is possible to construct some type of theory which addresses duality in a generic sigma model (M, ds^2, B). In our construction, the group $O_Q(2n)$ will appear but in a different manner. The reader is reminded of the analogy presented in Table 1.1. In these notes we will address a certain "restricted" type of duality which leads to a well-posed mathematical problem. The more general discussion requires the use of the full machinery of exterior differential systems and is beyond the scope of these lectures [14]. Remember that the circular duality transformation may viewed as an induced map on paths arising from an affine map from (x,p)-space to $(\tilde{x},\tilde{p})$-space. In this section we address whether it is possible to have a similar phenomenon arise when we have a generic target space. We will see that in certain situations duality can arise as an ordinary map from a certain $2n$-manifold to another $2n$-manifold.

The starting point of our discussion will be generating functions. Assume we have two sigma models (M, ds^2, B) and $(\widetilde{M}, \widetilde{ds}^2, \widetilde{B})$. We would like to know if there is duality transformation between these models. Let us postulate that the generating function has the form

$$G[x,\tilde{x}] = \int \left(u_\mu(x(\sigma),\tilde{x}(\sigma))\frac{dx^\mu}{d\sigma} + v_\mu(x(\sigma),\tilde{x}(\sigma))\frac{d\tilde{x}^\mu}{d\sigma}\right) d\sigma\ . \tag{5.1}$$

This generating function is a generalization to arbitrary manifolds of the one postulated for $SU(2)$ in [2]. Such a generating function has many desirable properties. It is reparametrization invariant. The integrand is the pullback of the 1-form $u_\mu(x,\tilde{x})dx^\mu + v_\mu(x,\tilde{x})d\tilde{x}^\mu$ on $M \times \widetilde{M}$. Since we are interested in theories where the dynamical variables are paths on a manifold we see that (5.1) is very natural. It is simply the integral of a 1-form along a curve in the product manifold $M \times \widetilde{M}$. A brief computation shows that

$$\begin{aligned}\pi_\mu(\sigma) &= a_{\mu\nu}(x(\sigma),\tilde{x}(\sigma))\frac{dx^\nu}{d\sigma} + b_{\mu\nu}(x(\sigma),\tilde{x}(\sigma))\frac{d\tilde{x}^\nu}{d\sigma}\ ,\\ \tilde{\pi}_\mu(\sigma) &= c_{\mu\nu}(x(\sigma),\tilde{x}(\sigma))\frac{dx^\nu}{d\sigma} + d_{\mu\nu}(x(\sigma),\tilde{x}(\sigma))\frac{d\tilde{x}^\nu}{d\sigma}\ .\end{aligned}$$

Solving for the domain and range variables we get

$$\frac{d\tilde{x}^\mu}{d\sigma} = A^\mu{}_\nu(x(\sigma),\tilde{x}(\sigma))\frac{dx^\nu}{d\sigma} + B^{\mu\nu}(x(\sigma),\tilde{x}(\sigma))\pi_\nu(\sigma) \,, \tag{5.2}$$

$$\tilde{\pi}_\mu(\sigma) = C_{\mu\nu}(x(\sigma),\tilde{x}(\sigma))\frac{dx^\nu}{d\sigma} + D_\mu{}^\nu(x(\sigma),\tilde{x}(\sigma))\pi_\nu(\sigma) \,. \tag{5.3}$$

This canonical transformation linear in $dx/d\sigma$ and π is very suggestive of target space duality. The reason is that the hamiltonian is quadratic in $dx/d\sigma$ and π and thus gets mapped into something which is quadratic in $d\tilde{x}/d\sigma$ and $\tilde{\pi}$. The difficulty is that the new metric and anti-symmetric tensor might not be functions of only $\tilde{x}$. *This is the obstruction to duality under Ansatz* (5.1). Part of our discussion will be to try to understand whether there are local obstructions to the existence of duality transformations. It is worthwhile stating this again explicitly.

> Do canonical transformations (5.2) and (5.3) lead to a new $\widetilde{\mathcal{H}}$ such that $\widetilde{ds}^2$ and $\widetilde{B}$ are only functions of $\tilde{x}$? In general $\widetilde{\mathcal{H}}$ will be a non-local function of the variables.

Finally, a more complicated Ansatz for (5.1) which might include terms such as $d^2x/d\sigma^2$ or $\sqrt{(dx/d\sigma)^2}$ will generally not lead to transformation equations which are linear in $dx/d\sigma$ and π as exemplified in (5.2) and (5.3).

We proceed to attack these issues. Note that equations (5.2) and (5.3), being ordinary differential equations (ODE), are always integrable. To explicitly see this, assume we are given $(x(\sigma),\pi(\sigma))$ then we insert them into equation (5.2) which we integrate[1] to obtain $\tilde{x}(\sigma)$. Subsequently we insert $x(\sigma),\pi(\sigma),\tilde{x}(\sigma)$ into (5.3) to get $\tilde{\pi}(\sigma)$. What we are going to do is to replace equations (5.2) and (5.3) by an equivalent set of equations (up to ambiguities involving constants of integration).

Introduce new variables p and $\tilde{p}$ and define $p(\sigma)$ and $\tilde{p}(\sigma)$ by

$$\frac{dp}{d\sigma} = \pi(\sigma) \,, \tag{5.4}$$

$$\frac{d\tilde{p}}{d\sigma} = \tilde{\pi}(\sigma) \,. \tag{5.5}$$

It can be shown [14] that the space with coordinates (x,p) is a manifold $T^\sharp M$ isomorphic[2] to T^*M. I do not understand the relationship between the use of the cotangent bundle here and in the interesting work presented in [18, 19]. Canonical transformation equations (5.2) and (5.3) may be rewritten as

$$\frac{d\tilde{x}^\mu}{d\sigma} = A^\mu{}_\nu(x(\sigma),\tilde{x}(\sigma))\frac{dx^\nu}{d\sigma} + B^{\mu\nu}(x(\sigma),\tilde{x}(\sigma))\frac{dp_\nu}{d\sigma} \,, \tag{5.6}$$

$$\frac{d\tilde{p}_\mu}{d\sigma} = C_{\mu\nu}(x(\sigma),\tilde{x}(\sigma))\frac{dx^\nu}{d\sigma} + D_\mu{}^\nu(x(\sigma),\tilde{x}(\sigma))\frac{dp_\nu}{d\sigma} \,. \tag{5.7}$$

---

[1] Of course there are arbitrary constants of integration.

[2] In the case of toroidal target spaces we saw that (x,p) space was a $2n$-torus rather than $T^*(\mathbf{T}^n)$ due to some global and quantum properties. All our discussions are local thus we will not see such phenomena.

These are integrable because they are ODE's. An advantage of the above over (5.2) and (5.3) is that the above treat x and p symmetrically.

Is is worthwhile to make a small mathematical digression. Given a manifold N, let $PN = \{\gamma : \mathbf{R} \to N\}$ be the space of all paths in N. Every map $f : N \to \tilde{N}$ induces a map $f_* : PN \to P\tilde{N}$. The opposite is not true. Not all maps from PN to $P\tilde{N}$ arise as maps from N to $\tilde{N}$. This is illustrated in equation (5.6) which is a map from $P(T^\sharp M)$ to $P(T^\sharp \widetilde{M})$. Here we explicitly see that the map depends not only on the point $(x, \tilde{x}) \in T^\sharp M \times T^\sharp \widetilde{M}$ but also on the "velocities" $dx/d\sigma$ and $dp/d\sigma$. In general equations (5.6) and (5.7) do not define a map from $T^\sharp M$ to $T^\sharp \widetilde{M}$. We are interested in looking for a phenomenon similar to that discussed after equation (3.12) where the map between the phase spaces is induced by a map between $T^\sharp M$ and $T^\sharp \widetilde{M}$.

In these lectures we address the question of when do equations (5.6) and (5.7) arise as maps from $T^\sharp M$ to $T^\sharp \widetilde{M}$ in such a way that we get a dual sigma model. We proceed naively by performing a formal mathematical manipulation on (5.6) and (5.7): let us multiply both sides of the equations by $d\sigma$ and obtain the *exterior differential system* (EDS) which will schematically be written as

$$d\tilde{x} = A dx + B dp \,, \tag{5.8}$$

$$d\tilde{p} = C dx + D dp \,. \tag{5.9}$$

This exterior differentials system is equivalent to the partial differential equations (PDE) $\partial\tilde{x}/\partial x = A$, $\partial\tilde{x}/\partial p = B$, $\partial\tilde{p}/\partial x = C$ and $\partial\tilde{p}/\partial p = D$. In general PDE's have no solutions. There are certain integrability conditions which must be satisfied in order for the system to be integrable. Roughly, we have to be able to integrate simultaneously along $2n$ independent directions. This is very different from our original ODE system which has no integrability conditions. System (5.6) and (5.7) is equivalent to finding one dimensional integrable manifolds of (5.8) and (5.9). What we are proposing is that we look for $2n$-dimensional integrable manifolds of (5.8) and (5.9). If such an integrable manifold exists then we have a map from $T^\sharp M$ to $T^\sharp \widetilde{M}$. The integrability conditions for EDS (5.8) and (5.9) will describe local geometric conditions which must be satisfied on $T^\sharp M$ and $T^\sharp \widetilde{M}$ in order for a solution to exist.

5.2 Details

We now turn to more detailed study of what was proposed in the last section. Assume we are given a sigma model (M, ds^2, B). Let Γ be the Levi-Civita connection associated with metric ds^2. We introduce new variables p and relate them to π via

$$\pi_\mu = \frac{dp_\mu}{d\sigma} - \frac{dx^\lambda}{d\sigma}\Gamma_\lambda{}^\nu{}_\mu p_\nu = \frac{\nabla}{d\sigma} p_\mu \,. \tag{5.10}$$

This definition is different than the one previously given by equation (3.11). The definition given above is much better because p transforms covariantly with respect to coordinate transformations on the base space M. Just as before one can show that the space with coordinates (x, p) is a manifold $T^\sharp M$ isomorphic to the cotangent bundle T^*M. For any covariant vector v_μ define the covariant variation $\delta_\Gamma v$ by

$$\delta_\Gamma v_\mu = \delta v_\mu - \delta x^\lambda \Gamma_\lambda{}^\nu{}_\mu v_\nu \,. \tag{5.11}$$

A brief computation shows that

$$\delta_\Gamma \pi_\mu = \frac{\nabla}{d\sigma}\delta_\Gamma p_\mu - R^\nu{}_{\mu\lambda\rho}\delta x^\lambda \frac{dx^\rho}{d\sigma} p_\nu \,. \tag{5.12}$$

The symplectic structure (2.3) may be rewritten as

$$\begin{aligned} &\int d\sigma\, \delta x^\mu \wedge \delta\pi_\mu \\ &= \frac{1}{2}\int d\sigma \begin{pmatrix} \delta x & \delta_\Gamma p \end{pmatrix} \wedge \begin{pmatrix} 0 & I_n \\ I_n & 0 \end{pmatrix} \begin{pmatrix} \nabla/d\sigma & 0 \\ b & \nabla/d\sigma \end{pmatrix} \begin{pmatrix} \delta x \\ \delta_\Gamma p \end{pmatrix}, \end{aligned} \tag{5.13}$$

where $b_{\mu\nu} = p_\lambda (dx^\rho/d\sigma) R^\lambda{}_{\rho\mu\nu}$. Note that our old friend Q appears in the above. What is surprising is that

$$\begin{pmatrix} \nabla/d\sigma & 0 \\ b & \nabla/d\sigma \end{pmatrix} \tag{5.14}$$

is the unique torsion free pseudo-Riemannian connection on $T^\sharp M$ associated with the Q-metric

$$ds_Q^2 = dx^\mu \otimes (dp_\mu - dx^\lambda \Gamma_\lambda{}^\nu{}_\mu p_\nu) + (dp_\mu - dx^\lambda \Gamma_\lambda{}^\nu{}_\mu p_\nu) \otimes dx^\mu \tag{5.15}$$

on $T^\sharp M$. A short computation shows that (5.14) is a skew-adjoint operator with respect to the Q-metric.

We see that some natural geometric structures related to the pseudo-Riemannian geometry of the Q-metric are beginning to appear.

5.3 A digression and an analogous problem

We expand on the analogy discussed in Table 1.1. Assume we are in Euclidean space $\mathbf{R}^n$. The Euclidean group $E(n)$ is the group of isometries of Euclidean space. Since Felix Klein we have understood that Euclidean geometry is the study of the properties of figures which are invariant under the action of the Euclidean group (the isometry group of Euclidean space). In a similar fashion one can define hyperbolic and elliptic geometries. Klein's ideas seem to fail when one considers Riemannian geometry: on a generic Riemannian manifold M there are no isometries. E. Cartan realized that this was not fatal and that the orthogonal group played a very important role in Riemannian geometry. Cartan reformulated Euclidean geometry using the observation that $\mathbf{R}^n = E(n)/O(n)$. In modern language, $E(n)$ is a principal $O(n)$-bundle over $\mathbf{R}^n$. Cartan studied the properties of $\mathbf{R}^n$ in terms of the properties of $E(n)$. In doing so he realized that the Maurer-Cartan equations for the group $E(n)$ contain all the information necessary to extract both the properties of $\mathbf{R}^n$ and $O(n)$. Cartan now attacked the problem of Riemannian geometry by observing that on every Riemannian manifold (M, ds^2) one could construct a larger space $O(M)$ called the bundle of orthonormal frames. For a Euclidean space, the bundle of orthonormal frames may be identified with $E(n)$. Cartan was able to write down a generalization of the Maurer-Cartan equations on the bundle $O(M)$. These equation are known as the first and second structural equations of the space. What Cartan discovered was that in Riemannian geometry there was a group action of $O(n)$ on the bundle of frames $O(M)$ rather than on the base space M. The base space $M = O(M)/O(n)$ and all properties of M may be understood in terms of the properties of $O(M)$. This is the starting point for modern Riemannian geometry.

Assume we have an isometry from a Riemannian manifold (M, ds^2) to a Riemannian manifold $(\widetilde{M}, \widetilde{ds}^2)$. This leads to a system of non-linear PDE's given by

$$\tilde{g}_{\mu\nu}(\tilde{x}) \frac{\partial \tilde{x}^\mu}{\partial x^\rho} \frac{\partial \tilde{x}^\nu}{\partial x^\sigma} = g_{\rho\sigma}(x) \,. \tag{5.16}$$

Cartan realized that there was a better and more geometric way to formulate these equations. Cartan constructed local orthonormal coframes $\omega^m = e^m{}_\mu dx^\mu$ on M, and $\tilde{\omega}$ on $\widetilde{M}$ where the metrics may be written as $ds^2 = \omega^m \otimes \omega^m$ and $\widetilde{ds}^2 = \tilde{\omega}^m \otimes \tilde{\omega}^m$. An isometry requires the existence of an orthogonal matrix-valued function R such that $\tilde{\omega}^m = R^m{}_n \omega^n$. Cartan now observed that this leads us to a first order EDS $\tilde{e}^m{}_\mu d\tilde{x}^\mu = R^n{}_m e^n{}_\nu dx^\nu$. Cartan went much further. He realized that one could promote R to a new independent variable. This is similar to introducing a new variable $u = dy/dx$ in a second order ODE and writing the original equation as a pair of first order ODEs. Instead of working on M, Cartan worked on a space with coordinates (x, R). In modern language, this is the bundle of orthonormal frames. Also, the structural equations are globally defined on the bundle of orthonormal frames whereas (5.16) are only valid locally. This is what led to his invention of the theory of G-structures (principal sub-bundles of the frame bundle) and generalized geometries.

In the modern viewpoint, an isometry may be defined in the following way. Assume we have a map $f : M \to \widetilde{M}$; the differential of this map $df : TM \to T\widetilde{M}$ naturally acts on the bundle of *all* frames, *i.e.*, $\frac{\partial}{\partial x^\nu} = \frac{\partial \tilde{x}^\mu}{\partial x^\nu} \frac{\partial}{\partial \tilde{x}^\mu}$. If df lifts to a bundle map from the *orthonormal frame bundle* $O(M)$ to the *orthonormal frame bundle* $O(\widetilde{M})$ then f is an isometry. We have to be careful with the converse. A bundle map $O(M) \to O(\widetilde{M})$ is fiber-preserving and thus induces a map $M \to \widetilde{M}$ on the base spaces. In general, the bundle map from $O(M)$ to $O(\widetilde{M})$ will not be an isometry because it will not be the lift of a map between the bases.

When is a bundle map $F : O(M) \to O(\widetilde{M})$ the lift of a map $f : M \to \widetilde{M}$? Recall that frame bundles are endowed with a globally defined $\mathbf{R}^n$-valued canonical 1-form. Let θ and $\tilde{\theta}$ be the respective canonical 1-forms on $O(M)$ and $O(\widetilde{M})$. If $F^*\tilde{\theta} = \theta$ then F is the lift of a map between the bases. This means that an isometry can be defined as a bundle map between orthogonal frame bundles which preserves the canonical 1-forms.

The formulation just presented is global whereas the formulation in terms of PDE's (5.16) is local. The question of the existence of an isometry between Riemannian manifolds is a difficult one. There are both local and global issues which must be addressed. In this report we only discuss local issues. Roughly, the local existence of an isometry is guaranteed if there exists coordinate systems on M and $\widetilde{M}$ such that the curvature and its higher derivatives agree up to a certain order. Global issues are much more difficult [20].

The key point is contained in the following scenario.

Assume we have a pseudo-isometry from $(T^\sharp M, ds_Q^2)$ to $(T^\sharp M, \widetilde{ds_Q^2})$. It is relatively easy to show that not only does the pseudo-isometry preserve the metric but also the connection. Consequently, the induced map on paths preserves the symplectic form (5.13). This means that the pseudo-isometry between the finite dimensional spaces $T^\sharp M$ and $T^\sharp \widetilde{M}$ induces a canonical transformation from the infinite dimensional phase space $(x(\sigma), \pi(\sigma))$ to the infinite dimensional phase space $(\tilde{x}(\sigma), \tilde{\pi}(\sigma))$.

The above requires some qualifications because we have ignored global issues[3]. The construction just described justifies the analogy presented in Table 1.1.

5.4 Back to duality

We will now perform the mathematical construction we require. We begin with a sigma model (M, ds^2, B). Using the metric we construct a local orthonormal coframe $\{\omega^m\}$. We put coordinates on the "fake cotangent bundle" $T^\sharp M$ by noting that a 1-form may be written as $p_m \omega^m$. This allows us to define 1-forms $\mu_m = dp_m - \omega^l \Gamma_l{}^n{}_m p_n$. These 1-forms define the horizontal distribution on $T^\sharp M$ associated with the Levi-Civita connection. The Q-metric on $T^\sharp M$ is given by

$$ds_Q^2 = \omega^m \otimes \mu_m + \mu_m \otimes \omega^m \ . \tag{5.17}$$

A pseudo-isometry is a map from $T^\sharp M$ to $T^\sharp \widetilde{M}$ such that

$$\begin{pmatrix} \tilde{\omega} \\ \tilde{\mu} \end{pmatrix} = S \begin{pmatrix} \omega \\ \mu \end{pmatrix} , \tag{5.18}$$

where $S \in O_Q(2n)$. We now rephrase the above in the language of bundles.

Given the metric ds_Q^2 on $T^\sharp M$ we construct the bundle of Q-orthonormal frames, denoted by $O_Q(T^\sharp M)$. This bundle has base space $T^\sharp M$ and fiber $O_Q(2n)$. A pseudo-isometry between $(T^\sharp M, ds_Q^2)$ and $(T^\sharp \widetilde{M}, \widetilde{ds_Q^2})$ is given by a bundle map between $O_Q(T^\sharp M)$ and $O_Q(T^\sharp \widetilde{M})$ which preserves the canonical 1-forms. Once we have a pseudo-isometry we can construct the desired canonical transformation. This is the full story as far as the "canonical structure" is concerned. This is only half the problem because we have to worry about the hamiltonian.

6 The hamiltonian structure

The hamiltonian is related to a positive definite metric on $T^\sharp M$ defined by

$$ds_{\mathcal{H}}^2 = (\mu_m - B_{mn}\omega^n) \otimes (\mu_m - B_{ml}\omega^l) + \omega^m \otimes \omega^m \ . \tag{6.1}$$

Note that for any curve $(x(\sigma), p(\sigma))$ on $T^\sharp M$, the evaluation of the above on the tangent vector to the curve yields $2\mathcal{H}$ where $\mathcal{H}$ is defined by (2.5).

We are now in a position to see what is required to have duality within our scenario. Assume we have a Q-pseudo-isometry $f : T^\sharp M \to T^\sharp \widetilde{M}$. If in addition, f

[3]A discussion of global issues in duality may be found in [21].

preserves the metric $ds^2_{\mathcal{H}}$ then the sigma model (M, ds^2, B) will be dual to the sigma model $(\widetilde{M}, \widetilde{ds}^2, \widetilde{B})$. We are interested in maps from $T^\sharp M$ to $T^\sharp \widetilde{M}$ which preserve two different symmetric 2-tensors, ds^2_Q and $ds^2_{\mathcal{H}}$, of different type. We will call such maps K-isometries. To formulate this problem in terms of bundles we have to use the bundle consisting of frames which are simultaneously orthonormal with respect to ds^2_Q and $ds^2_{\mathcal{H}}$. The fiber of this bundle is isomorphic to $K = O_Q(2n) \cap O(2n) \approx O(n) \times O(n)$, the maximal compact subgroup of $O_Q(2n)$. We can now state the main result of these lectures. Let $K(T^\sharp M, ds^2_Q, ds^2_{\mathcal{H}})$ be the bundle consisting of frames which are simultaneously orthonormal with respect to ds^2_Q and $ds^2_{\mathcal{H}}$.

Theorem 6.2

Assume there exists a bundle map between the frame bundles $K(T^\sharp M, ds^2_Q, ds^2_{\mathcal{H}})$ and $K(T^\sharp \widetilde{M}, \widetilde{ds}^2_Q, \widetilde{ds}^2_{\widetilde{\mathcal{H}}})$ which preserves the canonical 1-forms; then the sigma model (M, ds^2, B) is dual to the sigma model $(\widetilde{M}, \widetilde{ds}^2, \widetilde{B})$. Said differently, the sigma model (M, ds^2, B) is dual to the sigma model $(\widetilde{M}, \widetilde{ds}^2, \widetilde{B})$ if there exists a K-isometry between $T^\sharp M$ and $T^\sharp \widetilde{M}$.

To relate these ideas to concepts familiar from toroidal target space duality we discuss in more detail the bundle $K(T^\sharp M, ds^2_Q, ds^2_{\mathcal{H}})$. A general Q-orthonormal frame *at a point* in $T^\sharp M$ may be written as

$$S \begin{pmatrix} \omega \\ \mu \end{pmatrix} \tag{6.3}$$

where $S \in O_Q(2n)$. First we decompose S in a way that explicitly exhibits the hamiltonian. The Lie algebra of $O_Q(2n)$ may be written in a block decomposition as

$$\begin{pmatrix} \delta & \beta \\ \gamma & -\delta^t \end{pmatrix}, \tag{6.4}$$

each entry is an $n \times n$ matrix and β, γ are skew. A matrix in the Lie algebra of K is of the form

$$\begin{pmatrix} \alpha & \beta \\ \beta & \alpha \end{pmatrix}, \tag{6.5}$$

where α, β are skew. This suggests a decomposition

$$\begin{pmatrix} \delta & \beta \\ \gamma & -\delta^t \end{pmatrix} = \begin{pmatrix} \alpha & \beta \\ \beta & \alpha \end{pmatrix} + \begin{pmatrix} \sigma & 0 \\ 0 & -\sigma \end{pmatrix} + \begin{pmatrix} 0 & 0 \\ \epsilon & 0 \end{pmatrix}, \tag{6.6}$$

where σ is symmetric and ϵ is skew. Near the identity one can rewrite (6.3) in the form

$$T \begin{pmatrix} e^{\sigma} & 0 \\ 0 & e^{-\sigma} \end{pmatrix} \begin{pmatrix} I & 0 \\ -B & I \end{pmatrix} \begin{pmatrix} \omega \\ \mu \end{pmatrix} = T \begin{pmatrix} e^{+\sigma}\omega \\ e^{-\sigma}(\mu - B\omega) \end{pmatrix}, \tag{6.7}$$

where $T \in K$ and B is skew. Note that $e^{\pm\sigma}$ are non-singular symmetric matrices. Since T is orthogonal, we see that we have essentially computed a square root for (2.5). We have explicitly found the B field and the e^{σ} term is roughly the square root of ds^2. This shows that at a point on $T^\sharp M$ the moduli space for $ds^2_{\mathcal{H}}$ is given

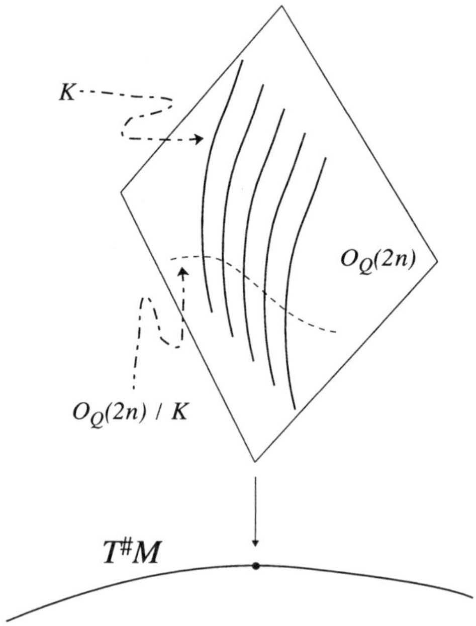

Figure 2. The quadrilateral represents the fiber $O_Q(2n)$ over a point in $T^\sharp M$. The coset space $O_Q(2n)/K$ arises from a fibration of $O_Q(2n)$ with the fibers isomorphic to K. The leaves of the fibration are depicted by the heavy vertical curves. We schematically denote the orbit space $O_Q(2n)/K$ by the dashed curve.

by $O_Q(2n)/K$. The moduli space depends on n^2 real parameters. The reader should compare this with the standard discussion in toroidal target spaces.

It is now worthwhile to consider a second decomposition which is reminiscent of both the Iwasawa decomposition and the Euler angle decomposition. Observe that there exists an $R \in O(n)$ such that $\sigma = R^{-1}\Delta R$ where Δ is diagonal. What we do is insert this into (6.7), absorb an R^{-1} into T, redefine B and push R all the way to the right.

$$T'\begin{pmatrix} e^{+\Delta} & 0 \\ 0 & e^{-\Delta} \end{pmatrix}\begin{pmatrix} I & 0 \\ -B' & I \end{pmatrix}\begin{pmatrix} R & 0 \\ 0 & R \end{pmatrix}\begin{pmatrix} \omega \\ \mu \end{pmatrix} = T'\begin{pmatrix} e^{+\Delta}\omega' \\ e^{-\Delta}(\mu' - B'\omega') \end{pmatrix}. \tag{6.8}$$

In the above $T' \in K$, B' is skew, $\omega' = R\omega$ and $\mu' = R\mu$. Many explicit examples of target space duality directly give a hamiltonian which can be put into this final form. This decomposition depends crucially on the details of how the subgroup K sits inside of $O_Q(2n)$.

The symplectic structure on the infinite dimensional phase space is associated with ds_Q^2. This gives us a frame bundle with fiber $O_Q(2n)$ as in Figure 2. Fix a point in $T^\sharp M$. Each orbit of K in $O_Q(2n)$ determines a hamiltonian at that point. A choice of a K-orbit on the fiber over each point in $T^\sharp M$ is same as a choice of a sub-bundle with reduced structure group K. In general this bundle will not be associated with a hamiltonian. Remember that earlier we were careful to state that the moduli space for the hamiltonian *at a point* in $T^\sharp M$ was $O_Q(2n)/K$. If (x, p) are local coordinates on $T^\sharp M$ then observe that $g_{\mu\nu}(x)$ and $B_{\mu\nu}(x)$ are only functions of x. They do not depend on p. Bundles of type $K(T^\sharp M, ds^2_Q, ds^2_{\mathcal{H}})$ are a subset of the set of bundles of K-orthonormal frames. This has important consequences when studying the principal K-bundles over $T^\sharp M$ because the bundles $K(T^\sharp M, ds^2_Q, ds^2_{\mathcal{H}})$ are very special. If one writes the structural equations for a general principal K-bundle one finds many curvature and torsion terms. The fact that ds^2 and B "come from M" imposes many constraints on the structural equations which are discussed in [14]. The structural equations contain the information necessary to understand the local obstructions to the existence of K-isometries.

7 A CHALLENGE TO MATHEMATICIANS

The content of these lectures suggest a very interesting mathematical problem. Can one discover examples of geometrical data which would lead, between $T^\sharp M$ and $T^\sharp \widetilde{M}$, to nontrivial *local* K-isometries? The reason for the requirement of locality is that in these lectures we did not discuss global issues except in the case of toroidal target spaces. The existence of local isomorphisms between the principal bundles $K(T^\sharp M, ds^2_Q, ds^2_{\mathcal{H}})$ and $K(T^\sharp \widetilde{M}, \widetilde{ds}^2_Q, \widetilde{ds}^2_{\widetilde{\mathcal{H}}})$ which preserve the canonical 1-forms is intimately related to Cartan's problem of equivalence [13]. There one finds that the existence of such local isomorphisms is codified in the curvature and torsion of the bundles. Explicit formulas will appear in [14]. The only explicit examples known of this phenomenon are the toroidal target spaces. It is not clear whether the non-abelian duality examples fall in this "restricted" class. Global existence theorems are much more complicated [20].

There are well established procedures for constructing the dual theory when the sigma model admits a continuous group action [5, 6, 8, 9, 11]. Is there a systematic procedure which can be invoked to look for the existence of dual target spaces when there are no symmetries?

ACKNOWLEDGMENTS

I would like to thank L. Baulieu, V. Dotsenko, V. Kazakov, M. Picco and P. Windey for the terrific job they did in organizing a very stimulating and dynamic summer institute at Cargèse. I would like to thank J. Sanchez-Guillen and the Theoretical Physics Group of the University of Santiago for their hospitality during my visit. I am indebted to the Iberdrola Foundation for the visiting professorship which made my visit to Santiago possible. This research was supported in part by The National Science Foundation, Grant PHY–9507829.

Many colleagues have contributed to my understanding of target space duality. I am most indebted to Robert Bryant for explaining to me the many mysteries of exterior differential systems and Cartan's method of equivalence. He has patiently and generously answered my questions. Is Singer always clears my mathematical confusions with clear and succinct explanations. Chien-Hao Liu has been a fantastic sounding board, always asking important probing questions which have helped clarify my ideas. I would like to thank Thom Curtright for sharing some of his ideas with me. Joaquin Sanchez-Guillen patiently listened and provided feedback during our many discussions at Santiago. I would like to thank Paul Windey for many discussions at conception of this program for studying target space duality. I would like to thank Chien-Hao Liu and Paul Watts for their critical comments on this manuscript.

References

[1] A. Giveon, E. Rabinovici and G. Veneziano, "Duality in string background space", Nucl. Phys. B322 (1989) 167.

[2] T. Curtright and C. Zachos, "Currents, charges, and canonical structure of pseudodual chiral models", Phys. Rev. D49 (1994) 5408–5421, `hep-th/9401006`.

[3] E. Alvarez, L. Alvarez-Gaumé and Y. Lozano, "A canonical approach to duality transformations", Phys. Lett. B336 (1994) 183–189, `hep-th/9406206`.

[4] Y. Lozano, "Non-abelian duality and canonical transformation", Phys.Lett. B355 (1995) 165–170, `hep-th/9503045`.

[5] A. Giveon, M. Porrati and E. Rabinovici, *Target space duality in string theory*, Phys. Reports 244 (1994) 77–202, `hep-th/9401139`.

[6] X. de la Ossa and F. Quevedo, "Duality symmetries from nonabelian isometries in string theory", Nucl. Phys. B403 (1993) 377, `hep-th/9210021`.

[7] B.E. Fridling and A. Jevicki, "Dual representations and ultraviolet divergences in nonlinear sigma models", Phys. Lett. 134B (1984) 70.

[8] E. Alvarez, L. Alvarez-Gaumé and Y. Lozano, "On Non-abelian Duality", Nucl. Phys. B424 (1994) 155–183, `hep-th/9403155`.

[9] A. Giveon and M. Roček, "On Nonabelian Duality", Nucl. Phys. B421 (1994) 173, `hep-th/9308154`.

[10] A. Giveon and M. Roček, "Introduction to duality", `hep-th/9406178`.

[11] M. Roček and E. Verlinde, "Duality, quotients, and currents", Nucl. Phys. B373 (1992) 630, `hep-th/9110053`.

[12] Yau, S.T., *Essays on Mirror Manifolds*, edited by S.T. Yau, International Press, 1992.

[13] R.B. Gardner, "The method of equivalence and its applications", CBMS-NSF Regional Conference Series in Applied Mathematics, Vol. 58, SIAM, 1989.

[14] O. Alvarez, "The classical geometry of target space duality", Part I and Part II, in preparation.

[15] R. Bryant, S.S. Chern, R.B. Gardner, H.L. Goldschmidt, P.A. Griffiths, *Exterior Differential Systems*, Mathematical Sciences Research Institute publication 18, Springer-Verlag, 1991.

[16] R. Abraham and J.E. Marsden, *Foundations of mechanics*, 2nd ed., Benjamin/Cummings Publ., 1978.

[17] V.I. Arnold, *Mathematical methods of classical mechanics*, 2nd ed., GTM 60, Springer-Verlag, 1989.

[18] C. Klimčík and P. Ševera, "Strings in spacetime cotangent bundle and T-duality", Mod. Phys. Lett. A10 (1995) 323–330, `hep-th/9411003`.

[19] C. Klimčík and P. Ševera, "Dual and non-abelian duality and the Drinfeld double", Phys. Lett. B351 (1995) 455–462, `hep-th/9502122`.

[20] M. Gromov, *Partial Differential Relations*, Springer-Verlag, 1986.

[21] E. Alvarez, L. Alvarez-Gaumé, J. Barbón and Y. Lozano, "Some global aspects of duality in string theory", Nucl. Phys. B415 (1994) 71–100, `hep-th/9309039`.

(PERTURBED) CONFORMAL FIELD THEORY APPLIED TO 2D DISORDERED SYSTEMS: AN INTRODUCTION

DENIS BERNARD[1]

SERVICE DE PHYSIQUE THÉORIQUE DE SACLAY
F-91191, GIF-SUR-YVETTE, FRANCE
AND
INSTITUT DES HAUTES ÉTUDES SCIENTIFIQUES
F-91440, BURES-SUR-YVETTE, FRANCE

1 INTRODUCTION

These lectures will probably look too elementary to experts in disordered systems. It will perhaps be more appropriate as an introduction to this field for conformal field theorists.

The plan of these notes is the following:

1-INTRODUCTION. We introduced the types of problems we will be concerned with. We mention a few open problems; the most challenging being the description of the quantum Hall transition.

2-A FEW BASIC TECHNIQUES. In this section we have gathered the basic tools which are usually used. This includes the supersymmetric method and the replica method.

3-THE DIRECT METHOD. There, we show that due to specific properties of two-dimensional field theories, it is (sometimes) possible to compute directly the disorder dependence of the correlation functions and then to average over the disorder. Examples are provided by WZW models coupled to random vector potentials, a particular case being the massless Dirac model coupled to a random vector potential. We discuss the Abelian and non-Abelian cases and show in both cases that the disorder can be factorized. At large impurity density, the disordered WZW model is conformally invariant being equivalent to the product of the WZW model over the

[1]Member of the CNRS

Low Dimensional Applications of Quantum Field Theory
Edited by L. Baulieu *et al.*, Plenum Press, New York, 1997

non-compact space $\mathcal{G}^C/\mathcal{G}$ times an arbitrary number of copies of the pure version of the original WZW model.

4-THE SUPERSYMMETRIC METHOD. We take the random Ising model as the simplest example for illustrating the supersymmetric method. We show that it can be formulated as a Gross-Neveu type model but over the Lie superalgebra $OSp(2N|2N)$, and we use this formulation to recover its properties usually obtained by the replica method. We discuss its connection with the random Dirac theory: a model which has been introduced for describing the quantum Hall transition. These examples illustrate the fact that non-unitary conformal field theories based on affine Lie superalgebras with zero superdimension provide candidates for critical theories of gaussian disordered systems.

5-THE REPLICA METHOD. We take the random phase sine-Gordon model as example. Its large distance behavior in the low temperature phase is still controversial. We present the approach based on the renormalization group and the symmetric replica trick. We also discuss a large N version of this model.

6-THE VARIATIONAL METHOD. We again use the random phase sine-Gordon model to illustrate this method. This approach consists in finding a good variational Ansatz at fixed disorder and then averaging. It provides exact lower bound to the free energy.

Besides stochastic differential equations, cf. Polyakov's lecture in this volume, there are at least two classes of disordered problems which we can study.

(i) The first ones concern statistical systems with random coupling constants, cf e.g. [1]. Imagine having a physical system with two kinds of degree of freedom with two very different relaxation times. The degrees of freedom, called spin variables, which have the shortest relaxation time will thermalize well before the other degrees of freedom, which we call impurities. The impurities are also called quenched variables because they are not in thermal equilibrium with the spin variables. We consider a large number of such physical systems with different realizations of impurity configurations. Since the spin variables are in thermal equilibrium but not the impurities we are interested in the average over the impurity configurations of the free energy of the thermalized spin systems.

A classical example is provided by the Ising model with random bond coupling constants. The thermalized spin variables σ_i, defined on the vertices of a lattice, take two values $\sigma_i = \pm$. The effect of the impurities is represented by the randomness of the interaction J_{ij} which couple neighborhood spins. At fixed disorder, its partition function is:

$$Z[J] = \sum_{\{\sigma_i\}} \exp\left(- \sum_{\langle i,j \rangle} J_{ij}\sigma_i\sigma_j \right)$$

A probability distribution $P[J]$ is given to the random coupling constants. The average thermal properties will then be given by the average of the logarithm of the partition function: $\overline{\log Z[J]}$.

In the continuum limit and close to the critical point, *i.e.* in its scaling limit, it is known from prehistory that the Ising model is described by a massive real Majorana fermion with mass $\overline{m} \sim \tau = \frac{(T_C - T)}{T_C}$ with T_C the critical temperature. In presence of randomness in the bond interaction, the mass becomes a random function of the space position. Therefore, at fixed disorder, the action describing the random bond Ising model in its continuous limit is:

$$S[m(x)] = \int \frac{d^2x}{4\pi} \left(\overline{\Psi} i\partial\!\!\!/ \Psi + m(x)\overline{\Psi}\Psi \right)$$

with Ψ a Majorana fermion. The random mass $m(x)$ is coupled to the energy operator as it should be in order to represent the randomness of the bond interaction. This action is of the form:

$$S[g(x)] = S_* + \int d^2x g(x)\Phi(x) \tag{1.1}$$

where S_* is the action at the renormalization group fixed point describing the pure system, and $\Phi(x)$ a scalar field. As formulated in eq.(1.1), the problem can also be interpreted as a random field problem.

The Ising model with random bond interactions provides one of the simplest example of disordered system in 2d [2, 3, 4]. We will study it in details in the following (using the supersymmetric method). Besides the disordered WZW models, we will also study another example (using the replica method): the random phase sine-Gordon model. It has been introduced for describing a large variety of two dimensional disordered physical systems [5, 6, 7, 8]. Contrary to the random bond Ising model which is now well understood, there is up to now no consensus on the large distance behavior of the random phase sine-Gordon model. We will present the various suggestion which have been made concerning this behavior.

(ii) The second class of problems deals with particles moving in random potential, cf e.g. [1]. The random potential then represents the impurities that the particles, e.g. the electrons, encountered while traveling in a metal. The property of the medium will be characterized by the behavior of the electrons in this disordered surrounding. For example, its conductivity properties will depends on whether the averaged wave functions are localized or not. These problems are generically called localization problems .

We are thus interested in the averaged properties of hamiltonians of the form:

$$H = H_0 + V$$

where H_0 is a fixed hamiltonian, and V a random perturbation. A standard example is: H_0 a pure kinetic term, $H_0 = -\partial_x^2$ and $V = V(x)$ a potential representing the interaction with the underlying medium.

Quantities of interest would be e.g. the averaged density of states which is given by the average of the trace of the Green function:

$$\overline{\rho(E)} = \frac{1}{\pi} \overline{\mathrm{Im}\,\mathrm{tr}\, \left(\frac{1}{H - E - i\epsilon} \right)} \tag{1.2}$$

More generally, we are interested in computing averages of products of Green functions. Indeed, computing the conductivity tensor $\sigma_{\mu\nu}$ using the Kubo formula requires computing the average of the product of an advanced and a retarded Green function.

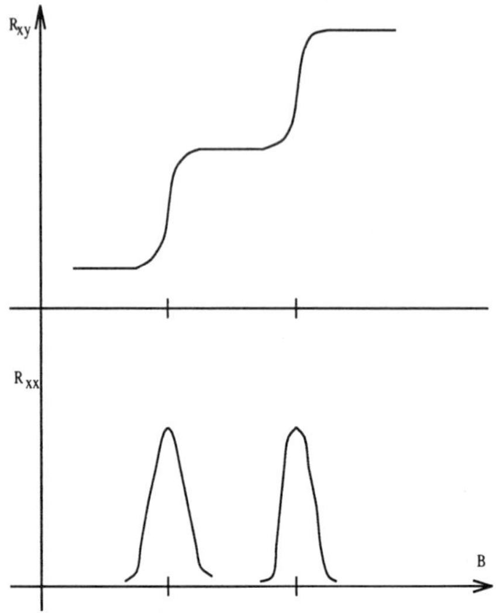

Figure 1. The transverse and longitudinal Hall resistances R_{xy} and R_{xx} as a function of the magnetic field B.

Let me introduce a recent localization/delocalization problem (which actually was the motivation for these lectures). It concerns the transition between the plateaus of the integer quantum Hall effect, cf [9, 10] and references therein. Without going into any details, we recall that the quantum Hall effect is characterized by the quantization of the transverse Hall resistance R_{xy} to inverse integer values in unit of (h/e^2). This integer quantization remains valid when the magnetic field varies on some domain, which are called plateaus. On these plateaus the longitudinal resistance R_{xx} vanishes, but it becomes non-zero in the regime separating two plateaus. The values of the resistance R_{xy} and R_{xx} as a function of the magnetic field is represented in Fig.1.

The integer quantum Hall effect (IQHE) is correctly described by considering a 2d gas of non-interacting electrons subject to a transverse magnetic field. As is well known, the energy spectrum is then given by the Landau levels, which are highly degenerate. In a pure system without any impurities, the Hall resistance R_{xy} will be a step-function as a function of the Fermi level. The plateaus occur when the Fermi level of the electrons gas is between two Landau levels.

However, since the degeneracy of a landau level is field dependent, (this degeneracy is ϕ/ϕ_0 where ϕ is the magnetic flux through the system and $\phi_0 = h/e$ the flux quantum), the Hall resistance in a pure system is a linear function of the magnetic field B. More precisely:

$$R_{xy}^{pure} = (\frac{h}{e^2})\frac{\phi}{N_e\phi_0} = (\frac{h}{e^2})\frac{1}{\nu}$$

where N_e is the number of electrons in the system and $\nu = \frac{N_e\phi_0}{\phi}$ is called the filling factor. The explanation of the plateaus for R_{xy} as a function of the magnetic field B

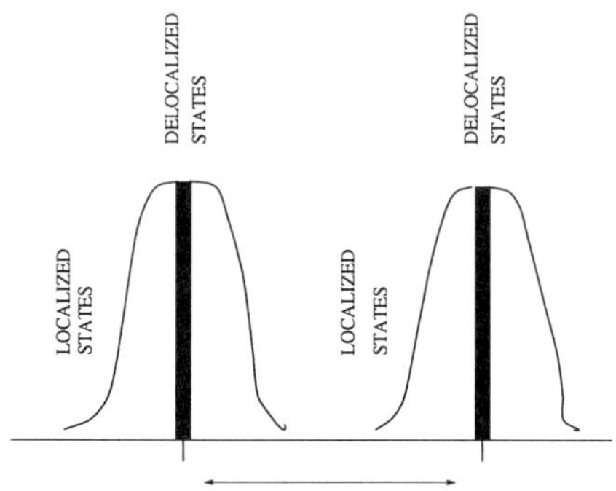

Figure 2. The density of states in two consecutive Landau bands.

and not of the Fermi level requires considering the effect of the impurities [11, 12]. The impurity potential splits the degeneracy of the Landau levels, which now form energy bands. The states whose energy is in the border of these bands are localized while those which are in the center of the bands are delocalized. The localized states then serve as reservoir for the electrons which stabilized the Fermi level between two Landau bands while the magnetic field varies. This stabililization of the Fermi level give rise to the plateaus. It is expected that the delocalized states are present only at the center of the band; that is there is not a band of delocalized states. The presence of delocalized states is necessary for having a non-zero conductivity.

Between two plateaus the Fermi level is (almost) in the middle of a Landau band. Since the delocalized states are in the middle of this band, the transition between two plateaus of the integer quantum Hall effect is thus a transition between localized and delocalized states. (De)localized states are characterized by a (in)finite localization length. Near this transition, this localization length $\xi(E)$ behaves as follows:

$$\xi(E) \sim |E - E_c|^{-\nu} \tag{1.3}$$

with E_c the band center energy. The density of states is non-singular at the transition point [13]. Various measurements, as well as numerical simulations, gave access to the exponent ν: $\nu \sim 2.3 \pm 0.1 \sim 7/3$ [14, 15, 9]. The behavior (1.3) has to be compared to the standard Anderson localization in two dimensions which predicts that any tiny amount of disorder will localize all the states.

What is the field theory describing the quantum Hall transition is still an open question, cf ref.[10] for a recent discussion. The problem consists in determining the appropriate infrared fixed point, and therefore seems to be a "simple" exercise in conformal field theory. Unfortunately, this infrared fixed point is a strong coupling problem in all the models introduced so far; it is therefore difficult to attack. A sigma model approach was introduced in ref.[16, 17]. A very elegant network model was developed in ref.[18]. A supersymmetric spin chain, which arises as an anisotropic limit of the network model, was introduced in ref.[10]. A model closely related to the random bond Ising model, namely the random Dirac model, has also been proposed for describing the IQHE localization/delocalization transition [19]. We will present it in section 4.

2 A FEW BASIC TECHNIQUES

These problems can be analyzed with similar techniques. The two standard ways of studying them use either the replica trick, cf eg [1] or the supersymmetric method [20]. In this section we present a very brief introduction to these methods. (The similarity of the techniques used to study disordered statistical model and localization problems is apparent at a first naive level, but more refined studies of both subjects reveal their differences.)

As it will become clear below, the supersymmetric method applies to system which are gaussian at fixed disorder, while the replica trick has a larger domain of applicability since it is not restricted to free theory. Perturbatively the replica and supersymmetric methods are equivalent. However, a few non-perturbative results have been obtained using the supersymmetric method. Clearly, once these disordered problems have been reformulated in a field theoretical languages, all the standard field theoretical methods (e.g renormalization group, mean field approximation, variational approximation, large N expansion, etc.) are available.

In the following sections we will illustrate these methods on two examples: the random bond Ising model and the random phase sine-Gordon model.

2.1 The supersymmetric method

The supersymmetric method [20] applies to models which are gaussian at fixed disorder. We will present it in the random Hamiltonian problem. As above let us consider a random hermitian hamiltonian

$$H = H_0 + V,$$

where H_0 is a fixed hamiltonian and V a random potential. We are interested in computing averages of product of advanced or retarded Green functions:

$$\overline{\left(\frac{1}{H_E - i\epsilon}\right)\left(\frac{1}{H_{E'} + i\epsilon}\right)\cdots}$$

with $\epsilon \to 0^+$. We have set $H_E = H - E$. The supersymmetric method is based on a gaussian integral representation of the Green functions. E.g.:

$$\frac{1}{H_E - i\epsilon} = \frac{1}{Det[H_E - i\epsilon]} \int d\psi \psi \psi^* \exp\left(\pm i \psi^* (H_E - i\epsilon)\psi\right) \tag{2.1}$$

with ψ complex Grassmannian variables. In eq.(2.1), we have the freedom to choose the $\pm$ sign. There are no convergence problem in this fermionic integral. To be able to average over the disorder, we need to re-exponentiate the determinant $Det[H_E - i\epsilon]$. The inverse of this determinant can be written as a gaussian integral over complex bosonic variables:

$$\frac{1}{Det[H_E - i\epsilon]} = \int d\phi exp\left(-i\phi^* (H_E - i\epsilon)\phi\right) \tag{2.2}$$

The prefactor $(-i)$ in the exponential is chosen by requiring the convergence of the gaussian integral. Gathering these gaussian integrals, we obtain the supersymmetric representation of the Green functions:

$$\left(\frac{1}{H_E - i\epsilon}\right) = \int d\psi d\phi \psi\psi^* \exp\left(-i\psi^*(H_E - i\epsilon)\psi - i\phi^*(H_E - i\epsilon)\phi\right) \tag{2.3}$$

$$\begin{aligned}\left(\frac{1}{H_E - i\epsilon}\right)\left(\frac{1}{H_{E'} + i\epsilon}\right) &= \int d\psi d\phi \quad \psi_R\psi_R^*\psi_A\psi_A^* \\ &\quad \exp\left(-i\psi_R^*(H_E - i\epsilon)\psi_R - i\phi_R^*(H_E - i\epsilon)\phi_R\right. \\ &\quad \left.-i\psi_A^*(H_{E'} + i\epsilon)\psi_A + i\phi_A^*(H_{E'} + i\epsilon)\phi_A\right)\end{aligned} \tag{2.4}$$

etc.

In the equation representing product of two Green functions, the indices A, R refers to the advanced or retarded Green functions. Note that due to the convergence requirement, the sign in front of the $(i\epsilon)$ factor differs in the advanced and retarded sector. More generally, products of N Green functions are represented by integrals over N couples of superpartners (ψ, ϕ) with appropriate choice of the sign factor according to the convergence condition.

In the form (2.4), the average over the disorder can be easily done. Assume for simplicity that the random potential has a gaussian distribution:

$$P[V] = \exp\left(-\frac{1}{2\sigma} tr(V^2)\right) dV$$

Consider for example the average of one Green function. The term coupled to the random potential is $\frac{1}{2\sigma} tr(V^2) + itr(V(\phi\phi^* - \psi\psi^*))$. After integration over V, the averaged Green functions becomes:

$$\overline{\left(\frac{1}{H - i\epsilon}\right)} = \int d\psi d\phi \psi\psi^* \exp\left(-S_{eff}\right) \tag{2.5}$$

with

$$S_{eff} = i\psi^*(H_0 - E - i\epsilon)\psi + i\phi^*(H_0 - E - i\epsilon)\phi - \frac{\sigma}{2} tr\left(\phi\phi^* - \psi\psi^*\right)^2$$

The bosonic and fermionic sectors are coupled by the disorder. This effective action is easily generalized when considering products of higher number of Green functions.

Notice that if H is a quantum hamiltonian for a physical system in d spatial dimensions, *i.e.* a $(d+1)$ system, than the supersymmetric action describes a d dimensional quantum field theory. In particular, since the quantum Hall effect is a $(2+1)$ system its localization/delocalization transition would be described by a 2d field theory.

2.2 *The replica method*

Let us now recall briefly the starting point of the replica method, cf. [1]. We are interested in computing the average of the free energy, or of correlation functions.

Since the free energy is proportional to the logarithm of the partition function $Z[J]$, we have to compute the average of $\log Z[J]$. The replica method is based on the identity:

$$\log Z[J] = \lim_{n\to 0} \frac{Z^n[J]-1}{n} = \frac{d}{dn} Z^n[J]\Big|_{n=0} \tag{2.6}$$

In other words, we compute the quenched averages of one disordered system by introducing n copies of this system, in all of which the random variables take the same values, and then computing the averaged partition function of this new system at $n = 0$.

This method leads immediately to a criteria selecting the systems in which the disorder is relevant. This criteria is called the Harris criteria.

Let us describe a disordered system in the continuum limit by the action:

$$S[g(x)] = S_* + \sum_a \int d^2x g^a(x)\Phi_a(x) \tag{2.7}$$

where S_* is the action at the renormalization group fixed point describing the pure system at its critical point, and $\Phi_a(x)$ are scalar fields. When the coupling constant $g^a(x)$ are independent of x and not random, the action (2.7) describes the behavior of the statistical model near the critical point in its scaling limit. In this case the field $\Phi_a(x)$ is relevant, and therefore will influence the large distance behavior of the pure system, if its scaling dimension is less than two: $\dim(\Phi_a) < 2$. This is changed when disorder is present.

Assume for simplicity that the random variables $g^a(x)$ are gaussian variables with one and two-point functions:

$$\overline{g^a(x)} = 0 \quad and \quad \overline{g^a(x)g^b(y)} = \delta^{ab}\sigma_a\delta(x-y) \tag{2.8}$$

Before averaging over the disorder, the partition function is

$$Z[g(x)] = \int DX \exp(-S[g(x)])$$

and the n replicated one becomes:

$$Z^n[g(x)] = \int \prod_r DX_r \exp\left(-\sum_r S_*^r - \sum_a \int d^2x g^a(x) \sum_r \Phi_a^r(x)\right)$$

where the index r runs from 1 to n, S_*^r is the action for the rth replic and $\Phi_a^r(x)$ is the field $\Phi_a(x)$ in the rth copy. To have introduced the replicated copies allows us to easily average over the disorder. Assuming the gaussian measure (2.8) for $g^a(x)$, we obtain:

$$\overline{Z^n[g(x)]} = \int \prod_r DX_r \exp(-S_{eff}) \tag{2.9}$$

with

$$S_{eff} = \sum_r S_*^r + \frac{1}{2}\sum_a \sigma_a \int d^2x \sum_{r,s} \Phi_a^r(x)\Phi_a^s(x) \tag{2.10}$$

The replica are now coupled by the disorder. The field perturbing the uncorrelated action $\sum_r S_*^r$ is quadratic in terms of the $\Phi_a(x)$'s. It will be relevant if $2\dim(\Phi_a(x)) < 2$. (A more careful study is needed to analyze the short distance singularity in $\Phi_a^r(x)\Phi_a^s(x)$ when $r = s$). Hence:

$$\dim \Phi_a(x) < 1 \Rightarrow \textit{relevant disorder.}$$

This is the Harris criteria in two dimensions. Note that for a field to be relevant in a quenched system its dimension has to be half what is required in the pure system. If these dimensions depend on parameters the Harris criteria provides a first glance to the phase diagram.

The quenched correlation functions are defined by:

$$\overline{\langle \mathcal{O}(x)\cdots\rangle\langle\cdots\rangle} = \overline{\left(\frac{1}{Z[g(x)]}\int DXe^{-S[g(x)]}\mathcal{O}(x)\cdots\right)\cdots}$$

The replica method can also be used to compute the averaged correlation functions. It provides a way to re-exponentiate the inverse partition function. The rules are the following:

$$\begin{aligned}
\overline{\langle \mathcal{O}(x)\rangle} &= \langle \mathcal{O}^r(x)\rangle|_{n=0} \\
\overline{\langle \mathcal{O}(x)\mathcal{O}(y)\rangle} &= \langle \mathcal{O}^r(x)\mathcal{O}^r(y)\rangle|_{n=0} \\
\overline{\langle \mathcal{O}(x)\rangle\langle \mathcal{O}(y)\rangle} &= \langle \mathcal{O}^r(x)\mathcal{O}^s(y)\rangle|_{n=0}, \quad r\neq s \\
&\cdots
\end{aligned} \tag{2.11}$$

In particular, the average of the connected Green functions are represented by the correlation function of $\frac{1}{n}\sum_r \mathcal{O}^r(x)$. The simplest way to prove eq.(2.11) consists in introducing sources in the quenched partition function.

3 THE DIRECT METHOD: THE RANDOM VECTOR POTENTIAL MODEL

As a warm up we introduce a very simple gaussian model, which evidently can be completely solved. (We will later use a few of its properties in our study of the more interesting random phase sine-Gordon model.)

In two dimensions it is sometimes possible to compute exactly field theory partition functions and correlation functions with sources. In this section, we also described how this property can be applied to very particular disordered 2d systems. The example we choose consists in a non-Abelian generalization of the gaussian model based on the WZW models. It provides one of the rare disordered models which can be solved directly without relying on the supersymmetric or replica method. This solution relies on the fact that the disorder dependence can be factorized.

3.1 The gaussian model

The bosonic form of its action is defined to be:

$$\begin{aligned}
S_0[\Phi, A] &= \int \frac{d^2x}{4\pi}\left(\frac{1}{2}(\partial_\mu\Phi)^2 + iA_\mu\epsilon_{\mu\nu}\partial_\nu\Phi\right), \\
&= \int \frac{d^2x}{\pi}\left(\frac{1}{2}\partial_z\Phi\partial_{\bar z}\Phi + i(A_{\bar z}\partial\Phi + A_z\partial_{\bar z}\Phi)\right).
\end{aligned} \tag{3.1}$$

where A_μ is a random quenched variable with measure:

$$P[A] = \exp\left[-\frac{1}{2\sigma}\int \frac{d^2x}{4\pi} A_\mu A_\mu\right] = \exp\left[-\frac{1}{\sigma}\int \frac{d^2x}{\pi} A_z A_{\bar{z}}\right] \tag{3.2}$$

In eq.(3.1) we introduced the complex coordinate $z = x + iy$, $\bar{z} = x - iy$ on the plane.

Since it is a gaussian model, it can be easily solved without using the replica trick or the supersymmetric formulation, but by directly computing the quenched averages. Let us introduce the Hodge decomposition of A_μ: $A_\mu = \partial_\mu \xi + \epsilon_{\mu\nu}\partial_\nu \eta$. The fields ξ and η decouple in the measure (3.2):

$$P[A] = P[\xi;\eta] = \exp\left[-\frac{1}{2\sigma}\int \frac{d^2x}{4\pi}\left((\partial_\mu\xi)^2 + (\partial_\mu\eta)^2\right)\right] \tag{3.3}$$

The action $S_0[\Phi, A]$ is independent of ξ and therefore the field ξ is irrelevant. This fact was expected since the field ξ represents a pure gauge whereas the physically relevant quantity is the field strength $F = \epsilon_{\mu\nu}\partial_\mu A_\nu = (\partial_\mu\partial_\mu)\eta$. Moreover, the field η can be absorbed into a translation of Φ:

$$S_0[\Phi, A_\mu = \epsilon_{\mu\nu}\partial_\nu\eta] = S_0[\Phi + i\eta, A = 0] + \int \frac{d^2x}{4\pi}\frac{1}{2}(\partial_\mu\eta)^2. \tag{3.4}$$

The fact that the field η can be absorbed into a shift of Φ does not mean that the quenched correlation functions are identical to those in the pure system. Using eq.(3.4), we have

$$\langle\prod_n e^{i\alpha_n\Phi(x_n)}\rangle_{A_\mu=\epsilon_{\mu\nu}\partial_\nu\eta} = e^{\sum_n \alpha_n\eta(x_n)}\ \langle\prod_n e^{i\alpha_n\Phi(x_n)}\rangle_{A=0}.$$

It factorizes into the product of the correlation functions in the pure system times a simple function of the impurities. However, the average of this function is not irrelevant since the variables η have long-range correlations: $\overline{\eta(x)\eta(y)} = -\pi\sigma(\partial_z\partial_{\bar{z}})^{-1}(x,y) = -\sigma\log|x-y|^2$. In particular, it changes the values of the critical exponents.

Averages of products of correlation functions can then be computed. One has:

$$\begin{aligned}\overline{\langle\prod_n e^{i\alpha_n\Phi(x_n)}\rangle_A \cdots \langle\prod_m e^{i\beta_m\Phi(y_m)}\rangle_A} &= \langle\langle e^{\sum_n \alpha_n\eta(x_n)+\cdots+\sum_m \beta_m\eta(y_m)}\rangle\rangle \times \\ &\times\langle\prod_n e^{i\alpha_n\Phi(x_n)}\rangle_{A=0}\cdots\langle\prod_m e^{i\beta_n\Phi(y_m)}\rangle_{A=0}\end{aligned}$$

where $\langle\langle\cdots\rangle\rangle$ refers to the η-correlation functions with the free field measure (3.3). All the correlators are therefore given by gaussian integral. We get:

$$\begin{aligned}\overline{\langle\prod_n e^{i\alpha_n\Phi(x_n)}\rangle_A \cdots \langle\prod_m e^{i\beta_m\Phi(y_m)}\rangle_A} &= \prod_{n,m}|x_n - y_m|^{-2\sigma\alpha_n\beta_m} \times \\ &\times \prod_{n<n'}|x_n - x_{n'}|^{2\alpha_n\alpha_{n'}}\cdots\prod_{m<m'}|y_m - y_{m'}|^{2\beta_m\beta_{m'}}\end{aligned} \tag{3.5}$$

Clearly, conformal invariance is unbroken in the random abelian case. The dimensions Δ of the vertex operators $\exp(i\alpha\Phi(x))$ in the quenched theory are:

$$\Delta^\alpha_{quenched} = \Delta^\alpha_{pure} - \sigma\alpha^2.$$

It is interesting to look at the connected Green functions of the current $J_z = i\partial\Phi$. It turns out that all these quenched connected correlations are chiral; in particular, they are holomorphic and only depend on $z = x + iy$. For example, we have:

$$\begin{aligned}\overline{\langle i\partial\Phi(z)\rangle} &= 0,\\ \overline{\langle i\partial\Phi(z)i\partial\Phi(w)\rangle_{conn.}} &= 2\pi\left(\partial\frac{1}{\overline{\partial}}\right)_{z,w} = \frac{1}{(z-w)^2}\end{aligned}$$

More generally, one verifies that:

$$\overline{\langle i\partial\Phi(z_1)\cdots i\partial\Phi(z_P)\rangle_{conn.}} = \langle i\partial\Phi(z_1)\cdots i\partial\Phi(z_P)\rangle_{A=0}. \tag{3.6}$$

In other words, the averages of these connected correlation functions are unaffected by the disorder. This property is not true for the average of disconnected correlation functions of $J_z = i\partial\Phi$.

A more complete derivation of these Ward identities which follows from a $U(1)$ symmetry is presented in the section devoted to the study of the random phase sine-Gordon model.

It is an interesting exercise to decipher the operator product algebra of these models.

3.2 *The non-Abelian case: factorization*

We now turn to the non-Abelian case which provides an exceptional quenched model which can be solved directly, without using any replica or supersymmetric method. This is done by factorizing the disorder dependence. We derive it directly from the definition of the quenched correlation functions. This result was also found in ref.[21] using the supersymmetric method, (it appears while we were preparing these notes).

The non-Abelian model appears naturally when generalizing the previous model, after fermionization, to a N-component model. Indeed, using standard bosonization/fermionization rules, $\overline{\Psi}\gamma_\mu\Psi = \epsilon_{\mu\nu}\partial_\nu\Phi$, and $\overline{\Psi}i\partial\!\!\!/\Psi = \frac{1}{2}(\partial_\mu\Phi)^2$, we can rewrite the action (3.1) in an equivalent fermionic form:

$$S_0 = \int \frac{d^2x}{\pi}\left(\overline{\Psi}(i\partial\!\!\!/ + A\!\!\!/)\Psi\right).$$

It describes a Dirac fermion coupled to a random magnetic field.

The non-Abelian generalization describes N massless Dirac fermions minimally coupled to a random non-abelian gauge field [22, 23, 21]. Let us introduce the components of the fermions $(\overline{\psi}^j_+, \psi^j_+)$ and $(\overline{\psi}_{-;j}, \psi_{-;j})$, $j = 1, \cdots, N$. The action is defined as:

$$S[A] = \int \frac{d^2x}{\pi}\left(\psi_{-;j}\left(\partial_{\overline{z}}\delta^j_k + A^j_{\overline{z},k}\right)\psi^k_+ + \overline{\psi}_{-;j}\left(\partial_z\delta^j_k + A^j_{z,k}\right)\overline{\psi}^k_+\right) \tag{3.7}$$

where $A^j_{z,k} = i\sum_a A^a_z(t^a)^j_k$, with $(A^a_z)^* = A^a_{\overline{z}}$, is the gauge field. Here the hermitian matrices t^a form the N-dimensional representation of $SU(N)$. We denote by f^{abc} the $SU(N)$ structure constants: $[t^a, t^b] = if^{abc}t^c$. The Dirac fermions take values in this

N-dimensional representation. The gauge field is assumed to be a quenched variable with the gaussian measure:

$$P[A] = \exp\left[-\frac{1}{\sigma}\int \frac{d^2x}{\pi}\sum_a A_z^a A_{\bar{z}}^a\right] \tag{3.8}$$

At fixed disorder the partition function is:

$$Z[A] = e^{W[A]} = \int D\psi e^{-S[A]} = Det[i\partial\!\!\!/ + A\!\!\!/] \tag{3.9}$$

It can be expressed in terms of the WZW action [24], see below eq.(3.32).

The way to factorize the disorder consists in implementing a chiral gauge transformation. So we parametrized the vector potential by an element G of the complexified group $SU(N)^C$ as follows:

$$A_{\bar{z}} = G^{-1}(\partial_{\bar{z}}G) = -A_z^* \tag{3.10}$$

This is always possible on the sphere provided the connexion A is regular enough. This parametrization is unique up to left multiplication by a constant group element. Notice that the number of degree of freedom is preserved by this transformation: in the A or G parametrization there are $2dimSU(N)$ degree of freedom. We then have $\psi_-(\partial_{\bar{z}} + A_{\bar{z}})\psi_+ = (\psi_- G^{-1})\partial_{\bar{z}}(G\psi_+)$ and $\overline{\psi}_-(\partial_z + A_z)\overline{\psi}_+ = (\overline{\psi}_- G^*)\partial_z(G^{-1\,*}\overline{\psi}_+)$. We can therefore absorb the dependence on A by transforming the fermions as:

$$\begin{aligned} \hat{\psi}_- = \psi_- G^{-1} \quad &; \quad \hat{\overline{\psi}}_- = \overline{\psi}_- G^* \\ \hat{\psi}_+ = G\psi_+ \quad &; \quad \hat{\overline{\psi}}_+ = G^{-1\,*}\overline{\psi}_+ \end{aligned} \tag{3.11}$$

This is the chiral gauge transformation. It maps the action (3.7) into the free Dirac action for the gauge transformed fermions. We denote it $S_{free}[\hat{\psi}]$. The Jacobian for the transformation $\psi \to \hat{\psi}$ is non-trivial but given by the chiral anomaly:

$$\left|\frac{D\psi}{D\hat{\psi}}\right| = \frac{Det[i\partial\!\!\!/ + A\!\!\!/]}{Det[i\partial\!\!\!/]} = Z[A]$$

The extra factor $Det[i\partial\!\!\!/]$ is irrelevant when computing the quenched correlation function but relevant in the evaluation of the conformal anomaly. The crucial point is that this Jacobian is equal to the partition function at fixed disorder. Therefore, the partition and the Jacobian simplifies when computing the correlation functions at fixed disorder, and we get:

$$\begin{aligned} \langle\psi_- \cdots\rangle_A &= \frac{1}{Z[A]}\int D\psi e^{-S[A]}\psi_- \cdots \\ &= \int D\hat{\psi} e^{-S_{free}[\hat{\psi}]}(\hat{\psi}_- G)\cdots \end{aligned} \tag{3.12}$$

This is the announced factorization property: the correlation functions at fixed disorder factorize into the product of a correlation function in the free Dirac theory times an explicitly known functional of the vector potential.

We now have to average over the disorder. For this it is convenient, if not necessary, to change variable from A to G. The Jacobian is known, see e.g. [25]:

$$DA = DG \exp\left(2h^v S_{wzw}(GG^*)\right) \tag{3.13}$$

where DG is the Haar measure on $SU(N)^C$, $S_{wzw}(GG^*)$ is the WZW action and h^v is the dual Coxeter number, equals to N is the $SU(N)$ case. Since only the product GG^* enters into the Jacobian, we decompose $SU(N)^C$ as $(SU(N)^C/SU(N)) \times SU(N)$. For G this means:

$$G = HU \quad with \quad H \in SU(N)^C/SU(N), U \in SU(N)$$

The Haar measure DG factorizes on the Haar measure on $(SU(N)^C/SU(N))$ and $SU(N)$. Thus, we finally obtain:

$$DA = DUDH \exp\left(2h^v S_{wzw}(HH^*)\right) \tag{3.14}$$

In eq. (3.14) the part related to the compact space $SU(N)$ only involves the Haar measure, while the part related to the non-compact symmetric space $(SU(N)^C/SU(N))$ involves the WZW action. Although the WZW action seems to appears with a 'wrong' sign in eq.(3.14), the non-compact sigma model is well defined since the metric on $(SU(N)^C/SU(N))$ is negative definite.

For an arbitrary value of the disorder strength σ we also have to include the factor (3.8) into the measure for A. This factor couples U and H. But things simplify at $\sigma = \infty$. For this value of the coupling constant, we have:

$$P_{\sigma=\infty}[A]DA = DUDH \exp\left(2h^v S_{wzw}(HH^*)\right) \tag{3.15}$$

The compact and non-compact sectors are now independent. Moreover, the field U of the compact sector completely decouples from the quenched correlation functions of the local spinless neutral operators as e.g. $(\psi_- \overline{\psi}_+)$. Indeed, we have:

$$\overline{\langle(\psi_-\overline{\psi}_+)\cdots\rangle_A\cdots}_{\sigma=\infty} = \int DHD\hat{\psi} e^{2h^v S_{wzw}(HH^*) - S_{free}[\hat{\psi}]} \, (\hat{\psi}_- \, HH^* \, \overline{\hat{\psi}_+}) \cdots \tag{3.16}$$

Notice that a necessary condition for the cancelation of the U-dependence in eq.(3.16) is to consider $SU(N)$ scalar operators *i.e.* we have to sum over the color indices of the fermions ψ_- and $\overline{\psi}_+$ in order to multiply U and U^* to get $UU^* = 1$.

The free Dirac action plus the non-compact sigma model are conformally invariant. This is clear for the free massless Dirac theory. The sigma-model on the non-compact space $(SU(N)^C/SU(N))$ defined by the WZW action was studied in ref.[25] in connexion with path integral construction of the coset models. There it is was shown that it is a non-unitary conformal field theory which carries a representation of the affine Kac-Moody algebra of negative level $\tilde{k} = -2h^v$. Therefore, the Virasoro central charge of the non-compact sector is:

$$C = \frac{(-2h^v) dim\, su(N)}{(-2h^v) + h^v} = 2 dim\, SU(N) = 2(N^2 - 1) \tag{3.17}$$

The scaling dimension of the primary fields ϕ_R which belong to a representation R are given as usual by:

$$dim[\phi_R] = 2\frac{Cas(R)}{2((-2h^v)+h^v)} = -2\frac{Cas(R)}{2h^v} \tag{3.18}$$

with $Cas(R)$ the Casimir operator in R. The negative value of this dimension is an echo of the non-unitary character of the theory.

As an application of eq.(3.16) we can derive the scaling dimension of the operator $(\psi_-\overline{\psi}_+)$ in the quenched correlation functions. It is the sum of the scaling dimensions of the operator $(\hat{\psi}_-\hat{\overline{\psi}}_+)$ in the free Dirac theory plus the one of the operator (HH^*) in the vector representation ($\square$) in the non-compact WZW model. Explicitely:

$$dim_{\sigma=\infty}[\overline{(\psi_-\overline{\psi}_+)}] = 1 - 2\frac{Cas(\square)}{(-2N)} = 1 - \frac{N^2-1}{N^2}$$

This agrees with ref.[22, 21].

In summary, we have shown that at $\sigma = \infty$ we have the following equivalence, valid in the local spinless neutral sector:

$$\overline{Dirac_{A\sigma=\infty}} = (Free\ Dirac)^{\mathcal{N}} \times (SU(N)^C/SU(N))_{\tilde{k}=-2h^v}$$

with $\mathcal{N}$ the number of correlations we are averaging. This equivalence is a direct consequence of the factorization property eq.(3.15,3.16) of the random vector potential model.

3.3 Quenched current correlation functions

The previous factorization property is really useful only at infinite coupling σ. Here we describe how averages of current correlation functions can be computed for any value of σ. We also show that, as in the Abelian gaussian model, the averages of the chiral connected current correlation functions are unaffected by the disorder [23].

We are thus interested in computing the quenched average of the correlation functions of the currents $J^a_\mu = (\overline{\Psi}\gamma_\mu t^a\Psi)$. Explicitly, their components are:

$$\begin{aligned} J^a_z &= \psi_{-;j}(t^a)^j_k\psi^k_+ \\ J^a_{\bar{z}} &= \overline{\psi}_{-;j}(t^a)^j_k\overline{\psi}^k_+ \end{aligned} \tag{3.19}$$

While these currents are conserved in the pure system, they are not conserved once the disorder has been turned on. However, as explained in ref.[23], the quenched theory still possesses a $su(N)$ symmetry. The currents generating this symmetry are represented by the insertion of the following operators $\mathcal{J}^a_\mu$ in the quenched averages:

$$\mathcal{J}^a_\mu = \pi\frac{\delta}{\delta A^a_\mu}$$

Although the fields J^a_z are not conserved, their quenched correlation functions can be computed. This relies on the Polyakov-Wiegmann (PW) formula [24] for the effective action $W[A]$ defined in eq.(3.9). It can be exactly computed by integrating its anomalous transformation under a chiral gauge transformation. Indeed, let $G_\mu = \sum_a t^a G^a_\mu[A]$ with $G^a_\mu[A] = \langle J^a_\mu \rangle_A = \pi \frac{\delta W[A]}{\delta A^a_\mu}$ be the generating functions of the connected current Green function in the pure system. They satisfy the anomalous Ward identities [24],

$$\begin{aligned} \partial_z G_{\bar z} + \partial_{\bar z} G_z + [A_z, G_{\bar z}] + [A_{\bar z}, G_z] &= 0, \\ \partial_z G_{\bar z} - \partial_{\bar z} G_z + [A_z, G_{\bar z}] - [A_{\bar z}, G_z] &= 2F_{z\bar z}[A], \end{aligned} \tag{3.20}$$

with $F_{z\bar z}[A] = \partial_z A_{\bar z} - \partial_{\bar z} A_z + [A_z, A_{\bar z}]$. Eqs.(3.20) completely specify $G_\mu[A]$. The solution of eqs.(3.20) can written as:

$$G_z[A] = A_z - \frac{1}{\partial_{\bar z} + ad.A_{\bar z}} \partial_z A_{\bar z} \tag{3.21}$$

$$G_{\bar z}[A] = A_{\bar z} - \frac{1}{\partial_z + ad.A_z} \partial_{\bar z} A_z \tag{3.22}$$

Here $ad.A_z$ denoted the adjoint action of A_z. Notice that $G_z[A]$ is local in A_z but non-local in $A_{\bar z}$.

Knowing explicitly the correlation functions for any impurity sample, it is a priori possible to take the quenched average. Let us first concentrate on the average of the chiral correlation functions involving only currents of the same chirality; e.g. involving only J^a_z. Consider first the average of products of one-point functions. Since the quenched average is defined by $\overline{A^a_{\bar z}(x) A^b_{\bar z}(y)} = \sigma\pi\delta^{ab}\delta^{(2)}(x-y)$ and since $\langle J^a_z \rangle_A$ are linear in A_z, these quenched correlations can be computed using Wick's theorem applied on A. For example, the two-point and three-point functions are:

$$\begin{aligned} \overline{\langle J^a_z(z_1)\rangle_A \langle J^b_z(z_2)\rangle_A} &= 2\sigma\pi\delta^{ab} \left(\frac{1}{\partial_{\bar z}}\partial_z\right)_{(z_1,z_2)} = \frac{2\sigma\delta^{ab}}{(z_1-z_2)^2} \\ \overline{\langle J^a_z(z_1)\rangle_A \langle J^b_z(z_2)\rangle_A \langle J^c_z(z_3)\rangle_A} &= i(\pi\sigma)^2 f^{abc} \left(\frac{1}{\partial_{\bar z}}\partial_z\right)_{(z_1,z_2)} \\ &\quad \times \left[\left(\frac{1}{\partial_{\bar z}}\right)_{(z_3,z_1)} - \left(\frac{1}{\partial_{\bar z}}\right)_{(z_3,z_2)}\right] \\ &\quad + (\text{cyclic permutation}) \\ &= 3\sigma^2 \frac{i f^{abc}}{(z_1-z_2)(z_1-z_3)(z_2-z_3)} \end{aligned}$$

More generally, the average of products of one-point functions is the sum of connected correlations which can be expressed in terms of the correlation functions of the pure system:

$$\left[\overline{\langle J^{a_1}_z(z_1)\rangle_A \cdots \langle J^{a_M}_z(z_M)\rangle_A}\right]^{con.} = M\sigma^{M-1} \langle J^{a_1}_z(z_1) \cdots J^{a_M}_z(z_M)\rangle_0 \tag{3.23}$$

Here, $\langle \cdots \rangle_0$ denote the pure correlation functions. They are known exactly [26]. The relation between the quenched correlations and their connected parts is the usual one.

More interesting is the average of products of correlations with insertion of the conserved currents $\mathcal{J}_\mu^a$ since they encode the underlying symmetry algebra. We find:

$$\begin{aligned}
&\overline{\mathcal{J}_z^{n_1}(z_1)\cdots\mathcal{J}_z^{n_M}(z_M)\langle J_z^{a_1}(w_1)J_z^{b_1}(\xi_1)\cdots\rangle_A^c\cdots\langle J_z^{a_P}(w_P)J_z^{b_P}(\xi_P)\cdots\rangle_A^c} \\
= &\ \langle J_z^{n_1}(z_1)\cdots J_z^{n_M}(z_M)J_z^{a_1}(w_1)J_z^{b_1}(\xi_1)\cdots\rangle_0\cdots\langle J_z^{a_P}(w_P)J_z^{b_P}(\xi_P)\cdots\rangle_0 \\
+ &\ \sum_{j=1}^{M}\langle J_z^{n_1}(z_1)\cdots\widehat{J_z^{n_j}(z_j)}\cdots J_z^{n_M}(z_M)J_z^{a_1}(w_1)J_z^{b_1}(\xi_1)\cdots\rangle_0\times \\
&\qquad\times\langle J_z^{n_j}(z_j)J_z^{a_2}(w_2)J_z^{b_2}(\xi_2)\cdots\rangle_0\cdots\langle J_z^{a_P}(w_P)J_z^{b_P}(\xi_P)\cdots\rangle_0 \\
+ &\ \cdots \\
+ &\ \langle J_z^{a_1}(w_1)J_z^{b_1}(\xi_1)\cdots\rangle_0\cdots\langle J_z^{n_1}(z_1)\cdots J_z^{n_M}(z_M)J_z^{a_P}(w_P)J_z^{b_P}(\xi_P)\cdots\rangle_0
\end{aligned} \tag{3.24}$$

The hatted fields have to be omitted. Here, we assumed that there is no insertion of one-point functions. The formula (3.24) is actually simpler in words: it is obtained by distributing the currents $J_z^{n_1}(z_1),\cdots,J_z^{n_M}(z_M)$ among the pure correlators in all possible way, each counted only once. Notice that all the chiral quenched correlation functions are purely algebraic, without any logarithmic correction.

Eq.(3.24) with no insertion of $\mathcal{J}_z^n$ shows that the connected correlation function of the fields J_z^a are unaffected by the disorder.

From eq.(3.24) we read the operator product expansion of the fields. The currents $\mathcal{J}_z^a$ satisfy:

$$\mathcal{J}_z^{n_1}(z_1)\mathcal{J}_z^{n_2}(z_2)=\frac{\delta^{n_1n_2}}{(z_1-z_2)^2}+\frac{if^{n_1n_2n_3}}{z_1-z_2}\mathcal{J}^{n_3}(z_2)+reg. \tag{3.25}$$

Therefore, the quenched conserved currents satisfy the commutation relations of a Kac-Moody algebra, exactly as the currents in the pure system do.

The operator product expansion between the conserved currents $\mathcal{J}_a^z$ and the correlators $\langle J_z^{a_1}(w_1)J_z^{a_2}(w_2)\cdots\rangle_A$ are:

$$\begin{aligned}
\mathcal{J}_z^n(z)\langle J_z^{a_1}(w_1)J_z^{a_2}(w_2)\cdots\rangle_A^c \sim &\sum_j\frac{\delta^{na_j}}{(z-w_j)^2}\langle J_z^{a_1}(w_1)\cdots\widehat{J_z^{a_j}(w_j)}\cdots\rangle_A^c \\
&+\sum_j\frac{if^{na_ja'_j}}{z-w_j}\langle J_z^{a_1}(w_1)\cdots J_z^{a'_j}(w_j)\cdots\rangle_A^c \\
&+reg.
\end{aligned} \tag{3.26}$$

These operator product are unusual in conformal field theory with Kac-Moody symmetry. In particular, they imply that the fields $\langle J_z^{a_1}(w_1)J_z^{a_2}(w_2)\cdots\rangle_A^c$ are *not* associated to highest weight vector representations. We don't know to which category of representations they correspond to.

Contrary to the chiral quenched correlation functions which are easy to compute, the averages of correlation functions involving fields of opposite chiralities are difficult, if not impossible, to evaluate directly. This is due to the fact that the generating functions $G_z[A]$ and $G_{\bar z}[A]$ are non-local in $A_{\bar z}$ and A_z respectively. A naive perturbative expansion is spoiled by untractable divergences. However, since the model is asymptotically free in the infrared, one can used renormalization group techniques to evaluate the large distance behavior of a few two-point functions [23].

3.4 Generalization to any WZW model

The previous discussion generalizes easily to any WZW models. So let us consider a WZW model over the Lie group $\mathcal{G}$ at level k. Its action is [37, 26]:

$$S_{wzw}(g) = k\int \frac{d^2x}{8\pi} tr(\partial_z g^{-1}\partial_{\bar{z}} g) + k\Gamma(g)$$

where $\Gamma(g)$ is the WZW topological term. We define the action of the disordered WZW model to be the WZW action in presence of sources. Namely:

$$\begin{aligned} S_{wzw}(g;A) &= S_{wzw}(g) \\ &-k\int \frac{d^2x}{2\pi}\Big(tr(A_z J_{\bar{z}} + A_{\bar{z}} J_z) + tr(g A_z g^{-1} A_{\bar{z}}) - tr(A_z A_{\bar{z}})\Big) \end{aligned} \quad (3.27)$$

where $J_z, J_{\bar{z}}$ are the WZW currents and $A_z, A_{\bar{z}}$ the random variables. We assume that the probability distribution $P[A]$ is given by eq.(3.8).

As a functional of g and A the action (3.27) satisfies the "Polyakov-Wiegmann" relation:

$$S_{wzw}(g;A) = S_{wzw}(g^h;A^h) - S_{wzw}(hh^*;A^h), \quad (3.28)$$

with

$$\begin{aligned} A^h_{\bar{z}} &= hA_{\bar{z}}h^{-1} + h\partial_{\bar{z}}h^{-1} \\ g^h &= hgh^* \end{aligned} \quad (3.29)$$

for any element h of the complexified group $\mathcal{G}^C$.

If we parametrize the impurity connexion as in previous section by $A_{\bar{z}} = G^{-1}\partial_{\bar{z}}G$ with $G \in \mathcal{G}^C$, this action action has been cooked up such that:

$$S_{wzw}(g;A) = S_{wzw}(g^G) - S_{wzw}(GG^*) \quad (3.30)$$

with

$$A_{\bar{z}} = G^{-1}\partial_{\bar{z}}G = -A^*_z \quad (3.31)$$

This follows from eq.(3.28) and the fact that $A^G = 0$.

This has two consequences. First we can explicitly evaluate the partition function for any sample of the disorder. Namely, we have:

$$Z[A] = \int Dg e^{-S_{wzw}(g;A)} = \exp(S_{wzw}(GG^*)) \quad (3.32)$$

In the last equation we used the fact that for the Haar measure on $\mathcal{G}$ we have: $\int dg f(g_1 g g_2^*) = \int dg f(g)$ for any $g_1, g_2 \in \mathcal{G}^C$ and any function regular enough.

The second consequence concerns the correlation functions of the $\mathcal{G}$-scalar local operators at fixed disorder. For example consider the operator

$$\Phi_R(z,\bar{z}) = \sum_i \overline{\phi}_{\overline{R};i}(\bar{z})\phi^i_R(z)$$

where $\phi^i_R(z)$ the primary fields in the representation R. These operators are invariant under the global $\mathcal{G}$ symmetry. In the path integral formulation, they are represented by the insertion of $tr(\rho_R(g))$, ie:

$$\Phi_R(z,\bar{z}) \equiv tr(\rho_R(g)) \tag{3.33}$$

where ρ_R denotes the representation R of $\mathcal{G}$. We then have:

$$\begin{aligned}\langle \mathrm{tr}\ \rho_R(g)\cdots\rangle_A &= \int \frac{Dg}{Z[A]} e^{-S_{wzw}(g;A)} tr(\rho_R(g))\cdots \\ &= \int \frac{Dg}{Z[A]} e^{-S_{wzw}(g^G)+S_{wzw}(GG^*)} tr(\rho_R(g))\cdots\end{aligned}$$

where we have used eq.(3.30). Now we see from eq.(3.32) that the partition function $Z[A]$ cancels against $\exp(S_{wzw}(GG^*))$. Therefore, using the invariance of the Haar measure, we obtain:

$$\langle tr(\rho_R(g))\cdots\rangle_A = \int Dg e^{-S_{wzw}(g)} tr(\rho_R(g)\rho_R(GG^*)^{-1})\cdots \tag{3.34}$$

We have used the cyclicity of the trace in order to reconstitute the product GG^*. In other words, we have used the fact that the operator $tr(\rho_R(g))$ is $\mathcal{G}$ invariant. This is important because it proves that the G-dependence of these correlation functions projects from $\mathcal{G}^C$ to $\mathcal{G}^C/\mathcal{G}$. Indeed, let us factorized $G \in \mathcal{G}^C$ as:

$$G = HU \quad with \quad H \in \mathcal{G}^C/\mathcal{G} \quad and \quad U \in \mathcal{G} \tag{3.35}$$

Then since $UU^* = 1$, we have:

$$\langle \mathrm{tr}\ \rho_R(g)\cdots\rangle_A = \int Dg e^{-S_{wzw}(g)} tr(\rho_R(g)\rho_R(HH^*)^{-1})\cdots \tag{3.36}$$

We thus have proved that the correlation functions of the neutral local operators factorize into the WZW correlation functions times an explicitely known function of the $\mathcal{G}^C/\mathcal{G}$ component of the impurity connexion. This is analogue to the factorization of the previous section.

Using the factorization (3.34) we can now average over A. As before, it is convenient to change variable from A to $G = HU$. The Jacobian is given in eq.(3.14). In particular at $\sigma = \infty$, we obtain:

$$\overline{\langle \mathrm{tr}\ \rho_R(g)\cdots\rangle_A\cdots} = \int DHDg e^{2h^v S_{wzw}(HH^*)-S_{wzw}(g)} tr(\rho_R(g)\rho_R(HH^*)^{-1})\cdots \tag{3.37}$$

Therefore, we have proved the following equivalence, valid in the $\mathcal{G}$-neutral sector:

$$\overline{WZW^{(k)}_{\mathcal{G}}}_A = \left(WZW^{(k)}_{\mathcal{G}}\right)^{\mathcal{N}} \times \left(WZW^{(\tilde{k}=-2h^v)}_{\mathcal{G}^C/\mathcal{G}}\right) \tag{3.38}$$

where $\mathcal{N}$ is the number of correlation functions we are averaging. These are conformal field theories. The conformal dimensions of the operators in the disordered theory can be evaluate using the formula (3.18).

4 THE SUPERSYMMETRIC METHOD: THE RANDOM BOND ISING MODEL

In this section, we apply the supersymmetric method to study the random bond Ising model and the closely related Dirac theory coupled to a random potential and a random mass. These models have been analysed using the replica method in ref.[2, 3, 4].

As we will soon explain, the randomness of the bond interaction has a marginal effect on the large distance behavior of the Ising model. This allows us to use renormalization group techniques to study the infrared behavior of this random model. At criticality, the disorder only induces logarithmic corrections to the pure system. For example, the averages of the spin correlation functions at criticality are [3, 4]:

$$\overline{\langle\sigma(R)\sigma(0)\rangle^N} \sim \left(\frac{a}{R}\right)^{N/4} (\log(R/a))^{N(N-1)/8}$$

These have to be compared to the pure correlation function which are:

$$\langle\sigma(R)\sigma(0)\rangle^N \sim \left(\frac{a}{R}\right)^{N/4}.$$

Close to criticality, the specific heat of the Ising model possesses a logarithmic divergency: $C_v = \log(\tau)$ with $\tau \sim (T_C - T)/T_C$. In the disordered model, the behavior of the specific heat near criticality reads [2]:

$$C_v \sim \frac{1}{g_a}\log\left(1 + g_a \log(\frac{1}{\tau})\right)$$

where g_a represents the strength of the disorder. This $\log(\log\tau)$ behavior has to be compared to the $(\log\tau)$ behavior in the pure system.

The perturbative study of the Dirac theory with a random potential and a random mass is very similar to the perturbative study of the random mass Ising model. However, the crucial difference is that contrary to the randomness of the mass, the randomness of the potential is marginally relevant.

Although the supersymmetric method is only applicable to models which are gaussian at fixed disorder, it is certainly a good starting point for disentangling properties of a large class of disordered conformal field theories. The study of random bond Ising model clearly reveals that an appropriate algebraic framework for studying gaussian disordered systems at criticality should be based on affine Lie superalgebra with zero superdimension. In the case of the random Ising model, this algebra is $OSp(2N|2N)$. The fact that the algebra has zero superdimension, *i.e.* has an equal number of bosonic and fermionic generators, ensures that the Virasoro central charge of the Sugawara stress-tensor is zero. This is needed by construction for a disordered system. The vanishing of the central charge does not imply the triviality of the theory since it is non-unitary. The general framework should probably be based on the Wess-Zumino-Witten models, or their cosets, on Lie superalgebras which, like $OSp(2N|2N)$, have equal numbers of bosonic and fermionic generators. The relevance of affine Lie superalgebras was independently realized in ref.[21]

4.1 The model

In its scaling limit (near criticality) the Ising model is described by a massive real Majorana fermion with mass $\overline{m} \sim \tau = \frac{(T_C - T)}{T_C}$ with T_C the critical temperature. In presence of disorder, the mass becomes a function of the space position. The random Ising model is defined by the action ($z = x + iy$):

$$S[m(x)] = \int \frac{d^2x}{4\pi} \left(\psi \partial_{\bar{z}} \psi + \bar{\psi} \partial_z \bar{\psi} + i m(x) \bar{\psi} \psi \right) \tag{4.1}$$

where ψ are grassmanian variables. The mass $m(x)$ is coupled to the energy operator: $\epsilon(x) = i\bar{\psi}(x)\psi(x)$. It is choosen to be a random quenched variable with a gaussian measure:

$$P[m] = \exp\left(-\frac{1}{4g} \int \frac{d^2x}{2\pi} (m(x) - \overline{m})^2 \right) \tag{4.2}$$

The energy operator $\epsilon(x)$ has dimension one. The Harris criteria thus tell us that randomness in the bond interaction is marginal in the 2d Ising model. As we will see, it is not exactly marginal but only marginally irrelevant.

Since, at fixed disorder the action (4.1) is a free quadratic action, we can use the supersymmetric formulation to study the quenched model. In order to compute averages of product of correlation functions, we need to introduce an arbitrary number of copies of the fermions and of their supersymmetric partners. For simplicity, let us restrict to the case of two copies. The generalization to an arbitrary number of copies will be given later. Let us denote by ψ_1 and ψ_2 the two real Majorana fermions, and introduce the complex fermions $\psi_\pm$ by:

$$\psi_\pm = \frac{1}{\sqrt{2}} (\psi_1 \pm i\psi_2) \tag{4.3}$$

At fixed disorder, the fermionic action is:

$$\begin{aligned} S_F &= S_1[m(x)] + S_2[m(x)] \\ &= \int \frac{d^2x}{4\pi} \left(\psi_1 \partial_{\bar{z}} \psi_1 + \psi_2 \partial_{\bar{z}} \psi_2 + \bar{\psi}_1 \partial_z \bar{\psi}_1 + \bar{\psi}_2 \partial_z \bar{\psi}_2 + i m(x) (\bar{\psi}_1 \psi_1 + \bar{\psi}_2 \psi_2) \right) \\ &= \int \frac{d^2x}{2\pi} \left(\psi_- \partial_{\bar{z}} \psi_+ + \bar{\psi}_- \partial_z \bar{\psi}_+ + i \frac{m(x)}{2} (\bar{\psi}_- \psi_+ - \psi_- \bar{\psi}_+) \right) \end{aligned} \tag{4.4}$$

By definition this action can be used to compute products of two Ising correlation functions. Namely:

$$\langle A(z) \cdots \rangle_{m(x)} \langle B(w) \cdots \rangle_{m(x)} = \frac{1}{Z[m(x)]} \int D\psi_1 D\psi_2 A_1(z) \cdots B_2(w) \cdots e^{-S_F}$$

where $\langle \cdots \rangle_{m(x)}$ refers to the Ising correlation function at fixed desorder. $A_1(z)$ and $B_2(w)$ are the expressions of the Ising operators $A(z)$, $B(w)$ in the first and second copies. The partition function is a determinant:

$$Z[m(x)] = \int D\psi_\pm e^{-S_F} = Det \begin{pmatrix} -i\frac{m}{2} & \partial_{\bar{z}} \\ \partial_z & i\frac{m}{2} \end{pmatrix} = Det\ H_{Dirac} \tag{4.5}$$

Its inverse can be represented as a path integral over bosonic complex fields η and γ:

$$\frac{1}{Z[m(x)]} = \int D\eta D\gamma e^{-S_B}$$

with

$$S_B = \int \frac{d^2x}{2\pi} \left(\eta \partial_{\bar{z}} \gamma + \bar{\eta} \partial_z \bar{\gamma} + i \frac{m(x)}{2} (\bar{\eta}\gamma - \eta\bar{\gamma}) \right) \tag{4.6}$$

Since the Dirac Hamiltonian H_D, as defined in (4.5), is purely imaginary, $H_D^\dagger = -H_D$, the integral over the bosonic variables is an integral of an imaginary gaussian if we defined the complex conjugasion by $\eta^* = \bar{\gamma}, \gamma^* = \bar{\eta}$. To insure absolute convergence, we could add the term $\epsilon \int \frac{d^2x}{\pi} (\bar{\eta}\gamma + \eta\bar{\gamma})$ in the action S_B.

The total action is $S_{tot} = S_F + S_B$. By construction, products of Ising correlation functions at fixed disorder can be rewritten as:

$$\langle A(z) \cdots \rangle_{m(x)} \langle B(w) \cdots \rangle_{m(x)} = \int D\psi_\pm D\eta D\gamma A_1(z) \cdots B_2(w) \cdots e^{-S_{tot}}$$

The action S_{tot} is supersymmetric. The supersymmetric transformation can be defined as:

$$\begin{aligned} \delta\psi_- &= \eta \quad , \quad \delta\psi_+ = 0 \\ \delta\eta &= 0 \quad , \quad \delta\gamma = -\psi_+ \end{aligned}$$

Similarly for the $\overline{\psi}_\pm, \cdots$. It is easy to check that the kinetic term in S_{tot} is susy invariant: $\delta(\psi_- \partial_{\bar{z}} \psi_+ + \eta \partial_{\bar{z}} \gamma) = 0$. Moreover the field coupled to the mass terms in S_{tot} is susy exact:

$$\left(\bar{\psi}_- \psi_+ - \psi_- \bar{\psi}_+ + \bar{\eta}\gamma - \eta\bar{\gamma} \right) = \delta \left(\bar{\psi}_- \gamma - \psi_- \bar{\gamma} \right)$$

This insures that the partition function defined by S_{tot} is independent of the random mass $m(x)$, exactly as BRS symmetry ensures gauge invariance in gauge theories. We actually have a stronger result: S_{tot} is susy-exact. This is similar to topological theory; however, the difference with topological theory is that we are interested in correlation functions of non susy-closed operators. As we will explain below, this choice of the supersymmetric transformation is not unique.

4.2 The effective action

We now set $\overline{m} = 0$, *i.e.* we consider the effect of the disorder at criticality. Averaging over the quenched variables with the gaussian measure (4.2) leads to the effective action:

$$\begin{aligned} S_{eff} &= \int \frac{d^2x}{2\pi} \left(\psi_- \partial_{\bar{z}} \psi_+ + \bar{\psi}_- \partial_z \bar{\psi}_+ + \eta \partial_{\bar{z}} \gamma + \bar{\eta} \partial_z \bar{\gamma} \right) + \frac{g}{8} \int \frac{d^2x}{\pi} \Phi_{pert} \\ &= S_* + \frac{g}{8} \int \frac{d^2x}{\pi} \Phi_{pert} \end{aligned} \tag{4.7}$$

with

$$\begin{aligned}\Phi_{pert} &= \left(\bar{\psi}_-\psi_+ - \psi_-\bar{\psi}_+ + \bar{\eta}\gamma - \eta\bar{\gamma}\right)^2 \\ &= 2(\bar{\psi}_1\psi_1)(\bar{\psi}_2\psi_2) + (susy\ partners).\end{aligned} \tag{4.8}$$

Clearly this action is supersymmetric since the total action S_{tot} was supersymmetric for any values of the disorder before averaging.

This action can be viewed as a perturbation of the (non-unitary) conformal field theory specified by the action S_*. This fixes the normalization of the fields to be:

$$\psi_-(z)\psi_+(w) \sim \frac{1}{z-w} \quad , \quad \gamma(z)\eta(w) \sim \frac{1}{z-w} \tag{4.9}$$

The Virasoro algebra is the standard one. The Virasoro central charge is zero. Note that since the fermions $\psi_\pm$ have dimension half, the perturbing field Φ_{pert} has dimension two. It is therefore marginal.

The conformal field theory specified by S_* is invariant under a supersymmetric algebra whose conserved currents are:

$$\begin{aligned} G_\pm(z) = \eta(z)\psi_\pm(z) \quad &, \quad \hat{G}_\pm(z) = \gamma(z)\psi_\pm(z) \\ K(z) = \eta^2(z) \quad &, \quad \hat{K}(z) = \gamma^2(z) \\ J(z) =: \psi_-(z)\psi_+(z) : \quad &, \quad H(z) =: \gamma(z)\eta(z) : \end{aligned} \tag{4.10}$$

There are four fermionic currents, which are generators of supersymmetric transformations, and four bosonic ones. They form a representation of the affine $OSp(2|2)$ current algebra at level one. The root system of $OSp(2|2)$ is given below:

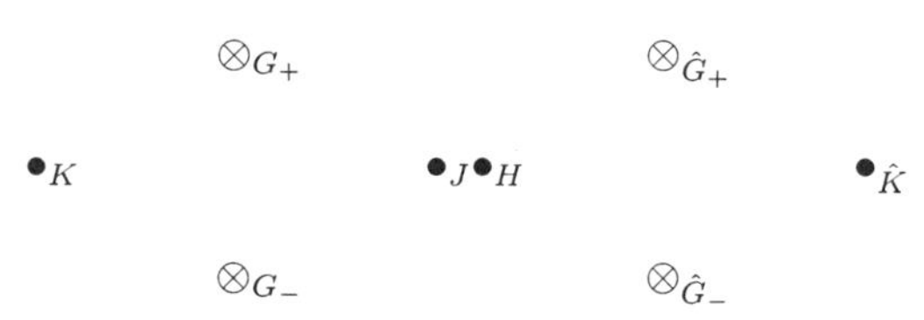

The black dots refer to the bosonic generators, while the symbols $\otimes$ represent the fermionic roots. The non-trivial operator product expansions of the currents are the following:

$$\begin{aligned} J(z)J(w) \sim \frac{1}{(z-w)^2} \quad &; \quad H(z)H(w) \sim \frac{-1}{(z-w)^2} \\ J(z)G_\pm(w) \sim \frac{\pm 1}{(z-w)}G_\pm(w) \quad &; \quad J(z)\hat{G}_\pm(w) \sim \frac{\pm 1}{(z-w)}\hat{G}_\pm(w) \\ H(z)G_\pm(w) \sim \frac{1}{(z-w)}G_\pm(w) \quad &; \quad H(z)\hat{G}_\pm(w) \sim \frac{-1}{(z-w)}\hat{G}_\pm(w) \\ H(z)K(w) \sim \frac{2}{(z-w)}K(w) \quad &; \quad H(z)\hat{K}(w) \sim \frac{-2}{(z-w)}\hat{K}(w) \\ \hat{G}_\pm(z)G_\mp(w) \quad \sim \quad & \frac{1}{(z-w)^2} + \frac{1}{(z-w)}(H(w) \pm J(w)) \end{aligned} \tag{4.11}$$

$$\hat{K}(z)K(w) \sim \frac{2}{(z-w)^2} + \frac{4}{(z-w)}H(w)$$

$$G_-(z)G_+(w) \sim \frac{1}{(z-w)}K(w) \quad ; \quad \hat{G}_-(z)\hat{G}_+(w) \sim \frac{1}{(z-w)}\hat{K}(w)$$

$$K(z)\hat{G}_\pm(w) \sim \frac{-2}{(z-w)}G_\pm(w) \quad ; \quad \hat{K}(z)G_\pm(w) \sim \frac{2}{(z-w)}\hat{G}_\pm(w)$$

The perturbing field can be written in terms of these fields:

$$\Phi_{pert} = 2\left[\overline{J}J - \overline{H}H + \frac{1}{2}(\overline{K}\hat{K} + \overline{\hat{K}}K) + \overline{G}_-\hat{G}_+ - \overline{\hat{G}_-}G_+ + \overline{G}_+\hat{G}_- - \overline{\hat{G}_+}G_-\right]$$

In other words, the perturbation (4.8) is a current-current perturbation. It preserves the $OSp(2|2)$ symmetry.

The generalization to $2N$ copies is obvious. We introduce $2N$ copies of Majorana fermions ψ_α, $\alpha = 1, \cdots, 2N$, or N copies of Dirac fermions $\psi_\pm^a$, $a = 1, \cdots, N$, and $2N$ copies of their supersymmetric partners. The perturbing field is similarly obtained by a gaussian integral from the free random theory:

$$\Phi_{pert} = 2\sum_{\alpha<\beta}(\bar{\psi}_\alpha\psi_\alpha)(\bar{\psi}_\beta\psi_\beta) + (susy\ partners) \tag{4.12}$$

The symmetry algebra is now extended to the affine $OSp(2N|2N)$ current algebra. The perturbing field is still bilinear in the currents. Hence, the random Ising model is described by a Gross-Neveu model on the Lie superalgebra $OSp(2N|2N)$. In particular it is an integrable model, with factorizable scattering, Yangian symmetry, etc.

4.3 Renormalization group computation

The action S_{eff} describes an interacting theory. Although the perturbing field has dimension two, and therefore is naively marginal, it breaks conformal invariance. To see it requires computing the beta function. As explained in the Appendix, the beta function at one loop is encoded in the operator product expansion (OPE) of the perturbing field. Since S_* is a free gaussian theory, this OPE is easily computed using Wick's theorem and eqs. (4.9). We obtain:

$$\Phi_{pert}(z)\Phi_{pert}(w) = \frac{8}{|z-w|^2}\Phi_{pert}(w) + \text{irrelevantterms.}$$

Using the formula (8.3) of the Appendix, we get the beta function:

$$\dot{g} = \beta(g) = -g^2 + \cdots \tag{4.13}$$

The dots refer to higher loop corrections. We recognize the beta function for asymptotically free theory. It is easily integrated, giving the coupling constant flow:

$$g_R = \frac{g_a}{1 + g_a \log(R/a)}$$

where g_a is the value of the coupling constant at the lattice cut-off a. Since g_a is positive by definition (it is the strengh of the disorder), g_R decreases at large distances R. In other words, the theory is asymptotically free in the infrared regime.

The beta function is independent of the number of copies, *i.e.* of N.

4.4 Quenched correlation functions

The fact that the random Ising model is asymptotically free in the infrared regime is important since it allows us to compute the large distance behavior of two point quenched correlation functions using the renormalization group, (using formula (8.4) of the Appendix for the anomalous dimensions).

We recall the OPE between the energy operator and the spin field. The fusion rules are $\epsilon \times \sigma = \sigma$. More precisely, we have:

$$\epsilon(z)\sigma(w) = i(\bar{\psi}\psi)(z)\sigma(w) = \frac{C_{\epsilon\sigma\sigma}}{|z-w|}\sigma(w) + \cdots$$

with $C^2_{\epsilon\sigma\sigma} = 1/4$.

Let us consider the quenched average of products of two point functions of the spin field, *i.e.* $\overline{\langle\sigma(R)\sigma(0)\rangle^N}$. Consider first the case $N = 1$. In the supersymmetric effective theory, the quenched correlation function $\overline{\langle\sigma(R)\sigma(0)\rangle}$ is represented by the two point function of the spin field σ_1 in the first copy. Its OPE with the perturbing field is :

$$\Phi_{pert}(z)\sigma_1(w) = \frac{1}{|z-w|}\mathcal{O}(w) + \cdots$$

where $\mathcal{O}(w)$, an operator of dimension $(1/8 + 1)$, is irrelevant compared to σ_1 which has dimension $1/8$. Therefore, the anomalous dimension of σ_1 at one loop is $\gamma^{(1)} = 1/8 + \cdots$, up to irrelevant terms. Hence:

$$\overline{\langle\sigma(R)\sigma(0)\rangle} \sim \left(\frac{1}{R}\right)^{1/4} \tag{4.14}$$

It has the same infrared behavior as the spin-spin correlation function in the pure system at the critical temperature.

Consider now the general case N arbitrary. In the supersymmetric effective action, the quenched correlation function $\overline{\langle\sigma(R)\sigma(0)\rangle^N}$ is represented by the two point function of the operator $\mathcal{O}^{(N)}(z)$, product of the spin fields in the N first copies:

$$\mathcal{O}^{(N)}(z) = \sigma_1(z)\cdots\sigma_N(z)$$

When computing the OPE between the perturbing field Φ_{pert} and $\mathcal{O}^{(N)}(z)$ only the first term in eq.(4.12), $2\sum_{\alpha<\beta}(\bar{\psi}_\alpha\psi_\alpha)(\bar{\psi}_\beta\psi_\beta)$, gives a relevant contribution. When computing the OPE between this term and $\mathcal{O}^{(N)}(z)$ we can contract any pair of products of $(\bar{\psi}_\alpha\psi_\alpha)(\bar{\psi}_\beta\psi_\beta)$ with the pair of product of spin fields $\sigma_\alpha\sigma_\beta$. E.g.

$$(\bar{\psi}_1\psi_1)(\bar{\psi}_2\psi_2)\sigma_1(z)\sigma_2(z)\cdots\sigma_N(z)$$

Thus we have:

$$\Phi_{pert}(z)\mathcal{O}^{(N)}(w) = -\frac{N(N-1)C^2_{\epsilon\sigma\sigma}}{|z-w|^2}\mathcal{O}^{(N)}(w) + \textit{irrelevant terms}.$$

Therefore, the anomalous dimension of $\mathcal{O}^{(N)}(w)$ is

$$\gamma^{(N)} = \frac{N}{8} - \frac{N(N-1)C^2_{\epsilon\sigma\sigma}}{4} g + \cdots$$

Recall that $C^2_{\epsilon\sigma\sigma} = 1/4$. It implies:

$$\begin{aligned}\overline{\langle\sigma(R)\sigma(0)\rangle^N}_{g_a} &= \exp\left(-2\int_{g_a}^{\sigma_R} \frac{\gamma^{(N)}}{\beta} d\sigma\right) \overline{\langle\sigma(a)\sigma(0)\rangle^N}_{g_R} \\ &\sim \left(\frac{a}{R}\right)^{N/4} (\log(R/a))^{N(N-1)/8} \end{aligned} \tag{4.15}$$

There are logarithmic corrections to the pure system.

4.5 The specific heat near criticality

In order to compute the behavior of the mean specific heat near the critical point we consider a non zero value of $\overline{m}$:

$$\overline{m} = \frac{T_C - T}{T_C} = \tau$$

The specific heat C_v per unity of volume $C_v = \frac{1}{Vol.}\left(\frac{\partial E}{\partial T}\right)$ can be expressed in terms of the connected correlation function of the energy operator $\epsilon(x)$:

$$C_v = \int d^2x \overline{\langle\epsilon(x)\epsilon(0)\rangle^{conn.}}$$

Near criticality, its behavior is given by the infrared singular behavior of $\overline{\langle\epsilon(x)\epsilon(0)\rangle^{conn.}}$ at zero mass. The infrared cut-off is specified by the mass scale $1/\overline{m}$. Thus:

$$C_v \sim \int^{1/\overline{m}} d^2x \overline{\langle\epsilon(x)\epsilon(0)\rangle^{conn.}}_{\overline{m}=0} \tag{4.16}$$

In the supersymmetric effective theory, the connected two-point correlation function for the energy operator is represented by the two-point function of the operator $\epsilon'(x) = \epsilon_1(x) - \epsilon_2(x)$ where $\epsilon_{1,2}(x)$ refer to the energy operators in the first and second copies:

$$2\overline{\langle\epsilon(x)\epsilon(0)\rangle^{conn.}} = \langle\epsilon'(x)\epsilon'(0)\rangle$$

To evaluate the behavior of C_v as $\overline{m} \to 0$ requires evaluating the infrared behavior of $\langle\epsilon'(x)\epsilon'(0)\rangle$. This can be done using the renormalization group by computing the anomalous dimension of $\epsilon'(x)$. The OPE between $\epsilon'(x)$ and the perturbing field Φ_{pert} follows from the fusion rule, $\epsilon \times \epsilon = 1 + \epsilon$, in the Ising model:

$$\Phi_{pert}(z)\epsilon'(w) = \frac{2}{|z-w|^2}\epsilon'(w) + \cdots$$

Formula (8.4) of the Appendix gives:

$$\gamma_{\epsilon'}(g) = 1 + \frac{1}{2} g + \cdots$$

Therefore, at large distances we have:

$$\langle \epsilon'(R)\epsilon'(0)\rangle \sim \left(\frac{a}{R}\right)^2 \frac{1}{(1+g_a\log(R/a))}$$

This gives the critical behavior of the specific heat:

$$\begin{aligned} C_v &\sim \int^{1/\tau} d^2x\overline{\langle \epsilon(x)\epsilon(0)\rangle^{conn.}} \\ &\sim \frac{1}{g_a}\log\left(1+g_a\log(\frac{1}{\tau})\right) \end{aligned} \tag{4.17}$$

This $\log(\log\tau)$ behavior has to be compared to the $(\log\tau)$ behavior in the pure system.

4.6 Miscellaneous remarks on the random Dirac theory

A model[2] very closely connected to the random bond Ising model has been introduced in connection with the quantum Hall transition [19]. It is a model of Dirac fermions coupled to a random potential, or more generally, to a random vector potential, a random mass and a random scalar potential. Its action is:

$$\begin{aligned} S &= \int \frac{d^2x}{2\pi}\Big(\psi_-(\partial_{\bar z}+A_{\bar z})\psi_+ + \bar\psi_-(\partial_z+A_z)\bar\psi_+ \\ &\quad + i\frac{m(x)}{2}(\bar\psi_-\psi_+ - \psi_-\bar\psi_+) + i\frac{V(x)}{2}(\bar\psi_-\psi_+ + \psi_-\bar\psi_+)\Big) \end{aligned}$$

The random variables A, m and V have a gaussian distribution with width g_A, g_M and g_V. Let us denote by Φ_A, Φ_M and Φ_V the perturbing fields coupled to the constants g_A, g_M and g_V after averaging over the disorder. In the two copy sector, we can write them in terms of the $OSp(2|2)$ currents as follows:

$$\begin{aligned} \Phi_A &= (\overline{H}-\overline{J})(H-J) \\ \frac{1}{2}(\Phi_V+\Phi_M) &= 2\overline{G}_-\hat G_+ - 2\overline{\hat G_+}G_- + \overline{K}\hat K + \overline{\hat K}K \\ \frac{1}{2}(\Phi_V-\Phi_M) &= 2\overline{H}H - 2\overline{J}J + 2\overline{\hat G_-}G_+ - 2\overline{G}_+\hat G_- \end{aligned}$$

It is an interesting exercise to compute the beta function at one-loop for these coupling constants. We find:

$$\begin{aligned} \beta_A = \dot g_A &= 32 g_M g_V + \cdots \\ \beta_V - \beta_M = \dot g_V - \dot g_M &= 8(g_V+g_M)^2 + \cdots \\ \beta_V + \beta_M = \dot g_V + \dot g_M &= 8(g_V+g_M)(g_V-g_M) + 8g_A(g_V+g_M) + \cdots \end{aligned} \tag{4.18}$$

For generic value of the initial coupling constants this describes complicated flows. In the particular case with $g_A = g_V = 0$, we recover the beta function of the random bond Ising model.

[2]The remarks of this section arose from discussions with Martin Zirnbauer.

For $g_A = g_M = 0$, we have $\dot{g}_V = 8g_V^2$. This means that a random potential is marginally relevant in the Dirac theory [19]. The coupling constant g_V grows at large distances. The infrared fixed is expected to describe the quantum Hall effect transition. However, being a strong coupling problem no concrete description of it has been proposed.

Note that in this perturbative computation we put all the epsilons which distinguished between advanced and retarded sectors equal to zero. In other words, we did not distinguished whether we are computing averages of product of only advanced Green functions, or averages of product of advanced and retarded Green functions. It is likely that the infared behavior will not be the same in these two cases. It is tempting to conjecture that it will be trivial when we consider only advanced Green functions but non-trivial when mixed products of advanced and retarded Green functions are considered.

Being formulated as current-current perturbation of a first order free theory, we can implement a Hubbard-Stratonovitch transformation on the random Dirac theory in order to produce a dual theory whose coupling constant is $1/g$. The transformation goes as follows: one first represents the current-current interaction by a gaussian integral over an auxiliary gauge connexion A, and then integrate over the free fermions and bosons. This produces a dual theory whose fundamental fields are the connexion A, or equivalently the group element G such that $A_{\bar{z}} = G^{-1}\partial_{\bar{z}}G$, and whose fundamental coupling constant is $1/g$. However, in the mixed sector with both advanced and retarded Green functions, this naive transformation seems to be spoiled by divergences in the gaussian integrals.

5 The replica method: the random phase sine-Gordon model

In this section, we apply the replica method to the random phase sine-Gordon model. It is a random version of the 2d XY model. It has also been used to describe other random physical systems; e.g. the 2d XY model in a random field [5, 6], interfacial roughening transition [7], randomly pinned flux lines in supraconductors [8], etc.

Let us first introduce the model from the XY point of view. Recall that the partition function of the XY model is defined as:

$$Z = \sum_{\{h_i\}} \exp\left(-\frac{K}{2}\sum_{\langle i,j\rangle}(h_i - h_j)^2\right)$$

It describes the thermodynamics of a fluctuating surface. The variables h_i are interpreted as the height of the surface above a base plane. In the pure system the h_i take integer values. The disordered model is defined by the same partition function but the height variables h_i take the values $h_i = d_i + n_i$ with n_i integers and d_i some random variables. Similarly as for the standard XY model, we can formulate this problem in an alternative form by using the Poisson summation formula: $\sum_n f(n) = \sum_m \int d\Phi f(\Phi) e^{i2\pi m\Phi}$. Introducing the variables Φ_i, one gets:

$$Z = \int D\Phi \sum_{\{m_i\}} \exp\left(-\frac{K}{2}\sum_{\langle i,j\rangle}(\Phi_i - \Phi_j)^2 + i2\pi\sum_j m_j(\Phi_j - d_j)\right)$$

In the continuum limit, the variables Φ_i are replaced by a field $\Phi(x)$ and the discrete Laplacian becomes a continuous one. Near the critical point, only the first harmonic of the local potential is relevant. Keeping only this first harmonics gives the action:

$$S = \int \frac{d^2x}{4\pi} \left(\frac{K}{2} (\partial_\nu \Phi)^2 - \Delta \cos(\Phi(x) - d(x)) \right)$$

with $d(x)$ a random quenched fields. This is the model we will use to illustrate the replica method.

As we explain below, this model as two different phases: a high temperature phase in which the disorder is irrelevant and a low temperature phase in which it is relevant. In the high temperature phase nothing interesting happens. In the low temperature phase, the large distance behavior of the system could a priori be different than in the pure system. However, a precise description of this behavior is still missing. For example, the behavior of the quenched correlation functions of $\Phi(x)$ is still controversial. One finds that at large distance we have:

$$\overline{\langle [\Phi(x) - \Phi(0)]^2 \rangle} = A(\log|x|) + B(T - T_c)^2 (\log|x|)^2$$

However, renormalization group computations predict a non-vanihsing $B \neq 0$, while variational approaches give $B = 0$. In the sequel we will present both approaches.

5.1 The model

Let us first rewrite the action with more appropriate notations. We also need to slightly generalize it by introducting a random potential in addition to the random phase. The action reads:

$$S(\Phi|A_\mu;\xi) = \int \frac{d^2x}{4\pi} \left(\frac{K}{2} (\partial_\nu \Phi)^2 - A_\nu \partial_\nu \Phi - \xi(x) e^{i\Phi} - \xi^*(x) e^{-i\Phi} \right) \tag{5.1}$$

The constant K is proportional to the inverse temperature, $K \propto 1/T$. In the formula, $A_\nu(x)$ and $\xi(x)$ are random quenched variables with gaussian measure:

$$\begin{aligned} P[A] &= \exp\left[-\frac{1}{2g} \int \frac{d^2x}{4\pi} A_\mu A_\mu \right] \\ P[\xi] &= \exp\left[-\frac{1}{2\sigma} \int \frac{d^2x}{4\pi} \xi \xi^* \right] \end{aligned} \tag{5.2}$$

In absence of disorder, the dimension of the vertex operator $\exp(i\Phi)$ is: $\dim(e^{i\Phi}) = \frac{1}{K}$. The Harris criteria then tell us that there are two different phases:

low temperature phase $(K > K_c = 1) \Rightarrow$ *disorder is relevant*
high temperature phase $(K < K_c = 1) \Rightarrow$ *disorder is irrelevant*

Recall that $K \propto 1/T$. At high temperature, the large distance behavior is identical to that of the pure system. At the critical temperature $K = K_c$, the disorder has a marginal effect. At low temperature, the large distance behavior is a priori different.

5.2 Symmetries

Similarly as for the random gaussian model, the random phase sine-Gordon model possesses a $U(1)$ symmetry whose Noether current corresponds to insertion of $\partial_\mu \Phi$ in the quenched connected correlation functions. This symmetry in particular implies that all the quenched connected correlation functions of $\partial_\mu \Phi$ are unaffected by the disorder.

Just as for the Gaussian model, we can decompose A_μ as $A_\mu = \partial_\mu \Lambda + \epsilon_{\mu\nu}\partial_\nu \zeta$. Then the field ζ decouples from the action and from the measure. So we can set it to zero, and we have:

$$P[A] = \exp\left[-\frac{1}{2g}\int \frac{d^2x}{4\pi}(\partial_\mu \Lambda)^2\right], \quad with \quad A_\mu = \partial_\mu \Lambda \tag{5.3}$$

We will denote by $S(\Phi|\Lambda;\xi)$ the action (5.1) with $A_\mu = \partial_\mu \Lambda$.

The simplest way to visualize this symmetry consists in introducing sources for $\partial_\mu \Phi$ in the action (5.1):

$$S(\Phi|\Lambda;\xi) \to S(\Phi|J_\mu;\Lambda;\xi) = S(\Phi|\Lambda;\xi) - \int \frac{d^2x}{4\pi} J_\mu \partial_\mu \Phi$$

Let us decompose J_μ as $J_\mu = \partial_\mu \rho + \epsilon_{\mu\nu}\partial_\nu \eta$. The field η decouples and only the field ρ is relevant. The $U(1)$ symmetry follows form the fact that the ρ and Λ dependence can be absorbed into a shift of Φ. Namely:

$$S(\Phi|J_\mu;\Lambda,\xi) = S(\Phi - \frac{\rho+\Lambda}{K}|J_\mu = \Lambda = 0,\hat{\xi}) - \frac{1}{2K}\int \frac{d^2x}{4\pi}(\partial_\mu(\rho+\Lambda))^2 \tag{5.4}$$

with $\hat{\xi} = \xi e^{i(\rho+\Lambda)/K}$ and $J_\mu = \partial_\mu \rho$. As a consequence, the generating function of the connected correlation functions of $\partial_\mu \Phi$, which is $\log Z[J_\mu, \Lambda, \xi]$, can be written as:

$$\log Z[J_\mu, \Lambda, \xi] = -\frac{1}{2K}\int \frac{d^2x}{4\pi}(\partial_\mu(\rho+\Lambda))^2 + \log Z[J_\mu = \Lambda = 0, \hat{\xi}] \tag{5.5}$$

Now, notice that $\hat{\xi}$ and ξ have identical distribution. Therefore, averaging (5.5) over the disorder implies:

$$\overline{\log Z[J_\mu, \Lambda, \xi]} = -\frac{1}{2K}\int \frac{d^2x}{4\pi}(\partial_\mu \rho)^2 + const.$$

where the constant is independent of J_μ. This shows that the quenched average of the connected correlation functions of the $U(1)$ current are identical to the correlation functions of this current in absence of disorder. In other words, the connected correlation functions are protected from the disorder by the $U(1)$ symmetry.

It is worth noticing that the presence of the disorder restore the $U(1)$ symmetry which is absent in the pure sine-Gordon model.

In connection with this $U(1)$ symmetry, the model possesses the remarkable property that the g-dependence of the correlation functions of the vertex operators can be factorized. Consider eq.(5.4) in absence of source: $J_\mu = 0$. The fact that the Λ-dependence can be absorbed in a translation of Φ implies for the correlation functions that:

$$\langle e^{i\alpha_1\Phi(x_1)}\cdots\rangle_{\Lambda,\xi} = \left(\prod_p e^{i\frac{\alpha_p}{K}\Lambda(x_p)}\right)\langle e^{i\alpha_1\Phi(x_1)}\cdots\rangle_{\Lambda=0,\hat{\xi}}$$

Let $G_{\alpha_1,\cdots}(x_1,\cdots|g,\sigma)$ be the quenched correlations of the vertex operator $\exp(i\alpha\Phi)$:

$$G_{\alpha_1,\cdots}(x_1,\cdots|g,\sigma) = \overline{\langle e^{i\alpha_1\Phi(x_1)}\cdots\rangle_{\Lambda,\xi}\langle\cdots\rangle_{\Lambda,\xi}} \tag{5.6}$$

Integrating over Λ using the free field gaussian measure (5.3), and using the fact that $\hat{\xi}$ and ξ have the same measure, we deduce:

$$G_{\alpha_1,\cdots}(x_1,\cdots|g,\sigma) = \prod_{p<q}|x_p - x_q|^{2g\alpha_p\alpha_q/K^2}G_{\alpha_1,\cdots}(x_1,\cdots|g=0,\sigma) \tag{5.7}$$

Equivalently,

$$\partial_g G_{\alpha_1,\cdots}(x_1,\cdots|g,\sigma) = \left(\sum_{p<q}\frac{\alpha_p\alpha_q}{K^2}\log(|x_p - x_q|^2)\right)G_{\alpha_1,\cdots}(x_1,\cdots|g,\sigma) \tag{5.8}$$

This identity will be useful for analyzing the renormalization group equations.

5.3 *The effective action*

Since the action (5.1) is not free we cannot rely on the supersymmetric method but only on the replica trick. Therefore, as explained in the introduction, we consider n copies of the system with the same disorder and then average over the disorder. This gives the following effective action:

$$S_{eff} = \int\frac{d^2x}{4\pi}\left(\frac{K}{2}\sum_r(\partial_\mu\Phi^r)^2 - \frac{g}{2}\sum_{r,s}(\partial_\mu\Phi^r)(\partial_\mu\Phi^s) - 2\sigma\sum_{r\neq s}\exp(i(\Phi^r - \Phi^s))\right) \tag{5.9}$$

Let us first look at the kinetic term. It is of the form $\int\frac{d^2x}{4\pi}\frac{1}{2}(\partial_\mu\Phi^r)G_{rs}(\partial_\mu\Phi^s)$ with

$$G_{rs} = K\delta_{rs} - g = (K-g)\delta_{rs} - g(1-\delta_{rs})$$

Since the interaction only involves the difference of the replicated fields, it is convenient to decompose the kinetic terms as:

$$\frac{1}{2}(\partial_\mu\Phi^r)G_{rs}(\partial_\mu\Phi^s) = \frac{(K-ng)}{2n}\left(\partial_\mu(\sum_r\Phi^r)\right)^2 + \frac{K}{4n}\sum_{r\neq s}(\partial_\mu(\Phi^r - \Phi^s))^2 \tag{5.10}$$

Note that the field $(\sum_r\Phi^r)$ decouples. Its correlation functions are therefore unaffected by the disorder. Since these correlation functions represent the averages of the connected correlation functions of the $U(1)$ current, we recover the previous result obtained from the $U(1)$ Ward identities. In particular, $\partial_\mu(\sum_r \Phi^r)$ is a $U(1)$ conserved current. We could also use this decomposition to derive eq.(5.8).

The kinetic term (5.10) fixes the normalization of the vertex operators. In particular, we find the dimension of the perturbing field: $\dim\left(e^{i(\Phi^r-\Phi^s)}\right) = \frac{2}{K}$. It is independent of σ. It is relevant for $K > 1$: we thus recover the Harris criteria.

This decomposition has a Lie algebraic interpretation. It corresponds to the decomposition of $U(n) = SU(n) \times U(1)$. The second term in (5.10) can be rewritten as $\frac{K}{2}(\partial_\mu \vec{\varphi})^2$ where $\vec{\varphi}$ takes values in the Cartan subalgebra of $SU(n)$. In terms of $\vec{\varphi}$, the perturbing field in (5.9) reads: $\sum_{\vec{\alpha}} \exp(i\vec{\alpha}.\vec{\varphi})$ where the sum extends over all the $SU(n)$ roots. At $K = K_c$, the fields $\exp(i\vec{\alpha}.\vec{\varphi})$ have dimension two, and the perturbing field is a current-current interaction. The action (5.9) at $K = K_c$ is therefore equivalent to the $SU(n)$ Gross-Neveu model. We will later see that at this point the model is asymptotically free in the infrared. Hence at the critical temperature $K = K_c$, the disorder only induces logarithmic corrections.

5.4 Renormalization group

The renormalization group allows to perturbatively analyse the behavior of the system in the low temperature phase. We will do a one loop computation, which is valid close to the critical temperature, *i.e.* $\frac{K-K_c}{K_c} \ll 1$.

As we are already familiar with, the one-loop beta functions are encoded in the OPE of the fields. Let us introduce the following notation:

$$\mathcal{O}_1 = \sum_{r\neq s} \exp(i(\Phi^r - \Phi^s)) \tag{5.11}$$

$$\mathcal{O}_2 = \frac{1}{2n}\sum_{r,s} i\partial_z(\Phi^r - \Phi^s) i\partial_{\bar{z}}(\Phi^r - \Phi^s) \tag{5.12}$$

The field $\mathcal{O}_1$ is the perturbing field associated to the coupling constant σ. The field $\mathcal{O}_2$ is one of the kinetic field. Notice that $\mathcal{O}_2 = (i\partial_z\vec{\varphi})(i\partial_{\bar{z}}\vec{\varphi})$. It is necessary to introduce it since it is generated from $\mathcal{O}_1$ by OPE. Indeed we have:

$$\begin{aligned}
\mathcal{O}_1(z)\mathcal{O}_1(0) &= \frac{2(n-2)}{|z|^{2/K}}\mathcal{O}_1(0) + \frac{2n}{|z|^{2/K-2}}\mathcal{O}_2(0) + \textit{irrelevant terms}, \quad (5.13)\\
\mathcal{O}_2(z)\mathcal{O}_1(0) &= \frac{2/K^2}{|z|^2}\mathcal{O}_1(0) + \textit{irrelevant terms},\\
\mathcal{O}_2(z)\mathcal{O}_2(0) &= \textit{irrelevant terms}.
\end{aligned}$$

For $n = 1$ these are the familiar OPE of the Kosterlitz-Thouless transition. Note the change of behavior between $n < 2$ or $n > 2$. The first OPE indicates how σ is renormalized, but the second equation shows that the kinetic term needs also to be renormalized.

Using the formula (8.3) of the Appendix, we get the beta functions:

$$\begin{aligned}
\beta_\sigma &= 2\left(\frac{K-K_c}{K}\right)\sigma + (n-2)\sigma^2 + \cdots\\
\beta_K &= \frac{n}{2}\sigma^2 + \cdots
\end{aligned}$$

This shows that the coupling to A_μ would have been generated at one loop even if we did not start with it. Since the field $(\sum_r \Phi^r)$ decouples, the coupling $(K - ng)$ is unrenormalized (at any order in perturbation theory). Thus: $\beta_K = n\beta_g$. Setting $n = 0$ as required by the replica trick, we get:

$$\begin{aligned}
\beta_\sigma &= 2\left(\frac{K-K_c}{K}\right)\sigma - 2\sigma^2 + \cdots \\
\beta_g &= \frac{1}{2}\sigma^2 + \cdots \\
\beta_K &= 0
\end{aligned} \tag{5.14}$$

So, K is unrenormalized at $n = 0$. It may appears surprising that the coupling g is renormalized although we know exactly the g-dependence of any correlation functions. We will later see that this is not in contradiction with the renormalization group equations.

From equation (5.14), we immediatly see that in the low temperature phase ($K > K_c$), the beta function β_σ possesses a non trivial infrared zero at σ_*:

$$\sigma_* = \left(\frac{K-K_c}{K_c}\right) + \cdots \qquad for \ \left(\frac{K-K_c}{K_c}\right) \ll 1. \tag{5.15}$$

Notice the fact the other beta function β_g does not vanish at σ_*:

$$\beta_g(\sigma_*) = \frac{1}{2}\left(\frac{K-K_c}{K_c}\right)^2 + \cdots \tag{5.16}$$

Hence, even at σ_* the coupling g will continue to flow. We may characterize such pseudo-fixed point as a "run away fixed point". This is a particularity of the model which has direct consequences on the correlation functions.

5.5 Quenched correlation functions

Let us now analyse the renormalization group equations. As we already said, the field A_μ would have been generated at one loop if not present initially. But on other hand we know that the dependence on g can be completely disentangled. This apparent conflict should have an effect on the renormalization group equations.

Consider the correlation functions $G_{\alpha_1,\cdots}(x_1,\cdots|g,\sigma)$ defined in eq.(5.6). They satisfy the renormalization group equations:

$$\left[\sum_p x_p^\nu \frac{\partial}{\partial x_p^\nu} + \sum_p \gamma_p(g,\sigma) - \beta_\sigma(\sigma)\frac{\partial}{\partial\sigma} - \beta_g(\sigma)\frac{\partial}{\partial g}\right] G_{\alpha_1,\cdots}(x_1,\cdots|g,\sigma) = 0$$

where $\gamma_p(g,\sigma)$ are the anomalous dimensions. But the g-dependence is explicitly known, therefore using eq.(5.8) we get:

$$\left[\sum_p x_p^\nu \frac{\partial}{\partial x_p^\nu} + \sum_p \gamma_p(g,\sigma) - \beta_\sigma(\sigma)\frac{\partial}{\partial\sigma} - \beta_g(\sigma)\sum_{p<q}\frac{\alpha_p\alpha_q}{K^2}\log(|x_p - x_q|^2)\right] G_{\alpha_1,\cdots}(\cdots) = 0$$

Note that it is now possible to set g equal to zero. In particular at the infrared fixed point σ_*, in which $\beta_\sigma(\sigma_*) = 0$, we get:

$$\left[\sum_p x_p^\nu \frac{\partial}{\partial x_p^\nu} + \sum_p \gamma_p(\sigma_*) - \beta_g(\sigma_*) \sum_{p<q} \frac{\alpha_p \alpha_q}{K^2} \log(|x_p - x_q|^2)\right] G^*_{\alpha_1,\dots}(\cdots) = 0 \quad (5.17)$$

with $\beta_g(\sigma_*)$ given in eq.(5.16). The effect of the g-flow is to add the extra logarithmic term in the renormalization group equations (5.17).

This can be used to compute two-point functions at the infrared fixed point. Consider,

$$\begin{aligned} G_1(x) &= \overline{\langle \exp(i\alpha(\Phi(x) - \Phi(0)))\rangle}_{\sigma_*} \\ G_2(x) &= \overline{\langle \exp(i\alpha\Phi(x))\rangle \langle \exp(-i\alpha\Phi(0))\rangle}_{\sigma_*} \end{aligned}$$

Let $\gamma_{1,2}(\sigma_*)$ be their anomalous dimensions at the infrared fixed point. The renormalization group equation (5.17) gives:

$$G_{1,2}(x) = |x|^{-2\gamma_{1,2}(\sigma_*)} \exp\left(-\frac{\alpha^2 \beta_g(\sigma_*)}{2K^2}(\log|x|)^2\right) \quad (5.18)$$

Notice the $(\log|x|)^2$ correction which arises from the renormalization of g. This term is independent of the anomalous dimension.

The anomalous dimensions are even in α and vanishes at $\alpha = 0$, therefore:

$$\gamma_{1,2}(\sigma_*) = \frac{\alpha^2}{K}\rho_{1,2}(\sigma_*) + \mathcal{O}(\alpha^4), \quad \textit{with} \quad \rho_1(\sigma_*) = 1 + \mathcal{O}\left(\frac{K - K_c}{K_c}\right)$$

Expanding in power of α^2, gives the two point functions of Φ:

$$\overline{\langle [\Phi(x) - \Phi(0)]^2\rangle} = \frac{2\rho_1(\sigma_*)}{K}\log|x| + \frac{\beta_g(\sigma_*)}{2K^2}(\log|x|)^2 \quad (5.19)$$

$$\overline{[\langle \Phi(x) - \Phi(0)\rangle]^2} = \frac{2\rho_2(\sigma_*)}{K}\log|x| + \frac{\beta_g(\sigma_*)}{2K^2}(\log|x|)^2 \quad (5.20)$$

Note that the $(\log|x|)^2$ cancels in the connected correlation function $\overline{\langle [\Phi(x) - \Phi(0)]^2\rangle_c}$ as it should be, since this connected correlation function is unaffected by the disorder.

There is a crossover from a $(\log|x|)$ to a $(\log|x|)^2$ behavior. For $|x| < R_{cross}$ we have a usual $(\log|x|)$ behavior, while for $|x| > R_{cross}$ we have a $(\log|x|)^2$ behavior. The crossover length is R_{cross} with

$$\log R_{cross} \sim \frac{2K\rho_1(\sigma_*)}{\beta_g(\sigma_*)} \sim \frac{K_c^2}{(K - K_c)^2}, \qquad \textit{for} \quad (K - K_c) \ll 1.$$

R_{cross} is exponentially large close to the phase transition. However eqs.(5.19,5.20) are true to all order in perturbation theory if we can rely on the renormalization group in the replica symmetric approach.

We can use eqs.(5.19,5.20) to find an estimate of the width of the surface for a system of finite length L, *i.e.* to find an estimate of $\overline{\langle \Phi(0)^2\rangle_L}$. We define it as the integral of the Fourier transform $\Gamma(q)$ of (5.19) using $1/L$ as ultraviolet cutoff. At short momenta $q \ll 1/R_{cross}$, the propagator $\Gamma(q)$ is dominated by the Fourier transform of the $(\log|x|)^2$ term; *i.e.* $\Gamma(q) \sim \beta_g(\sigma_*)\left(\frac{\log q^2}{q^2}\right)$. Thus:

$$\overline{Q(L)} = \overline{\langle\Phi(0)^2\rangle_L} = \int_{1/L} d^2q\Gamma(q) \sim \beta_g(\sigma_*)\,(\log L)^2 \tag{5.21}$$

for $L \gg R_{cross}$. This has to be compare with the pure case which gives a $(\frac{1}{K}\log L^2)$ behavior.

5.6 A large N model

The formula (5.19, 5.20) are still controversial, theoretically as well as numerically. Variational approaches, (part of which we will describe below), predict a $(\log|x|)$ behavior [27, 28, 29]. The RG flows was also found to be unstable again assymmetric replica perturbations [30, 31]. The numerical verifications of (5.19) are also not settled: a $(\log|x|)$ behavior was found in ref.[32, 33], while more recent simulations [34] seem to indicate a $(\log|x|)^2$ behavior. In view of this conflict, and since the variational approaches are argued to be exact for systems with a large number of components [35], in ref.[36] we studied a large N version of the model (5.1). Using RG computations based on the (a priori symmetric) replica trick, we find that our large N model possesses a non-trivial infrared fixed point in which the correlation functions have the form (5.19, 5.20) but with the $(\log|x|)^2$ term suppressed by a factor $(1/N^3)$ compared to the $(\log|x|)$ term.

To introduce the large N version of (5.1), it is convenient to fermionize it. The fermionic form of the random phase sine-Gordon model is a massless Thirring model coupled to a quenched potential A_μ and a random phase ξ. To define its large N version, we need to introduce N Dirac fermions with components $\psi^k_\pm$ and $\overline{\psi}^k_\pm$ with $k = 1, \cdots, N$. Let $z = x + iy$ and $\bar{z} = x - iy$ be the complex coordinates on the plane. The action is:

$$\begin{aligned} S^{(N)} \;&=\; \int \frac{d^2x}{\pi}\left(\sum_{k=1}^{N}(\psi_{-;k}\partial_{\bar{z}}\psi^k_+ + \overline{\psi}_{-;k}\partial_z\overline{\psi}^k_+) - \frac{a}{N}(\sum_{k=1}^{N}\psi_{-;k}\psi^k_+)(\sum_{k=1}^{N}\overline{\psi}_{-;k}\overline{\psi}^k_+)\right) \\ &-\int \frac{d^2x}{\pi}\left(A_{\bar{z}}(\sum_{k=1}^{N}\psi_{-;k}\psi^k_+) + A_z(\sum_{k=1}^{N}\overline{\psi}_{-;k}\overline{\psi}^k_+) + \xi(\sum_{k=1}^{N}\overline{\psi}_{-;k}\psi^k_+) + \xi^*(\sum_{k=1}^{N}\psi_{-;k}\overline{\psi}^k_+)\right) \end{aligned} \tag{5.22}$$

In absence of disorder, it is conformally invariant. The random potential $A_\mu = (A_{\bar{z}}, A_z)$ is coupled to the $U(1)$ currents $J_z = \sum_{k=1}^N \psi_{-;k}\psi^k_+$ and $\overline{J}_{\bar{z}} = \sum_{k=1}^N \overline{\psi}_{-;k}\overline{\psi}^k_+$ of the unperturbed theory. At $\xi = 0$, the random potential does not break conformal invariance.

There are a priori many ways to generalize the action (5.1) to a large N version. The action (5.22) has been designed in such way as (i) to keep the number of disordered variables fixed, (ii) to preserve the exact conformal invariance in absence of disorder, and (iii) to parallel as much as possible standard properties of large N models.

The fermionic action (5.22) can be bosonized back using non-Abelian bosonization [37]. As usual, since the pure system describes N Dirac fermions, the pure bosonized theory will be described by a $su(N)$ Wess-Zumino-Witten (WZW) model at level one plus a massless free field. The $su(N)$ WZW model at level one possesses primary fields taking values in the $(N-1)$ fundamental representations of $su(N)$. Let $\phi_\square^k$ and $\phi_{\bar\square;k}$ be the chiral WZW primary fields which take values in the defining representation of $su(N)$ and in its complex conjugate. Their conformal weights are both equal to $\left(\frac{N-1}{2N}\right)$. Let us denote by φ the gaussian free field. The original fermions $\psi_\pm^k$ can be written as the product of these WZW primary fields by a vertex operator of the gaussian model. Namely, $\psi_+^k = \phi_\square^k e^{\frac{i}{\sqrt{N}}\varphi}$ and $\psi_{-;k} = \phi_{\bar\square;k} e^{-\frac{i}{\sqrt{N}}\varphi}$, and similarly for the other chiral components $\overline{\psi}_+^k$ and $\overline{\psi}_{-;k}$. The bosonic form of the action (5.22) is:

$$\begin{aligned} S^{(N)} &= S_{wzw} + \frac{K}{2}\int \frac{d^2x}{4\pi}(\partial_\nu\varphi)^2 \\ &- \int \frac{d^2x}{4\pi}\left(A_\nu(x)\partial_\nu\varphi + \xi(x)\Phi e^{\frac{i}{\sqrt{N}}\varphi} + \xi^*(x)\Phi^* e^{-\frac{i}{\sqrt{N}}\varphi}\right) \end{aligned} \tag{5.23}$$

where S_{wzw} refers to the $su(N)_1$ WZW action. We have introduced the composite fields $\Phi(z,\bar z) = \sum_{k=1}^N \phi_\square^k(z)\overline{\phi_{\bar\square;k}}(\bar z)$ and $\Phi^* = \sum_{k=1}^N \phi_{\bar\square;k}(z)\overline{\phi_\square^k}(\bar z)$. Equivalently, $\Phi = tr_\square(G)$ with G the group valued field of the WZW model. For $N=1$ the WZW terms are absent and we recover the action (5.1) of the random phase sine-Gordon model.

The renormalization group computation done in ref.[36] predicts a non-trivial infrared fixed point in the low temperature phase with correlation function of the form (5.19) but with $\beta_g(\sigma_*)$ given by:

$$\beta_g(\sigma_*) = \frac{1}{8N^3}\left(\frac{K-K_c}{K_c}\right)^2 + \cdots, \quad for N \gg 1, \quad \frac{K-K_c}{K_c} \ll 1$$

The occurence of this factor $1/N^3$ could explain why the $(\log|x|)^2$ term does not manifest itself in the variational approaches.

6 The variational method: the random phase sine-Gordon model

Various variational approaches to disorder systems have been proposed. They can be applied after or before disorder averaging. In the first case, one first averages over the disorder using the replica trick, and then implements a variational method on the replicated model, cf. e.g. [35]. In the second case, one applies a variational method on any sample of fixed disorder, and then averages over the disorder. We will present the second method using the random phase sine-Gordon model as example, following ref.[29]. But it can clearly be applied to other models. The variational method at fixed disorder leads to an exact bound for the averaged free energy.

Consider the partition function $Z[\xi(x)] = \int D\Phi \exp(-S(\Phi|\xi))$, where $S(\Phi|\xi)$ is the action (5.1) at $A_\mu = 0$, which we recall:

$$S(\Phi|\xi) = \int \frac{d^2x}{4\pi}\left(\frac{K}{2}(\partial_\nu\Phi)^2 - \xi(x)e^{i\Phi} - \xi^*(x)e^{-i\Phi}\right)$$

At fixed disorder, we approximate the partition function by a gaussian action:

$$S_0(\Phi|G) = \frac{1}{2}\int d^2x d^2y \Phi(x) G^{-1}(x-y)\Phi(y) = \frac{1}{2}\int \frac{d^2k}{(2\pi)^2}\hat{\Phi}(k)\hat{G}^{-1}(k)\hat{\Phi}(k) \quad (6.1)$$

Here the hatted quantities refer to the Fourier transforms. Notice that the Ansatz (6.1) is choosen to be translation invariant, and that the gaussian is centered around the origin. We choose a gaussian anstaz otherwise the computations are undoable. The kernel $G(x-y)$ is the variational parameter.

The approximated partition function is $Z_0 = \int D\Phi \exp(-S_0)$. Let F and F_0 be the respective free energies, *i.e.* $Z = e^{-F}$ and $Z_0 = e^{-F_0}$. For any realization of the disorder and for any choice of S_0, the following inequality holds:

$$F \leq F_0 + \langle (S - S_0) \rangle_0 \quad (6.2)$$

where $\langle \cdots \rangle_0$ refers to the expectation values with the measure S_0. This is proved using the following inequality:

$$\frac{Z}{Z_0} = \langle e^{-(S-S_0)} \rangle_0 \geq e^{-\langle (S-S_0)\rangle_0}$$

Therefore, the best approximated action S_0 is found from:

$$\frac{\delta}{\delta \hat{G}(k)}\Big(F_0 + \langle (S - S_0)\rangle_0 \Big) = 0. \quad (6.3)$$

The inequality (6.2) gives a upper bound to the free energy.

Let us apply this general setup to our example. As it is formulated the method is more appropriate to study the free energy than the correlation functions. Thus we consider the system in a box of finite volume of size $Vol. = L^2$ and look for the L-dependence of the free energy. That is, we use the variational free energy to analyse the finite size effects.

Since S_0 is Gaussian, F_0 and $\langle (S - S_0)\rangle_0$ are easily computed. The free energy F_0 per unit of volume is given by the logarithm of a determinant:

$$\frac{F_0}{Vol.} = -\frac{1}{2}\int d^2k \log \hat{G}(k)$$

while for the expectation value of $(S - S_0)$ per unit of volume we have:

$$\frac{\langle (S - S_0)\rangle_0}{Vol.} = \frac{\pi K}{2}\int \frac{d^2k}{(2\pi)^2} k^2 \hat{G}(k) - \frac{e^{-Q/2}}{Vol.}\int \frac{d^2x}{4\pi}(\xi(x) + \xi^*(x)) \quad (6.4)$$

with

$$Q = \langle \Phi^2(0) \rangle = G(0) = \int \frac{d^2k}{(2\pi)^2}\hat{G}(k) \quad (6.5)$$

This parameter is naturally interpreted as the width of the system.

Notice that the expectation value (6.4) only depends on a particular moment of $\xi(x)$. It does not depend on all the details of the disorder configuration. It is this particular fact which makes the variational approach doable.

The variational equations (6.3) are simple to compute. They determine the kernel $\hat{G}(k)$:

$$\hat{G}(k) = \frac{4\pi}{K}\left(\frac{1}{k^2+M^2}\right) \tag{6.6}$$

where the effective mass is:

$$M^2 = \frac{1}{\pi K}\frac{e^{-Q/2}}{(Vol.)^{1/2}}D_0 \quad with \quad D_0 = \frac{1}{(Vol.)^{1/2}}\int \frac{d^2x}{4\pi}(\xi(x)+\xi^*(x)) \tag{6.7}$$

We have introduced the prefactor $(Vol.)^{1/2}$ in the definition of D_0 to make it scale invariant. Since the effective mass M^2 depends on Q, which is a functional of the kernel $\hat{G}(k)$, eqs. (6.5, 6.6) form a set of non linear coupled equations for $\hat{G}(k)$ which we rewrite below:

$$Q = \frac{4\pi}{K}\int \frac{d^2k}{(2\pi)^2}\left(\frac{1}{k^2+M^2}\right) \quad with \quad M^2 = \frac{1}{\pi K}\frac{e^{-Q/2}}{(Vol.)^{1/2}}D_0 \tag{6.8}$$

The effective mass depends on the disorder through its moment D_0. This moment can be either positive or negative, with a symmetric probability distribution around the origin. Thus we have to study separately the two cases: $D_0 > 0$ or $D_0 < 0$. This analysis was done in ref.[29].

For $D_0 > 0$, the effective mass is real and the Green function $\hat{G}(k)$ has no pole. Therefore, for Q we get:

$$Q = \int_{1/L}^{1/a} \frac{d^2k}{(2\pi)^2}\hat{G}(k) = \frac{1}{K}\log\left(\frac{a^{-2}+M^2}{L^{-2}+M^2}\right) \tag{6.9}$$

where $k_{UV} = 1/a$ and $k_{IR} = 1/L$ are the ultraviolet and infrared cutoff. Recall that the volume is $Vol. = L^2$. Inserting this expression in the definition (6.7) of the effective mass gives:

$$M^2 = \frac{D_0}{\pi K}\frac{1}{L}\left(\frac{a^{-2}+M^2}{L^{-2}+M^2}\right)^{-1/2K} \tag{6.10}$$

The effective mass vanishes as $L\to\infty$, but with different power of L for $K < K_c = 1$ or $K > K_c = 1$. These differences arise from the different behavior of the last term in eq.(6.10). For $K < K_c$, we have $M^2 \ll (1/L)^2$, and therefore the last term in eq.(6.10) behaves like $(1/L)^{1/K}$. For $K > K_c$, the mass term dominates, $M^2 \gg (1/L)^2$, and therefore, the last term in (6.10) behaves like $M^{1/K}$. Hence, for $D_0 > 0$ we obtain:

$$M^2(L) \sim \left(\frac{1}{L}\right)^{1+\frac{1}{K}} \quad ; \quad Q(L) \sim \frac{1}{K}\log L^2 \qquad for \quad K < K_c \tag{6.11}$$

$$(M^2(L))^{1-\frac{1}{2K}} \sim \left(\frac{1}{L}\right) \quad ; \quad Q(L) \sim \frac{1}{2K-1}\log L^2 \qquad for \quad K > K_c \tag{6.12}$$

For $D_0 < 0$, since the situation is quite different since the effective mass square is negative. The Green function $\hat{G}(k)$ now possesses a pole at $k^2 = -M^2$. In order to analyse the effect of this pole we have to remember that for a system in a box of length L, the momenta are quantized to discret values: $(k_x, k_y) = (\frac{2\pi n_x}{L}, \frac{2\pi n_y}{L})$, with (n_x, n_y) integers. So in the key equation (6.8) the integral is actually a discret sum: $\int \frac{d^2k}{(2\pi)^2} \to \frac{1}{L^2}\sum_{n_x,n_y}$. In eq. (6.8), we seperate the first terms which correspond to $(n_x = \pm, n_y = \pm)$ from the others which we approximate by an integral. We obtain:

$$\begin{aligned} Q(L) &= \frac{4\pi}{KL^2}\frac{4}{\left(\frac{2\pi}{L}\right)^2 + M^2} + \frac{4\pi}{K}\int_{1/L}^{1/a}\frac{d^2k}{(2\pi)^2}\left(\frac{1}{k^2+M^2}\right) \\ &= \frac{4\pi}{KL^2}\frac{4}{\left(\frac{2\pi}{L}\right)^2 + M^2} + \frac{1}{K}\log(L^2/a^2) \end{aligned} \tag{6.13}$$

where we have neglected the M^2 dependence in the last integral. Once again, the effective mass behaves differently as $L \to \infty$ for $K < K_c$ and $K > K_c$. These different behaviors are distinguished by the relative importance between the two terms in eq.(6.13). Indeed, suppose that $M^2 \ll 1/L^2$. Then the first term in (6.13) is irrelevant and therefore $Q(L) \sim \frac{1}{K}\log L^2$. However, inserting this value of Q in the definition of M^2 as a function of Q, cf eq.(6.8), we deduce then that $M^2(L) \sim (1/L)^{1+\frac{1}{K}}$. Thus consistency of the hypothesis $M^2 \ll 1/L^2$ requires $K < 1$. When M^2 becomes of order $1/L^2$, the first term in (6.13) dominates and M^2 remains frozen to this values. The expression of M^2 as a function of Q then tell us that $Q(L) \sim \log L^2$. Summarizing, for $D_0 < 0$ we get:

$$M^2(L) \sim \left(\frac{1}{L}\right)^{1+\frac{1}{K}} \quad ; \quad Q(L) \sim \frac{1}{K}\log L^2 \qquad for \quad K < K_c \tag{6.14}$$

$$M^2(L) \sim \left(\frac{1}{L}\right)^{2} \quad ; \quad Q(L) \sim \log L^2 \qquad for \quad K > K_c \tag{6.15}$$

Notice that in the high temperature phase $K < K_c$, the behavior of $Q(L)$ is the same for D_0 positive or negative, while this behavior is different in the low temperature phase.

These behaviors have been obtained at fixed disorder. We can now average over the disorder. Since the effective mass was only a function of D_0 which is symmetrically distributed around the origin, the average value of $Q(L)$ is half of the sum of its values for D_0 positive and negative. Hence, at large L, we have:

$$\overline{Q(L)} \sim \frac{1}{K}\log L^2 \qquad for \quad K < K_c \tag{6.16}$$

$$\overline{Q(L)} \sim \frac{K}{2K-1}\log L^2 \qquad for \quad K > K_c \tag{6.17}$$

The averaged behavior at high temperature is the same as in absence of disorder. That is, the disorder is irrelevant in the high temperature phase as we found in the previous section using the replica approach. But the disorder is relevant in the low temperature phase. However the result obtained for $\overline{Q(L)}$ in the low temperature phase disagree with the result obtained sing the symmetric replica trick, cf eq.(5.21).

This variational method gives poor results for the correlation functions. Indeed, since we choose the Gaussian anstaz (6.1) to be centered around the orgin, the variational one-point functions vanishes: $\langle\Phi(x)\rangle_0 = 0$. But the connected correlation are unaffected by the disorder since it is protected by the $U(1)$ symmetry. Therefore, the variational two-point function is also unaffected by the disorder, which is probably not realistic. A more appropriate Ansatz could be to choose a gaussian action not centered around the origin.

Also, this variational approach does not take one-loop effect into account. This could be the origin of the disagrement between the renormalization group and the variational approaches.

7 Replica symmetry breaking or not?

In this section we very shortly describe how replica symmetry breaking is incorporated in the renormalization group perturbative approach based on the replica method. The aim is not to present all the details and subtilities of the replica symmetry breaking, (there already exist extensive reviews on the subject), but only to introduce the main steps.

As example we choose the minimal conformal models perturbed by random bond interaction. The replica symmetric was studied in [38], while the study of the theory with replica symmetry breaking was done in [39]. The basic examples are the random bond Ising or Potts models.

The simplest minimal model is the Ising model, whose disordered version was studied above in the supersymmetric approach. In the scaling limit, random bond interaction is represented by a perturbation by the energy operator $\epsilon(x)$ with a random coupling constant. In the Kac classification this operator is the Φ_{21} operator. Thus we consider the random models:

$$S[g(x)] = S_* + \int d^2x g(x)\Phi_{21}(x)$$

where S_* represents the action of the corresponding minimal conformal model, e.g the Ising or Potts models. As we have seen, in the Ising model the disorder is marginal and only induces logarithmic corrections. The dimension of Φ_{21} in the Potts model is $\dim(\Phi_{21}) = \frac{4}{5} < 1$. It is therefore a strictly relevant disordered perturbation.

This models were studied in ref.[38] using perturbative renormalization group cpmputation in the a priori symmetric replicated theory. After replica, the effective action is:

$$S_{eff} = \sum_{z} S_*^r + \sigma \int d^2x \sum_{r\neq s} \Phi_{21}^r(x)\Phi_{21}^s(x) \tag{7.1}$$

In eq.(7.1), one chooses to restrict the sum to $r \neq s$ since in the OPE of two $\Phi_{21}(x)$ operators, which is given by the fusion rule $\Phi_{21} \times \Phi_{21} = 1 + \Phi_{31}$, only the identity arises with a singular term and the field Φ_{31} is irrelevant.

Replica symmetry breaking is incorporated in two steps [40]. One first promotes the coupling constant σ to a matrix σ_{rs}, and then consider σ_{rs} at $n = 0$ has a hierarchical Parisi matrix. This amounts to parametrize it by a diagonal element and a function $\sigma(x)$ with $x \in [n, 1]$, $n \to 0$:

$$\sigma \to \sigma_{rs} \to (\tilde{\sigma}, \sigma(x)) \tag{7.2}$$

The multiplication law of two Parisi matrices parametrized by

$$(\tilde{\sigma}_1, \sigma_1(x)) \text{ and } (\tilde{\sigma}_2, \sigma_2(x))$$

is then defined by :

$$(\tilde{\sigma}_1, \sigma_1(x)) \cdot (\tilde{\sigma}_2, \sigma_2(x)) = (\tilde{h}, h(x)) \tag{7.3}$$

with

$$\begin{aligned} h &= \tilde{\sigma}_1\tilde{\sigma}_2 - \overline{\sigma_1\sigma_2} & (7.4) \\ h(x) &= -n\sigma_1(x)\sigma_2(x) + (\tilde{\sigma}_1 - \overline{\sigma}_1)\sigma_2(x) + (\tilde{\sigma}_2 - \overline{\sigma}_2)\sigma_2(x) \\ &\quad - \int_n^x dy(\sigma_1(x) - \sigma_1(y))(\sigma_2(x) - \sigma_2(y)) & (7.5) \end{aligned}$$

where $\overline{f} = \int_n^1 dx f(x)$.

One can now study the consequences of this anstaz in the renormalization group. This is done by first computing the renormalization group equations with the matrix σ_{rs} and then implementing Parisi's Ansatz. As usual the one loop beta functions $\beta_{rs} = \dot{\sigma}_{rs}$ are encoded in the operator product expansions. However, non replica symmetric fixed points in the random mininal conformal models only appear at two loops.

The two-loop beta function was computed in ref.[39] using an epsilon expansion. The central charge is parametrized as:

$$c = 13 - 6(\alpha_+^2 + \alpha_+^{-2}) \quad with \quad \alpha_+^2 = \frac{4}{3} - \epsilon$$

The case $\epsilon = 0$ corresponds to the Ising model $c = \frac{1}{2}$. The Potts model corresponds to $\epsilon = \frac{2}{15}$. The dimension of the Φ_{21} primary field is then:

$$\dim(\Phi_{21}) = 1 - \frac{3}{2}\epsilon$$

For $\epsilon = 0$ the disorder is marginal, while for $\epsilon \ll 1$ the disorder is slightly relevant. This allows to implement an ϵ-expansion, and in particular to determine the non-trivial fixed point in an ϵ-expansion. Since we assumed that the diagonal matrix elements of σ_{rs} vanish, in Parisi's Ansatz the matrix σ_{rs} is represented only be a function $\sigma(x)$. The vanishing of the beta functions then determines the possible fixed point function $\sigma_*(x)$. According to ref.[39], the fixed point equation reads:

$$3\epsilon\sigma_*(x) - 2\overline{\sigma_*}\sigma_*(x) - \int_0^x dy(\sigma_*(x) - \sigma_*(y))^2 + \sigma_*^3(x) + \overline{\sigma_*^2}\sigma_*(x) = 0 \tag{7.6}$$

with $\overline{\sigma_*} = \int_0^1 dx\sigma_*(x)$.

The replica symmetric solution corresponds to $\sigma_*(x) = const$:

$$\sigma_*(x) = const. = \frac{3}{2}\epsilon + \frac{9}{4}\epsilon^2 + \cdots$$

The non replica symmetric solution is given by a solution for which $\sigma_*(x)$ growth linearly for $0 < x < x_1$ and then remains constant for $x_1 < x < 1$. This is called a one-step replica symmetry breaking. It is found by deriving eq.(7.6) with respect to x as many times as necessary. Explicitly, one has:

$$\sigma_*(x) = \begin{cases} \frac{1}{3}x, & \text{if } 0 < x < x_1, \\ \frac{1}{3}x_1, & \text{if } x_1 < x < 1. \end{cases} \quad with \quad x_1 = \frac{9}{2}\epsilon + \frac{27}{2}\epsilon^2 + \cdots.$$

This solution is not present at one-loop. It has been checked that it is a stable solution of eq.(7.6).

The symmetric and non-symmetric solution can be distinguished by analysing the anomalous dimensions at the corresponding infrared fixed points. For example, the dimension of the energy operator at the infrared fixed point differs in the two solutions by two percent up to $\mathcal{O}(\epsilon^3)$. There is up to now no numerical evidence in favor of the non-symmetric solution [41].

ACKNOWLEDGEMENTS: It is a pleasure to thank H. Orland and M. Zirnbauer for very useful discussions.

8 APPENDIX: RENORMALIZATION GROUP AND OPE.

In this appendix we gather standard informations about the renormalization group in two dimension, cf e.g. [42] [43].

Consider partition functions and correlation functions computed with the measure $\int D\phi \ \exp(-S)$ with a perturbed action:

$$S = S_* + \sum_i g^i \int d^2x \Phi_i(x) \tag{8.1}$$

where $\Phi_i(x)$ are relevant primary operators of dimension h_i. Suppose that these fields satisfy the following operator product expansion:

$$\Phi_i(x)\Phi_j(y) = \frac{C_{ij}^k}{|x-y|^{h_i+h_j-h_k}}\Phi_k(y) + \cdots \tag{8.2}$$

Then, the beta function at one loop is:

$$\dot{g}^i = \beta^i(g) = (2-h_i)g^i - \pi \sum_{jk} C_{jk}^i g^j g^k + \cdots \tag{8.3}$$

No summation in the first term but summation over j, k in the second. The summation is over all the relevant fields generated by the product operator expansion.

In the same way, if $\mathcal{O}_\alpha$ is a set of operators with OPE with the perturbing field Φ_i given by the structure constant $C_{i\alpha'}^\alpha$. Then the matrix of anomalous dimensions ($\gamma = -a\partial_a \log Z$) is:

$$\gamma_{\alpha'}^\alpha = h_\alpha \delta_{\alpha'}^\alpha + 2\pi \sum_j C_{j\alpha'}^\alpha g^j + \cdots \tag{8.4}$$

Note that we have : $\gamma^i_j = 2\delta^i_j - \partial_i\beta^j$.

The renormalization group equations are:

$$\left[\sum_a x^\nu_a \frac{\partial}{\partial x^\nu_a} + \sum_a \gamma_a(g) - \sum_j \beta^j(g)\frac{\partial}{\partial g^j}\right] \langle \mathcal{O}_1(x_1)\cdots\mathcal{O}_P(x_P)\rangle = 0 \qquad (8.5)$$

For the two point functions, the integrated version of the RG equation reads:

$$\langle \mathcal{O}(R)\mathcal{O}(0)\rangle_{g(a)} = \langle \mathcal{O}(a)\mathcal{O}(0)\rangle_{g(R)} \exp\left(-2\int_{g(a)}^{g(R)} d\overline{g}\frac{\gamma_\Phi(\overline{g})}{\beta(\overline{g})}\right) \qquad (8.6)$$

where γ_Φ is the γ-function for Φ.

References

[1] Cf. e.g. : *Ill condensed matter*, Les Houches, Eds. R.Balian, R. Maynard and G. Toulouse, North-Holland 1978.

[2] V. Dotsenko and Vl. Dotsenko, Adv. Phys. 32 (1983) 129.

[3] A. Ludwig, Nucl. Phys. B330 (1990) 639.

[4] R. Shankar, Phys. Rev. Let. 58 (1987) 2466

[5] J.L. Cardy and S. Ostlund, Phys. Rev. B 25 (1982) 6899.

[6] J. Villain and J.F. Fernandez, Z. Phys. B54 (1984) 139.

[7] D. Cule and Y. Shapir, Phys. Rev. Lett. 74 (1995) 114.

[8] M.P.A. Fisher, Phys. Rev. Lett. 62 (1989) 1415.

[9] B. Huckestein, condmat/9501106.

[10] M. Zirnbauer, Ann. Physik 3 (1994) 513.

[11] R. Laughlin Phys. Rev. B23 (1981) 5632

[12] B. Halperin, Phys. Rev. B.25 (1982) 2185

[13] E. Brézin, C. Itzykson and D. Gross, Nucl. Phys. B235 (1984) 24.

[14] S. Koch, R. Haug, K. von Klitzing and K. Ploog, Phys.Rev.Let. 67 (1991) 883.

[15] H.Wei, D.Tsui, A.Paalanen and A. Pruisken, Phys.Rev.Let. 61 (1988) 1294.

[16] A. Pruisken, Nucl. Phys. B235 (1984) 277.

[17] H. Weidenmuller, Nucl. Phys. B290 (1987) 87.

[18] J. Chalker and P. Coodington, J. Phys. C21 (1988) 2665.

[19] A.Ludwig, M.Fisher, R.Shankar and G. Grinstein, Phys.Rev. B50 (1994) 7526.

[20] K. Efetov, Adv. Phys. 32 (1983) 53.

[21] C. Mudry, C. Chamon and X.G. Wen, condmat/9509054.

[22] A.Nersesyan, A.Tsvelik and F.Wegner, Phys.Rev.Let 72 (1994) 2628.

[23] D. Bernard, Nucl. Phys. B441 (1995) 471.

[24] A.Polyakov and P.Wiegman, Phys.Lett. 131B (1983) 121.

[25] K. Gawedzki and A. Kupianen. Nucl. Phys. B320 (1989) 625.

[26] V.Knizhnik and A.Zamolodchikov, Nulc. Phys. B247 (1984) 83.

[27] S.E. Korshunov, Phys. Rev. B 48 (1993) 3969.

[28] T. Giamarchi and P. Le Doussal, Phys. Rev. Lett. 72 (1994) 1530.

[29] H. Orland and Y. Shapir, Europhys. Lett. 30 (1995) 103.

[30] T. Giamarchi and P. Le Doussal, Phys. Rev. Lett. 74 (1995) 606.

[31] J. Kierfeld, J. Phys. I (France) 5 (1995) 379

[32] G. Batrouni and T. Hwa, Phys. Rev. Lett. 72 (1994) 4133.

[33] H. Rieger, cond-mat/9502092.

[34] E. Marinari, R. Monasson and J. Ruiz-Lorenzo, J. Phys. A28 (1995) 3975.

[35] M. Mézard and G. Parisi, J. Phys. A 23 (1990) L1229.

[36] M. Bauer and D. Bernard, SPhT-95-073; condmat/9506141.

[37] E. Witten, Comm. Math. Phys. 92 (1984) 455.

[38] A. Ludwig, Nucl. Phys. B330 (1990) 639.

[39] V. Dotsenko, Vk. Dotsenko, M. Picco and P. Pujol, condmat/9502134.

[40] cf e.g. M. Mezard, G. Parisi and M. Virasoro, "Spin-glass thery and beyond" World scientific, 1987.

[41] M. Picco, Phys. Rev. B54 (1996) 14930.

[42] J. Cardy, in Les Houches, Eds. E. Brézin and J. Zinn-Justin, North-Holland, 1988.

[43] A. Zamolodchikov, Sov. J. Nucl. Phys. 46 (1987) 1090.

DOES COUPLING TO GRAVITY PRESERVE INTEGRABILITY?

ADEL BILAL

CNRS - LABORATOIRE DE PHYSIQUE THÉORIQUE
DE L'ECOLE NORMALE SUPÉRIEURE
24 RUE LHOMOND, 75231 PARIS CEDEX 05, FRANCE

1 INTRODUCTION

These notes are based on results obtained in collaboration with Ian Kogan. They have already been briefly announced in our previous articles [1]. Here I will give some more details.

When a renormalisable quantum field theory is coupled to gravity three things may happen: the coupled system is more difficult or even non-renormalisable, it is simpler, or about the same. The first option is of course the generic one in four dimensions. In two dimensions, things tend to be better. We know of examples where adding conformal matter and gravity leads to a topological field theory, *i.e.* a much simpler system. In two dimensions, there are also many exactly integrable models. What happens to them when they are coupled to two-dimensional gravity? To try to get an idea, Ian Kogan and I have investigated the case of the integrable chiral Gross-Neveu model. We did neither prove nor disprove integrability of this model when coupled to gravity, but we obtained some encouraging indications that integrability might indeed be preserved.

In these notes, after introducing the model, I will describe these indications: we checked, up to the second non-trivial order whether the two-particle S-matrix remains elastic in the presence of gravity. It turned out that this is the case.

2 THE CHIRAL GROSS-NEVEU MODEL AND THE BETHE ANSATZ

The action of the chiral Gross-Neveu model is given by

$$S \sim \frac{1}{\sqrt{2}}\int \mathrm{d}t\mathrm{d}x \left[-i\bar{\psi}^j\gamma^a\partial_a\psi^j + \frac{1}{2}g^2(\bar{\psi}^j\psi^j)^2 - \frac{1}{2}g^2(\bar{\psi}^j\gamma_5\psi^j)^2\right]$$

Low Dimensional Applications of Quantum Field Theory
Edited by L. Baulieu *et al.*, Plenum Press, New York, 1997

$$= \int \mathrm{d}t\mathrm{d}x \left[i(\psi_+^j)^* \partial_- \psi_+^j + i(\psi_-^j)^* \partial_+ \psi_-^j + \sqrt{2} g^2 (\psi_+^i)^* (\psi_-^j)^* \psi_-^i \psi_+^j \right]. \quad (2.1)$$

(I will detail my conventions in the next section.) One has N colours of Dirac fermions ψ^j ($j = 1, \ldots N$), and ψ_-^j and ψ_+^j refer to right and left-moving components (Weyl fermions).[1] It is often convenient to rewrite the ψ self-couplings using two auxiliary scalar fields σ and π with unit propagator and coupling to the fermions as $\sim g\sigma\bar{\psi}^j\psi^j$ and $\sim ig\pi\bar{\psi}^j\gamma_5\psi^j$. Gross and Neveu [2] computed the effective potential for these scalar fields. To leading order in $\frac{1}{N}$, the effective potential for σ (at $\pi = 0$ *e.g.*) is given by all fermion one-loop diagrams with an arbitrary number of σ's attached. Each individual diagram is IR divergent, but summing them up leads, after renormalisation, to $V_{\text{eff}}(\sigma) = \frac{1}{2}\sigma^2 + \frac{1}{4\pi} g^2 N \sigma^2 \left[\ln\left(\frac{\sigma}{\sigma^*}\right) - 3 \right]$, where σ^* is some fixed subtraction value. This effective potential has a non-trivial minimum at non-zero σ. Hence, σ develops a non-vanishing expectation value, $<\sigma>= \sigma^* e^{1-\pi/Ng^2}$ which in turn, through the $\sigma\bar{\psi}\psi$-interaction, induces a mass $M \sim e^{-\pi/Ng^2}$ for the fermions. Clearly, this dynamical mass generation is non-perturbative.

An alternative way to understand the mass generation is via the Bethe Ansatz which, of course, also proves the integrability of the model [3]. Let me very briefly sketch the procedure. One starts with the Hamiltonian that corresponds to the action (2.1). This Hamiltonian is then diagonalized in a n-particle Fock space. At this point there is *no* Dirac sea. The diagonalisation of the Hamiltonian provides the energy levels of the empty Dirac sea (cut off at some large negative energy since n is finite). The eigenstates are pseudoparticles and correspond to the fermions of (2.1) with zero mass. One has a many-body theory of massless pseudoparticles. Then one fills this (cut-off) Dirac sea with the pseudoparticles up to some Fermi level. The physical spectrum is given by particle-hole type excitations. Suppose one takes one of the pseudoparticles out of the filled sea. This creates a hole. But in addition, all other pseudoparticles in the sea are affected by this hole creation. One can show that all $\frac{n}{2}$ right-moving pseudoparticles have their energy and momentum changed by $\Delta E = \Delta k \sim \frac{a}{n}$, while all left-moving ones suffer a change $\Delta E = -\Delta k \sim \frac{b}{n}$ for some constants a, b. Parametrizing the latter as $a = Me^{\theta}$ and $b = Me^{-\theta}$ the total change then is

$$\begin{aligned} \Delta E_{\text{tot}} &= \frac{n}{2}\left(\frac{a}{n} + \frac{b}{n}\right) = M\cosh\theta \\ \Delta k_{\text{tot}} &= \frac{n}{2}\left(\frac{a}{n} - \frac{b}{n}\right) = M\sinh\theta. \end{aligned} \quad (2.2)$$

This is precisely the dispersion relation for a massive excitation.[2]

Using the Bethe Ansatz, it has been shown [3, 4] that the S-matrix for pseudoparticle scattering is factorisable and elastic. Factorisability means that any $m \to l$ scattering process can be described by products of $2 \to 2$ S-matrix elements. Elasticity means that in any such scattering process only internal quantum numbers (*e.g.*

[1] In ref. [1], we mistakenly called ψ_- left-moving and ψ_+ right-moving.

[2] Of course, there is also the hole energy and momentum which still obey a massless dispersion relation and which decouple [4].

colour) are exchanged between the particles while they keep their individual momenta and energies. Since physical excitations are made up from pseudoparticle configurations, it could also be shown [3] that the S-matrix for the scattering of the physical excitations, *i.e.* of the physical (massive) particles, also is elastic and factorisable, as a consequence of these same properties for the pseudoparticle S-matrix.

3 Coupling to gravity

To couple the Gross-Neveu model to two-dimensional gravity, we replace the action (2.1) by its generally covariant form[3]

$$S_{\mathrm{M}}=\frac{1}{\sqrt{2}}\int \mathrm{d}t\mathrm{d}x(\det e)\Big[-i\bar{\psi}^j\gamma^a e_a{}^\mu\left(\partial_\mu+\frac{1}{4}\omega_\mu{}^{cd}\gamma_{cd}\right)\psi^j+\frac{1}{2}g^2(\bar{\psi}^j\psi^j)^2-\frac{1}{2}g^2(\bar{\psi}^j\gamma_5\psi^j)^2\Big] \tag{3.1}$$

where $e_a{}^\mu$ is the inverse zweibein, $\omega_\mu{}^{cd}$ the spin-connection and $\gamma_{cd}=\frac{1}{2}[\gamma_c,\gamma_d]$. It is well-known that for two-dimensional Majorana spinors the spin-connection drops out. Here however, the ψ^j are Dirac spinors and we need to introduce the $\omega_\mu{}^{cd}$. We also add the gravity action

$$S_{\mathrm{G}}=-\frac{\gamma}{16\pi}\int \mathrm{d}t\mathrm{d}x\sqrt{-g}R\frac{1}{\nabla^2}R+\mu\int \mathrm{d}t\mathrm{d}x\sqrt{-g}. \tag{3.2}$$

The latter is the well-known induced gravitational action as first derived by Polyakov [5] (see also section 2 of the second ref. [1]). It can be interpreted as due to the metric dependence of the ghost and matter measures in the functional integral. The value of the prefactor γ will be given below.

In conformal gauge the gravitational action (3.2) reduces to the Liouville action. Here instead, we will work in chiral (or light-cone) gauge where the metric and zweibein take the form

$$\begin{aligned} \mathrm{d}s^2 &= -2\mathrm{d}x^+\mathrm{d}x^- - 2h_{++}(\mathrm{d}x^+)^2 \\ e_{+-}e_{-+} &= 1\,, \quad e_{--}=0\,, \quad \frac{e_{++}}{e_{+-}}=h_{++}. \end{aligned} \tag{3.3}$$

We are free to make a convenient choice of the Lorentz phase such that $e_{+-}=-1$. Then $e_{-+}=-1$, too, and $e_{++}=-h_{++}$. It is then straightforward to compute the spin-connection from $\omega_\mu{}^{ab}=-\omega_\mu{}^{ba}$ and $\partial_{[\mu}e_{\nu]}{}^a+\omega_{[\mu}{}^{ab}e_{\nu]}{}^c\eta_{bc}=0$. One readily finds $\omega_+{}^{+-}=\partial_-h_{++}$ and $\omega_-{}^{+-}=0$, so that $\frac{1}{4}\omega_+{}^{cd}\gamma_{cd}=\frac{1}{2}\partial_-h_{++}\gamma_5$ while $\frac{1}{4}\omega_-{}^{cd}\gamma_{cd}=0$.

[3] Our conventions are: $x^\pm=\frac{1}{\sqrt{2}}(x^0\pm x^1)$, $\partial_\pm=\frac{1}{\sqrt{2}}(\partial_0\pm\partial_1)$, $x_ay^a=x_+y^++x_-y^-$. The Minkowski metric is $\eta_{00}=-1$, $\eta_{11}=1\Rightarrow\eta_{+-}=\eta^{+-}=-1$, and $g_{\mu\nu}=e_{a\mu}e^a{}_\nu$, and $e_a{}^\mu e^a{}_\nu=\delta^\mu_\nu$ where, as usual, Lorentz indices $(a,b,\ldots)$ are raised and lowered with η_{ab} while $g_{\mu\nu}$ is used for curved space indices $(\mu,\nu,\ldots)$. One defines $\bar{\chi}=\chi^+\gamma^0=\chi\gamma^0$ and $\gamma^0\gamma^1=\gamma_5$ so that $\bar{\chi}\gamma^0=-\chi$ and $\bar{\chi}\gamma^1=\chi\gamma_5$. Furthermore $(\gamma^0)^+=-\gamma^0,(\gamma^1)^+=\gamma^1,\gamma_5^+=\gamma_5$ and $\chi_\pm=\frac{1}{2}(1\pm\gamma_5)\chi$, so that $\bar{\chi}\gamma^\pm=\frac{1}{\sqrt{2}}\bar{\chi}(\gamma^0\pm\gamma^1)=-\sqrt{2}\chi_\mp$. Finally note that for a 2×2-matrix M one has $\det M=M_{00}M_{11}-M_{01}M_{10}=M_{++}M_{--}-M_{+-}M_{-+}$.

Finally the matter action takes the form

$$S_{\rm M} = \int {\rm d}t{\rm d}x\Big[\ \ i{\psi_-^j}^*\left(\partial_+ - h_{++}\partial_- - \tfrac{1}{2}\partial_- h_{++}\right)\psi_-^j + i{\psi_+^j}^*\partial_-\psi_+^j$$
$$+\sqrt{2}g^2(\psi_+^i)^*(\psi_-^j)^*\psi_-^i\psi_+^j\Big]. \tag{3.4}$$

The contribution from the spin connection is of course just what one needs to be able to rewrite this action in the more symmetric form

$$S_{\rm M} = \int {\rm d}t{\rm d}x\Big[\ \ i{\psi_-^j}^*\partial_+\psi_-^j + i{\psi_+^j}^*\partial_-\psi_+^j + \tfrac{i}{2}h_{++}\left(\partial_-{\psi_-^j}^*\psi_-^j - {\psi_-^j}^*\partial_-\psi_-^j\right)$$
$$+\sqrt{2}g^2(\psi_+^i)^*(\psi_-^j)^*\psi_-^i\psi_+^j\Big]. \tag{3.5}$$

The gravity action (3.2) in chiral gauge becomes

$$S_{\rm G} = \frac{\gamma}{8\pi}\int {\rm d}t{\rm d}x(\partial_-^2 h_{++})\frac{1}{\partial_-(\partial_+ - h_{++}\partial_-)}\partial_-^2 h_{++} + \mu\int {\rm d}t{\rm d}x. \tag{3.6}$$

From these expressions one sees that the matter action splits into a h_{++}-independent part that is exactly the same as in eq. (2.1) plus a term $\sim h_{++}\left(\partial_-{\psi_-^j}^*\psi_-^j - {\psi_-^j}^*\partial_-\psi_-^j\right)$ which provides an interaction between the right-moving fermions and the "graviton" (h_{++}). The left-moving fermions ψ_+^j do not interact with gravity. In the gravitational action the second term does not depend on h_{++} and hence does not contribute to the dynamics. So we will drop it in the following. The first term, however, does depend on h_{++} in a non-polynomial way. If we use $\frac{1}{\gamma}$ as a formal expansion parameter (*i.e.* $h_{++}\sim\frac{1}{\sqrt{\gamma}}$) one can expand the $(\partial_+ - h_{++}\partial_-)^{-1}$ in a power series leading to

$$S_{\rm G} = \frac{\gamma}{8\pi}\int {\rm d}t{\rm d}x h_{++}\partial_-\frac{1}{\partial_+}\left[1 + \partial_+^{-1}h_{++}\partial_- + (\partial_+^{-1}h_{++}\partial_-)^2 + \ldots\right]\partial_-^2 h_{++}. \tag{3.7}$$

from which one can read of the h_{++} ("graviton")-propagator, as well as the various vertices involving an arbitrary number of "gravitons".

The graviton propagator is given by

$$< h_{++}(-k)h_{++}(k) >= -\frac{4\pi i}{\gamma}\frac{k_+}{k_-^3} \tag{3.8}$$

with an appropriate $i\epsilon$-prescription to be discussed below, while the vertex *e.g.* between three gravitons of momenta p, q and $k = -p-q$ is

$$i\frac{\gamma}{4\pi}\left[p^2q^2 + p^2k^2 + q^2k^2\right]. \tag{3.9}$$

From the matter action (3.5) one reads the fermion propagators $\frac{i}{k_-}$ for the left-moving fermion (ψ_+) and $\frac{i}{k_+}$ for the right-moving fermion (ψ_-). Most important for us here is the vertex between an (incoming) right-moving fermion of momentum p, an (outgoing) right-moving fermion of momentum p' and an (incoming) graviton of momentum $k = p' - p$ which is

$$-\frac{i}{2}(p_- + p'_-). \tag{3.10}$$

Of course, there is also the four-fermion vertex that was already there before coupling to gravity. It is $i\sqrt{2}g^2$ between an (incoming) left-moving fermion of colour j, an (incoming) right-moving fermion of colour i and an (outgoing) left-moving fermion of colour i and an (outgoing) right-moving fermion of colour j.

4 The vanishing of the gravitational corrections to the S-matrix

Our goal now is to compute the two pseudoparticle S-matrix in the presence of gravity and show that it is still elastic. As explained above, the pseudoparticles are the massless fermions described by the action (3.5), while the physical particles are the massive fermions. The pseudoparticles correspond to an *empty* Dirac sea. This means that one cannot create a pseudoparticle-antipseudoparticle pair. Accordingly, the pseudoparticle propagators $\frac{i}{k_\pm}$ read from (3.5) do *not* have the standard Feynman $i\epsilon$-prescription which would be $\frac{ik_\mp}{k_+k_-+i\epsilon}$. Instead, we must only use the retarded propagator

$$< \psi_{\mp}^{l}{}^{*}(-k)\psi_{\mp}^{m}(k) > = \delta_{lm}\frac{i}{k_\pm + i\epsilon} = \delta_{lm}\frac{ik_\mp}{k_+k_- + i\epsilon\,\mathrm{sgn}k_\mp}. \tag{4.1}$$

An important consequence of the appearence of the retarded fermion propagators only, and of the structure of the four-fermion vertex, is that the diagrams of Fig. 1 do contribute to pseudoparticle scattering amplitudes, while those of Figs. 2 and 3 do not exist or give vanishing contribution.

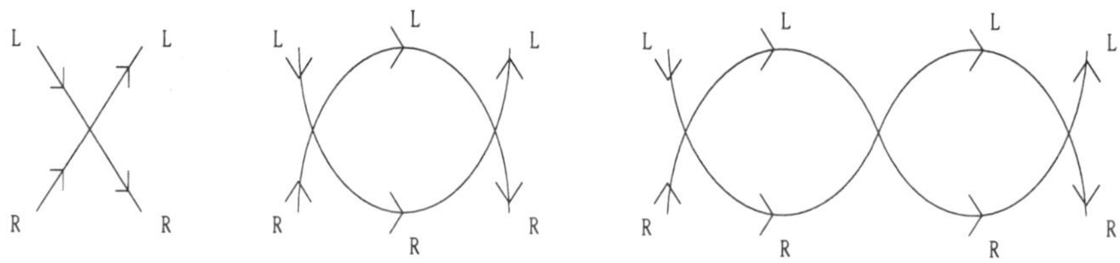

Figure 1. Diagrams contributing to the left-right scattering.

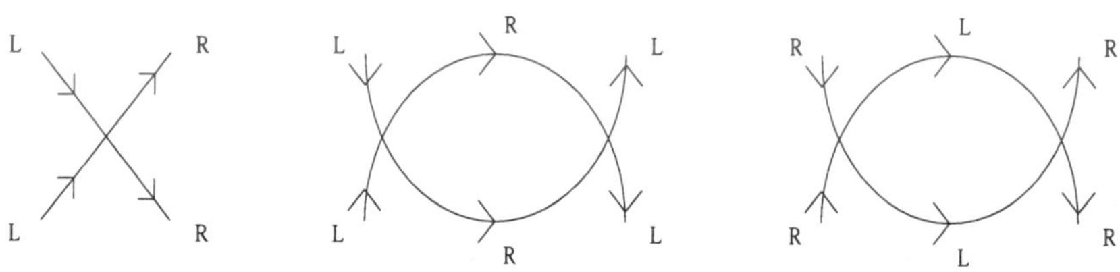

Figure 2. Left-left and right-right scattering diagrams giving vanishing contributions.

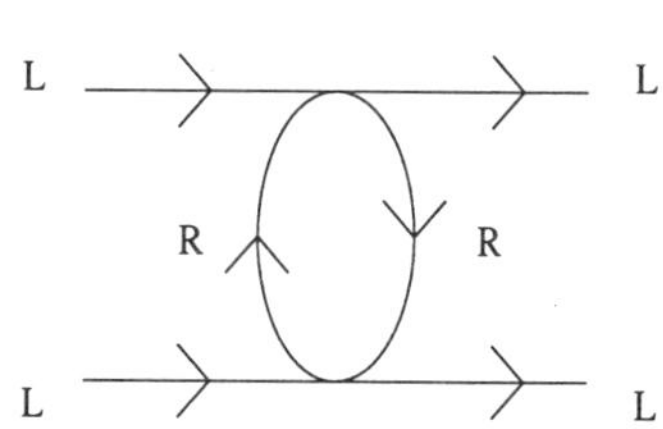

Figure 3. Another left-left scattering diagram giving a vanishing contribution.

Thus, without gravity, there is no left-left and no right-right scattering. The only non-trivial scattering is left-right $\to$ left-right, corresponding to diagrams as in Fig.

1. How does the latter manage to be elastic, *i.e.* preserve the individual momenta? This is a simple consequence of energy-momentum conservation and the mass-shell condition. Indeed, let the initial momenta be p and q and the final momenta p' and q' with p, p' for the left-moving and q, q' for the right-moving fermions. The conservation condition then is

$$p_+ + q_+ = p'_+ + q'_+ \ , \quad p_- + q_- = p'_- + q'_- . \tag{4.2}$$

The mass-shell conditions are

$$p_- = p'_- = 0 \ , \quad q_+ = q'_+ = 0. \tag{4.3}$$

Combining both equations gives

$$p_+ = p'_+ \ , \quad q_- = q'_- \tag{4.4}$$

which together with (4.3) shows that the individual momenta are unchanged and the two pseudoparticle S-matrix is elastic in the left-right $\to$ left-right chanel. Of course, this applies only to the S-matrix; general off-shell amplitudes do not preserve the individual momenta.

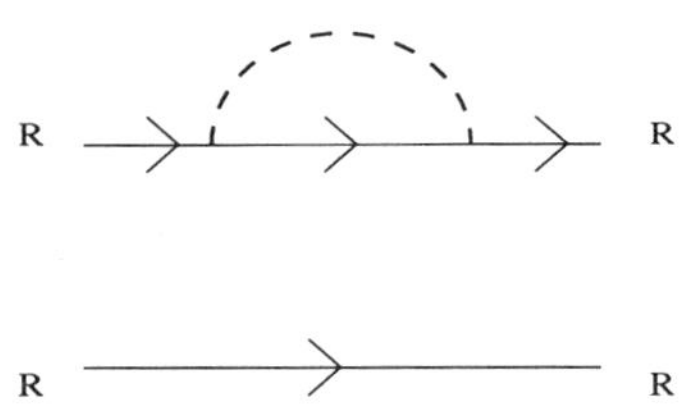

Figure 4: Disconnected diagram contributing only to the gravitational self-energy of the right fermions

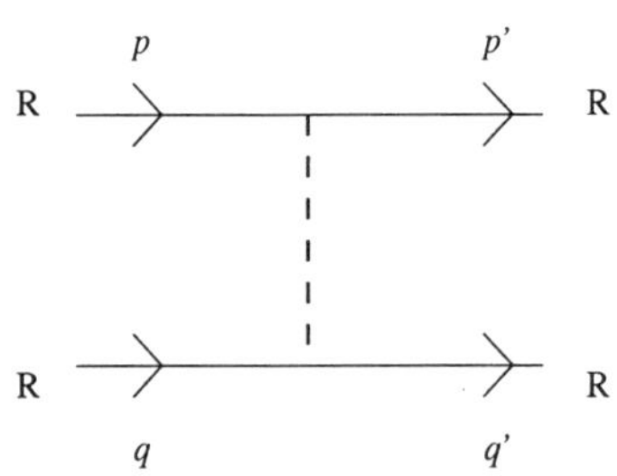

Figure 5. Gravitational tree-level scattering of two right fermions.

What changes when we couple this system to gravity? Now the right-moving fermions couple to the graviton with the vertex (3.10). This will affect the left-right $\to$ left-right correlation functions, but by the above kinematic argument the corresponding S-matrix element will remain elastic. Also, the left-left $\to$ left-left scattering is insensitive to gravity and thus remains elastic, too. The only non-trivial case is the right-right $\to$ right-right scattering. There is no kinematic reason for this to remain elastic, and the fermions do couple to gravity via diagrams as shown in Figs. 4 - 7. Of course, the diagram of Fig. 4 cannot lead to any non-trivial scattering. The first diagram that could do so is the tree-diagram of Fig. 5 involving a one-graviton exchange between two right-moving fermions. Recall from eq. (3.8) that the graviton propagator is $-\frac{4\pi i}{\gamma}\frac{k_+}{k_-^3}$. Contrary to the fermions, it should be the Feynman propagator and have the standard $i\epsilon$-prescription, *i.e.*

$$-\frac{4\pi i}{\gamma}\frac{k_+}{k_-^3} \rightarrow -\frac{4\pi i}{\gamma}\frac{k_+^4}{(k_+k_- + i\epsilon)^3} = -\frac{4\pi i}{\gamma}\frac{k_+}{(k_- + i\epsilon\,\mathrm{sgn}k_+)^3}. \tag{4.5}$$

Let's first compute the diagram of Fig. 5. Actually, we will compute the amputated diagram on shell as relevant for the S-matrix. It is given by

$$\frac{i\pi}{\gamma}(p_- + p'_-)(q_- + q'_-)\frac{(p_+ - p'_+)}{(p_- - p'_-)^3}. \tag{4.6}$$

Putting the external momenta on shell, $p_+ = p'_+ = q_+ = q'_+ = 0$ for right-moving fermions, the amplitude (4.6) vanishes. It is clear from the structure of the graviton propagator that one always gets zero as long as it is coupled to two on-shell right-moving fermions. It then immediately follows that both diagrams of Fig. 6 also vanish on shell. So, up to order $\frac{1}{\gamma^2}$, we are left with the two diagrams of Fig. 7 only.

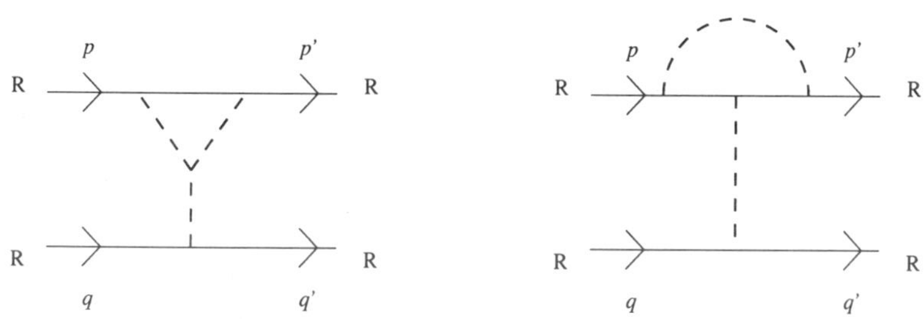

Figure 6. Vertex corrections to the diagram of Fig. 5.

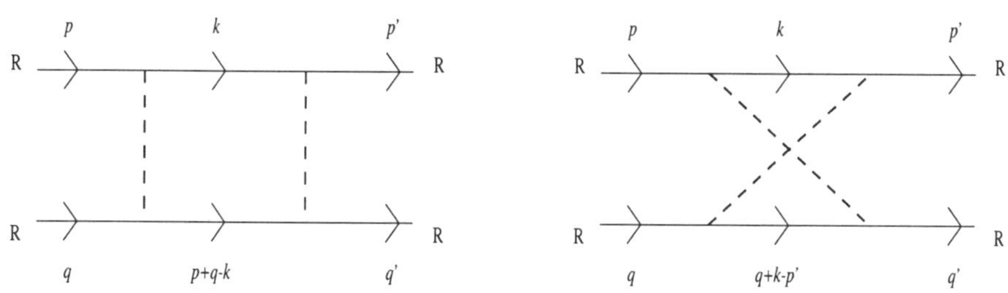

Figure 7. The box and crossed box diagrams contributing to the one-loop gravitational scattering of two right fermions.

They yield

$$\frac{1}{4\gamma^2}\int d^2k\,\frac{(p_-+k_-)(p'_-+k_-)(p_+-k_+)(k_+-p'_+)}{k_+(p_--k_-)^3(k_--p'_-)^3} \times\left[\frac{(2q_-+p_--k_-)(2q'_-+p'_--k_-)}{q_++p_+-k_+}+\frac{(2q_--p'_-+k_-)(2q'_--p_-+k_-)}{q_+-p'_++k_+}\right]. \tag{4.7}$$

Again, we only want to evaluate it on-shell where it reduces to

$$-\frac{(q_-+q'_-)}{2\gamma^2}\int d^2k\,\frac{(k_-+p_-)(k_-+p'_-)(2k_--p_--p'_-)}{(k_--p_-)^3(k_--p'_-)^3}. \tag{4.8}$$

Although it looks as if the integrand does not depend on k_+, this is not true. The k_+-dependence comes in through the $i\epsilon$-term in the graviton propagators. Let's integrate over k_- first. The integrand falls off fast enough at infinity so that one can close the contour by a semicircle in the upper or lower half-plane. We have two third-order poles (assuming $p_- \neq p'_-$ *i.e.* $p \neq p'$; if $p = p'$ we have a pole of order six and the integral obviously vanishes). According to the $i\epsilon$-prescription (4.5) they are at $k_- = p_- - i\epsilon\,\mathrm{sgn}(k_+ - p_+)$ and $k_- = p'_- - i\epsilon\,\mathrm{sgn}(k_+ - p'_+)$. If we choose the semicircle in the lower half-plane, we will pick up residues from the first pole only if $k_+ > p_+$ and from the second pole only if $k_+ > p'_+$. One gets

$$\begin{aligned}&-\frac{2\pi i}{\gamma^2}\frac{(q_-+q'_-)(p_-+p'_-)}{(p_--p'_-)^3}\int d^2k\left[\theta(k_+-p_+)-\theta(k_+-p'_+)\right]\\ &=\frac{2\pi i}{\gamma^2}\frac{(q_-+q'_-)(p_-+p'_-)(p_+-p'_+)}{(p_--p'_-)^3}.\end{aligned} \tag{4.9}$$

Had we chosen the semi-circle in the upper half-plane, the result would have been the same, of course. Thus the one-loop result (on shell) is, up to the factor $\frac{2}{\gamma}$, identical to the tree-level result, and hence vanishes for the same reason ($p_+ = p'_+ = 0$ on shell).

Let me stress that the vanishing of the scattering amplitude at one loop was in no way obvious a priori. One has to add both diagrams of Fig. 7 and use the on shell condition. Then, it is only through the subtlety of the $i\epsilon$-prescription that one gets the factor $(p_+ - p'_+)$ which makes the amplitude vanish on shell. It is very tempting to speculate that this sort of mechanism will persist at all orders in $\frac{1}{\gamma}$.

5 Conclusions

We have seen that, in general, the coupling to gravity does modify the right-right $\to$ right-right correlation functions. However, the gravitational corrections to the *on shell* right-right $\to$ right-right scattering amplitude vanish at tree and one-loop level, *i.e.* up to and including order $\frac{1}{\gamma^2}$. We thus conjectured [1] that this might be true to all orders in $\frac{1}{\gamma}$. Let's assume here that this is indeed the case, and see what it implies.

First of all, one might object the use of perturbation theory in $\frac{1}{\gamma}$. The constant γ is given by [5]

$$\gamma = \frac{1}{12}\left(c - 13 - \sqrt{(c-1)(c-25)}\right) \tag{5.1}$$

where c is the total matter central charge. Of course, gravity is well understood for $c \leq 1$ where γ is real. This is the weak coupling regime. Indeed, as $c \to -\infty$ one has $\gamma \sim \frac{c}{6}$. Since $\frac{1}{\gamma}$ is the gravitational coupling constant, $c \to -\infty$ is the gravitational weak-coupling limit. A perturbation expansion in $\frac{1}{\gamma}$ could be expected to be reasonable as long as c is large and negative. In the Gross-Neveu model this is certainly not the case. Moreover, in ref. [1] we have shown that, in the presence of gravity, certain fermion correlation functions have a diverging expansion in $\frac{1}{\gamma}$ for all γ, and that the Borel-resummed perturbation series exhibits a typically non-perturbative behaviour. Here, however, the situation is different: the perturbative expansion of the right-right $\to$ right-right on shell scattering amplitude, having only zero coefficients, converges everywhere in the complex $\frac{1}{\gamma}$-plane, yielding zero for all γ. Although one cannot exclude a non-perturbative contribution $\sim e^{-a\gamma}$ to the S-matrix, this does not seem to be very likely.

So if the gravitational corrections to the right-right $\to$ right-right (pseudoparticle) S-matrix elements do indeed vanish, as they do for the left-left $\to$ left-left elements, the only corrections are to the left-right $\to$ left-right (pseudoparticle) S-matrix elements. But as already noted earlier, the latter nevertheless remain elastic, and one would conclude that the two pseudoparticle scattering S-matrix remains elastic in all chanels when the coupling to gravity is included. One could then go on and speculate that all S-matrix elements for multi-pseudoparticle scattering factorize, and hence reduce to products of two-pseudoparticle scattering S-matrices, which are all elastic. Thus the complete S-matrix for pseudoparticle scattering would be factorisable and elastic. Since the physical S-matrix for the scattering of the physical (massive) fermions is obtained from the pseudoparticle S-matrix, it probably would turn out to be elastic and factorisable, too. If this chain of hypothesis goes through, the chiral Gross-Neveu model would indeed remain integrable when coupled to gravity. However, there is still a long way to go.

Acknowledgments

It is a great pleasure to thank Ian Kogan for the enjoyable collaboration that led to the results presented here. We also had profited from discussions with D. Gross and A. Polyakov.

References

[1] A. Bilal and I.I. Kogan, *Gravitationally dressed conformal field theory and emergence of logarithmic operators*, Princeton University preprint PUPT-1482, hep-th/9407151; A. Bilal and I.I. Kogan, Nucl. Phys. B (4) 49 1995 569 , hep-th/9503209.

[2] D.J. Gross and A. Neveu, Phys. Rev. D (1) 0 1974 3235.

[3] N. Andrei and J.H. Lowenstein, Phys. Rev. Lett. 4 (3) 1979 1698.

[4] J.H. Lowenstein, *Introduction to the Bethe Ansatz approach in* (1 + 1)-*dimensional models in*: Les Houches 1982 (Elsevier).

[5] A. Polyakov, Mod. Phys. Lett. A (2) 1987 893 ; V. Knizhnik, A. Polyakov and A. Zamolodchikov, Mod. Phys. Lett. A (3) 1988 819.

SELF-AVOIDING RANDOM MANIFOLDS

FRANÇOIS DAVID*

SERVICE DE PHYSIQUE THÉORIQUE†, CEA SACLAY
F-91191 GIF-SUR-YVETTE CEDEX, FRANCE

1 TETHERED SURFACES AND RANDOM MANIFOLDS

Several important developments of theoretical physics in the last 15 years come from the extension of the concept of random walk to fluctuating extended objects. This has been very fruitful both in high energy physics, where the quantum fluctuations of strings ($1+1$-dimensional objects) and of p-branes ($p+1$-dimensional) in Minkowski space are considered, and in condensed matter physics, where the thermal fluctuations of 2-dimensional films or membranes in Euclidean 3-dimensional space are a fascinating subject (see for instance [1]).

It is known that a 2-dimensional surface, with an intrinsic metric $g_{\alpha\beta}$, embedded in flat d-dimensional target space (the embedding being described by the mapping $x \to \vec{r}(x)$) is characterized by its extrinsic metric

$$h_{\alpha\beta} = \partial_\alpha \vec{r} \partial_\beta \vec{r} \tag{1.1}$$

Dimensional analysis shows that the relevant terms involving $\vec{r}$ in the action may involve one derivative $\partial_\alpha \vec{r}$ (tangent vectors) and two derivatives $\partial_{\alpha\beta}\vec{r}$ (extrinsic curvatures) of the embedded surface. Two very different classes of models exist, characterized by the coupling between the intrinsic and the extrinsic metric.

(1) The intrinsic metric is proportional to the extrinsic one, *i.e.* is coupled to the embedding by the constraint $g_{\alpha\beta} = \partial_\alpha \vec{r} \partial_\beta \vec{r}$. This is the case for the "rigid string" model of Polyakov, and for the equivalent Canham-Helfrich model of fluid membranes with bending rigidity. The action is

$$S[\vec{r}] = \int d^2x \sqrt{g} \left[\tau + \frac{\kappa}{2} (\Delta \vec{r})^2 \right] \tag{1.2}$$

*Physique Théorique CNRS

†Laboratoire de la Direction des Sciences de la Matière du CEA

Low Dimensional Applications of Quantum Field Theory
Edited by L. Baulieu *et al.*, Plenum Press, New York, 1997

with τ the bare string (or surface) tension, and κ the bending rigidity. Renormalization group calculations show that at large distances, the bending rigidity becomes irrelevant, and that the effective action is nothing but that of the Polyakov string, with an effective string tension τ_{eff} dynamically generated [2, 3].

(2) The intrinsic metric $g_{\alpha\beta}$ does not fluctuate (or it has a dynamics decoupled from that of the extrinsic metric). This class of models is not very useful for high energy physics, but is relevant in statistical physics to describe the so-called "tethered membranes", which generalize the concept of flexible 1-dimensional chains (polymers) to 2-dimensional networks [4]. It is this class of models that I shall discuss in these lectures.

Let me consider a flexible two dimensional regular triangular network fluctuating in 3-dimensional space. A simple discrete action for the model is

$$S = -\kappa \sum_{\substack{\text{neighbouring} \\ \text{triangles } t,t'}} \vec{n}_t \cdot \vec{n}_{t'} + \tau \sum_{\text{links } i,j} (\vec{r}_i - \vec{r}_j)^2 \tag{1.3}$$

where $\vec{n}_t$ is the normal vector to the triangle t. The first term is the extrinsic curvature term, with κ the bending rigidity, the second term is a Gaussian elastic term. Numerical simulations and analytical arguments indicate a very interesting behavior for such a model. If κ is small enough, the bending rigidity is irrelevant at large distances, and the surface is in a "crumpled", or "collapsed", phase, with $\langle \vec{n}_t \rangle = 0$. This means that O(3) rotational invariance is not broken and that the surface has no average orientation. If κ is large enough, the surface is in a flat phase, characterized by a non-zero average orientation $\langle \vec{n} \rangle \neq 0$, and a spontaneous breakdown of O(3) invariance. These two phases are separated by a crumpling transition at some κ_c. Numerical simulations indicate that (at $d = 3$) this transition is continuous, and characterized by non-trivial critical exponents.

An continuous model à la Landau-Ginzburg to describe tethered surfaces is given by the effective action [5]

$$S = \int d^D x \left[\frac{\kappa}{2} (\Delta \vec{r})^2 + \frac{t}{2} h_{\alpha\alpha} + \frac{K}{2} (h_{\alpha\alpha})^2 + \mu \left(h_{\alpha\beta} - \frac{\delta_{\alpha\beta}}{D} h_{\gamma\gamma} \right)^2 \right] \tag{1.4}$$

with $h_{\alpha\beta} = \partial_\alpha \vec{r} \partial_\beta \vec{r}$ the extrinsic metric. D is the internal dimension of the surface (the model describes in fact D-dimensional tethered manifolds), $\Delta = \partial_\alpha \partial_\alpha$ is the Laplacian, κ the bending rigidity. t, K and μ are the effective elastic moduli (t corresponds to a tension, K and μ are related to the so-called Lamé coefficients). Neglecting the fluctuations, the minimization of this action shows that if $t > 0$, $\langle \partial_\alpha \vec{r} \rangle = 0$ and $\langle h_{\alpha\beta} \rangle = 0$, so that the surface is crumpled, while if $t < 0$ $\langle \partial_\alpha \vec{r} \rangle \neq 0$ and the surface is flat. $\langle h_{\alpha\beta} \rangle$ vanishes and there is a continuous crumpling transition at $t = 0$. This analysis neglects the effect of fluctuations and is valid only if the internal dimension of the manifold is $D > 4$. For $D < 4$ fluctuations become important, and their effect can be estimated by an ϵ-expansion for $D = 4 - \epsilon$ [5], or by a large d expansion [6, 7, 8] (where d is the dimension of the target space). It might seems surprising that for 2-dimensional manifolds ($D = 2$) the flat phase phase still exists, since it is characterized by a spontaneous breakdown of the continuous rotational O(d) symmetry, which should be forbidden by the Mermin-Wagner-Coleman

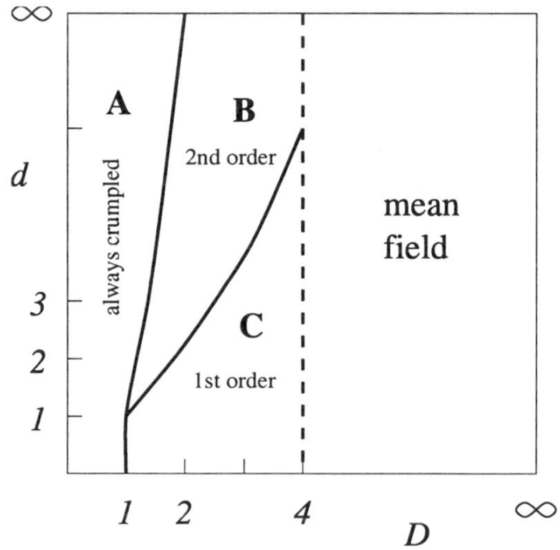

Figure 1. Nature of the crumpling transition for phantom manifolds as a function of D and d.

theorem. Such a crumpling transition is indeed forbidden for $D = 1$ (polymers): for any dimension $d \geq 1$ of target space, infinite semi-flexible polymers with a non-zero bending rigidity κ are always crumpled at large distance, and the flat phase does not exist. In fact there is no contradiction for $D = 2$. The transverse degrees of freedom of the manifolds $\vec{r}_\perp$ (undulations) are coupled to the longitudinal degrees of freedom $\vec{r}_\parallel$ (phonons), so that the global symmetry of the model is not the compact group O(d) (rotations), but the non-compact group of Euclidean displacements E(d) (translations+rotations). There is a non-trivial coupling between phonons (longitudinal modes) and undulations (transverse modes) which generates effective long range interactions between these transverse modes. In the presence of such long range interactions the Mermin-Wagner theorem does not apply.

Let me give a tentative picture which emerge from the analytical and numerical studies of the crumpling transition. For $D > 4$ the crumpling transition is continuous and its critical exponents are given by mean field theory. For $D = 4 - \epsilon$, ϵ small, one loop calculations indicates that the crumpling transition is continuous for $d > d_c$ large enough, but becomes discontinuous (fluctuation induced first-order) for $d < d_c$ small. Large d calculations shows that there is a second order crumpling transition for $D \geq D_1(d) < 2$ with $D_1(\infty) = 2$. For $D < D_1(d)$ there is no flat phase and no crumpling transition. Thus we expect that the domain $1 \geq D \geq 4$ in the (d, D) plane will be separated into three regions. A domain A for small D where the manifold is always crumpled, a domain B where there is a second order crumpling transition, and a domain C where the crumpling transition is always first order. Numerical simulations indicates that the point $(D = 2, d = 3)$ is in B [9]. Recent studies of the folding problem show that the point $(D = 2, d = 2)$ is in C [10].

The crumpled phase is the simplest to characterize. It corresponds to $t > 0$ in

(1.4), and at large distance only the quadratic t term is relevant. Therefore the action for a crumpled manifold is Gaussian, and is nothing but the massless free field action

$$S = \int d^D x \, \frac{1}{2} (\nabla_x \vec{r})^2 \tag{1.5}$$

The properties of such manifolds are easy to compute. For instance, if one considers a finite manifold with an internal extent L, its average squared size $\langle R^2 \rangle$ in target space scales as

$$\langle R^2 \rangle \propto \begin{cases} L^{(2-D)} & \text{if } D < 2 \\ \ln(L) & \text{if } D = 2 \\ \text{constant} & \text{if } D > 2 \end{cases} \tag{1.6}$$

which implies that the fractal dimension of a crumpled Gaussian manifold is $d_f = 2D/(2-D)$ if $D < 2$, and is infinite if $D \geq 2$.

2 Self-avoiding crumpled Manifolds

The above considerations apply to "phantom manifolds", which are free to intersect themselves. Indeed, the action (1.4) takes into account only local couplings in the internal space. Such local couplings are the only relevant one for strings, but for physical tethered networks self-avoiding interactions, which involve elements of the manifold which are close in target space but are arbitrarily far apart in internal space, are relevant. It is expected that such interactions will change the scaling properties of the manifold in the crumpled phase and at the crumpling transition. For instance, in the crumpled phase, the average squared size will now scale as

$$\langle R^2 \rangle \propto L^{2\nu} \qquad 0 < \nu < 1 \tag{2.1}$$

with a critical exponent $\nu \geq \nu_0 = \sup[(2-D)/2, 0]$. If $\nu_0 < \nu < 1$ the manifold will be swollen by self-avoidance, but still crumpled. This is the case for polymers $D = 1$ for $1 < d < 4$, with for instance $\nu = 3/4$ for $d = 2$. The effect of self-avoidance may even be so strong that the manifold stays flat and that the crumpling transition disappear, in this case $\nu = 1$. Such a behavior has been observed in several numerical simulations of self-avoiding tethered membranes in three dimensions.

A simple analytical model to describe crumpled self-avoiding manifolds has been introduced in [11, 12]. It is a simple extension of the continuous Edwards model for polymers. The action S (the free energy for a configuration $\vec{r}$) is the sum of a Gaussian elastic energy and of a 2-body repulsive interaction, proportional to the coupling constant b:

$$S[\vec{r}] = \int d^D x \, \frac{1}{2} (\nabla_x \vec{r})^2 + b \int d^D x \int d^D y \; \delta^d(\vec{r}(x) - \vec{r}(y)). \tag{2.2}$$

The internal dimension D may be taken as a continuous parameter, interpolating between polymers ($D = 1$) and membranes ($D = 2$). The issue is to compute the critical exponents describing the scaling behavior of large manifolds, for instance the exponent ν (related to the fractal dimension d_f of the manifold by $\nu = D/d_f$), and

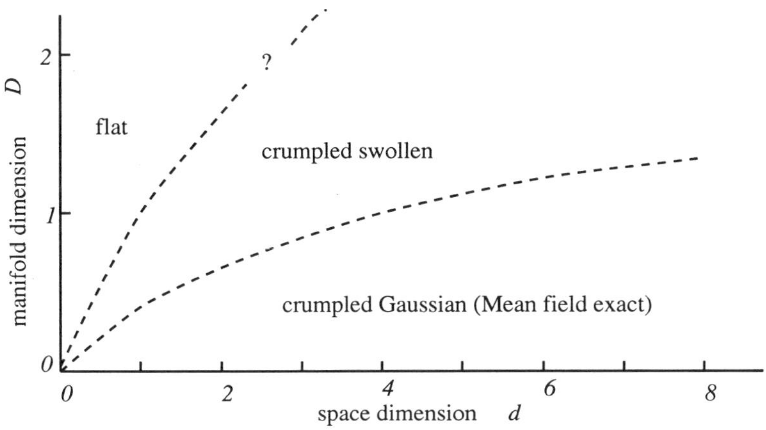

Figure 2. Self-avoiding manifolds as a function of d and D.

the configuration exponent γ, related to the scaling of the partition function Z of a finite manifold with internal extent L by

$$Z \propto L^{\gamma-1} \text{ constant}^{L^D}. \tag{2.3}$$

The mean field exponents are obtained by setting $b = 0$. One recovers the free Gaussian action and the exponents $\nu_0 = (2 - D)/2$ and $\gamma_0 = 1 - d(2 - D)/2$ (if D is not integer).

Dimensional analysis shows that the mean field theory is invalid if the engineering dimension of b, ϵ, is positive

$$[b] = \epsilon = 2D - d(2 - D)/2 > 0. \tag{2.4}$$

In this case, we expect that $\nu > \nu_0$. For small enough d, and certainly for $d \leq D$, we expect that the manifold is flat. The general picture of the expected behavior as a function of the internal dimension D and of the external dimension d is presented in Fig.2.

A natural idea is to compute the corrections to mean field by a ϵ-expansion à la Wilson-Fisher. This has been done for polymers, first by using the so-called de Gennes trick [13]: in the scaling limit the self-avoiding walk can be mapped (by a Legendre transform) onto a local field theory with $O(n)$ symmetry defined in the d-dimensional target space, in the limit $n \to 0$. It is then equivalent to study the limit of a single very long polymer and the massless limit of the $n = 0$ theory, for which standard renormalization group theory is appliquable. Unfortunately no such equivalence exists for manifolds, beyond the case $D = 1$. Another renormalization scheme used for polymers is the so-called direct renormalization scheme [14]. It has been applied to self-avoiding manifolds by Aronowitz & Lubensky [11] and by Kardar & Nelson [12], who performed calculations to first order in ϵ. The basic idea of this method is to perform explicit perturbative calculations for a finite manifold and for $\epsilon > 0$. Perturbation theory is then UV and IR finite, but has UV divergences when $\epsilon \to 0$. It appears that these poles in $1/\epsilon$ can be removed by reexpressing the observables in terms of adequate dimensionless renormalized quantities, such as the second virial coefficient. The internal size L of the manifold plays the role of the inverse of a renormalization mass scale, and renormalization group equations can be obtained by considering the L dependence of the renormalized theory. At first order in perturbation theory, the consistency of these renormalization group equations has been checked explicitly by Duplantier, Hwa & Kardar [15]. However, at that time it was not clear whether this direct renormalization approach could be justified beyond first order (except for $D = 1$, where the de Gennes trick is used to show the equivalence between direct renormalization and the standard minimal subtraction scheme).

3 Renormalization for multi-local Theories:

Recently it has become possible to prove the consistency of this approach, and the renormalizability of the model (2.2) directly in the internal D-dimensional space, despites the fact that this model is a non-local field theory in D dimensions [16]. Let me present the general idea for the proof, which is due to B. Duplantier, E. Guitter and myself. The perturbation theory for this model is obtained by expanding the observables as power series in b. The bi-local "interaction vertex" (in field theoretic language I call it a bilocal operator) is written in Fourier transform as

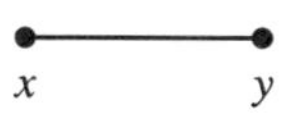

$$= \delta^d(\vec{r}(x) - \vec{r}(y)) = \int d^d\vec{k}\ e^{i\vec{k}(\vec{r}(x)-\vec{r}(y))}. \tag{3.1}$$

It can be viewed in a Coulomb gas representation as the integral over the "charge" $\vec{k}$ of a neutral "dipole" with charge $+\vec{k}$ at x and charge $-\vec{k}$ at y. The term of order b^K in the perturbative expansion of the partition function (as well as of other observables) involves K dipoles $(x_1, y_1), \ldots, (x_K, y_K)$. The integration over the charges $\vec{k}_1, \ldots \vec{k}_K$ gives an integral over the positions of the dipoles of the determinant of the "dipole energy" quadratic form Q

$$\int \cdots \int \prod_{i=1}^{K} d^Dx_i\, d^Dy_i\ \det\left[Q[x_i, y_i]\right]^{-d/2}. \tag{3.2}$$

Q is a $K \times K$ matrix such that $\sum_{i,j=1}^{K} \vec{k}_i Q_{ij} \vec{k}_j$ is the Coulomb energy (in D-dimensions) of the K dipoles. Each Q_{ij} is a linear combination

$$Q_{ij} = G_0(x_i, x_j) + G_0(y_i, y_j) - G_0(x_i, y_j) - G_0(x_j, y_i). \tag{3.3}$$

of the Coulomb potentials G_0 between the endpoints of pairs of dipoles i and j

$$G_0(x, x') = \langle r(x)r(x')\rangle_0 = \frac{\Gamma((D-2)/2)}{4\pi^{D/2}}\,|x - x'|^{2-D}. \tag{3.4}$$

This Coulomb potential is properly defined for $0 < D < 2$ by analytic continuation in D. For $0 < D < 2$ it is negative, but it vanishes for $x = x'$ (while for $D > 2$ it

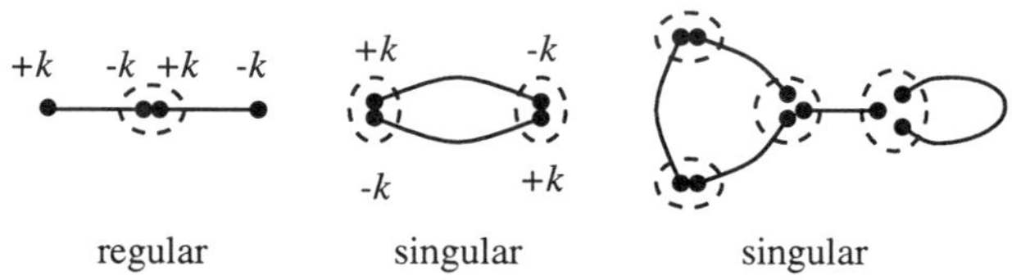

Figure 3. A non-neutral regular configuration and two neutral UV singular configurations.

diverges), and it decreases at large distances (as for $D > 2$). The integration over the $2K$ points in a non-integer D-dimensional space can also be defined properly by analytic continuation in D and the use of distance geometry. This amounts to replace the integration over the $2K \times D$ coordinates of the $2K$ points by an integration over the $K \times (2K-1)$ scalar distances between these points.

One can show that when the determinant $\det[Q]$ vanishes short distance UV singularities occur in the integrals. This occurs if and only if some of the end-points of (not necessarily the same) dipoles coincide, so that the end-points form "atoms", while the dipoles form "molecules", and if moreover one can assign non-zero charges $\vec{k}_i$ to the dipoles while each atom stays globally neutral. This condition is more easily depicted graphically on Figure 3.

The associated singularities of these integrals are related to the behavior at short distance of the expectation value (with respects to the free Gaussian model) of products of bilocal operators as given by Eq. 3.1. One can show that this short distance behavior is encoded in a multilocal operator product expansion (MOPE) , which generalizes Wilson's operator product expansion. Let me give two examples:

When the two points x and y of the bi-local interaction operator tend towards a single point, this operator can be expanded in terms of local operators involving derivatives of the field $\vec{r}$. The first terms of the expansion are explicitly (not writing explicitly the D and d dependence of the coefficients)

$$= c_0\, |x-y|^{\epsilon-2D}\, \mathbf{1} + c_1\, |x-y|^{\epsilon-D-2}(x^\alpha - y^\alpha)(x^\beta - y^\beta)\ :\nabla_\alpha \vec{r}\, \nabla_\beta \vec{r}: + \cdots \tag{3.5}$$

$\mathbf{1}$ is the identity operator (its expectation value is 1), the : : in the operator $:\nabla_\alpha \vec{r}\, \nabla_\beta \vec{r}:$ denotes the normal ordering subtraction prescription required to deal properly with the UV singularities contained in $\nabla_\alpha \vec{r}\, \nabla_\beta \vec{r}$.

The second example is less simple, and shows that when the end-points of two bilocal operators tend pairwise towards two different points, this generates again bilocal operators

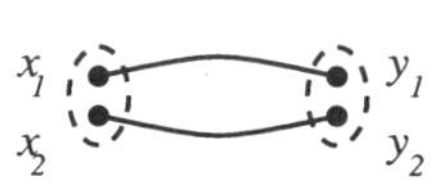

$$= d_0\, \left[|x_1 - x_2|^{2-D} + |y_1 - y_2|^{2-D} \right]^{-d/2}$$

$$+ \cdots \tag{3.6}$$

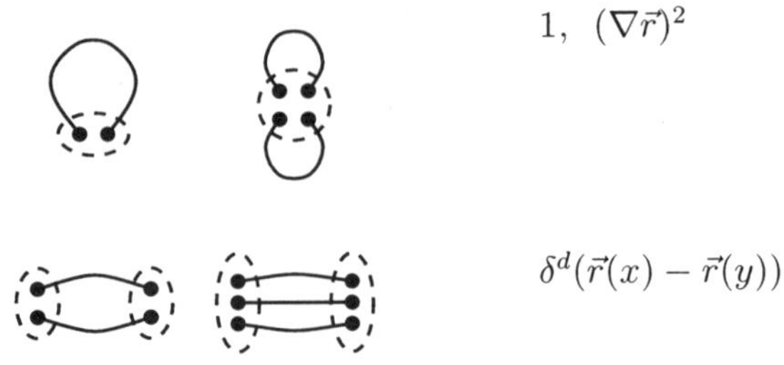

Figure 4. UV divergent configurations and the associated relevant operators.

This structure is generic, and products of local and bilocal operators generate multilocal operators of the general form

$$\Phi\{x_1, \cdots, x_P\} \ = \ \int d^d\vec{r}_0 \prod_{i=1}^{P} \left[(\nabla_{\vec{r}_0})^{m_i} \ \delta^d(\vec{r}_0 - \vec{r}(x_i)) \ A_i(x_i) \right] . \tag{3.7}$$

where the $A_i(x_i)$ are local operators, which can be decomposed into products of multiple x-derivatives of $\vec{r}$. The m_i are integers. For $P = 1$ and $m = 0$ one recovers local operators $A(x)$ ($m > 0$ gives 0). For $P = 2$, $m_1 = m_2 = 0$ and $A_1 = A_2 = \mathbf{1}$ one recovers the bilocal interaction operator, etc... These operators have a very special form: they can be viewed as a local convolution in the target d-dimensional $\vec{r}$ space of a non-local product (in the internal D-dimensional space) of the P local operators A_i.

The MOPE implies that the formalism of renormalization theory and of renormalization group equations, which has been developed for local quantum field theories, can be adapted for this model. One is in fact interested in the IR scaling behavior of the lattice model, when some length scale L goes to ∞. This lattice model is described by the action (2.2), with a short distance lattice cut-off a. To study this IR limit it is equivalent to look at the UV continuum limit of the model when the physical length scale L is kept fixed, while the UV cut-off a goes to 0. In this limit one can construct, via renormalization, a finite renormalized theory with $a = 0$, which obeys renormalization group equations. From these equations, one recovers the large distance behavior of the lattice model we started from. The procedure works well in perturbation theory when one is close to the upper critical dimension, *i.e.* for ϵ small, and it leads to the ϵ-expansion.

In our case, the MOPE can be used to determine, by power counting, which multilocal operators are relevant and give UV singularities (poles in $1/\epsilon$). Then one can also show that these poles can be subtracted by adding to the action (2.2) counterterms proportional to the marginally relevant multilocal operators, leading to the UV finite renormalized theory. For the model of self-avoiding manifolds, this analysis shows that the UV divergences are associated only with local and bilocal operators, as depicted on Fig. 4, and that only three operators are relevant: the identity operator $\mathbf{1}$, the elastic energy operator $(\nabla\vec{r})^2$ and the bilocal operator $\delta^d(\vec{r}(x) - \vec{r}(y))$. $\mathbf{1}$ is strongly relevant, and gives power-like UV divergences proportional to a^{-D} (a being a short-distance cut-off). The two other operators are superficially relevant,

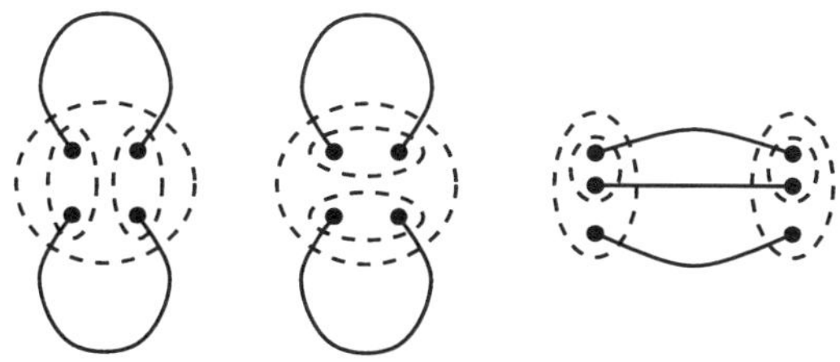

Figure 5. Examples of nested singular configurations

they give logarithmic UV divergences or equivalently poles in $1/\epsilon$ at $\epsilon = 0$.

The fact that the so-called superficial divergences, associated to a global contraction of points towards a singular configuration, can be subtracted by counterterms is a consequence of the MOPE. A complete proof of the renormalizability of the theory is possible, but much more delicate. It requires a control of the subdivergences coming from successive contractions associated to nested singular configurations, such as those depicted of Fig. 3.

4 Scaling for infinite self-avoiding Manifold

A first application of this formalism is the derivation of scaling laws. Since the model is renormalizable (at least perturbatively), it can be made UV finite (for $\epsilon \simeq 0$) by introducing two counterterms in the action. The new renormalized action is of the form

$$S[\vec{r}] = \frac{Z}{2} \int d^D x \, (\nabla \vec{r})^2 + b_R \, \mu^\epsilon \, Z_b \int\int d^D x \, d^D y \; \delta(\vec{r}(x) - \vec{r}(y)). \qquad (4.1)$$

b_R is the dimensionless renormalized coupling constant (the perturbative expansion in b_R is UV finite order by order). Z is a wave-function renormalization factor and Z_b a coupling constant renormalization factor, both are perturbative series in b_R, with poles up to degree $1/\epsilon^{K-1}$ at order K. μ is the renormalization momentum scale. As for ordinary local theories, such as the Landau-Ginzburg-Wilson Φ^4 model, one can change b_R and $\vec{r}$ in Eq. 4.1 into bare quantities in order to rewrite the renormalized Hamiltonian as a bare action given by Eq. 2.2. The renormalization group β-function and the anomalous dimension γ of the field $\vec{r}$ are defined in the standard way

$$\beta(b_R) = \mu \frac{\partial}{\partial \mu} b_R \Big|_{\text{bare}} \quad ; \quad \gamma(b_R) = -\frac{1}{2} \, \mu \frac{\partial}{\partial \mu} \ln Z \Big|_{\text{bare}} . \qquad (4.2)$$

The β-function is found to be of the form

$$\beta(b_R) = -\epsilon \, b_R + \mathrm{C} \, b_R + \mathcal{O}(b_R^2) \quad ; \quad \mathrm{C} = \mathrm{C}(\mathrm{D}) \text{ positive constant} , \qquad (4.3)$$

and therefore there is, at least for small $\epsilon > 0$, an IR attractive fixed point $b_r^\star = \mathcal{O}(\epsilon)$, which governs the scaling behavior of self-avoiding polymerized surfaces at large distance. The existence of this fixed point ensures the universality of this non-trivial scaling for $\epsilon > 0$, and that no new interactions, possibly non-local in external space, are generated by the RG transformations.

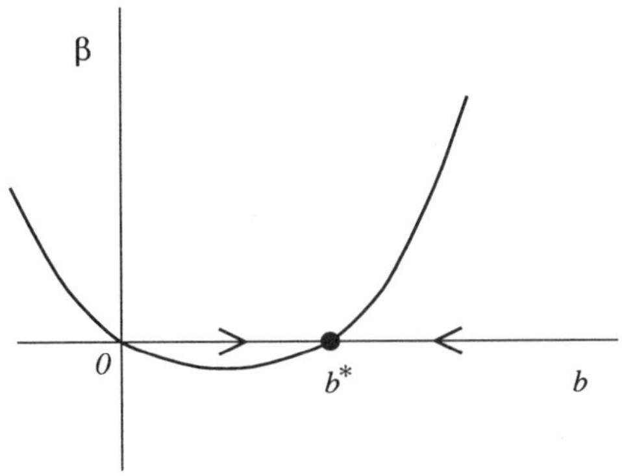

Figure 6. The β-function and the RG flow for $\epsilon > 0$.

The explicit calculation for the scaling exponents ν and γ leads to the same results for the scaling exponents ν and γ than the direct renormalization method at first order in ϵ. With this method higher order calculations are feasible, but technically quite difficult. In particular, already at second order the RG functions cannot be expressed analytically, and numerical integration methods have to be developed. Work is in progress to compute the scaling exponents at order ϵ^2.

5 Finite Size Scaling and direct Renormalization

The model given by Eq. 2.2 describes an infinite manifold with flat internal metric, corresponding to an infinite and regular flexible lattice. Finite manifolds are described by a similar model, but the D-dimensional manifold M is now embodied with a fixed non-trivial Riemannian metric $g_{\alpha\beta}(x)$ (examples are closed manifolds with the topology of the sphere $\mathcal{S}_D$ or the torus $\mathcal{T}^D$), and may have a boundary ∂M (open manifold with the topology of the disk for instance). A similar analysis can be performed for such models, and the MOPE structure of short distance singularities is still valid, but new local operators $A(x)$, which depend on the internal metric on M and on the boundary ∂M, such as the scalar curvature R, appear in the MOPE and in Eq. (3.7). The renormalized action now contains at least five operators and five independent renormalization factors Z

$$\begin{aligned} S[\vec{r}] \; = \; & \int_M Z\,1 + \int_M Z\,(\nabla\vec{r})^2 + \int\!\!\int_M Z\,b\,\delta^d(\vec{r}-\vec{r}) \\ & + \int_M Z\,R + \int_{\partial M} Z\,1. \end{aligned} \tag{5.1}$$

The curvature operator $\int_M R$ is superficially relevant only for $D=2$ and the boundary operator $\int_{\partial M} 1$ only for $D=1$. When these additional terms are not relevant, the first three renormalization factors Z are the same for finite curved manifolds than for the infinite flat plane. This property is analogous to the renormalization property of local field theories in finite geometries, which justifies the finite scaling laws for critical systems in finite geometries, and it has two very important consequences:

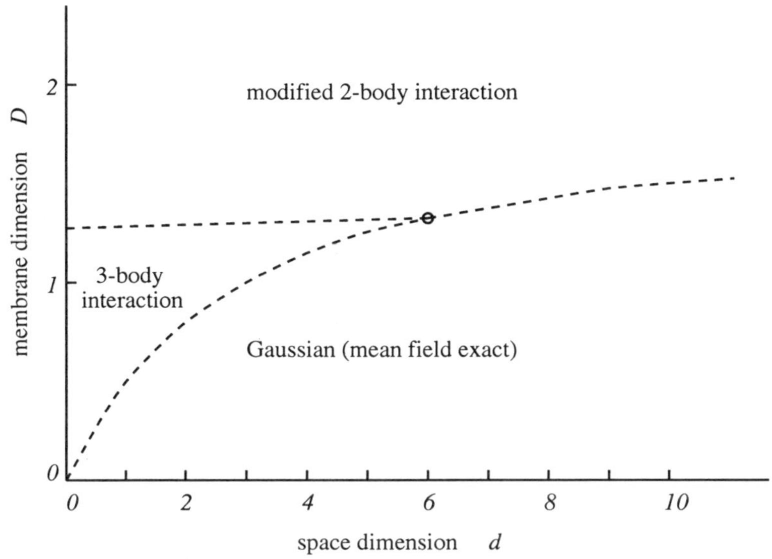

Figure 7. Relevant interactions for the Θ-point in the *d*-*D* plane.

(i) The scaling hypothesis at the basis of the direct renormalization approach, which relies explicitly on calculations with finite manifolds, is shown to be valid to all orders in perturbation theory; *(ii)* for "abstract" manifolds with dimension $D < 2$, with the only exception of open polymers ($D = 1$ open surface), the following hyperscaling relation [17] relating the configuration and the ν exponents holds:

$$\gamma = 1 - \nu d. \tag{5.2}$$

6 Self-avoiding Manifold at the tricritical Θ-point:

Finally, let me briefly discuss recent results obtained with K. Wiese on the scaling behavior of polymerized membranes at the Θ-point [18]. This point separates the swollen phase, where the self-avoidance repulsive forces that I considered previously dominate, from the dense collapsed phase, where short ranged attractive forces dominate. At the Θ-point the effective two body repulsive coupling b vanishes, and two different interactions may become relevant. The first one is the 3-body contact repulsion, which is usually considered for polymers

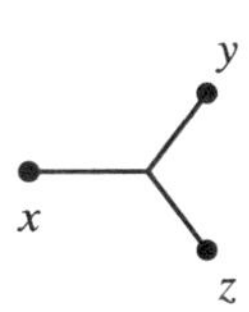

$$= \iiint d^Dx\, d^Dy\, d^Dz\; \delta^d(\vec{r}(x) - \vec{r}(y))\, \delta^d(\vec{r}(x) - \vec{r}(z)). \tag{6.1}$$

The second one is a modified 2-body interaction, repulsive at short range but attractive at larger range ($\Delta_{\vec{r}}$ is the d-dimensional Laplacian)

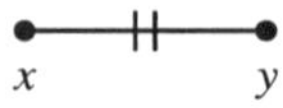

$$= - \int\!\!\int d^D x \, d^D y \; \Delta_{\vec{r}} \, \delta^d(\vec{r}(x) - \vec{r}(y)). \tag{6.2}$$

Calculations at first order are not feasible analytically, and already require numerical evaluations of complicated integrals. The results of such one loop calculations are schematically depicted on Fig. 6, where the domains where the 3-body and modified 2-body terms are respectively relevant are shown. This indicates that the last modified 2-body term is the relevant one for 2-dimensional manifold in any external dimension d. There is also a quite interesting and non-trivial crossover between the two terms around $D = 4/3$ $d = 6$, which must be studied by a double ϵ-expansion.

7 Conclusion

The theoretical study of the scaling behavior of polymerized flexible membranes leads to the development of new multilocal continuum field theories, and to new applications of renormalization group methods. I hope that these methods will lead to a quantitative progress in the understanding of the behavior of real 2-dimensional polymerized membranes. This requires results beyond first order in the ϵ-expansion (recall that $D = 2$ correspond to $\epsilon = 4$), and a better understanding of the relation between this RG approach and more heuristic or approximate methods, such as variational methods or approximate recursion relations. The sophisticated renormalization theory for multilocal models presented here should hopefully also find applications in other problems of statistical physics, or in other areas of theoretical physics.

References

[1] *Statistical Mechanics of Membranes and Surfaces*, Proceedings of the Fifth Jerusalem Winter School for Theoretical Physics (1987), D. R. Nelson, T. Piran and S. Weinberg Eds., World Scientific, Singapore (1989).

[2] L. Peliti and S. Leibler, Phys. Rev. Lett. 54 (1985) 1690.

[3] A. M. Polyakov, Nucl. Phys. B268 (1986) 406.

[4] Y. Kantor, M. Kardar and D. R. Nelson, Phys. Rev. Lett. 57 (1986) 791; Phys. Rev. A 35 (1987) 3056.

[5] M. Paczuski, M. Kardar and D. R. Nelson, Phys. Rev. Lett. 60 (1988) 2639.

[6] F. David and E. Guitter, Europhys. Lett. 5 (1988) 709.

[7] J. A. Aronovitz and T. C. Lubensky, Phys. Rev. Lett. 60 (1988) 2634.

[8] M. Paczuski and M. Kardar, Phys. Rev. A 39 (1989) 6086.

[9] Y. Kantor and D. R. Nelson, Phys. Rev. Lett. 58 (1987) 2774; Phys. Rev. A 36 (1987) 4020.

[10] P. Di Francesco and E. Guitter, Europhys. Lett. 26 (1994) 455; Phys. Rev. E50 (1994) 4418.

[11] J. A. Aronowitz and T. C. Lubensky, Europhys. Lett. 4 (1987) 395.

[12] M. Kardar and D. R. Nelson, Phys. Rev. Lett. 58 (1987) 1289.

[13] P.-G. de Gennes, Phys. Lett. 38 A (1972) 339.

[14] J. des Cloizeaux, J. de Physique 42 (1981) 635.

[15] B. Duplantier, T. Hwa and M. Kardar, Phys. Rev. Lett. 64 (1990) 2022.

[16] F. David, B. Duplantier and E. Guitter, Phys. Rev. Lett. 72 (1994) 311.

[17] B. Duplantier, Phys. Rev. Lett. 58 (1987) 2733; and in [1].

[18] K. Wiese and F. David, Nucl. Phys. B450 (1995) 495.

FOLDING, MEANDERS AND ARCHES

P. DI FRANCESCO, O. GOLINELLI, AND E. GUITTER

SERVICE DE PHYSIQUE THÉORIQUE
C.E.A. SACLAY, F91191 GIF SUR YVETTE CEDEX

1 INTRODUCTION

The concept of folding has an important place in polymer physics. Typically one considers the statistical model of a polymer chain made of say n identical constituents, and which may be folded onto itself. The entropy of such a system is obtained by counting the number of inequivalent ways of folding the chain. The combinatorial problem of enumerating all the *compact* foldings of a closed polymer chain happens to be equivalent to another geometrical problem, that of enumerating meanders [1], *i.e.* configurations of a closed road crossing a river through n bridges. To our knowledge, this is still an open problem, which indeed has been adressed by very few authors. Apart from the above folding interpretation, the meander problem arises in such various domains as the classification of 3-manifolds [3], computer science [4] and fine arts [5].

In the present communication (which is an abridged version of ref. [2]), we study various aspects of this meander problem. In section 2, we define the meander problem itself, as well as a somewhat simpler semi-meander problem corresponding to the compact folding of an open polymer chain. We also gather in this section a number of data on exact enumeration of some meanders, as well as conjectural asymptotics for some of these data. A meander may be viewed as a particular gluing of two arch configurations, representing the configuration of the road respectively above an below the river. Section 3 is devoted to the derivation of exact results for the statistical distribution of arches in arch configurations. In section 4, we introduce an exact recursion relation for constructing all semi-meanders, for which we develop a mean field approximation leading to an estimate of the entropy of folding of open polymers. Section 5 presents an alternative description of the meander problems in terms of random matrix models, and gathers a few consequent results. In section 6, we address the question of irreducible meanders [1], *i.e.* systems of several roads

Low Dimensional Applications of Quantum Field Theory
Edited by L. Baulieu *et al.*, Plenum Press, New York, 1997

crossing a river, which are interlocked in an irreducible way: in this case, an exact solution is derived for both irreducible meander and semi-meander numbers, leading to upper bounds on the meander and semi-meander numbers.

2 Definitions and generalities

2.1 Meanders

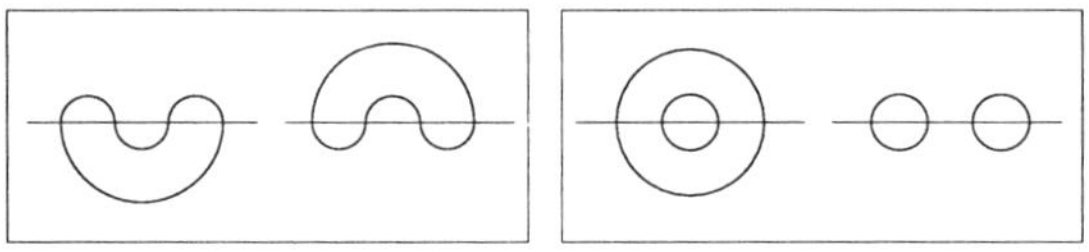

(a) **(b)**

FIGURE 1. The $M_2 = 2$ meanders of order 2 (a), and the $M_2^{(2)} = 2$ two-component meanders of order 2. The infinite river is represented as a horizontal line.

A *meander* is defined as follows. Let us consider an infinite straight line (river). A meander of order n is a closed self-avoiding connected loop (road) which intersects the line through $2n$ points[1] (bridges). A meander of order n can clearly be viewed as a compact folding configuration of a closed chain of $2n$ constituents (in one-to-one correspondence with the $2n$ bridges), by putting a hinge on each section of road between two bridges. Note also that the road and the river play indeed similar roles, by appropriately cutting the road and closing the river.

In the following, we will study the number M_n of inequivalent meanders of order n (by inequivalent we mean meanders which cannot be smoothly deformed into each other without changing the order of the bridges). We will also be interested in the numbers $M_n^{(k)}$ of inequivalent meanders of order n with k connected components, *i.e.* made of k closed connected non-intersecting but possibly interlocking loops, which cross the river through a total of $2n$ bridges. The number of connected components for a meander of order n may not exceed n, as each connected component of the meander must cross the river an even number of times, whereas the number of bridges is $2n$. Note that with this last definition, $M_n = M_n^{(1)}$. The $M_2 = M_2^{(1)} = 2$ meanders of order 2 and the $M_2^{(2)} = 2$ two-component meanders of order 2 are depicted in Fig.1 (a) and (b) respectively for illustration.

We display the first values of the multi-component meander numbers $M_n^{(k)}$ in Table I below (some of these numbers can be found in [1] and [8]).

[1]The number of intersections between a loop and an infinite line is necessarily an even number.

TABLE I. The numbers $M_n^{(k)}$ of inequivalent meanders of order n with k connected components, for $1 \leq k \leq n \leq 12$, obtained by exact enumeration on the computer

$k\backslash n$	1	2	3	4	5	6	7	8	9	10	11	12
1	1	2	8	42	262	1828	13820	110954	933458	8152860	73424650	678390116
2		2	12	84	640	5236	45164	406012	3772008	35994184	351173328	3490681428
3			5	56	580	5894	60312	624240	6540510	69323910	742518832	8028001566
4				14	240	3344	42840	529104	6413784	76980880	919032664	10941339452
5					42	990	17472	271240	3935238	54787208	742366152	9871243896
6						132	4004	85904	1569984	26200468	412348728	6230748192
7							429	16016	405552	8536890	161172704	2830421952
8								1430	63648	1860480	44346456	934582000
9									4862	251940	8356656	222516030
10										16796	994840	36936988
11											58786	3922512
12												208012

2.2 Folding a strip of stamps, semi-meanders

As mentioned in the introduction, the meander problem is equivalent to that of compact folding of a closed polymer chain. In this section, we will instead consider the case of an *open* polymer chain, moreover attached by one of its extremities. In another language, the problem is nothing but that of *folding a strip of stamps* [6] [7], and leads to a slightly different version of the meander problem, the *semi-meander* problem, which we describe now.

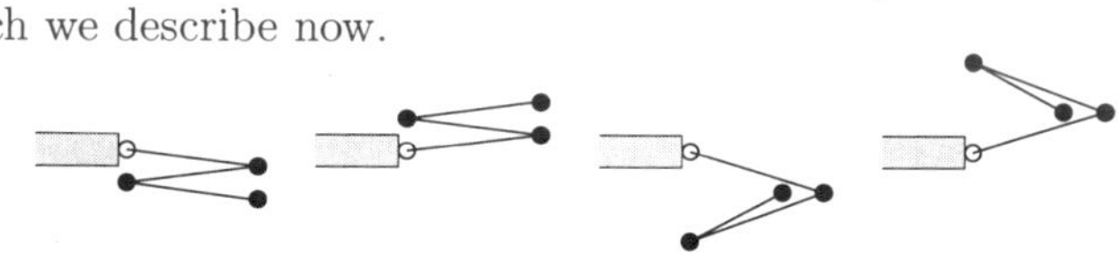

FIGURE 2. The 4 inequivalent foldings of a strip of 3 stamps. The fixed stamp is indicated by the empty circle. The other circles correspond to the edges of the stamps. The first stamp is fixed and attached to a support (shaded area).

The problem of *folding a strip of stamps* may be stated as follows. One considers a strip of $n-1$ stamps, the first of which (say the leftmost one) is fixed, and supposedly attached to some support, preventing the strip from winding around the first stamp. A folding of the strip is a complete piling of its stamps, which preserves the (non-intersecting) stamps and their connections, and only affects the relative positions of any two adjacent stamps: each stamp is folded either on top of or below the preceding one in the strip.

The number of inequivalent ways of folding a strip of $n-1$ stamps is denoted by S_n. In Fig.2, we display the $S_4 = 4$ inequivalent foldings of a strip of 3 stamps

(stamps are represented in side view, by segments, and also not completely folded to clearly indicate the succession of folds).

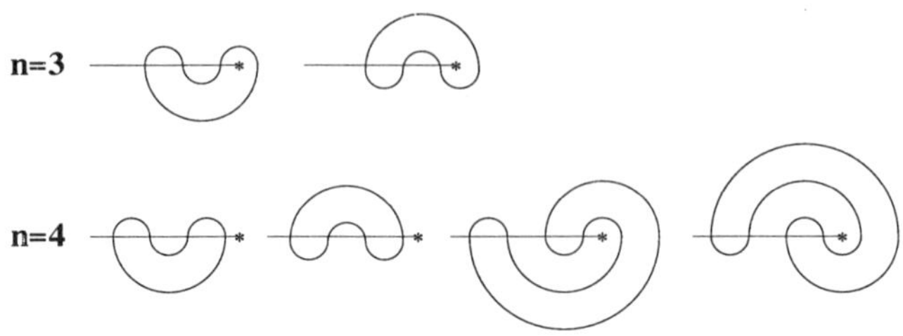

FIGURE 3. The $\bar{M}_3 = 2$ and $\bar{M}_4 = 4$ semi-meanders of order 3 and 4. The source of the corresponding semi-infinite river is indicated by an asterisk.

This folding problem turns out to be equivalent to a particular meander problem, which we will refer to as the *semi-meander* problem. Let us consider a half (semi-infinite) straight line (river) starting at a point (source). A semi-meander of order n is a closed self-avoiding connected loop which intersects the half-line through n points[2] (bridges). Let us denote by $\bar{M}_n$ the number of inequivalent semi-meanders of order n. In analogy with the definition of meanders with k connected components, we also denote by $\bar{M}_n^{(k)}$ the number of inequivalent semi-meanders of order n with k connected components. As before, we have in particular $\bar{M}_n^{(1)} = \bar{M}_n$. The $\bar{M}_3 = 2$ and $\bar{M}_4 = 4$ semi-meanders of order 3 and 4 are depicted in Fig.3.

Like in the meander case, the number k of connected components of order n semi-meanders cannot exceed n. There is actually only one semi-meander of maximal number of connected components: it corresponds to having n concentric circles, each crossing the half-river exactly once, hence

$$\bar{M}_n^{(n)} = 1 . \tag{2.1}$$

The first values of $\bar{M}_n^{(k)}$ are displayed in Table II below, for $1 \leq k \leq n \leq 14$ (some of these numbers can be found in [7] and [8]).

[2]The number n of bridges for a semi-meander need not be even, hence $n = 1, 2, 3, \ldots$

TABLE II. The numbers $\bar{M}_n^{(k)}$ of inequivalent semi-meanders of order n with k connected components, for $1 \leq k \leq n \leq 14$, obtained by exact enumeration on the computer

$k\backslash n$	1	2	3	4	5	6	7	8	9	10	11	12	13	14
1	1	1	2	4	10	24	66	174	504	1406	4210	12198	37378	111278
2		1	2	6	16	48	140	428	1308	4072	12796	40432	129432	413900
3			1	3	11	37	126	430	1454	4976	16880	57824	197010	675428
4				1	4	17	66	254	956	3584	13256	49052	179552	658560
5					1	5	24	104	438	1796	7238	28848	113518	444278
6						1	6	32	152	690	3028	12996	54812	228284
7							1	7	41	211	1023	4759	21533	95419
8								1	8	51	282	1451	7112	33721
9									1	9	62	366	1989	10227
10										1	10	74	464	2653
11											1	11	87	577
12												1	12	101
13													1	13
14														1

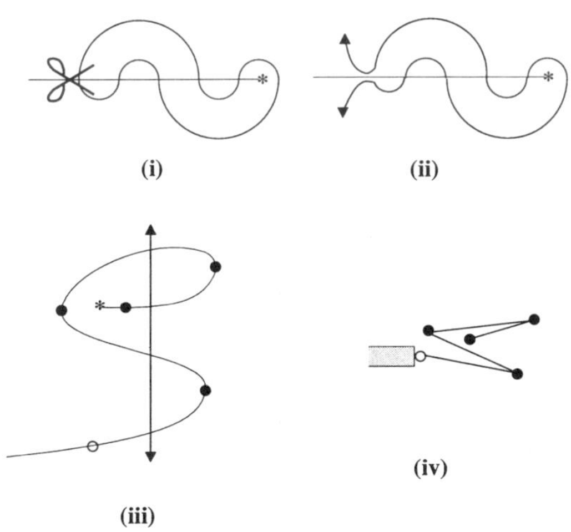

FIGURE 4. The mapping from semi-meanders to folded strips of stamps. (i) cut the leftmost arch of the semi-meander along the river. (ii) stretch the open circuit into a line. (iii) identify the segments of bent river with edges of stamps. (iv) draw the stamps according to the relative positions of crossings.

Let us prove that

$$S_n = \bar{M}_n \ . \tag{2.2}$$

Starting from a semi-meander of order n, let us construct a folding of the strip of $n-1$ stamps as follows. Cut the leftmost arch of the semi-meander along the river as indicated in Fig.4 (i), and stretch the (now open) circuit into an infinite line (ii). The river has been bent in this process, but the structure of its crossings with the line is preserved (except that the leftmost bridge has been erased). In a third step (iii), each segment of the river between two crossings of the infinite line is identified with an edge of stamp. In particular, the last edge of the strip of stamps is associated with the segment of river between its source and its first bridge, whereas the first edge of the fixed stamp corresponds to the infinite segment of river after the last bridge. This last choice singles out the first edge of the first stamp in such a way that no piece of the strip can wind around it. Finally (iv), the stamps can be drawn between the edges, the connection being indicated by the river. Note that in the process one of the bridges has been erased by the initial cut, henceforth we are left with $n-1$ bridges, hence $n-1$ stamps. This gives a one to one mapping from semi-meanders of order n to folded strips of $n-1$ stamps, thus proving (2.2).

An analogous construction allows one to relate the meander number M_n to that of foldings of a *closed* strip of $2n$ stamps. This construction is dual to the direct equivalence mentioned in the previous section, in the sense that the road and the river are exchanged.

2.3 Asymptotics

The data of the Tables I and II (see also [1] for M_n) enable one to evaluate numerically the asymptotic behaviour of M_n and $\bar{M}_n$ which read respectively

$$\begin{aligned} M_n &\sim \text{const}\ \frac{(12.25)^n}{n^{7/2}} \\ \bar{M}_n &\sim \text{const}\ \frac{(3.5)^n}{n^2} \end{aligned} \tag{2.3}$$

The exponent 7/2 for meanders was conjectured to be exact in [1].

Remarkably, the entropy of meanders

$$s = \lim_{n\to\infty} \frac{1}{n} \text{Log} M_n = \text{Log}\ R \sim 2.50 \tag{2.4}$$

is likely to be exactly twice that of semi-meanders

$$\bar{s} = \lim_{n\to\infty} \frac{1}{n} \text{Log} \bar{M}_n = \text{Log}\ \bar{R} \sim 1.25 \ . \tag{2.5}$$

Considering M_n (resp. $\bar{M}_n$) as the number of foldings of a closed (resp. open with a fixed end) strip of $2n$ (resp. $n-1$) stamps, general statistical mechanical considerations suggest that the leading behaviour of these numbers is exactly the same (*i.e.* $\propto \bar{R}^N$), when expressed in the total number of stamps $N = 2n$ (resp. $N = n-1$) for large N. This leads to the relation $\bar{s} = \frac{s}{2}$ ($=$ Log $\bar{R}$) between the corresponding thermodynamical entropies, *i.e.* $R = \bar{R}^2$ between the leading terms.

3 Arch statistics

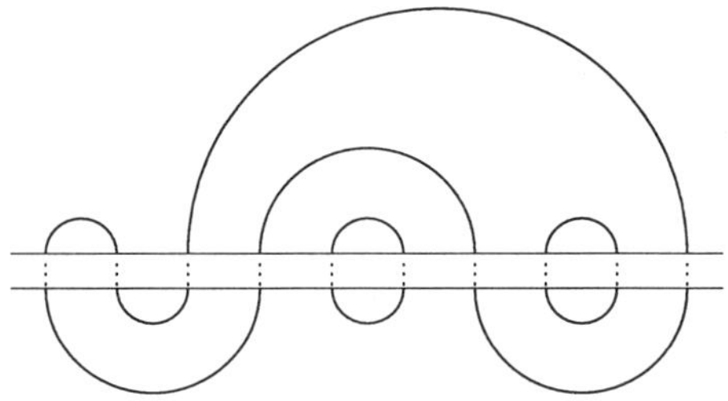

FIGURE 5. A generic meander is the superimposition of two arch configurations.

The most general meander of order n with arbitrary number of connected components is specified uniquely by its upper half (above the river) and lower half (below the river), as shown in Fig.5. Both halves form systems of n non-intersecting arches connecting $2n$ bridges by pairs. Any two arches are either disjoint or included, one into the other. Any such system of arches will be referred to as an ARCH CONFIGURATION of order n.

3.1 Catalan numbers

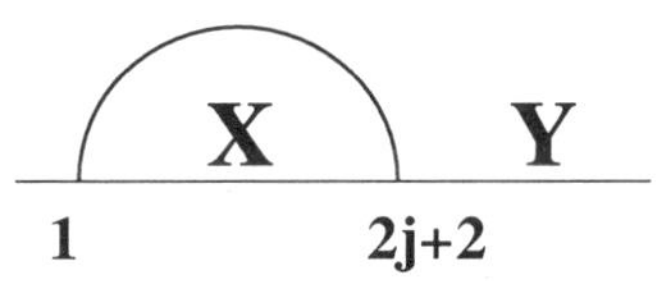

FIGURE 6. Recursion principle for arch configurations. One sums over all positions $2j+2$ of the right bridge of the leftmost arch, which separates the initial configuration into two configurations X of order j and Y of order $n-j-1$, respectively below the leftmost arch and to its right.

Let us first compute by recursion the number c_n of arch configurations of order n, linking n pairs of bridges. Starting from one such arch configuration, let us follow the arch linking the leftmost bridge (position 1) to another, say in position $2j+2$ (see Fig.6: this position has to be even). This arch separates the configuration into two sub-configurations of arches, X of order j (below the leftmost arch), and Y of order $n-j-1$ (to the right of the leftmost arch). Summing over the position of the right-hand bridge of the leftmost arch $(2j+2)$, we get the following simple recursion relation

$$c_n = \sum_{j=0}^{n-1} c_j \, c_{n-1-j} , \qquad n \geq 1 , \tag{3.1}$$

where we have set $c_0 = 1$. This rather simple example is nevertheless archetypical of a general type of reasoning used to count any relevant number associated to arch configurations. The scheme of Fig.6 is quite general.

The relation (3.1) is the defining recursion relation of the celebrated Catalan numbers c_n, which count, among other things, the numbers of parenthesings (with n opening and n closing parentheses) of words of $n+1$ letters. It follows immediately that

$$c_n = \frac{1}{2n+1}\binom{2n+1}{n} = \frac{(2n)!}{n!(n+1)!} \ . \tag{3.2}$$

In the following, we will need the generating function $C(x)$ of Catalan numbers,

$$\begin{aligned} C(x) &= \sum_{n=0}^{\infty} c_n\, x^n \\ &= 1 + x + 2x^2 + 5x^3 + 14x^4 + 42x^5 + 132x^6 + \cdots \end{aligned} \tag{3.3}$$

subject to the algebraic relation

$$xC(x)^2 = C(x) - 1 \ , \tag{3.4}$$

due to (3.1). One gets

$$C(x) = \frac{1-\sqrt{1-4x}}{2x} \ . \tag{3.5}$$

3.2 Arch configurations and meanders

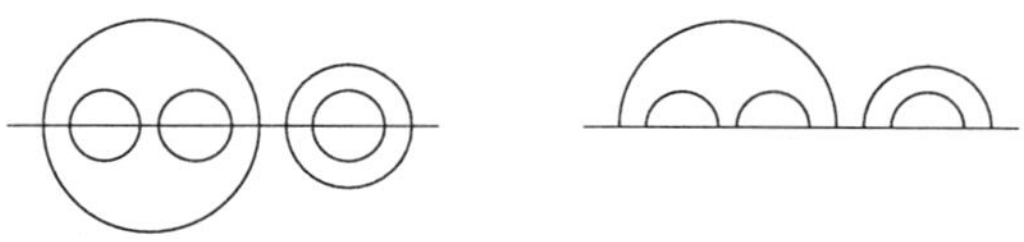

FIGURE 7. A particular 5-component meander of order 5, and the corresponding arch configuration.

Our interest in arch configurations is motivated by the fact that both meanders and semi-meanders can be built out of them. For instance, we get a direct one to one correspondence between n-component meanders of order n and arch configurations of order n: any arch configuration is completed by reflection wrt the river (see Fig.7). As a consequence, we have

$$M_n^{(n)} = c_n \ , \tag{3.6}$$

the total number of arch configurations of order n, in agreement with Table I.

More generally, any multi-component meander of order n is obtained by superimposing any two arch configurations of order n, one above the river, one below the river, and connecting them through the $2n$ bridges. As a consequence, we find the sum rule

$$\sum_{k=1}^{n} M_n^{(k)} = (c_n)^2 \ , \tag{3.7}$$

expressing the total number of multi-component meanders of order n as the total number of couples of (top and bottom) arch configurations of order n. This is readily checked on the data of Table I: the sums of numbers by columns are equal to the square of the lowest number ($M_n^{(n)} = c_n$) in each column.

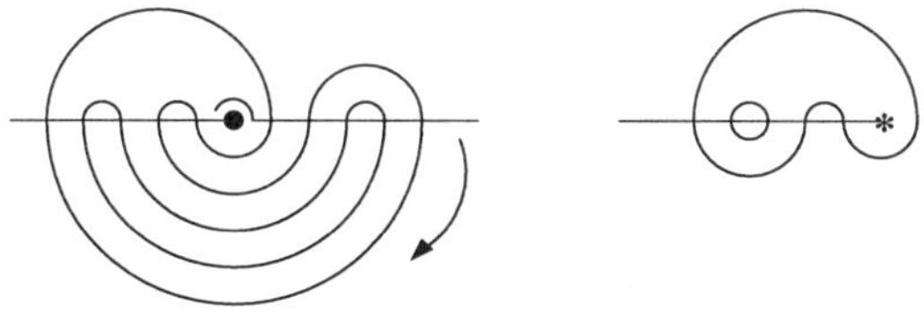

(a) **(b)**

FIGURE 8. Any semi-meander (b) is obtained from the superimposition of an arbitrary arch configuration (top of (a)) and a rainbow arch configuration (bottom of (a)) connected through the $2n$ bridges ($n = 5$ here), and by folding the river as indicated, thus identifying the bridges by pairs. The process (a) $\leftrightarrow$ (b) is clearly reversible.

It is instructive to note that any multi-component semi-meander of order n may also be obtained as the superimposition of one arbitrary arch configuration of order n above an infinite river, and a "rainbow" arch configuration below it (see Fig.8). The rainbow arch configuration of order n, denoted by $\mathcal{R}_n$, consists of n arches linking the opposite pairs of bridges: $(1, 2n)$, $(2, 2n-1)$, ...,$(n, n+1)$. Note that the number of bridges is doubled in this representation. To recover the semi-meander, one simply has to fold the infinite river into a semi-infinite one, identifying the $2n$ bridges by pairs according to the rainbow arches as indicated in Fig.8. As a consequence, we get the semi-meander version of the sum rule (3.7) for meanders

$$\sum_{k=1}^{n} \bar{M}_n^{(k)} = c_n , \tag{3.8}$$

expressing the total number of multi-component semi-meanders of order n as the total number c_n of arch configurations of order n (completed by the lower rainbow $\mathcal{R}_n$ to yield the semi-meanders). This sum rule is readily checked on the data of Table II: the sums of numbers by columns are the Catalan numbers.

The main difficulty in the meander and semi-meander problems is to find a direct way, for given arch configurations, to count the number of connected components of the resulting meander or semi-meander. This however is far beyond reach. Nevertheless, it is instructive to gather more refined statistical informations on the distribution of arches in random configurations, in view of a tentative generalization to the arch statistics of meanders with fixed number of connected components.

3.3 Statistics of exterior arches

Let us compute the distribution law of exterior arches in the arch configurations of order n. By exterior arches, we mean arches which have no other arch above them. For instance, the arch configuration of Fig.7 has two exterior arches, a rainbow $\mathcal{R}_n$ has only one exterior arch, etc. Let $E(n,k)$ denote the number of arch configurations of order n with exactly k exterior arches. A simple recursion relation can be obtained in the same spirit as for Catalan numbers, by following the general scheme of Fig.6. Starting from an order n arch configuration with k exterior arches, let us consider the leftmost arch, starting at the leftmost bridge. It is clearly an exterior arch. Let $2j+2$ be the position of the right bridge of this arch. Again, the arch separates the configuration into two arch configurations. The one below the arch is an arbitrary configuration among the c_j arch configurations of order j. The one to the right of the arch is an arch configuration of order $n-j-1$, with $k-1$ exterior arches. This leads to the recursion[3]

$$E(n,k) \;=\; \sum_{j=0}^{n-1} c_j \; E(n-j-1,k-1) \;, \tag{3.9}$$

for $k \geq 1$ and $n \geq 1$, with the initial condition $E(n,0) = \delta_{n,0}$ and $E(n,k) = 0$ for $k > n$. This defines the numbers $E(n,k)$ uniquely, and it is easy to prove that

$$E(n,k) \;=\; \frac{k}{2n-k}\binom{2n-k}{n} \;=\; \frac{k(2n-k-1)!}{n!(n-k)!} \;. \tag{3.10}$$

Another way of characterizing the distribution of arch configurations according to their number of exterior arches is through its factorial moments, defined as

$$\langle \binom{k}{l} \rangle_{\text{ext}} \;=\; \frac{\sum_{k=1}^{n}\binom{k}{l}E(n,k)}{\sum_{k=1}^{n}E(n,k)} \;, \tag{3.11}$$

where $\binom{k}{l} \equiv k(k-1)\ldots(k-l+1)/l!$. Thanks to the identity

$$\sum_{k=1}^{n}\binom{k}{l}E(n,k) \;=\; \frac{2l+1}{2n+1}\binom{2n+1}{n-l} \;=\; \binom{2n}{n-l} - \binom{2n}{n-l-1}, \tag{3.12}$$

we deduce that

$$\langle \binom{k}{l} \rangle_{\text{ext}} \;=\; (2l+1)\frac{n!(n+1)!}{(n-l)!(n+l+1)!} \;. \tag{3.13}$$

In particular, for $l=1$, the average number of exterior arches for arch configurations of order n reads

$$\langle k \rangle_{\text{ext}} \;=\; \frac{3n}{n+2} \;. \tag{3.14}$$

In the limit of infinite order, $n \to \infty$ (the thermodynamic limit), the lth factorial moment of the distribution of exterior arches tends to $2l+1$. In particular, the average number of exterior arches tends to 3.

4 Recursion relations for semi-meanders

In the following, we will mainly concentrate our efforts on semi-meanders. However, we will mention whenever possible the extensions of our results to meanders.

[3]One can also show that $E(n,k) = E(n,k+1) + E(n-1,k-1)$.

4.1 The main recursion

Let us present a simple algorithm for enumerating the semi-meanders of order n with k connected components. In this section, we will work in the infinite river/lower rainbow arch-framework for semi-meanders (see Fig.8 (a)), *i.e.* consider a semi-meander as the superimposition of the lower rainbow configuration $\mathcal{R}_n$, and some upper arch configuration.

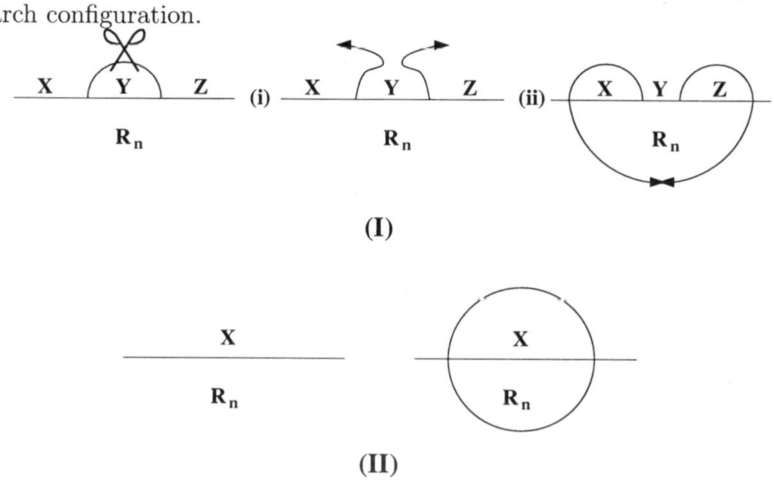

FIGURE 9. The construction of all the semi-meanders of order $n+1$ with arbitrary number of connected components from those of order n. Process (I): (i) pick any exterior arch and cut it (ii) draw its edges around the semi-meander and paste them below. The lower part becomes the rainbow configuration $\mathcal{R}_{n+1}$ of order $n+1$. The process (I) preserves the number of connected components. Process (II): draw a circle around the semi-meander of order n. The process (II) adds one connected component.

Starting from a semi-meander of order n with k connected components, let us construct several semi-meanders of order $n+1$ with k connected components in the following way, as indicated in Fig.9 (I) (i)-(ii).

(i) Pick any exterior arch of the upper arch configuration of the semi-meander and cut it.

(ii) Pull the two edges of the cut across the river (the left part of the exterior arch to the left, the right one to the right), and paste them around the lower rainbow, thus increasing the rainbow configuration by one arch ($\mathcal{R}_n \to \mathcal{R}_{n+1}$), and the number of bridges by 2.

The result is a certain semi-meander of order $n+1$ with the same number k of connected components. So for each semi-meander $\mathcal{M}$ of order n with k components, we can construct $E(\mathcal{M})$ semi-meanders of order $n+1$ with k components, where $E(\mathcal{M})$ denotes the number of exterior arches of $\mathcal{M}$. All these new semi-meanders are clearly distinct. There is however another way of generating more semi-meanders of order $n+1$, indicated in Fig.9 (II). Starting with any semi-meander of order n with $k-1$ connected components, one just adds an extra circular loop around it, which transforms the lower rainbow of order n into $\mathcal{R}_{n+1}$, and adds 2 bridges. The resulting semi-meander of order $n+1$ has clearly k connected components. Such a semi-meander cannot be obtained from some order n semi-meander by the procedure (I), because it has only one exterior upper arch, whereas the process (I) produces at least two exterior arches.

So all the order $n+1$ semi-meanders constructed by (I)-(II) are distinct. Conversely, given a semi-meander of order $n+1$ with k components, two cases may occur:

(a) it has only one exterior upper arch. In this case, it is surrounded by one circle, and therefore arises from the order n semi-meander with $k-1$ components inside the circle, through (II).

(b) it has at least two exterior upper arches. Cutting the lower exterior arch of the lower rainbow $\mathcal{R}_{n+1}$, pulling the edges of this arch above the upper configuration (the left edge by the left side, the right one by the right side, both suppressing the left and right-most bridges), and finally pasting the two edges on the upper side of the river (thus creating an exterior arch on the upper configuration), one gets an order n semi-meander with k components, which leads to the initial semi-meander of order $n+1$ through (I)-(i)-(ii), using the exterior arch constructed above.
This proves that the procedures (I)-(II) give a recursive algorithm for constructing *all* the semi-meanders of order $n+1$ with k connected components from the k and $k-1$-component meanders of order n. More precisely, these procedures can also be viewed as a recursive algorithm for constructing all the arch configurations of order n from those of order $n-1$. When completed by a lower rainbow configuration $\mathcal{R}_n$, so as to give multi-component semi-meanders, we have the interesting property that (I) preserves the number of connected components of the semi-meander, while (II) increases it by 1, which allows us in principle to follow this number of components throughout the construction.

The case $k = 1$ of connected semi-meanders is special, in the sense that the corresponding arch configurations are obtained by successive actions of the procedure (I) *only*. As a consequence, the number of connected semi-meanders of order $n+1$ is equal to the total number of exterior upper arches of all the semi-meanders of order n. Hence the average number of exterior arches in the connected semi-meanders of order n is

$$\langle \text{ext} \rangle_n = \frac{\bar{M}_{n+1}}{\bar{M}_n} \tag{4.1}$$

In particular, it tends to $\bar{R}$ in the large n limit: this links the thermodynamic entropy (2.5) of semi-meanders to their average thermodynamic number of exterior arches.

4.2 Mean field approximation

Starting from a particular arch configuration, let us generate its "descendents" by repeatedly using the processes (I) and (II) of Fig.9. After a given sequence of n ((I) or (II)) steps, let us denote by (m, e) respectively the *total* number of arch configurations generated and their average number of exterior arches. Then let us adopt the following "mean field type" recursive algorithm for an extra action by (I) or (II)

$$\begin{aligned} (m,e) &\underset{(I)}{\longrightarrow} (me,e) \\ (m,e) &\underset{(II)}{\longrightarrow} (m,1)\ . \end{aligned} \tag{4.2}$$

The second line of (4.2) is clear, as (II) builds one arch configuration out of each inital arch configuration, with only one exterior arch. The first line incorporates two suppositions. First, the number of arch configurations generated is approximated by a mean field value, namely the product of the initial number m of arch configurations by the *average* number e of exterior arches. Second, the average number of exterior arches is supposed to be unchanged.

Let us now choose as starting point $(m=1,e)$ the upper arch configuration of a semi-meander. Repeated actions of (I) only will generate its descendents which are themselves semi-meanders. Implicitly, (4.2) supposes a certain number of properties on the starting point of the recursion, like for instance the fact that repeated actions of (I) keep the average number of exterior arches e unchanged, while the number of semi-meanders obtained is multiplied by e each time. Thus in this scheme, e will be identified with $\bar{R}$, governing the large n behaviour of the number of semi-meanders $\bar{M}_n \sim \bar{R}^n$. The scheme (4.2) is only intended as an approximation and for that purpose the initial semi-meander is supposed to be very large, with a number e of exterior arches equal to the average $\bar{R}$. Let us try to evaluate e by identifying the total number $p_n(e)$ of configurations generated after all possible sequences of n actions of (I) or (II) with the total number $c_n \sim 4^n$ (this assumes that c_n gives the correct large n behaviour of the total number of descendents of a given configuration, independently of this configuration). We start with only one semi-meander, hence $p_0(e)=1$. After one step, according to eq. (4.2), we have generated $p_1(e)=e+1$ arch configurations. Let us decompose the number $p_n(e)$ into

$$p_n(e) \;=\; p_n^{(I)}(e) \;+\; p_n^{(II)}(e)\ , \tag{4.3}$$

where we make the distinction between the total number of arch configurations obtained by the process (I) (resp. (II)) in the last step (4.2) from the $p_{n-1}(e)$ previous ones. The algorithm (4.2) leads to the recursion relations

$$\begin{aligned} p_n^{(I)}(e) &= e\, p_{n-1}^{(I)}(e) \;+\; p_{n-1}^{(II)}(e) \\ p_n^{(II)}(e) &= p_{n-1}^{(I)}(e) \;+\; p_{n-1}^{(II)}(e). \end{aligned} \tag{4.4}$$

In terms of the vectors $P_n(e) = (p_n^{(I)}(e), p_n^{(II)}(e))^t$, this takes the matrix form $P_n(e) = M(e)P_{n-1}(e)$, with

$$M(e) \;=\; \begin{pmatrix} e & 1 \\ 1 & 1 \end{pmatrix}\ . \tag{4.5}$$

For the total number of arch configurations generated to behave like 4^n, we simply have to write that the largest eigenvalue of this matrix is 4, namely that the determinant of $M(e) - 4\,\mathbf{I}$ vanishes (this also fixes the other eigenvalue of $M(e)$ to be 2/3). This gives

$$e \;=\; \frac{11}{3} \;=\; 3.666\ldots \tag{4.6}$$

This estimate of $\bar{R}$ is close to the numerical estimate (2.3). But the sequence of approximations used to get (4.6) should certainly be refined.

A last remark is in order. One could wonder how much the above estimate depends on the starting point of the algorithm. In particular, starting with any semi-meander with a finite number k of connected components and a large order, we end up with the same estimate (4.6) for the average number of exterior arches. This in turn infers an estimate for $\bar{R}_k = \bar{R}_1 = \bar{R} = e = 11/3$ for the numbers $\bar{R}_k$ governing the large order behaviour of $\bar{M}_n^{(k)} \sim (\bar{R}_k)^n$. Moreover the average number of exterior arches for all the arch configurations (generated in our scheme by both (I) and (II)) reads, after n steps,

$$\langle \text{ext} \rangle_n \quad = \quad \frac{e.p_n^{(I)} + 1.p_n^{(II)}}{p_n^{(I)} + p_n^{(II)}} \; . \tag{4.7}$$

For large n, the vector $(p_n^{(I)}, p_n^{(II)})$ tends, up to a global normalization, to the eigenvector $(3,1)$ associated to the largest eigenvalue $\lambda_{\max} = 4$ of the matrix $M(11/3)$ (4.5). Therefore the average number of exterior arches for all arch configurations is estimated as

$$\langle \text{ext} \rangle \quad = \quad \frac{3e+1}{4} = 3 \; , \tag{4.8}$$

which coincides with the exact value, as given by (3.14).

5 Matrix model for meanders

Field theory, as a computational method, involves expansions over graphs weighted by combinatorial factors. In this section, we present a particular field theory which precisely generates planar graphs with a direct meander interpretation. The planarity of theses graphs is an important requirement, which ensures that the arches of the meander do not intersect each other, when drawn on a planar surface. The topology of the graphs in field theoretical expansions is best taken into account in matrix models, where the size N of the matrices governs a topological expansion in which the term of order N^{2-2h} corresponds to graphs with genus h. The planar graphs (with $h = 0$) are therefore obtained by taking the large N limit of matrix models (see for instance [9] for a review on random matrices).

5.1 *The matrix model as combinatorial tool*

Random matrix models are useful combinatorial tools for the enumeration of (connected) graphs [9]. Typically, one considers the following integral over Hermitian matrices of size $N \times N$

$$Z(g,N) \quad = \quad \frac{\int dM \; e^{-N\mathrm{Tr}(\frac{M^2}{2} - g\frac{M^4}{4})}}{\int dM \; e^{-N\mathrm{Tr}(\frac{M^2}{2})}} \tag{5.1}$$

where the integration measure is

$$dM \quad = \quad \prod_{i=1}^{N} dM_{ii} \prod_{i<j} d\mathrm{Re}M_{ij} d\mathrm{Im}M_{ij} \; . \tag{5.2}$$

The rules of Gaussian matrix integration are simple enough to provide us with a trick for computing the expression (5.1) as a formal series expansion in powers of g. For instance,

$$\langle M_{ij}M_{kl}\rangle_{\text{Gauss}} = \frac{\int dM\ e^{-N\text{Tr}(\frac{M^2}{2})}M_{ij}M_{kl}}{\int dM\ e^{-N\text{Tr}(\frac{M^2}{2})}} = \frac{\delta_{il}\delta_{jk}}{N} . \tag{5.3}$$

Expanding $Z(g, N)$ in powers of g, we are left with the computation of

$$Z(g, N) = \sum_{V=0}^{\infty} \frac{(Ng)^V}{V!} \langle (\text{Tr}\frac{M^4}{4})^V \rangle_{\text{Gauss}} . \tag{5.4}$$

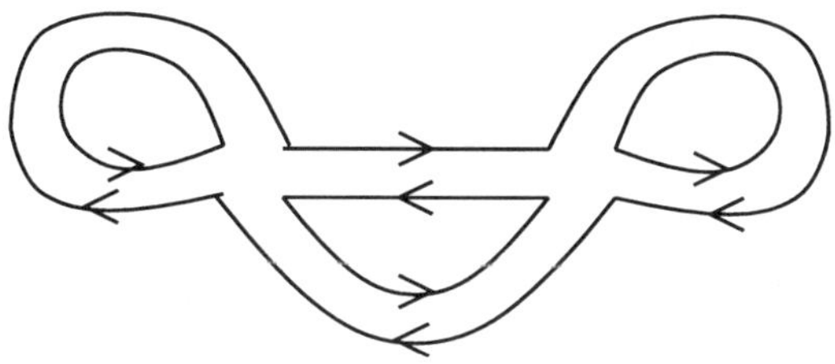

FIGURE 10. A ribbon graph with $V = 2$ vertices, $E = 4$ edges, and $L = 4$ oriented loops.

The result (5.3) suggests to represent any term in the small g expansion as a sum over graphs constructed as follows.

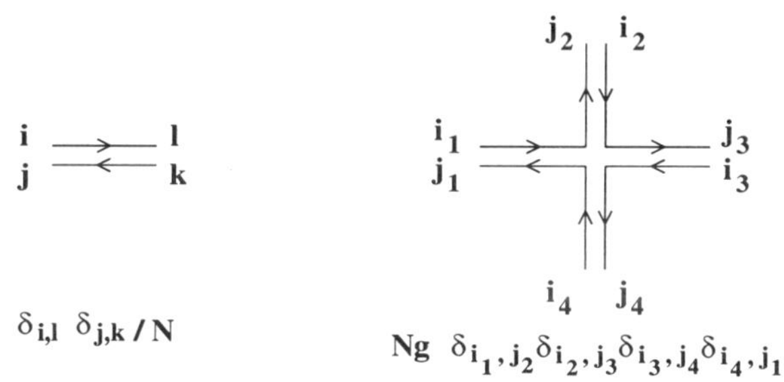

FIGURE 11. The Feynman rules for the one matrix model: the matrix indices are conserved along the oriented lines, which form closed loops. Edges receive a factor $1/N$, vertices (Ng), and all the matrix indices have to be summed over.

The graphs considered are ribbon graphs, *i.e.* with edges made of double-lines, oriented with opposite orientations (see Fig.10 for an example). These two oriented lines represent the circulation of matrix indices $i, j = 1, 2, \ldots, N$. Namely each end of an edge is associated with a matrix element M_{ij}, the index i being carried by the line pointing from (resp. j by the line pointing to) this end of the edge, as shown in Fig.11. An edge is therefore interpreted as the propagator between the states sitting at its extremities, according to (5.3), hence each edge will be weighed by a factor $1/N$ (hence an overall factor N^{-E}, where E is the total number of edges of the graph). To compute the term of order V in the small g expansion (5.4), one simply has to sum over all the ribbon graphs connecting the V four-valent vertices corresponding to the

V terms $\mathrm{Tr}M^4$ (an obvious 4-fold cyclic symmetry absorbs the factors 1/4). Each vertex has to be weighed by a factor Ng. These Feynman rules are summarized in Fig.11. An overall weight also comes from the summation over all the matrix indices $i = 1, 2, \ldots, N$ running on the oriented loops of the graph. This gives a global factor N^L for each graph, where L is the number of loops of the graph. For instance, the ribbon graph of Fig.10 receives a total weight $(Ng)^2 \times N^{-4} \times N^4 = (Ng)^2$.

The second trick is the fact that this sum can restricted to *connected* graphs only, by taking the logarithm of the function (5.1)

$$F(g,N) \;=\; \mathrm{Log}\, Z(g,N) \;=\; \sum_{\text{conn. graphs } \Gamma} g^V N^{V-E+L} \times \frac{1}{|\mathrm{Aut}(\Gamma)|} , \tag{5.5}$$

where $|\mathrm{Aut}(\Gamma)|$ denotes the order of the automorphism group of Γ, *i.e.* the number of permutations of its (supposedly labelled) vertices leaving the graph invariant. This symmetry factor results from the incomplete compensation of the factor $1/V!$ by the number of equivalent graphs (with different labelling of the vertices). Finally, we identify the power of N as the Euler-Poincaré characteristic of the graph Γ

$$\chi(\Gamma) \;=\; V - E + L \;=\; 2 - 2h , \tag{5.6}$$

which can be taken as the definition of the genus h of the graph. The number of oriented loops is indeed equal to that of faces F of the cellular complex induced by the graph, hence we can use the more standard definition of the Euler-Poincaré characteristic $\chi = V - E + F = V - E + L$.

Various techniques for direct computation of the integral (5.1) have made it possible to enumerate connected graphs with arbitrary genus, and derive many of their properties. In this section, we consider a matrix model adapted to the meander enumeration problem.

5.2 *The model*

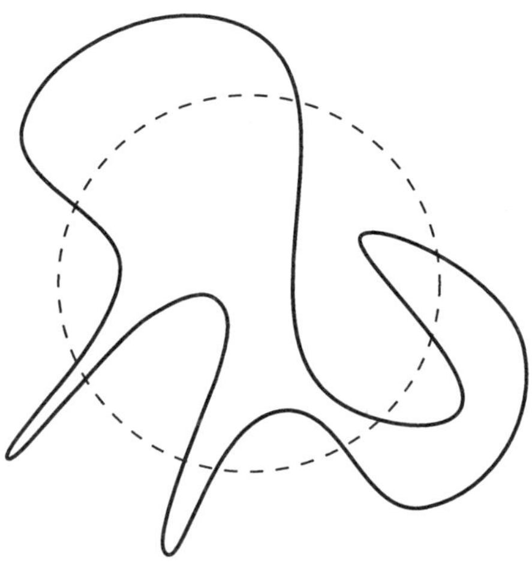

FIGURE 12. A sample black and white graph. The white loop is represented in thin dashed line. There are 10 intersections.

The enumeration of (planar) meanders is very close to that of 4-valent (genus 0) graphs made of two self-avoiding loops (say one black and one white), intersecting each other at simple nodes [1]. The white loop stands for the river, closed at infinity. The black loop is the road. Such a graph will be called a black and white graph. An example is given in Fig.12. The fact that the river becomes a loop replaces the order of the bridges by a cyclic order, and identifies the regions above the river and below it. Hence the number of meanders M_n is $2 \times 2n$ (2 for the up/down symmetry and $2n$ for the cyclic symmetry) times that of inequivalent black and white graphs with $2n$ intersections, weighed by the symmetry factor $1/|\mathrm{Aut}(\Gamma)|$. The same connection holds between $M_n^{(k)}$ and the black and white graphs where the black loop has k connected components.

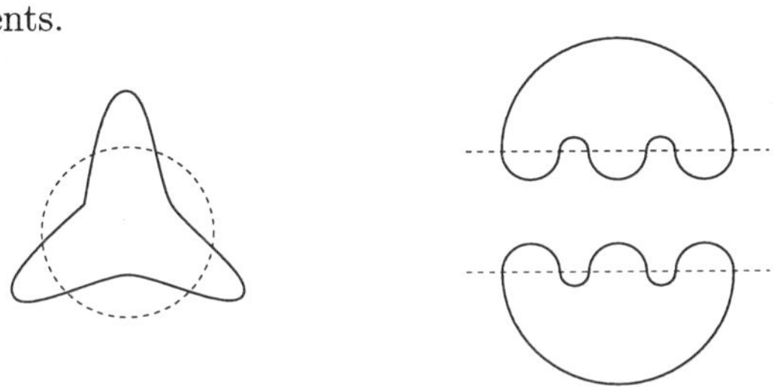

FIGURE 13. A particular black and white graph with 6 intersections, and its two associated meanders. The automorphism group of the black and white graph is $\mathbb{Z}_6$.

For illustration, we display a particular black and white graph Γ in Fig.13, together with its two corresponding meanders of order 3. The automorphism group of this black and white graph is $\mathbb{Z}_6$, with order $\mathrm{Aut}(\Gamma) = |\mathbb{Z}_6| = 6$. The two meanders come with an overall factor $1/(2 \times 6)$, hence contribute a total $2 \times 1/12 = 1/6$, which is precisely the desired symmetry factor.

In analogy with the ordinary matrix model (5.1), a simple way of generating black and white graphs is the use of the multi-matrix integral (with $m+n$ hermitian matrices of size N denoted by B and W)

$$Z(m,n,c,N) = \frac{1}{\kappa_N} \int \prod_{\alpha=1}^{m} dB^{(\alpha)} \prod_{\beta=1}^{n} dW^{(\beta)} e^{-N \mathrm{Tr}\, P(B^{(\alpha)},W^{(\beta)})} , \tag{5.7}$$

where the matrix potential reads

$$P(B^{(\alpha)},W^{(\beta)}) = \sum_{\alpha} \frac{(B^{(\alpha)})^2}{2} + \sum_{\beta} \frac{(W^{(\beta)})^2}{2} - \frac{c}{2} \sum_{\alpha,\beta} B^{(\alpha)} W^{(\beta)} B^{(\alpha)} W^{(\beta)} , \tag{5.8}$$

and the normalization constant κ_N is such that $Z(m,n,c=0,N) = 1$. In the following, the α and β indices will be referred to as color indices.

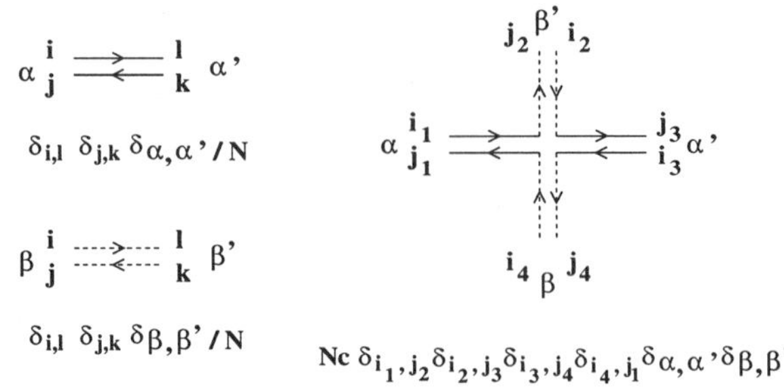

FIGURE 14. The Feynman rules for the black and white matrix model. Solid (resp. dashed) double-lines correspond to black (resp. white) matrix elements, whose indices run along the two oriented lines. An extra color index α (resp. β) indicates the number of the matrix in its class, $B^{(\alpha)}$, $\alpha = 1, 2, \ldots, m$ (resp. $W^{(\beta)}$, $\beta = 1, 2, \ldots, n$). The only allowed vertices are 4-valent, and have alternating black and white edges: they describe simple intersections of the black and white loops.

Like in the case (5.1), the logarithm of the function (5.7) can be evaluated pertubatively as a series in powers of c. A term of order V in this expansion is readily evaluated as a Gaussian multi-matrix integral. It can be obtained as a sum over 4-valent connected graphs, whose V vertices have to be connected by means of the two types of edges

$$\begin{aligned} \text{black edges} \quad \langle [B^{(\alpha)}]_{ij} [B^{(\alpha')}]_{kl} \rangle &= \frac{\delta_{il}\delta_{jk}}{N} \delta_{\alpha\alpha'} \\ \text{white edges} \quad \langle [W^{(\beta)}]_{ij} [W^{(\beta')}]_{kl} \rangle &= \frac{\delta_{il}\delta_{jk}}{N} \delta_{\beta\beta'} \ , \end{aligned} \tag{5.9}$$

which have to alternate around each vertex. The corresponding Feynman rules are summarized in Fig.14. This is an exact realization of the desired connected black and white graphs, except that any number of loops[4] of each color is allowed. In fact, each graph receives a weight

$$N^{2-2h}\ c^V\ m^b\ n^w\ , \tag{5.10}$$

where b (resp. w) denote the total numbers of black (resp. white) loops.

A simple trick to reduce the number of say white loops w to one is to send the number n of white matrices W to 0, and to retain only the contributions of order 1 in n. Hence

$$f(m, c, N) = \lim_{n \to 0} \frac{1}{n} \text{Log } Z(m, n, c, N) = \sum_{\substack{b.\&w.\ \text{conn. graphs } \Gamma \\ \text{with one } w \text{ loop}}} N^{2-2h}\ c^V\ m^b\ \frac{1}{|\text{Aut}(\Gamma)|} \ . \tag{5.11}$$

If we restrict this sum to the leading order N^2, namely the genus 0 contribution ($h = 0$), we finally get a relation to the meander numbers in the form

$$f_0(m, c) = \lim_{N \to \infty} \frac{1}{N^2} f(m, c, N)$$

[4]The reader must distinguish between these loops, made of double-lines of a definite color, from the oriented loops along which the matrix indices run.

$$= \sum_{p=1}^{\infty} \frac{c^{2p}}{4p} \sum_{k=1}^{p} M_p^{(k)} m^k \tag{5.12}$$

where the abovementioned relation between the numbers of black and white graphs and multi-component meanders has been used to rewrite the expansion (5.11).

5.3 Meander numbers as Gaussian averages of words

The particular form of the matrix potential (5.8) allows one to perform the exact integration over say all the W matrices (the dependence of P on W is Gaussian), with the result

$$Z(m,n,c,N) = \frac{1}{\theta_N} \int \prod_{\alpha=1}^{m} dB^{(\alpha)} \det [\mathbf{I} \otimes \mathbf{I} - c \sum_{\alpha} B^{(\alpha)\, t} \otimes B^{(\alpha)}]^{-n/2} e^{-N \mathrm{Tr} \sum_\alpha \frac{(B^{(\alpha)})^2}{2}} , \tag{5.13}$$

where $\mathbf{I}$ stands for the $N \times N$ identity matrix, $\otimes$ denotes the usual tensor product of matrices, and the superscript t stands for the usual matrix transposition. The prefactor θ_N is fixed by the condition $Z(m,n,c=0,N) = 1$. With this form, it is easy to take the logarithm and to let n tend to 0, with the result

$$\begin{aligned} f(m,c,N) &= -\frac{1}{2\theta_N} \int \prod_{\alpha=1}^{m} dB^{(\alpha)} \mathrm{Tr}(\mathrm{Log}[\mathbf{I} \otimes \mathbf{I} - c \sum_{\alpha} B^{(\alpha)\, t} \otimes B^{(\alpha)}]) e^{-N \mathrm{Tr} \sum_\alpha \frac{(B^{(\alpha)})^2}{2}} \\ &= \sum_{p=1}^{\infty} \frac{c^p}{2p} \langle \mathrm{Tr}(\sum_{\alpha=1}^{m} B^{(\alpha)\, t} \otimes B^{(\alpha)})^p \rangle_{\mathrm{Gauss}} \\ &= \sum_{p=1}^{\infty} \frac{c^p}{2p} \sum_{1 \le \alpha_1, \ldots, \alpha_p \le m} \langle |\mathrm{Tr}(B^{(\alpha_1)} \ldots B^{(\alpha_p)})|^2 \rangle_{\mathrm{Gauss}} , \end{aligned} \tag{5.14}$$

where we still use the notation $\langle \ldots \rangle_{\mathrm{Gauss}}$ for the multi-Gaussian average over the matrices $B^{(\alpha)}$, $\alpha = 1, 2, \ldots, m$. The modulus square simply comes from the hermiticity of the matrices $B^{(\alpha)}$, namely

$$\mathrm{Tr}(\prod B^{(\alpha_i)\, t}) = \mathrm{Tr}(\prod B^{(\alpha_i)\, *}) = \mathrm{Tr}(\prod B^{(\alpha_i)})^* . \tag{5.15}$$

Taking the large N limit of (5.12), it is a known fact [9] that correlations should factorize, namely

$$\langle |\mathrm{Tr}(\prod_{i=1}^{p} B^{(\alpha_i)})|^2 \rangle_{\mathrm{Gauss}} \underset{N \to \infty}{\longrightarrow} |\langle \mathrm{Tr}(\prod_{i=1}^{p} B^{(\alpha_i)}) \rangle_{\mathrm{Gauss}}|^2 . \tag{5.16}$$

By parity, we see that only even p's give non-vanishing contributions, and comparing with (5.12) we find a closed expression for the meander numbers of order n with k connected components

$$\sum_{k=1}^{n} M_n^{(k)} m^k = \sum_{1 \le \alpha_1, \ldots, \alpha_{2n} \le m} | \lim_{N \to \infty} \langle \frac{1}{N} \mathrm{Tr}(\prod_{i=1}^{2n} B^{(\alpha_i)}) \rangle_{\mathrm{Gauss}} |^2 . \tag{5.17}$$

This expression is only valid for integer values of m, but as it is a polynomial of degree n in m (with vanishing constant coefficient), the n first values $m = 1, 2, \ldots, n$

of m determine it completely. So we only have to evaluate the rhs of (5.17) for these values of m to determine all the coefficients $M_n^{(k)}$.

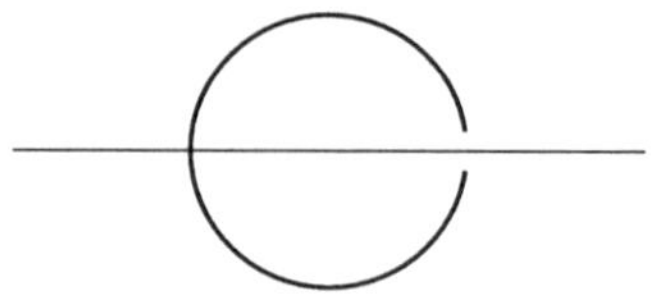

FIGURE 15. The connected toric meander of order 1: it has only 1 bridge.

The relation (5.17) suggests to introduce higher genus meander numbers, denoted by $M_p^{(k)}(h)$, with $M_{2n}^{(k)}(0) = M_n^{(k)}$ (note that the indexation is now by the number of intersections, or bridges), through the generating function

$$\sum_{h=0}^{\infty} \sum_{k=1}^{\infty} M_p^{(k)}(h) m^k N^{2-2h} = \sum_{1 \le \alpha_1, \ldots, \alpha_p \le m} \langle |\mathrm{Tr}(\prod_{i=1}^{p} B^{(\alpha_i)})|^2 \rangle_{\mathrm{Gauss}} , \tag{5.18}$$

which incorporates the contribution of all genera in the Gaussian averages. Note that the genus h is that of the corresponding black and white graph and not that of the river or the road alone. In particular, the river (resp. the road) may be contractible or not in meanders of genus $h > 0$. As an example the $M_1^{(1)} = 1$ toric meander is represented in Fig.15.

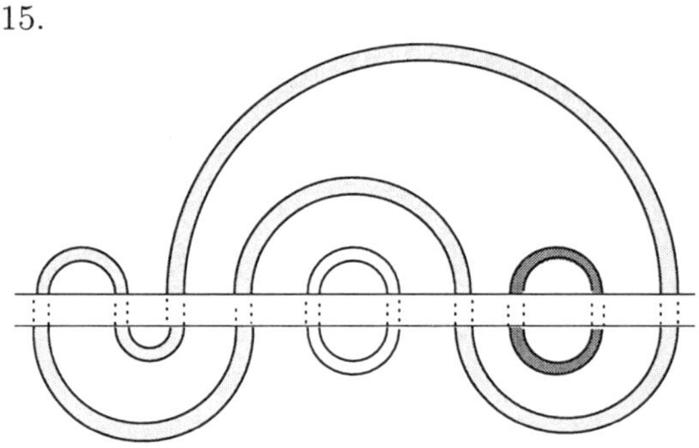

FIGURE 16. A typical graph in the computation of the rhs of (5.18). The two p-valent vertices corresponding to the two traces of words are represented as racks of p double legs ($p = 10$ here). The connected components of the resulting meander (of genus $h = 0$ on the example displayed here) correspond to loops of matrices $B^{(\alpha)}$. This is indicated by a different coloring of the various connected components. Summing over all values of α_i yields a factor m per connected component, hence m^3 here.

The relation (5.18) can also be proved directly as follows. Its rhs is a sum over correlation functions of the traces of certain words (products of matrices) with themselves. More precisely, using the hermiticity of the matrices $B^{(\alpha)}$, the complex conjugate of the trace $\mathrm{Tr}(\prod_{1 \le i \le 2n} B^{(\alpha_i)})$ can be rewritten as

$$\mathrm{Tr}(\prod_{1 \le i \le 2n} B^{(\alpha_i)})^* = \mathrm{Tr}(\prod_{1 \le i \le 2n} B^{(\alpha_{2n+1-i})}) , \tag{5.19}$$

i.e. in the form of an analogous trace, with the order of the B's reversed. According to the Feynman rules of the previous section in the case of only black matrices, such a correlation can be computed graphically as follows. The two traces correspond to two p-valent vertices, and the Gaussian average is computed by summing over all the graphs obtained by connecting pairs of legs (themselves made of pairs of oriented double-lines) by means of edges. Re-drawing these vertices as small racks of p legs as in Fig.16, we get a sum over all multi-component, multi-genera meanders (compare Fig.16 with Fig.5). More precisely, the edges can only connect two legs with the *same* matrix label α, which can be interpreted as a color: indeed, we have to sum over all colorings of the graph by means of m colors. But this coloring is constrained by the fact that the colors of the legs of the two racks have to be identified two by two (the color of both first legs is α_1,...,of both p-th legs is α_p). This means that each connected component of the resulting meander is painted with a color $\alpha \in \{1, 2, \ldots, m\}$. A graph of genus h comes with the usual weight N^{2-2h}. Summing over all the indices $\alpha_1, \ldots, \alpha_p = 1, 2, \ldots, m$, we get an extra factor of m for each connected component of the corresponding meander, which proves the relation (5.18).

In the genus 0 case, we must only consider planar graphs, which correspond to genus 0 meanders by the above interpretation. Due to the planarity of the graph, the two racks of $p = 2n$ legs each are connected to themselves through n edges each, and are no longer connected to each other: they form two disjoint arch configurations of order n. This explains the factorization mentioned in eq. (5.16), and shows that the genus 0 meanders are obtained by the superimposition of two arch configurations. The beauty of eq. (5.17) is precisely to keep track of the number of connected components k in this picture, by the m-coloring of the connected components.

This last interpretation leads to a straightforward generalization of (5.18) to semi-meanders and many meander-related numbers.

5.4 *Matrix expressions for semi-meanders and more*

In view of the above interpretation, we immediately get the generalization of eq. (5.17) to semi-meanders as

$$\begin{aligned} \sum_{k=1}^{n} \bar{M}_n^{(k)} m^k \quad &= \sum_{1\leq \alpha_1,\ldots,\alpha_n \leq m} \\ \lim_{N\to\infty} \frac{1}{N} \quad &\langle \mathrm{Tr}(B^{(\alpha_1)}B^{(\alpha_2)} \ldots B^{(\alpha_n)}B^{(\alpha_n)}B^{(\alpha_{n-1})} \ldots B^{(\alpha_1)})\rangle_{\mathrm{Gauss}} \end{aligned} \tag{5.20}$$

To get this expression, we have used the m-coloring of the matrices to produce the correct rainbow-type connections between the loops of matrices.

More generally, this provides a number of matrix integral identities for the following generalized semi-meanders. Let us label the arch configurations of order n by a permutation $\mu \in S_{2n}$, the symmetric group over $2n$ objects, in such a way that if we label the bridges of the arch configuration $1, 2, \ldots, 2n$, the permutation μ indicates the pairs of bridges linked by arches, namely, for any $i = 1, 2, \ldots, 2n$, $\mu(i)$ is the bridge linked to i by an arch. By definition, μ is made of n cycles of length 2, it is therefore an element of the class $[2^n]$ of S_{2n}. Note that an element of this class generally does not lead to an arch configuration, because the most general pairing of bridges has intersecting arches. A permutation $\mu \in [2^n]$ will be called ADMISSIBLE if it leads to an arch configuration. Let $\mathcal{A}_\mu$ be the arch configuration associated to some admissible $\mu \in [2^n]$. We can define some generalized semi-meander number $\bar{M}_n^{(k)}(\mathcal{A}_\mu)$ associated to $\mathcal{A}_\mu$ as the number of meanders of order n with k connected components whose lower arch configuration is $\mathcal{A}_\mu$. With this definition,

$$\bar{M}_n^{(k)}(\mathcal{R}_n) = \bar{M}_n^{(k)} , \tag{5.21}$$

where $\mathcal{R}_n$ is the rainbow configuration of order n, associated to the permutation $\mu(i) = 2n + 1 - i$, for $i = 1, 2, \ldots, 2n$. In other words, $\bar{M}_n^{(k)}(\mathcal{A}_\mu)$ is the number of closures by some arch configurations of order n of the lower arch configuration $\mathcal{A}_\mu$ which have k connected components. Eq. (5.20) extends immediately to

$$\sum_{k=1}^{n} \bar{M}_n^{(k)}(\mathcal{A}_\mu)\, m^k = \sum_{\substack{1\le\alpha_1,\ldots,\alpha_{2n}\le m \\ \alpha_i=\alpha_{\mu(i)},\ i=1,\ldots,2n}} \lim_{N\to\infty} \frac{1}{N} \langle \mathrm{Tr}(B^{(\alpha_1)}B^{(\alpha_2)}\ldots B^{(\alpha_{2n})})\rangle_{\mathrm{Gauss}} , \tag{5.22}$$

where the structure of the lower arch configuration $\mathcal{A}_\mu$ is encoded in the conditions $\alpha_i = \alpha_{\mu(i)}$, $i = 1, \ldots, 2n$, which identifies the colors of the arches according to $\mathcal{A}_\mu$.

Higher genus generalizations are straightforward, by simply removing the large N limit in the above expressions, namely

$$\sum_{k\ge1,h\ge0} \bar{M}_{2n}^{(k)}(\mathcal{A}_\mu, h)\, m^k\, N^{1-2h} = \sum_{\substack{1\le\alpha_1,\ldots,\alpha_{2n}\le m \\ \alpha_i=\alpha_{\mu(i)},\ i=1,\ldots,2n}} \langle \mathrm{Tr}(B^{(\alpha_1)}B^{(\alpha_2)}\ldots B^{(\alpha_{2n})})\rangle_{\mathrm{Gauss}} . \tag{5.23}$$

Note that the number $2n$ of bridges is even here, as we are considering arbitrary genus closures of a given arch configuration of order n. In genus $h = 0$, we recover the numbers $\bar{M}_{2n}^{(k)}(\mathcal{A}_\mu, 0) = \bar{M}_n^{(k)}(\mathcal{A}_\mu)$ defined by (5.22). In the particular case $\mathcal{A}_\mu = \mathcal{R}_n$, this defines higher genus semi-meander numbers $\bar{M}_{2n}^{(k)}(h) = \bar{M}_{2n}^{(k)}(\mathcal{R}_n, h)$, with the correspondence $\bar{M}_{2n}^{(k)}(0) = \bar{M}_n^{(k)}$. The resulting higher genus semi-meanders are obtained generically by allowing the 4 ends of two given arches to alternate along the river. As the arches cannot intersect each other, this requires increasing the genus of the graph. On the other hand, the lower rainbow configuration is contractible, hence the genus is also that of the contracted graph obtained by the folding process of Fig.8.

It is instructive to calculate the sum over all genera of these numbers, while keeping track of the numbers of connected components. We simply take $N = 1$, in which case the Gaussian average becomes an ordinary Gaussian average over real scalars $(b^{(1)}, \ldots, b^{(m)}) \in \mathbb{R}^m$

$$\begin{aligned} \sum_{k\ge1,h\ge0} \bar{M}_{2n}^{(k)}(\mathcal{A}_\mu, h)\, m^k &= \langle (\sum_{i=1}^{m} (b^{(i)})^2)^n \rangle_{\mathrm{Gauss}} \\ &= \lambda_m \int_0^\infty r^{m-1}\, r^{2n}\, e^{-\frac{r^2}{2}} dr \\ &= m(m+2)(m+4)\ldots(m+2n-2) , \end{aligned} \tag{5.24}$$

where the normalization constant has been fixed by the $n = 1$ case (the result is m). For $m = 1$, the above simply counts the total number of pairings between $2n$ legs, namely $(2n-1)!! = 1.3.5\ldots(2n-1)$. Note that the result (5.24) is independent of the lower arch configuration, it holds in particular for semi-meanders. The result (5.24) is a polynomial of degree n, with leading coefficient 1 corresponding to the only (genus 0) meander with n connected components, obtained by reflecting the lower arch configuration $\mathcal{A}_\mu$ wrt the river. For meanders, we simply get

$$\begin{aligned}\sum_{k\geq 1, h\geq 0} M_p^{(k)}(h)\, m^k &= \langle (\sum_{i=1}^{m} (b^{(i)})^2)^p \rangle_{\text{Gauss}} \\ &= m(m+2)(m+4)\ldots(m+2p-2)\ . \end{aligned} \tag{5.25}$$

Note that in this case the polynomial is of degree p, with leading coefficient 1, corresponding to the (genus 1) meander made of a collection of p loops intersecting the river only once.

5.5 *Computing averages of traces of words in matrix models*

As a warming up, let us first compute the rhs of eq. (5.20) in the case of one matrix $m = 1$, namely

$$\gamma_n = \lim_{N\to\infty} \frac{1}{N} \langle \text{Tr}(B^n) \rangle_{\text{Gauss}}\ . \tag{5.26}$$

By parity, we see that $\gamma_{2s+1} = 0$ for all integer s. A simple method usually applied for computing Gaussian averages uses the so-called *loop equations* of the matrix model. In the case of one matrix, these are obtained as follows. We write that the matrix integral of a total derivative vanishes, namely

$$\begin{aligned} 0 &= \int dB\, \frac{\partial}{\partial B_{ji}} \left[(B^{s+1})_{kl} e^{-N\text{Tr}\frac{B^2}{2}} \right] \\ \Rightarrow\quad 0 &= \langle -N(B^{s+1})_{kl} B_{ij} + \sum_{r=0}^{s} (B^r)_{kj} (B^{s-r})_{il} \rangle_{\text{Gauss}}\ . \end{aligned} \tag{5.27}$$

Taking $i = l$ and $j = k$, and summing over $i, j = 1, \ldots, N$, we finally get

$$\langle\ \text{Tr}(B^{s+2})\ \rangle_{\text{Gauss}} = \frac{1}{N} \sum_{r=0}^{s} \langle\ \text{Tr}(B^r)\ \text{Tr}(B^{s-r})\ \rangle_{\text{Gauss}}\ . \tag{5.28}$$

In the large N limit, due to the abovementioned factorization property, only the even powers of B contribute by parity, and setting $s + 2 = 2n$, this becomes

$$\gamma_{2n} = \sum_{r=0}^{n-1} \gamma_{2r} \gamma_{2n-2r-2}\ , \tag{5.29}$$

valid for $n \geq 1$, and $\gamma_0 = 1$. This is exactly the defining recursion (3.1) for the Catalan numbers, hence $\gamma_{2n} = c_n$, whereas $\gamma_{2n+1} = 0$. In this case, the equation (5.20) reduces therefore to the sum rule (3.8) for semi-meanders. Similarly, when $m = 1$, the equation (5.17) reduces to the sum rule (3.7) for meanders. An important remark is in order. It sould not be surprising that the general recursion principle for arches of Fig.6 resembles the large N limit of loop equations in matrix models. The role of the leftmost arch is played in the latter case by the differentiation wrt the matrix element B_{ji}: it can act at all the matrix positions in the word, which cut it into two even words, and can be graphically interpreted as just the right bridge position of the leftmost exterior arch in the pairings of matrices necessary to compute the Gaussian average of the trace of the word.

More generally, the loop equations for the Gaussian m-matrix model enable us to derive a general recursion relation for traces of words. The most general average of trace of word in m matrices in the large N limit is denoted by

$$\gamma^{(m)}_{p_1,p_2,\ldots,p_{mk}} = \lim_{N\to\infty} \frac{1}{N}\langle \mathrm{Tr}((B^{(1)})^{p_1}(B^{(2)})^{p_2}\ldots(B^{(m)})^{p_m}(B^{(1)})^{p_{m+1}}\ldots(B^{(m)})^{p_{mk}})\rangle_{\mathrm{Gauss}} \quad (5.30)$$

In the above, some powers p_j may be zero, but no m consecutive of them vanish (otherwise the word could be reduced by erasing the m corresponding pieces). Of course $2p = \sum_i p_i$ has to be an even number for (5.30) to be non-zero, by the usual parity argument. For $m = 1$, we recover $\gamma^{(1)}_p = \gamma_p$. If $\omega = \exp(2i\pi/m)$ denotes the primitive m-th root of unity, then we have the following recursion relation between large N averages of traces of words, for $m \geq 2$

$$\gamma^{(m)}_{p_1,p_2,\ldots,p_{mk}} = -\sum_{j=1}^{mk-1} \omega^j \gamma^{(m)}_{p_1,\ldots,p_j}\, \gamma^{(m)}_{p_{j+1},\ldots,p_{mk}}\,. \quad (5.31)$$

When j is not a multiple of m, it is understood in the above that the multiplets $(p_1,\ldots,p_j)$ and $(p_{j+1},\ldots,p_{km})$ have to be completed by zeros so a to form sequences of m-uplets. For instance, we write $\gamma^{(3)}_3 = \gamma^{(3)}_{3,0,0} = \gamma^{(3)}_{0,0,3}$. Note also that if only $q < m$ matrices are actually used to write a word, the corresponding $\gamma^{(m)}$ can be reduced to a $\gamma^{(q)}$ by erasing the spurious zeros (for instance, $\gamma^{(3)}_{3,0,0} = \gamma^{(1)}_3$). Together with the initial condition $\gamma^{(m)}_{0,\ldots,0,2n+1,0,\ldots,0} = \gamma_{2n+1} = 0$ and $\gamma^{(m)}_{0,\ldots,0,2n,0,\ldots,0} = \gamma_{2n} = c_n$, this gives a compact recursive algorithm to compute all the large N averages of traces of words in any multi-Gaussian matrix model.

A direct proof of eq. (5.31) goes as follows. Throughout this argument, we refer to the labels $\alpha = 1, 2, \ldots, m$ of the matrices as colors. The quantity $\gamma^{(m)}_{p_1,\ldots,p_{mk}}$ can be expressed as the sum over all possible planar pairings of matrices of the same color in the corresponding word, *i.e.* the sum over all arch configurations of order p (encoded in *admissible* permutations μ of $1, 2, \ldots, 2p$), preserving the color of the matrices (the matrices sitting at positions i and $\mu(i)$ have the same color α). Such a color-preserving arch configuration appears exactly once in the lhs of (5.31). We will show the relation (5.31) by proving that each such term also comes with a coefficient 1 in the rhs of (5.31). Let us evaluate the rhs of (5.31) in this language.

The index j may be viewed as the position of a "separator", which cuts the color preserving arch configurations into two disconnected pieces. The separator positions are labeled by the index $j = 1, 2, \ldots, mk - 1$. For a given color-preserving arch configuration in the lhs of (5.31), the only terms of the rhs of (5.31) contributing to it are those where the separator index j takes its values at positions inbetween the exterior arches linking various blocks of the same color. These positions will be referred to as available positions.

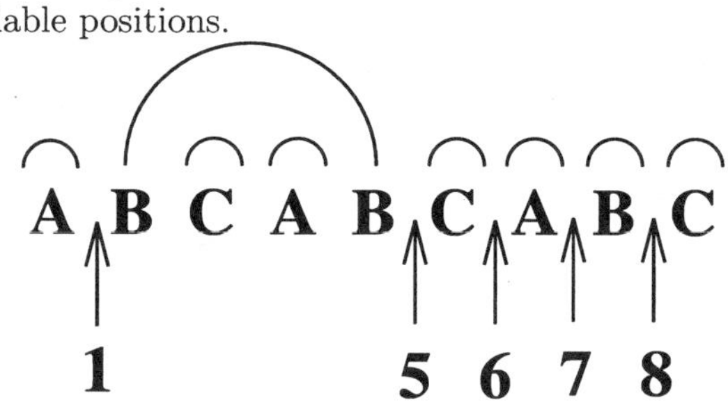

FIGURE 17. A sample color-preserving pairing of matrices for $m = 3$ matrices. Each block of matrices is denoted by a single letter A, B, C, according to its color 1, 2, 3. The precise pairing of matrices within blocks is not indicated for simplicity. The available separator positions are indicated by arrows. One checks that (i) the positions are consecutive modulo 3, and (ii) the number of available positions is $5 = 2 \times 3 - 1$.

Available positions satisfy the two following properties, illustrated in Fig.17.

(i) any two successive available positions are labelled by *successive* integers modulo m: if an available separator position sits between two blocks of colors α and $\alpha+1$, hence at a position of the form $j = sm + \alpha$, then the block of color $\alpha + 1$ is linked to other blocks of the same color, and the next available separator position sits between a block of color $\alpha + 1$ and a block of color $\alpha + 2$, hence at a position of the form $j = tm + \alpha + 1$. These two positions are consecutive modulo m.

(ii) the first available separator position is $j = ml + 1$: it sits to the right of the first set of related blocks of color 1. The last available separator position is of the form $j = qm + (m - 1)$, as it sits to the left of the rightmost set of related blocks of color m. Hence, thanks to property (i), the total number of available separator positions is of the form $rm - 1$.

Consequently, the total contribution of a given color-preserving admissible pairing in the rhs of (5.31) is of the form

$$(r-1)\sum_{j=1}^{m}(-\omega^s) + \sum_{j=1}^{m-1}(-\omega^s) \; = \; 1 \; . \tag{5.32}$$

So we have proved that the total contribution of each admissible pairing in the rhs of (5.31) is 1. This completes the proof of (5.31).

Let us show explicitly how to use the recursion (5.31) to compute the thermodynamic average of the trace of a particular word in $m = 3$ matrices. We wish to compute

$$\gamma^{(3)}_{2,1,2,0,1,0} \; = \; \lim_{N\to\infty} \frac{1}{N} \langle \mathrm{Tr}(A^2BC^2B) \rangle_{\mathrm{Gauss}} \; , \tag{5.33}$$

where we denote by A, B, C the matrices with respective colors 1, 2, 3. Applying (5.31) we get

$$\gamma^{(3)}_{2,1,2,0,1,0} \quad = \; -j\gamma^{(1)}_2\gamma^{(2)}_{2,2} - j^2\gamma^{(2)}_{2,1}\gamma^{(2)}_{2,1}$$

$$- \gamma^{(3)}_{2,1,2}\gamma^{(1)}_1 - j\gamma^{(3)}_{2,1,2}\gamma^{(1)}_1 - j^2\gamma^{(3)}_{2,1,2,0,1,0} \,, \tag{5.34}$$

where $j = \exp(2i\pi/3)$, and the various numbers m of matrices have been reduced to their minimal value. Regrouping the $\gamma^{(3)}_{2,1,2,0,1,0}$'s, and using the fact that $\gamma^{(1)}_{2s+1} = \gamma_{2s+1} = 0$ for integer s, we get

$$(1+j^2)\gamma^{(3)}_{2,1,2,0,1,0} \;=\; -j\gamma^{(1)}_2\gamma^{(2)}_{2,2} \,. \tag{5.35}$$

Using the recursion (5.31) for $m = 2$, we get

$$\gamma^{(2)}_{2,2} \;=\; \gamma^{(1)}_2\gamma^{(1)}_2 \;=\; (c_1)^2 \,. \tag{5.36}$$

Finally, we get

$$\gamma^{(3)}_{2,1,2,0,1,0} \;=\; (c_1)^3 \;=\; 1 \,. \tag{5.37}$$

This corresponds to the only way of pairing matrices of the same color in the sequence

5.6 Combinatorial expression using the symmetric group

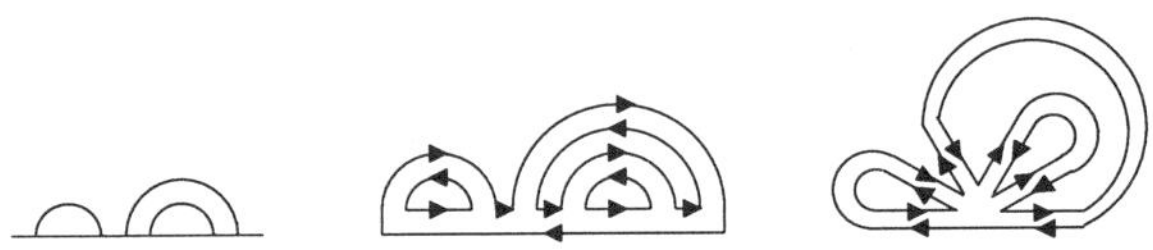

FIGURE 18. An arch configuration of order 3 and the corresponding interpretation as a ribbon graph, with $V = 1$ six-valent vertex and $E = 3$ edges. On the intermediate diagram, the arches have been doubled and oriented. These oriented arches indicate the pairing of bridges, *i.e.* represent the action of μ. Similarly, the oriented horizontal segments indicate the action of the shift permutation σ. Each oriented loop corresponds to a cycle of the permutation $\sigma\mu$.

ADMISSIBILITY CONDITION. In sect.5.4 above, we have seen how an arch configuration of order n could be encoded in an admissible permutation μ belonging to the class $[2^n]$ of S_{2n}. Let us write the admissibility condition explicitly. This condition states that arches do not intersect each other, namely that the ribbon graph (see Fig.18) with only one $2n$-valent vertex (the $2n$ bridges), whose legs are connected according to the arch configuration, is *planar*, *i.e.* of genus $h = 0$. This graph has $V = 1$ vertex, and $E = n$ edges (arches). Let us compute its number L of oriented loops in terms of the permutation μ. Let σ denote the "shift" permutation, namely $\sigma(i) = i+1$, $i = 1, 2, \ldots, 2n-1$ and $\sigma(2n) = 1$. Then an oriented loop in the ribbon graph is readily seen to correspond to a *cycle* of the permutation $\sigma\mu$. Indeed, the total number of loops is $L = \text{cycles}(\sigma\mu)$, the number of cycles of the permutation $\sigma\mu$. The admissibility condition reads

$$\begin{aligned} \chi \;=\; 2 = L - E + V \quad &= 1 - n + \text{cycles}(\sigma\mu) \\ \Leftrightarrow \quad \text{cycles}(\sigma\mu) \quad &= n+1 \,. \end{aligned} \tag{5.38}$$

Note that if we demand that the ribbon graph be of genus h, the above condition becomes

$$\text{cycles}(\sigma\mu) \; = \; n+1-2h \; . \tag{5.39}$$

CONNECTED COMPONENTS. Given an admissible permutation $\mu \in [2^n]$, let us now count the number of connected components of the corresponding semi-meander of order n. Let τ be the "rainbow" permutation $\tau(i) = 2n+1-i$. Note that τ changes the parity of the bridge label. On the other hand, the admissible permutation μ is readily seen to also change the parity of the bridge labels. As a consequence, the permutation $\tau\mu$ preserves the parity of bridge labels. In other words, even bridges are never mixed with odd ones. The successive iterations of the permutation $\tau\mu$ describe its cycles. The corresponding meander will be connected iff these cycles are maximal, namely μ has two cycles of length n (one for even bridges, one for odd bridges), i.e. $\tau\mu \in [n^2]$. We get a purely combinatorial expression for connected semi-meander numbers

$$\bar{M}_n \; = \; \text{card}\{\mu \in [2^n] \mid \text{cycles}(\sigma\mu) = n+1, \text{ and } \tau\mu \in [n^2]\} \; . \tag{5.40}$$

More generally, the semi-meander corresponding to μ will have k connected components iff $\tau\mu$ has exactly k pairs of cycles of equal length (one over even bridges, one over odd ones).

CHARACTER EXPRESSIONS. The above conditions on various permutations are best expressed in terms of the characters of the symmetric group. Denoting by $[i^{\nu_i}]$ the class of permutations with ν_i cycles of length i, and labelling the representations of S_{2n} by Young tableaux Y with $2n$ boxes as customary, the characters can be expressed as

$$\chi_Y([i^{\nu_i}]) \; = \; \det\left(p_{i+\ell_i-j}(\theta_.)\right)\Big|_{t_\nu} \; , \tag{5.41}$$

where the Young tableau has ℓ_i boxes in its i-th line, counted from the top, $t_\nu = \prod_i \frac{\theta_i^{\nu_i}}{\nu_i!}$, $p_m(\theta_.)$ is the m-th Schur polynomial of the variables $\theta_1, \theta_2, \ldots$

$$p_m(\theta_.) \; = \; \sum_{\substack{k_i \geq 0, i=1,2,\ldots \\ \Sigma i k_i = m}} \prod_i \frac{\theta_i^{k_i}}{k_i!} \; , \tag{5.42}$$

and we used the symbol $f(\theta_.)|_{t_\nu}$ for the coefficient of the monomial $\prod_i \frac{\theta_i^{\nu_i}}{\nu_i!}$ in the polynomial $f(\theta_.)$. As group characters, the χ_Y's satisfy the orthogonality relation

$$\sum_Y \chi_Y([\lambda])\chi_Y([\mu]) \; = \; \frac{(2n)!}{||[\lambda]||} \, \delta_{[\lambda],[\mu]} \; , \tag{5.43}$$

where the sum extends over all Young tableaux with $2n$ boxes, $[\lambda]$ denotes the class of a permutation $\lambda \in S_{2n}$, and $||[\lambda]||$ the order of the class. The order of the class $[i^{\nu_i}]$ is simply

$$||[i^{\nu_i}]|| \; = \; \frac{(2n)!}{\prod_i i^{\nu_i} \nu_i!} \; . \tag{5.44}$$

The orthogonality relation (5.43) provides us with a means of expressing any condition on classes of permutations in terms of characters. It leads to the following compact expression for the connected semi-meander numbers

$$\begin{aligned}\bar{M}_n &= \sum_{\substack{[i^{\lambda_i}]\in S_{2n}\\ \Sigma\lambda_i=n+1}} \sum_{\mu\in[2^n]} \delta_{[\sigma\mu],[i^{\lambda_i}]}\, \delta_{[\tau\mu],[n^2]} \\ &= \sum_{\substack{[i^{\lambda_i}]\in S_{2n}\\ \Sigma\lambda_i=n+1}} \sum_{\mu\in S_{2n}} \sum_{Y,Y',Y''} \frac{|[2^n]|\,|[i^{\lambda_i}]|\,|[n^2]|}{((2n)!)^3} \\ &\times \chi_Y([\mu])\chi_Y([2^n])\chi_{Y'}([\sigma\mu])\chi_{Y'}([i^{\lambda_i}])\chi_{Y''}([\tau\mu])\chi_{Y''}([n^2])\ . \qquad (5.45)\end{aligned}$$

Analogous expressions hold for (higher genus) semi-meanders with k connected components and for meanders as well. These make completely explicit the calculation of the various meander-related numbers. Unfortunately, the characters of the symmetric group are not so easy to deal with, and we were not able to use them in an efficient way to enhance our numerical data.

6 IRREDUCIBLE MEANDERS, EXACT RESULTS

A multi-component meander is said to be k-REDUCIBLE if one (proper) subset of its connected components can be detached from it by cutting the river k times between the bridges. A multi-component meander is said to be k-IRREDUCIBLE if it is *not* k-reducible, *i.e.* if no (proper) subset of its connected components can be detached from it by cutting the river k times between the bridges.

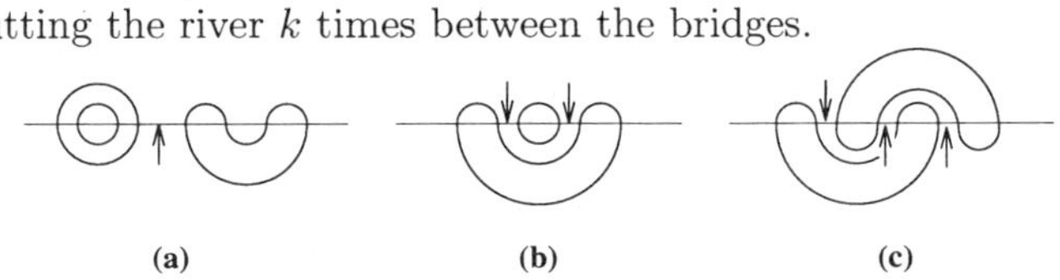

FIGURE 19. Reducibility and irreducibility for meanders. The cuts reducing the meanders are indicated by arrows. The meander (a) is 1-reducible, *i.e.* the succession of two meanders along the river. The meander (b) is 1-irreducible and 2-reducible. The meander (c) is 1-and 2-irreducible and 3-reducible.

In Fig.19, we give a few examples to illustrate the notion of reducibility and irreducibility of meanders. The same definition applies to the semi-meanders in the formulation with a semi-infinite river.

6.1 1-irreducible meanders and semi-meanders

A meander is 1-irreducible if it is not the succession along the river of at least two meanders. We can enumerate all the meanders by their growing number of 1-irreducible components. Denoting by $P_n^{(k)}$ the total number of 1-irreducible meanders of order

n with k connected components, we compute

$$M_n^{(k)} = \sum_{\substack{n_1+n_2+\ldots=n \\ k_1+k_2+\ldots=k}} P_{n_1}^{(k_1)} P_{n_2}^{(k_2)} \ldots , \tag{6.1}$$

hence the generating functions

$$\begin{aligned} M(z,x) &= \sum_{n\geq k\geq 1} M_n^{(k)} \, z^{2n} \, x^k \\ P(z,x) &= \sum_{n\geq k\geq 1} P_n^{(k)} \, z^{2n} \, x^k \end{aligned} \tag{6.2}$$

satisfy the equation

$$M(z,x) = P(z,x) + P(z,x)^2 + P(z,x)^3 + \ldots = \frac{P(z,x)}{1-P(z,x)} . \tag{6.3}$$

An analogous reasoning for semi-meanders leads to the relation

$$\frac{\bar{P}(z,x)}{1-P(z,x)} = \bar{M}(z,x) , \tag{6.4}$$

between the generating functions

$$\begin{aligned} \bar{M}(z,x) &= \sum_{n\geq k\geq 1} \bar{M}_n^{(k)} \, z^{n} \, x^k \\ \bar{P}(z,x) &= \sum_{n\geq k\geq 1} \bar{P}_n^{(k)} \, z^{n} \, x^k \end{aligned} \tag{6.5}$$

of respectively semi-meander and 1-irreducible semi-meander numbers of order n with k connected components, and $P(z,x)$ is defined in (6.2).

6.2 2-irreducible meanders and semi-meanders

The 2-irreducible meanders have been studied in [1] extensively (under the name of *irreducible meanders*). They use the following equivalent characterization: for any subset of connected components of a 2-irreducible meander, its set of bridges is not consecutive. Indeed, if it were not the case, one could cut the river before the first bridge and after the last one and detach the corresponding piece. One easily checks on Fig.19 (c) that the two sets of bridges of the two connected components of the meander are intertwined.

For completeness, we reproduce here their computation of the numbers q_n of 2-irreducible meanders of order n, by slightly generalizing their argument to include the numbers $Q_n^{(k)}$ of 2-irreducible meanders of order n with k connected components. The idea is to enumerate the $M_n^{(k)}$ meanders by focussing on their leftmost 2-irreducible component, namely the largest 2-irreducible subset of its connected components, containing the leftmost bridge.

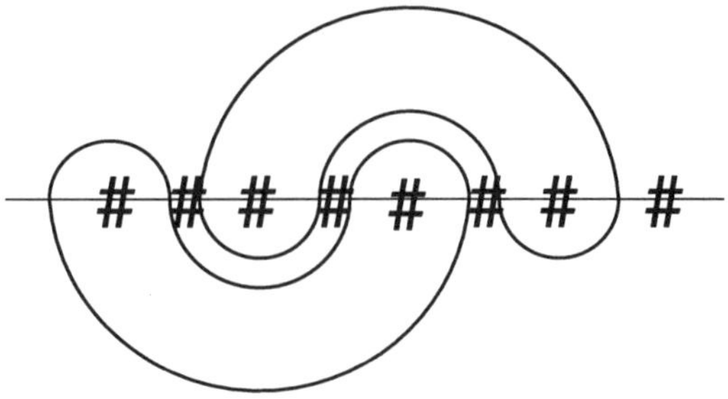

FIGURE 20. A general meander. The leftmost 2-irreducible component is represented in thick solid lines. The positions marked by # can be decorated with any meanders to get the most general meander with this leftmost 2-irreducible component.

In Fig.20, the leftmost 2-irreducible piece of a meander is depicted in thick solid line. Suppose this piece is of order p and has l connected components. The most general meander having this leftmost 2-irreducible piece can be obtained by decorating any segment of river between two consecutive bridges with arbitrary meanders of respective orders $n_1, n_2, \cdots, n_{2p}$, $p+n_1+\ldots+n_{2p}=n$ (there are $2p$ positions, indicated by the symbols # in Fig.20, for these possible decorations), and with respective numbers of connected components k_1, k_2, $\ldots k_{2p}$, with $l+\sum k_i = k$. We get the relation

$$M_n^{(k)} = \sum_{\substack{p+n_1+\cdots+n_{2p}=n \\ l+k_1+\ldots+k_{2p}=k}} Q_p^{(l)}\, M_{n_1}^{(k_1)}\, M_{n_2}^{(k_2)} \cdots M_{n_{2p}}^{(k_{2p})}, \tag{6.6}$$

with the convention that $M_0^{(0)} = 1$. In terms of the generating functions $M(z,x)$ of eq. (6.2) and

$$Q(z,x) = \sum_{n\geq k\geq 1} Q_n^{(k)}\, z^{2n}\, x^k , \tag{6.7}$$

the relation (6.6) reads

$$M(z,x) = Q(z(1+M(z,x)),x) . \tag{6.8}$$

In the special case $x=1$, if we denote by

$$\begin{aligned} B(z) &= 1+M(z,1) = \sum_{n=0}^{\infty}(c_n)^2 z^{2n} \\ q(z) &= 1+Q(z,1) = \sum_{n=0}^{\infty} q_n\, z^{2n} , \end{aligned} \tag{6.9}$$

then (6.8) reduces to

$$B(z) = q(zB(z)) . \tag{6.10}$$

The radius of convergence of the series $B(z)$ is $1/4$, due to the asymptotics of c_n, hence that of $q(z)$ is

$$z^* = \frac{1}{4}B(\frac{1}{4}) = \frac{1}{4}\sum_{n\geq 0} c_n^2\, 4^{-2n}$$

$$
\begin{aligned}
&= \frac{1}{4}C(\frac{x}{4})C(\frac{1}{4x})|_{x^0} \\
&= \frac{1}{4}\oint \frac{dx}{2i\pi x}C(\frac{x}{4})C(\frac{1}{4x}) \\
&= \oint \frac{dx}{2i\pi x}(1-\sqrt{1-x})(1-\sqrt{1-1/x}) \\
&= \oint \frac{dx}{2i\pi x}(-1+\sqrt{2-x-1/x}) \\
&= \frac{1}{2\pi}\int_0^{2\pi} d\theta(-1+2\sin\frac{\theta}{2}) \\
&= \frac{4-\pi}{\pi} \ .
\end{aligned}
\tag{6.11}
$$

Hence a leading behaviour for large n [1]

$$
q_n \sim \left(\frac{\pi}{4-\pi}\right)^{2n} \ . \tag{6.12}
$$

This gives an upper bound on the leading behaviour of M_n

$$
R \le \left(\frac{\pi}{4-\pi}\right)^{2} = 13.3923\ldots \tag{6.13}
$$

Let us extend these considerations to the semi-meander case. If $\bar{Q}_n^{(k)}$ denotes the number of 2-irreducible semi-meanders of order n with k connected components, let us enumerate the $\bar{M}_n^{(k)}$ meanders of order n with k connected components, focussing on their leftmost 2-irreducible piece, say of order p with l connected components.

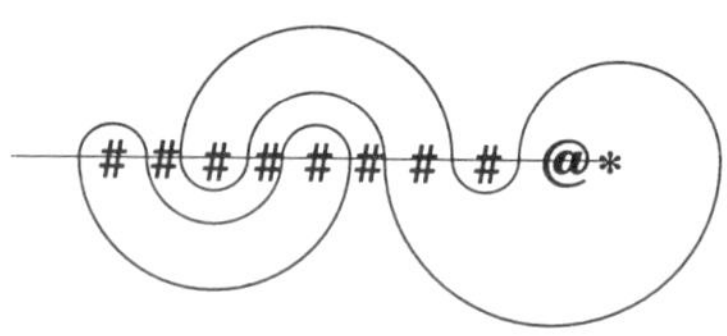

FIGURE 21. A general semi-meander. The leftmost 2-irreducible component is represented in thick solid lines. The positions marked by # can be decorated with any meanders to get the most general semi-meander with this leftmost 2-irreducible component. The last position, indicated by @, can be decorated with any semi-meander, possibly winding around the source of the river (*).

The most general semi-meander with given leftmost 2-irreducible component (as indicated by thick solid lines in Fig.21) is obtained by decorating the segments of river between any two consecutive bridges (there are $(p-1)$ such positions) with meanders of respective orders n_1, n_2,... n_{p-1}, with respectively k_1, ...,k_{p-1} connected components, and the segment of river between the source and the first bridge by any semi-meander of order q, with r connected components, such that $p+q+2\sum n_i = n$ and $l+r+\sum k_i = k$. This amounts to the following relation

$$
\bar{M}_n^{(k)} = \sum_{\substack{p+q+2(n_1+\cdots+n_{p-1})=n \\ l+r+k_1+\ldots+k_{p-1}=k}} \bar{Q}_p^{(l)} M_{n_1}^{(k_1)} \cdots M_{n_{p-1}}^{(k_{p-1})} \bar{M}_q^{(r)}, \tag{6.14}
$$

where the sum extends over non-negative values of n_i and k_i, with the convention that $M_0^{(0)} = 1 = \bar{Q}_0^{(0)} = \bar{M}_0^{(0)}$. This translates into a relation between the generating functions

$$\begin{aligned} \bar{M}(z,x) &= \sum_{n\geq k\geq 1} \bar{M}_n^{(k)}\, z^n\, x^k \\ \bar{Q}(z,x) &= \sum_{n\geq k\geq 1} \bar{Q}_n^{(k)}\, z^n\, x^k \end{aligned} \tag{6.15}$$

and $M(z,x)$ of eq. (6.2) (note that in (6.15) we have given a weight z per bridge in the *semi-infinite river* framework, while there would be twice that number of bridges in the *rainbow closing* framework)

$$\begin{aligned} \bar{M}(z,x) &= \bar{Q}(z(1+M(z,x)),x)\,\frac{1+\bar{M}(z,x)}{1+M(z,x)} \\ \Rightarrow \frac{\bar{M}(z,x)}{1+\bar{M}(z,x)}\,(1+M(z,x)) &= \bar{Q}(z(1+M(z,x)),x)\ . \end{aligned} \tag{6.16}$$

In the special case $x=1$, with the generating function

$$\bar{q}(z) = 1+\bar{Q}(z,1) = \sum_{n\geq 0} \bar{q}_n\, z^n, \tag{6.17}$$

where $\bar{q}_n = \sum_{k=1}^{n} \bar{Q}_n^{(k)}$ is the total number of 2-irreducible multi-component semi-meanders of order n, and with $C(z) = 1+\bar{M}(z,1)$, $B(z) = 1+M(z,1)$ defined respectively in (3.3) (6.9), the relation (6.16) reduces to

$$1+zB(z)C(z) = \bar{q}(zB(z))\ , \tag{6.18}$$

where we used the fact that $(C(z)-1)/C(z) = zC(z)$ (3.4). Reasoning as above, we find the convergence radius of the series $\bar{q}(z)$, $z^* = B(1/4)/4 = (4-\pi)/\pi$. Hence we have the asymptotics

$$\bar{q}_n \sim \left(\frac{\pi}{4-\pi}\right)^n\ . \tag{6.19}$$

This implies the upper bound on the leading behaviour of $\bar{M}_n$

$$\bar{R} \leq \frac{\pi}{4-\pi} = 3.659\ldots \tag{6.20}$$

This upper bound is below the mean field estimate (4.6) which therefore cannot be exact and needs to be improved.

7 Conclusion

We have studied the statistics of arch configurations, meanders and semi-meanders. This study emphasizes the role of exterior arches, instrumental in the recursive generation of both semi-meanders and arch configurations, and whose average number is directly linked to the corresponding entropies. A complete solution of these problems however requires the knowledge of the correlation between these exterior arches and those of a given depth. Discarding these correlations leads to a rough mean field result.

Using the alternative formulation (5.17)-(5.20) in the framework of random matrix models, we are left with the computation of particular Gaussian matrix averages of traces of words. Equation (5.31) provides a recursive way of computing any such average. As a by-product, we have derived an exact formula (5.40) expressing the semi-meander numbers in terms of the characters of the symmetric group.

Finally, we have presented the exact solutions for simpler meandric problems such as the 2-irreducible meander and semi-meander counting, providing interesting upper bounds on the entropy of meanders and semi-meanders.

In conclusion, we are still lacking of an efficient treatment of the main recursion relation for semi-meanders. Beyond the entropy problem, it would also be interesting to determine values of critical exponents in this problem, such as the exponent governing the subleading large order behaviour of the meander and semi-meander numbers (2.3). The latter are nothing but the usual α and γ configuration exponents of polymer chains. From this point of view, a fundamental issue is the determination of the universality class of the (self-avoiding) chain folding problem.

REFERENCES

[1] S. Lando and A. Zvonkin, *Plane and Projective Meanders*, Theor. Comp. Science 117 (1993) 227 and *Meanders*, Selecta Math. Sov. 11 (1992) 117.

[2] P. Di Francesco, O. Golinelli and E. Guitter, *Meander, folding and arch statistics*, to appear in Journal of Mathematical and Computer Modelling, special issue (1995).

[3] K.H. Ko, L. Smolinsky, *A combinatorial matrix in 3-manifold theory*, Pacific. J. Math 149 (1991) 319-336.

[4] K. Hoffman, K. Mehlhorn, P. Rosenstiehl and R. Tarjan, *Sorting Jordan sequences in linear time using level-linked search trees*, Information and Control 68 (1986) 170-184.

[5] A. Phillips, *La topologia dei labirinti*, in M. Emmer, ed. *L' occhio di Horus: itinerario nell'immaginario matematico*, Istituto della Enciclopedia Italia, Roma (1989) 57-67.

[6] J. Touchard, *Contributions à l'étude du problème des timbres poste*, Canad. J. Math. 2 (1950) 385.

[7] W. Lunnon, *A map-folding problem*, Math. of Computation 22 (1968) 193.

[8] N. Sloane, *The on-line encyclopedia of integer sequences*, e-mail: sequences@research.att.com

[9] E. Brézin, C. Itzykson, G. Parisi and J.-B. Zuber, *Planar Diagrams*, Commun. Math. Phys. 59 (1978) 35; see also P. Di Francesco, P. Ginsparg and J. Zinn-Justin, *2D Gravity and Random Matrices*, Phys. Rep. 254 (1995) 1, for a review and more references.

TOPOLOGICAL σ-MODELS AND THE LARGE-N TECHNIQUES

TOHRU EGUCHI

DEPARTMENT OF PHYSICS
FACULTY OF SCIENCE
UNIVERSITY OF TOKYO, TOKYO, JAPAN

1 INTRODUCTION

In our attempts at understanding the geometrical principles behind the string theory, the approach of topological field theories [1],[2] seems to offer an important clue. String theory exposes its geometrical structures in a most transparent manner in its topological formulation. Thus we may gain geometrical insights from the study of the topological version of string theories. When a string is compactified on a Kähler manifold K, the theory is described by an $N=2$ supersymmetric non-linear σ-model. After suitable twistings an $N=2$ σ-model yields a pair of topological field theories, topological σ-models (A-models) and B-models. B-models are essentially classical and easy to solve. Topological σ-model, on the other hand, is given by a sum over the holomorphic maps (instantons) from the Riemann surface to the target manifold and in general difficult to evaluate.

When the target space K is a Calabi-Yau manifold , a special situation arises; σ-model (A-model) may be replaced by the B-model associated with the mirror manifold of K and the genus=0 σ-model correlation functions are obtained from those of the B-model using the mirror map. Partition functions of the topological σ-models are also determined at higher genera $g=1,2$ using the method of holomorphic anomaly [3].

When we come to target manifolds K with a positive 1st Chern class $c_1(K)>0$ (Fano varieties), the situation becomes quite different from Calabi-Yau manifolds which are characterized by the scale invariance condition $c_1(K)=0$. It is well-known that the one-loop beta function of a supersymmetric non-linear σ-model is proportional to the 1st Chern class of K. When $c_1(K)>0$, the theory is asymptotically free and develops a mass gap due to dimensional transmutation. Examples of manifolds with $c_1(K)>0$ are given by complex projective spaces CP^n and Grassmannians.

Low Dimensional Applications of Quantum Field Theory
Edited by L. Baulieu *et al.*, Plenum Press, New York, 1997

Recently an algorithm based on the associativity of the operator product expansion has been used to recursively determine the instanton sum in the genus $g = 0$ free energy of the topological σ-model with the target manifolds $CP^2, CP^3, \cdots$ etc. [4, 5, 6, 7, 8]. This method is efficient and general, however, is limited to the genus $g = 0$ case at the moment.

In previous papers [9],[10] we have considered the CP^1, the simplest manifold with $c_1(K) > 0$ and constructed a matrix model which reproduces holomorphic maps from Riemann surfaces of arbitrary genera onto CP^1. A characteristic feature of the CP^1 model is the presence of the logarithmic terms in the action which implies the scaling violation of the theory. In these lectures I would like to present some basic materials of the topological σ-models and then discuss the topological CP^1-model and its possible generalizations. In section 2 I first briefly recall the basic facts about the topological σ-model and moduli space of holomorphic maps. In section 3 I introduce CP^1 model and describe its underlying integrable structure. In section 4 a large-N matrix model is constructed which reproduces holomorphic maps from Riemann surfaces onto CP^1. In section 5 I discuss a Landau-Ginzburg (LG) description of the CP^1 model and show that CP^1 correlation functions are reproduced with a superpotential $\cos X$. Hence the topological CP^1 model may be identified with the $N = 2$ supersymmetric sine-Gordon theory. The close relation between the CP^1 and $N = 2$ sine-Gordon models has been noted sometime ago based on the comparison of the particle spectra and S-matrices of these theories [11],[12]. I present an analysis of the saddle point equation in Section 6. Finally in Section 7 I discuss possible generalizations of our construction to other manifolds like CP^N.

2 The Topological σ Model

2.1 Lagrangian

It is well-known that when the target manifold of a supersymmetric non-linear σ model is Kähler, it is possible to construct a new supersymmetry generator by making use of the complex structure of the manifold and the theory acquires an extended $N = 2$ supersymmetry. Whenever a theory has an $N = 2$ supersymmetry, it is possible to twist the theory by redefining its spinor quantum numbers and convert it into a topological field theory [1],[2],[13]. After twisting, the spinor fields are turned into ghost fields, i.e. anti-commuting fields with integer spins. The topological σ models are obtained from twisting the $N = 2$ supersymmetric non-linear σ models. The Lagrangian of the topological σ model theory is given by the standard one of the non-linear σ model with spinor fields replaced by ghosts ρ, χ and the Dirac operator replaced by a suitable covariant derivative operator

$$\begin{aligned}
\mathcal{L} = t \int_\Sigma d^2z \Big(& \frac{1}{2} g_{i\bar{j}} (\partial_z \phi^i \partial_{\bar{z}} \phi^{\bar{j}} + \partial_{\bar{z}} \phi^i \partial_z \phi^{\bar{j}}) \\
& + g_{\bar{i}j} i \rho_z^{\bar{i}} D_{\bar{z}} \chi^j + g_{i\bar{j}} i \rho_z^i D_{\bar{z}} \chi^{\bar{j}} - R_{i\bar{i}j\bar{j}} \rho_{\bar{z}}^i \rho_z^{\bar{i}} \chi^j \chi^{\bar{j}} \Big) \\
= it \int_\Sigma & d^2z \{Q, V\} - it \times d,
\end{aligned}$$

$$V = \frac{1}{2} g_{i\bar{j}} \left(\rho_z^{\bar{i}} \partial_{\bar{z}} \phi^j + \rho_{\bar{z}}^j \partial_z \phi^{\bar{i}} \right),$$
$$d = i \int_\Sigma g_{i\bar{j}} (\partial_z \phi^i \partial_{\bar{z}} \phi^{\bar{j}} - \partial_{\bar{z}} \phi^i \partial_z \phi^{\bar{j}}). \tag{2.1}$$

Here d is the degree of the map $\phi : \Sigma \to K$. Q is the BRST operator given by the scalar component of the twisted $N = 2$ supercharge operator. BRST transformations of the fields are given by

$$\begin{aligned} &\delta\phi^i = i\epsilon\chi^i, \qquad \delta\phi^{\bar{i}} = i\epsilon\chi^{\bar{i}}, \\ &\delta\chi^i = 0, \qquad \delta\chi^{\bar{i}} = 0, \\ &\delta\rho_z^{\bar{i}} = -\epsilon\partial_z\phi^{\bar{i}} - i\epsilon\chi^{\bar{j}}\Gamma^{\bar{i}}_{\bar{j}\bar{m}}\rho_{\bar{z}}^{\bar{m}}, \\ &\delta\rho_{\bar{z}}^i = -\epsilon\partial_{\bar{z}}\phi^i - i\epsilon\chi^j\Gamma^i_{jm}\rho_{\bar{z}}^m. \end{aligned} \tag{2.2}$$

BRST variation δ is nilpotent, $\delta^2 = 0$, up to equations of motion. Thus the Lagrangian of the topological σ model is given by a BRST commutator (modulo a term proportional to the degree of the maps)

$$\text{Lagrangian} = \{Q, V\}. \tag{2.3}$$

When the Lagrangian is a BRST commutator, the WKB approximation of the path integral becomes exact and there are no higher-order corrections beyond the one-loop Gaussian fluctuations. In the case of the topological σ model, the saddle point of the action is found by saturating the inequality

$$\begin{aligned} &g_{i\bar{j}}(\partial_z\phi^i\partial_{\bar{z}}\phi^{\bar{j}} + \partial_{\bar{z}}\phi^i\partial_z\phi^{\bar{j}}) \\ &+g_{i\bar{j}}(- \partial_z\phi^i\partial_{\bar{z}}\phi^{\bar{j}} + \partial_{\bar{z}}\phi^i\partial_z\phi^{\bar{j}}) \\ &= 2g_{i\bar{j}}\partial_{\bar{z}}\phi^i\partial_z\phi^{\bar{j}} \geq 0. \end{aligned} \tag{2.4}$$

It is given by the holomorphic maps

$$\partial_{\bar{z}}\phi^i = \partial_z\phi^{\bar{i}} = 0. \tag{2.5}$$

Eq.(2.4) is analogous to the inequality in non-Abelian gauge theories

$$TrF^2 + TrFF^* = \frac{1}{2}Tr(F + F^*)^2 \geq 0 \tag{2.6}$$

which is saturated by the instantons, i.e. the anti-self-dual gauge fields. In this analogy the holomorphic maps (2.5) are called as world-sheet instantons. Thus the path-integral of topological σ model reduces to a sum over world-sheet instantons.

2.2 Instantons

For the sake of simplicity let us now consider instantons from a genus 0 Riemann surface, i.e. CP^1, onto the target space which is also CP^1. If we use local coordinates z and w for the world-sheet and target space CP^1, respectively, a degree-d instanton is described as

$$w = \frac{a_0 z^d + a_1 z^{d-1} + \cdots + a_d}{b_0 z^d + b_1 z^{d-1} + \cdots + b_d}. \tag{2.7}$$

The number of parameters of this map is given by

$$2(d+1)-1=2d+1 \tag{2.8}$$

where one parameter was subtracted since only the ratio matters in (2.7). When the system is coupled to gravity, we have to divide by the automorphism group $SL(2,\mathbf{C})$ of the sphere and hence the number of parameters reduces to

$$2d-2. \tag{2.9}$$

We can write the above result more formally as

$$\dim\mathcal{M}_{0,0}(CP^1;d)=2d-2 \tag{2.10}$$

where $\mathcal{M}_{g,s}(K;d)$ denotes the moduli space of degree-d instantons from genus g Riemann surfaces with s punctures onto a manifold K. A general formula is known for the dimensionality of the moduli space

$$\dim\mathcal{M}_{g,s}(K;d)=c_1(K)d+(1-g)(\dim K-3)+s \tag{2.11}$$

where $c_1(K)$ is the first Chern class of the Kähler manifold K. Note that $c_1(CP^1)=2$ and we recover (2.10).

2.3 Observables

Physical observables in the topological σ-model come from the de Rham classes of the target manifold. In fact given a closed differential n-form $f_{\mu_1,\cdots,\mu_n}(\phi)d\phi^{\mu_1}\wedge\cdots d\phi^{\mu_n}\in H^n(K;\mathbf{R})$ on K we can construct a BRST invariant operator $\mathcal{O}_f$

$$\begin{aligned}&\mathcal{O}_f=f_{\mu_1,\cdots,\mu_n}(\phi)\chi^{\mu_1}\cdots\chi^{\mu_n},\\&\delta\mathcal{O}_f=0.\end{aligned} \tag{2.12}$$

using the BRST transformation law (2.2). In the case of CP^1, there are two de-Rham classes 1 (identity) and ω (Kähler class) and the corresponding physical observables $\mathcal{O}$ are denoted as P and Q. In the case of CP^2 there are three de-Rham classes, $1,\omega,\omega^2$ and the corresponding physical operators are denoted as P,Q,R, respectively. These operators are fields on the world-sheet and depend on $z,\bar{z}$, however, coordinates are usually suppressed since correlation functions are position independent in topological field theories.

2.4 Ghost-Number Conservation

The ghost number of an observable $\mathcal{O}_i$ is given by $1/2$ of the degree of its corresponding differential form

$$\text{ghost number }(\mathcal{O}_i)=\frac{1}{2}\text{ degree of form }(\mathcal{O}_i). \tag{2.13}$$

Thus ghost number $(P) = 0$, ghost number $(Q) = 1$ in the CP^1 model and ghost number $(P) = 0$, ghost number $(Q) = 1$, ghost number $(R) = 2$ in the CP^2 model, etc. We introduce parameters $\{t_i\}$ in order to describe the coupling of the fields $\{\mathcal{O}_i\}$. Derivative of the free energy with respect to t_i gives the expectation value of the operator $\mathcal{O}_i$

$$\frac{\partial F}{\partial t_i} = \langle \mathcal{O}_i \rangle. \tag{2.14}$$

When the gravity is coupled to the the topological σ model, new BRST observables, gravitational descendants $\sigma_n(\mathcal{O}_i)$ $(n = 0, 1, \cdots)$ appear for each primary field $\mathcal{O}_i$. 0-th descendants $\sigma_0(\mathcal{O}_i)$ are identified as the primary fields $\mathcal{O}_i$ themselves. Descendant $\sigma_n(\mathcal{O}_i)$ carry a ghost number

$$\text{ghost number } (\sigma_n(\mathcal{O}_i)) = n + \alpha_i \tag{2.15}$$

where α_i is the ghost number of $\mathcal{O}_i$. Coupling parameters of the descendants $\sigma_n(\mathcal{O}_i)$ are denoted as $t_{n,i}$ $(t_{0,i} \equiv t_i)$. When only the primary couplings t_i are non-vanishing, the theory is said to be the small phase space. On the other hand, when descendant couplings are turned on, the theory is in the large phase space.

Now the ghost number conservation law reads as

$$\begin{aligned} &\langle \prod_{i=1}^{s} \sigma_{n_i}(\mathcal{O}_i) \rangle_{g,s} \neq 0 \\ &\Longleftrightarrow \sum_{i=1}^{s} (n_i + \alpha_i) = \dim \mathcal{M}_{g,s}(K; d) \\ &= c_1(K)d + (1-g)(\dim K - 3) + s. \end{aligned} \tag{2.16}$$

The above formula implies that the correlation function $\langle \prod_{i=1}^{s} \sigma_{n_i}(\phi_i) \rangle_{g,s}$ is non-zero and receives degree-d instanton contributions if one can solve the equation $\sum_{i=1}^{s}(n_i + \alpha_i) = c_1(K)d + (1-g)(\dim K - 3) + s$ for a non-negative integer d given the values of $\{n_i\}, \{\alpha_i\}, g, s$ and $c_1(K)$. We note the crucial difference between the case of Calabi-Yau manifolds with $c_1(K) = 0$ and the cases of Fano varieties $c_1(K) > 0$. In the former case the degree d does not enter in the right-hand-side of (2.16) and non-vanishing correlation functions receive contributions from instantons of all degrees. On the other hand, in the latter case contributions come from instantons of a definite degree. Instantons carry a weight proportional to $c_1(K)$ which is the coefficient of the 1-loop beta function.

2.5 Metric

Metric in the space of primary fields $\{\mathcal{O}_i\}$ in matter theory is given by the two-point functions on the sphere,

$$\eta_{ij} = \langle \mathcal{O}_i \mathcal{O}_j \rangle_{g=0}. \tag{2.17}$$

When gravity is coupled, the sphere with two-punctures becomes unstable and we should instead consider a three-point function

$$\eta_{ij} = \langle P \mathcal{O}_i \mathcal{O}_j \rangle_{g=0}. \tag{2.18}$$

Ghost number conservation (2.16) reads (only the "classical" piece $d = 0$ contributes to the metric)

$$\alpha_i + \alpha_j = \mathrm{dim} K \tag{2.19}$$

and thus the metric corresponds to the classical intersection pairing $\int_K \mathcal{O}_i \wedge \mathcal{O}_j = \eta_{ij}$. In the case of CP^1 model only non-vanishing element of the metric is given by $\langle PPQ \rangle = 1$. Thus the classical piece of the genus 0 free energy has the form, $F_0^{cl} = \frac{1}{2} t_P^2 t_Q$. In the case of CP^2 , $\langle PQQ \rangle = 1$, $\langle PPR \rangle = 1$ while others vanish. Classical free energy is given by $F_0^{cl} = \frac{1}{2} t_P t_Q^2 + \frac{1}{2} t_P^2 t_R$.

2.6 Instanton Contributions

Let us next evaluate the instanton contributions to the genus 0 free energy in the small phase space. If we consider the case of CP^1 and the correlation functions of the type $\langle P \cdots PQ \cdots Q \rangle$ where n P's and m Q's appear, the ghost number conservation law reads $m = 2d - 2 + n + m$. Hence n has to vanish for $d > 0$ and correlation functions with P insertion receive no contributions from instantons. On the other hand correlation functions of the type $\langle Q \cdots Q \rangle$ with any number of Q's receive a unique degree-1 instanton contribution and we obtain $\langle Q \cdots Q \rangle = \exp(t_Q)$. Thus we find the genus 0 free energy of CP^1 model

$$F_0 = F_0^{cl} + F_0^{qu} = \frac{1}{2} t_P^2 t_Q + e^{t_Q}. \tag{2.20}$$

In the case of CP^2 the situation becomes more complex. If we consider correlation functions of the type $\langle P \cdots PQ \cdots QR \cdots R \rangle$ in the small phase space with n P's, m Q's and ℓ R's, the ghost number conservation gives $m + 2\ell = 3d - 1 + n + m + \ell$. When $n = 0$, there exist non-vanishing correlation functions for all d with $\ell = 3d - 1$. Thus the genus 0 free energy has an expansion

$$\begin{aligned} F_0 = F_0^{cl} + F_0^{qu} &= \frac{1}{2} t_P t_Q^2 + \frac{1}{2} t_P^2 t_R + f(t_Q, t_R), \\ f(t_Q, t_R) &= \sum_{d=1} N_d \frac{t_R^{3d-1}}{(3d-1)!} e^{d t_Q} \end{aligned} \tag{2.21}$$

N_d is the number of degree-d instantons which pass through $3d - 1$ fixed points on CP^2. There is an algorithm which computes the numbers N_d recursively [4]. I will discuss about it section 6.

3 CP^1 Model

3.1 Constitutive Relation

The integrable structure of the CP^1 model is described by the basic 2-point functions

$$\langle PP \rangle = u, \qquad \langle PQ \rangle = v. \tag{3.1}$$

Combining (3.1) and (2.20) we note that

$$\langle QQ\rangle = e^{t_Q} = e^{\langle PP\rangle} \tag{3.2}$$

holds in the small phase space. Using the topological recursion relation [14] it is possible to show that (3.2) in fact holds in a large phase space where couplings $\{t_{n,P}, t_{n,Q}, n = 1, 2, \cdots\}$ of the gravitational descendants do not vanish [15]. Thus the relation

$$\langle QQ\rangle = e^{\langle PP\rangle} \tag{3.3}$$

holds in the large phase space of the theory and characterizes the CP^1 model (constitutive equation). It is then easy to see that the integrable structure behind the CP^1 model is given by the (1-dimensional) Toda hierarchy

$$\frac{\partial^2 u}{\partial t_Q^2} = \frac{\partial}{\partial t_Q}\frac{\partial}{\partial t_P}\langle PQ\rangle = \frac{\partial}{\partial t_P}\langle PQQ\rangle \frac{\partial^2}{\partial t_P^2} e^u. \tag{3.4}$$

(3.4) in fact is the Toda-lattice equation when we identify t_P as the continuum version of the index n of a Toda field $u_n = u(t_P)$.

3.2 Toda-hierarchy

The Lax formalism of the Toda-lattice hierarchy is well-known. We introduce a Lax operator (at genus $g = 0$ or dispersionless limit, it becomes a number rather than an operator)

$$L = p + v + e^u p^{-1} \tag{3.5}$$

and construct operators

$$B_n^Q = [L^n]_+, \quad n = 1, 2, \cdots \tag{3.6}$$

($+$ means to take terms with non-negative powers of p). (3.6) generate flows in the Toda times t_n ($n = 1, 2, \cdots$)

$$\frac{\partial L}{\partial t_n} = \{B_n^Q, L\}, \quad n = 1, 2, \ldots, \tag{3.7}$$

where $\{A, B\}$ denotes the Poisson bracket of the Toda theory defined by

$$\{A, B\} = p\Big(\frac{\partial A}{\partial p}\frac{\partial B}{\partial t_P} - \frac{\partial B}{\partial p}\frac{\partial A}{\partial t_P}\Big). \tag{3.8}$$

It is easy to identify the Toda times t_n as the descendant times $t_{n-1,Q}$ $n = 1, 2, \cdots$ of the operator Q. For instance, putting $n = 1, 2, \mathcal{O}_i = Q$, and $X = Y = P$ or $X = P$, $Y = Q$ in topological recursion relation [14]

$$\langle \sigma_n(\mathcal{O}_i)XY\rangle = \sum_j n\langle \sigma_{n-1}(\mathcal{O}_i)\mathcal{O}_j\rangle\langle \mathcal{O}^j XY\rangle \tag{3.9}$$

we obtain flow equations

$$\begin{aligned}
\frac{\partial u}{\partial t_{1,Q}} &= \langle \sigma_1(Q)PP\rangle = \langle QP\rangle\langle QPP\rangle + \langle QQ\rangle\langle PPP\rangle = \Big(\frac{1}{2}v^2+e^u\Big)', \\
\frac{\partial v}{\partial t_{1,Q}} &= \langle \sigma_1(Q)PQ\rangle = \langle QP\rangle\langle QPQ\rangle + \langle QQ\rangle\langle PPQ\rangle = (ve^u)', \\
\frac{\partial u}{\partial t_{2,Q}} &= \langle \sigma_2(Q)PP\rangle = 2\Big(\langle\sigma_1(Q)P\rangle\langle QPP\rangle + \langle\sigma_1(Q)Q\rangle\langle PPP\rangle\Big) = \Big(\frac{1}{3}v^3+2ve^u\Big)', \\
\frac{\partial v}{\partial t_{2,Q}} &= \langle \sigma_2(Q)PQ\rangle = 2\Big(\langle\sigma_1(Q)P\rangle\langle QPQ\rangle + \langle\sigma_1(Q)Q\rangle\langle PPQ\rangle\Big) = (v^2e^u+e^{2u})'.
\end{aligned} \tag{3.10}$$

Comparing the previous equation with

$$\begin{aligned}
\frac{\partial L}{\partial t_2} &= 2(ve^u)' + (v^2+2u)'e^u p^{-1}, \\
\frac{\partial L}{\partial t_3} &= 3(v^2e^u+e^{2u})' + (v^3+6ve^u)'e^u p^{-1},
\end{aligned}$$

which follow from (3.7), we find $t_2 = t_{1,Q}/2$, $t_3 = t_{2,Q}/3$. In general it is not difficult to check that the relation

$$t_n = \frac{1}{n} t_{n-1,Q}\,, \qquad n = 1,2,\cdots \tag{3.11}$$

holds.

3.3 Logarithmic Operators

Analysis of flows in the parameters $\{t_{n,P}, n = 1,2,\cdots\}$ is more involved. Flow equations can again be written down using the recursion relation (3.9). We have, for instance,

$$\begin{aligned}
\frac{\partial u}{\partial t_{1,P}} &= \langle \sigma_1(P)PP\rangle = (uv)', \\
\frac{\partial v}{\partial t_{1,P}} &= \langle \sigma_1(P)PQ\rangle = \Big(\frac{1}{2}v^2+(u-1)e^u\Big)', \\
\frac{\partial u}{\partial t_{2,P}} &= \langle \sigma_2(P)PP\rangle = (uv^2+2(u-2)e^u)', \\
\frac{\partial v}{\partial t_{2,P}} &= \langle \sigma_2(P)PQ\rangle = \Big(\frac{1}{3}v^3+2v(u-1)e^u\Big)'.
\end{aligned}$$

It is somewhat non-trivial to find generators for these flows. It turns out that operators involving logarithms of the Lax operator

$$\begin{aligned}
B_n^P &= [L^n(\log L - c_n)]_+, \quad n = 0,1,\ldots, \\
c_n &= \sum_{j=1}^{n} 1/j, \quad c_0 = 0
\end{aligned} \tag{3.12}$$

generate the flows in the $t_{n,P}$ variables,

$$\frac{\partial L}{\partial t_{n,P}} = 2\{B_n^P, L\}, \quad n = 0, 1, \cdots. \tag{3.13}$$

In (3.12), (3.13) the logarithm of L is defined by taking the average

$$\begin{aligned}\log L &= \log(p + v + e^u p^{-1}) \\ &= \frac{1}{2}\log p(1 + vp^{-1} + e^u p^{-2}) + \frac{1}{2}\log e^u p^{-1}(1 + ve^{-u}p + e^{-u}p^2) \\ &= \frac{u}{2} + \frac{1}{2}\log(1 + vp^{-1} + e^u p^{-2}) + \frac{1}{2}\log(1 + ve^{-u}p + e^{-u}p^2).\end{aligned} \tag{3.14}$$

Our operators $L^n(\log L - c_n)$ have a special "scaling" property

$$\frac{d}{dL}L^n(\log L - c_n) = nL^{n-1}(\log L - c_{n-1}). \tag{3.15}$$

We note in particular an important relation

$$\frac{\partial L}{\partial t_P} = 2\{B_0^P, L\} = 2\{[\log L]_+, L\}. \tag{3.16}$$

3.4 Hamiltonians

It is possible to recast the above Lax formulation into the Hamiltonian formalism. Hamilton's equations of motions are given by

$$\frac{\partial u}{\partial t_{n,Q}} = \{u, H_{n+1}^Q\}_1, \frac{\partial v}{\partial t_{n,Q}} = \{v, H_{n+1}^Q\}_1, \quad n = 0, 1, 2, \cdots \tag{3.17}$$

where the bracket is defined by the 1st Hamiltonian structure

$$\{u(x), v(y)\}_1 = \delta'(x - y), \{u, u\}_1 = \{v, v\}_1 = 0. \tag{3.18}$$

x, y in the above formula means the values of the "spatial" variable t_P. Hamiltonians are given by

$$\begin{aligned}H_1^Q &= \int (\frac{1}{2}v^2 + e^u)\, dx, H_2^Q = \frac{1}{2}\int (\frac{1}{3}v^3 + 2ve^u)\, dx, \\ H_3^Q &= \frac{1}{3}\int (\frac{1}{4}v^4 + 3v^2 e^u + \frac{3}{2}e^{2u})\, dx, \cdots.\end{aligned} \tag{3.19}$$

They can be represented as

$$\begin{aligned}H_n^Q &= \int \frac{1}{n(n+1)}\left(L^{n+1}\right)_0 dx \\ &= \frac{1}{n}\int \langle \sigma_n(Q)P\rangle\, dx, \quad n = 1, 2, \cdots\end{aligned} \tag{3.20}$$

where the suffix-0 denotes the terms proportional to p^0 in L^{n+1}.

Hamiltonians for the $\{t_{n,P}\}$ evolutions are defined similarly by

$$\frac{\partial u}{\partial t_{n,P}} = \{u, H_{n+1}^P\}_1, \frac{\partial v}{\partial t_{n,P}} = \{v, H_{n+1}^P\}_1, \quad n = 0, 1, 2, \cdots. \tag{3.21}$$

They are given by

$$\begin{aligned} H_1^P &= \int (uv)\, dx, H_2^P = \int (\frac{1}{2}uv^2 + (u-2)e^u)\, dx, \\ H_3^P &= \int (\frac{1}{3}uv^3 + 2v(u-2)e^u)\, dx, \cdots. \end{aligned} \tag{3.22}$$

For general n we can express them as

$$\begin{aligned} H_n^P &= \frac{2}{n} \int \Big(L^n(\log L - c_n)\Big)_0 dx \\ &= \frac{1}{n} \int \langle \sigma_n(P) P \rangle\, dx, \end{aligned} \tag{3.23}$$

Hamiltonians $\{H_n^Q\}$ and $\{H_m^P\}$ mutually commute with each other.

It is easy to show the existence of the 2nd Hamiltonian structure in CP^1 model which is a characteristic feature of completely integrable systems. A new Poisson bracket is defined by

$$\begin{aligned} &\{u(x), u(y)\}_2 = 2\partial_x \delta(x-y), \\ &\{u(x), v(y)\}_2 = \partial_x (v(x)\delta(x-y)), \\ &\{v(x), v(x)\}_2 = (2e^u(x)\partial_x + (\partial_x e^u))\delta(x-y), \\ &\{v(x), u(y)\}_2 = v(x)\partial_x \delta(x-y). \end{aligned} \tag{3.24}$$

They generate time evolutions $(n = 0, 1, 2 \cdots)$

$$\begin{aligned} \frac{\partial u}{\partial t_{n,Q}} &= \{u, H_n^Q\}_2, \\ \frac{\partial v}{\partial t_{n,Q}} &= \{v, H_n^Q\}_2, \\ \frac{\partial u}{\partial t_{n,P}} &= \{u, H_n^P\}_2, \\ \frac{\partial v}{\partial t_{n,P}} &= \{v, H_n^P\}_2. \end{aligned} \tag{3.25}$$

Note the shift in the suffices of the Hamiltonians from those of (3.17) and (3.21). Making use of the topological recursion relation it is possible to prove the existence of the 2nd Hamiltonian structure for arbitrary topological σ-models [16].

4 Matrix Integrals

4.1 Matrix Model

We now turn to the construction of a matrix model which reproduces the CP^1 model on Riemann surfaces of arbitrary genera. Let us first recall some basic formulas for the analysis of matrix models using orthogonal polynomials. We consider a one-matrix model with an action $\mathrm{Tr}\, V(M)$ which depends on the eigenvalues of a hermitian matrix M of size $N^* \times N^*$, $N^* = N \times t_P$. The orthogonal polynomials are defined by

$$\begin{aligned}\int d\lambda e^{NV(\lambda)}\psi_n(\lambda)\psi_m(\lambda) &= \delta_{n,m}h_m,\\ \psi_n(\lambda) &= \lambda^n + \text{lower terms}.\end{aligned} \tag{4.1}$$

The multiplication by λ of the orthogonal polynomials leads to a three-term recursion relation

$$\begin{aligned}\lambda\psi_n(\lambda) &= \sum_m Q_{nm}\psi_m(\lambda)\\ &= \psi_{n+1}(\lambda) + v_n\psi_n(\lambda) + R_n\psi_{n-1}(\lambda),\\ R_n &= h_n/h_{n-1} \equiv e^{N(\phi_n-\phi_{n-1})}.\end{aligned} \tag{4.2}$$

The free-energy of the system is given by $F = \log \Pi_0^{N^*-1} h_n = N^* \sum_0^{N^*-1} \phi_n$. The matrix Q has non-vanishing elements only on the diagonal and 1st off-diagonal lines and thus is a Jacobi matrix. In the continuum limit $N(\phi_n - \phi_{n-1}) \approx F''/N^2 = u$ and the matrix Q turns into the Lax operator

$$Q \approx L = p + v + e^u p^{-1}. \tag{4.3}$$

Q replaces L in the discussion of the CP^1 model at higher genera. Time evolution equations are now given by the commutation relations

$$\begin{aligned}\frac{n}{N}\frac{\partial Q}{\partial t_{n-1,Q}} &= [(Q^n)_+, Q], n = 1, 2, \cdots\\ \frac{1}{2N}\frac{\partial Q}{\partial t_{n,P}} &= [(Q^n(\log Q - c_n))_+, Q],\\ &\quad n = 0, 1, 2, \cdots,\end{aligned} \tag{4.4}$$

where $+$ means to take the upper-triangular part of a matrix including the diagonal line. It is easy to check that the matrix commutators reduce to the Poisson bracket (3.8) in the continuum limit and thus (4.4) reduce to (3.7) and (3.13) at genus $g = 0$.

The derivative operator $d/d\lambda$, on the other hand, becomes a lower-triangular matrix when acting on the orthogonal polynomials

$$\frac{d}{d\lambda}\psi_n(\lambda) = n\psi_{n-1}(\lambda) + \cdots = \sum_m P_{nm}\psi_m(\lambda). \tag{4.5}$$

By partial integration we have

$$\begin{aligned}
&\int d\lambda e^{NV(\lambda)} \frac{d}{d\lambda}\psi_n(\lambda)\psi_m(\lambda) \\
&= -\int d\lambda e^{NV(\lambda)} NV'(\lambda)\psi_n(\lambda)\psi_m(\lambda) \\
&= -NV'(Q)_{nm}h_m, \qquad n > m.
\end{aligned} \tag{4.6}$$

Hence

$$P = -NV'(Q)_-, \tag{4.7}$$

where $-$ means to take the lower-triangular part of the matrix. The well-known string equation then reads

$$[Q, P] = -N[Q, V'(Q)_-] = 1. \tag{4.8}$$

Let us now introduce our matrix model defined by an action

$$\begin{aligned}
\operatorname{Tr} V(M) = & - 2\operatorname{Tr} M(\log M - 1) \\
& + \sum_{n=1} 2t_{n,P} \operatorname{Tr} M^n(\log M - c_n) \\
& + \sum_{n=1} \frac{1}{n} t_{n-1,Q} \operatorname{Tr} M^n.
\end{aligned} \tag{4.9}$$

We then obtain

$$\begin{aligned}
N[V'(Q)_-, Q] &= -2N[(\log Q)_-, Q] \\
&\quad +N\sum 2nt_{n,P}[(Q^{n-1}(\log Q - c_{n-1}))_-, Q] \\
&\quad +N\sum t_{n-1,Q}[(Q^{n-1})_-, Q] \\
&= \frac{\partial Q}{\partial t_P} - \sum_{n=1} nt_{n,P}\frac{\partial Q}{\partial t_{n-1,P}} \\
&\quad -\sum_{n=2}(n-1)t_{n-1,Q}\frac{\partial Q}{\partial t_{n-2,Q}},
\end{aligned} \tag{4.10}$$

where we used (4.4) and the "scaling" property of our logarithmic potentials (3.15). (4.8) is now compactly expressed as

$$1 + \sum_{n=1}\sum_i nt_{n,i}\frac{\partial Q}{\partial t_{n-1,i}} = \frac{\partial Q}{\partial t_P}, (i = P, Q). \tag{4.11}$$

The diagonal element of (4.11) reads

$$\frac{\partial \langle PQ\rangle}{\partial t_P} = 1 + \sum_n\sum_i nt_{n,i}\langle\sigma_{n-1}(\mathcal{O}_i)PQ\rangle \tag{4.12}$$

in terms of the correlation functions. By integrating (4.12) we obtain

$$\begin{aligned}
u &= t_Q + \sum_{n=1}\sum_i nt_{n,i}\langle\sigma_{n-1}(\mathcal{O}_i)P\rangle, \\
v &= t_P + \sum_{n=1}\sum_i nt_{n,i}\langle\sigma_{n-1}(\mathcal{O}_i)Q\rangle.
\end{aligned} \tag{4.13}$$

These are the string equations of the CP^1 model [15].

The characteristic feature of our action (4.9) is the appearance of logarithms which generate additive terms under the scale transformation of the matrix M. This is the mechanism by means of which we reproduce the sum over instantons.

4.2 Virasoro Operators

It is easy to derive Ward identities of our model (4.9) by varying the matrix eigenvalues as $\lambda_i \to \lambda_i + \varepsilon\lambda_i^{m+1}$. We find

$$L_{-1}Z = \Big(-\frac{\partial}{\partial t_P} + \sum_{n=1} n t_{n,i}\frac{\partial}{\partial t_{n-1,i}} + N^2 t_P t_Q\Big)Z = 0, \tag{4.14}$$

$$\begin{aligned} L_0 Z \;=\; & \Big(\sum_{n=1} n t_{n-1,Q}\frac{\partial}{\partial t_{n-1,Q}} - \frac{\partial}{\partial t_{1,P}} \\ & + \sum_{n=1} n t_{n,P}\frac{\partial}{\partial t_{n,P}} - 2\frac{\partial}{\partial t_Q} \\ & + 2\sum_{n=1} n t_{n,P}\frac{\partial}{\partial t_{n-1,Q}} + N^2 t_P^2\Big)Z = 0, \end{aligned} \tag{4.15}$$

$$\begin{aligned} L_m Z \;=\; & \Big(-\frac{\partial}{\partial t_{m+1,P}} + \sum_{n=1} n t_{n,P}\frac{\partial}{\partial t_{n+m,P}} \\ & + \sum_{n=1}(n+m)t_{n-1,Q}\frac{\partial}{\partial t_{n+m-1,Q}} \\ & -2(m+1)c_{m+1}\frac{\partial}{\partial t_{m,Q}} \\ & +2\sum_{n=0}(n+m)(1+n(c_{n+m}-c_n))t_{n,P}\frac{\partial}{\partial t_{n+m-1,Q}} \\ & +\frac{1}{N^2}\sum_{r=1}^{m-1} r(m-r)\frac{\partial^2}{\partial t_{m-r-1,Q}\partial t_{r-1,Q}}\Big)Z = 0, \\ & m = 1, 2, \ldots \end{aligned} \tag{4.16}$$

The operators $\{L_m\}$ form a Virasoro algebra

$$[L_m, L_n] = (m-n)L_{m+n}, \qquad m, n \geq -1. \tag{4.17}$$

A special feature of the above operators is the presence of the mixing terms of the form $t_{n,P}\partial/\partial t_{n+m-1,Q}$ which arise due to the lack of scale invariance of the action (4.9). The factor 2 in front of these terms is identified as the 1st Chern class $c_1(CP^1) = 2$.

It is easy to derive a varieties of relations among correlation functions using Virasoro conditions (4.14), (4.15), (4.16) and compute intersection numbers of differential forms on the moduli space of instantons. I am not going into detailed discussions about intersection numbers except to quote a few data [10]

$$\begin{aligned}
\langle\sigma_{2m}(Q)\rangle_{0,m+1} &= \frac{(2m)!}{((m+1)!)^2},\\
\langle\sigma_{2m+1}(P)\rangle_{0,m+1} &= -2c_{m+1}\frac{(2m)!}{((m+1)!)^2},\\
\langle\sigma_1(P)\rangle_1 &= \frac{1}{24}\chi(CP^1) = \frac{1}{12},\\
\langle Q\rangle_1 &= -\frac{1}{24}\int_{CP^1}\omega = -\frac{1}{24}
\end{aligned} \tag{4.18}$$

which hold in the limit of zero couplings ($t_{n,i} = 0$ for all n, i) and are used in the following. Here $\langle\mathcal{O}\rangle_{0,d}$ denotes a genus-0 amplitude which receives a contribution from the degree d instantons and $\langle\mathcal{O}\rangle_1$ denote a genus-1 classical ($d = 0$) amplitude.

Instead I would like to point out that the L_0 equation for the CP^1 model can be generalized for a large class of manifolds as has been shown by Hori using the method of intersection theory [17]. His equation reads as

$$\begin{aligned}
&\sum_{n,i}(n+\alpha_i+(1-\dim K)/2)\tilde{t}_{n,i}\frac{\partial}{\partial t_{n,i}} + \sum_{n,i,j} C_\omega{}^i{}_j\tilde{t}_{n,i}\frac{\partial}{\partial t_{n-1,j}} \\
&+\frac{N^2}{2}\sum_{i,j}C_\omega{}^{ij}t_it_j + \frac{1}{24}\Big(\frac{(3-\dim K)}{2}\cdot\chi(K) - \int_K c_1(K)c_{\dim K-1}(K)\Big) = 0
\end{aligned} \tag{4.19}$$

where $\tilde{t}_{n,i} = t_{n,i} - \delta_{n,1}\delta_{i,P}$. $C_{\omega ij}$ denotes the classical intersection number among Kähler-class ω and differential forms $\mathcal{O}_i, \mathcal{O}_j$, $C_{\omega ij} = \int_K \omega\mathcal{O}_i\mathcal{O}_j$ and $\chi(K)$ is the Euler number. Forth terms in (4.19) comes from exceptional contributions at $g = 0, s = 3$ and $g = 1, s = 1$. Hori's equation together with the well-known string (L_{-1}) and dilaton equations form a basis for the analysis of topological σ models on general manifolds.

We should point out that there is a subtlety associated with the $g = 1$ free energy in the CP^1 model. From the definition of the partition function (4.9) and the fact that the size of the matrix M being equal to $(N \times t_P)^2$, it is possible to derive a relation

$$\Big(D - N\frac{\partial}{\partial N}\Big)F = 0, \tag{4.20}$$

where $F = \log Z$ and $D = \sum_{n=0}\tilde{t}_{n,i}\partial/\partial t_{n,i}$. If the free energy has an expansion without a term proportional to $\log N$, (4.20) would imply $\langle\sigma(P)\rangle_1 = 0$ at zero couplings in contradiction to the data (4.18). Thus F must contain a $\log N$ term as

$$F = \sum_{g=0} N^{2-2g}F_g - \frac{1}{12}\log N \tag{4.21}$$

in the $1/N$ expansion. (4.21) reproduces the value of $\langle\sigma(P)\rangle_1$.

The presence of the log term is related to the issue of whether the genus-1 free energy has an expression suggestive of a string field theory

$$F_1 = \frac{1}{24}\log\det u_{ij}, \tag{4.22}$$

where $u_{ij} = \partial^3 F_0/\partial t_P \partial t_i \partial t_j$. Unlike the case of minimal models where (4.22) holds [15, 18, 19] it may not be valid in the case of σ-models [20]. In the CP^1 model we have a relation

$$F_1 = \frac{1}{24} \log \det u_{ij} - \frac{1}{24} u, \quad (i, j = 1, 2) \tag{4.23}$$

where the extra term reproduces $\langle Q \rangle_1 = -\frac{1}{24}$ (4.18).

5 Landau-Ginzburg formulation

In the case of topological minimal models coupled to topological gravity at genus $g = 0$ Landau-Ginzburg (LG) description of the system has been developed where the scalar Lax operator of the dispersionless KP hierarchy plays the role of the superpotential [21]. In the following we shall show that it is also possible to develop a LG formulation in the case of the topological CP^1 model using the Lax operator of the Toda hierarchy as the superpotential. In this section we restrict ourselves to the small phase space and change our notation; p is replaced by z. The Lax operator is then given by

$$L = z + t_P + e^{t_Q} z^{-1}. \tag{5.1}$$

If one regards L as the superpotential, its extrema give the vacuum states

$$\partial_z L = 0 \implies z^2 = e^{t_Q}. \tag{5.2}$$

If the Kähler class Q is described by z, this gives the relation $Q \cdot Q = e^{t_Q}$ of the quantum cohomology ring of CP^1. We thus make an identification

$$P = 1, \qquad Q = z. \tag{5.3}$$

Note that the variable z itself is not the Landau-Ginzburg field since the Poisson bracket of the Toda theory (3.8) has an unusual form with an extra factor of z in the RHS. We may instead regard $X = \log z$ as the LG variable. The superpotential (5.1) is then rewritten as

$$L = \exp X + t_P + e^{t_Q} \exp(-X). \tag{5.4}$$

It is of the form of the sine-Gordon potential. Because of the change of variable $z \to X$ there appear extra Jacobian factors. The residue formula is now given by

$$\langle ABC \rangle = \oint dz \frac{A(z)B(z)C(z)}{z^2 \partial_z L}, \tag{5.5}$$

where the integration contour is taken around $z = \infty$. Note that the denominator equals

$$z^2 \partial_z L = z^2 - e^{t_Q} \tag{5.6}$$

which looks as if we were dealing with the A_1 minimal model. It is easy to check the flatness of the metric

$$\langle P \mathcal{O}_i \mathcal{O}_j \rangle = \eta_{ij}, \tag{5.7}$$

where $\eta_{PQ} = \eta_{QP} = 1,\ \eta_{PP} = \eta_{QQ} = 0$.

Our next step is to describe gravitational descendants $\sigma_n(\mathcal{O}_i),\ \mathcal{O}_i = P, Q$. In the case of the topological minimal models coupled to topological gravity it is possible to adopt the matter picture and express gravitational descendants using only the matter fields [22, 23]. It turns out that also in the case of CP^1 model it is possible to describe gravitational descendants using only the LG field. We postulate

$$\begin{aligned}\sigma_n(P) &= 2n[L^{n-1}(\log L - c_{n-1})z\partial_z L]_+ \ ,\\ \sigma_n(Q) &= [L^n z\partial_z L]_+ \ ,\end{aligned} \tag{5.8}$$

with $n = 0, 1, 2, \cdots$. Note that $\sigma_n(Q)$ is a polynomial of order $n+1$ in z, while $\sigma_n(P)$ is an infinite series in the variable z. We remark that $\sigma_0(Q) = z$ from (5.8) in agreement with (5.3). On the other hand, $\sigma_0(P)$ differs from 1, however, the difference is a BRST exact term.

In order to check our LG formulation we look at the flow equations of the CP^1 model at genus zero. Let us recall the flow equations of Section 3,

$$\begin{aligned}\frac{\partial L}{\partial t_{n,P}} &= 2\{[L^n(\log L - c_n)]_+, L\},\\ \frac{\partial L}{\partial t_{n,Q}} &= \frac{1}{n+1}\{[L^{n+1}]_+, L\}.\end{aligned} \tag{5.9}$$

Comparing the coefficients of z^{-1} on both sides we find

$$\begin{aligned}\langle\sigma_n(P)P\rangle &= 2\mathrm{res}(L^n(\log L - c_n)/z),\\ \langle\sigma_n(Q)P\rangle &= \frac{1}{n+1}\mathrm{res}(L^{n+1}/z).\end{aligned} \tag{5.10}$$

("res" means to take the coefficient of z^{-1}.) Thus we recover (3.20),(3.23). They are the analogues of the Gelfand-Dikii potentials of the KP hierarchy. Integrating (5.10) over t_P we find

$$\begin{aligned}\langle\sigma_n(P)\rangle &= \frac{2}{n+1}\mathrm{res}(L^{n+1}(\log L - c_{n+1})/z),\\ \langle\sigma_n(Q)\rangle &= \frac{1}{(n+2)(n+1)}\mathrm{res}(L^{n+2}/z).\end{aligned} \tag{5.11}$$

Let us examine 1-point functions at vanishing values of the coupling constants $t_P, t_Q = 0$. Using the binomial expansion of powers of $L = z + z^{-1}$ in (5.11) we immediately recover the previous result (4.18)

$$\langle\sigma_{2m}(Q)\rangle = \frac{(2m)!}{((m+1)!)^2}. \tag{5.12}$$

Similarly $\langle\sigma_{2m+1}(P)\rangle$ can be evaluated as

$$\langle\sigma_{2m+1}(P)\rangle = \frac{1}{m+1}\oint\frac{dz}{z}L^{2m+2}(\log L - c_{2m+2})$$

$$= \frac{1}{m+1} \oint \frac{dz}{z} L^{2m+2} (\log(1+z^{-2}) - c_{2m+2})$$
$$= \frac{1}{m+1} \oint \frac{dx}{(x-1)^{2m+2}} (\log x - c_{2m+2})$$
$$= \frac{2(2m+1)!}{(m+1)!(m+1)!} \oint dx \frac{x^{m+1}}{x-1} (\log x - c_{m+1}),$$

where we have made a change of variable $1 + z^{-2} = x$ and performed repeated partial integrations. Then by picking up a pole at $x = 1$ we reproduce (4.18)

$$\langle \sigma_{2m+1}(P) \rangle = -2c_{m+1} \frac{(2m+1)!}{((m+1)!)^2}. \tag{5.13}$$

As we have noted earlier, the superpotential of the CP^1 model has the form of the sine-Gordon theory when expressed in terms of the LG field X. This indicates the equivalence of the CP^1 and the $N = 2$ supersymmetric sine-Gordon theories at least in their topological versions. The correspondence between these theories have been known for some time based on the agreement of the particle spectra and scattering matrices [11, 12]. Our results seem to provide further evidence for their equivalence. It will be interesting to see if we can make the correspondence more precise and find out, in particucular, how the sine-Gordon theory may be turned into an asymptotically free theory at a special value (KT point) of the coupling constant.

The relation between our coupling constant t_Q and those of the physical theory $g^2/4\pi$, θ is given by

$$t_Q = -\Big(\frac{4\pi}{g^2} + i\theta\Big). \tag{5.14}$$

In asymptotically free theories the coupling constant is replaced by the Λ-parameter via the dimensional transmutation $e^{-4\pi/g^2} = \Lambda^{b_1}$ with b_1 being the coefficient of the one-loop beta function. For the $N = 2$ supersymmetric CP^n σ-model b_1 is equal to the 1st Chern class, $b_1 = n + 1$. Thus we have (putting $\theta = 0, n = 1$)

$$e^{t_Q} = \Lambda^2 \tag{5.15}$$

and the superpotential reads after shifting $X \to X + t_Q/2$

$$L = \Lambda(e^X + e^{-X}). \tag{5.16}$$

Clearly Λ corresponds to the sine-Gordon mass parameter.

6 Saddle Point Analysis

We consider the system in the small phase space and put $x = n/N = t_P$, $y = t_Q$. The eigenvalue density $\rho(\lambda)$ is constrained to satisfy

$$\int \rho(\lambda) d\lambda = x. \tag{6.1}$$

The large N saddle point equation for ρ is

$$2\mathrm{PP}\int \frac{\rho(\lambda')d\lambda'}{\lambda-\lambda'} = \log(\lambda^2) - y, \qquad \lambda \in \mathrm{supp}(\rho). \tag{6.2}$$

Here PP Denotes the principal part of the integral. To find the solution, it is convenient to introduce a holomorphic function

$$H(z) = \int \frac{\rho(\lambda)d\lambda}{z-\lambda}. \tag{6.3}$$

This function satifies the following properties:
(i) $H(z)$ is holomorphic on the z-plane cut along the support of ρ
(ii) $H(z) \to x/z$ as $z \to \infty$
(iii) $H(\lambda+i0) + H(\lambda-i0) = 2\mathrm{PP}\int \frac{\rho(\lambda')d\lambda'}{\lambda-\lambda'}$ for $\lambda \in \mathrm{supp}(\rho)$
(iv) $H(\lambda+i0) - H(\lambda-i0) = -2\pi i\rho(\lambda)$ for $\lambda \in \mathrm{supp}(\rho)$.
We thus look for a function $H(z)$ satisfying (i), (ii) and (iii) where the right hand side of (iii) is replaced by $\log(\lambda^2) - y$.

Let us first assume that the support of ρ is given by a segment $0 \le b \le \lambda \le a$ of the real line. Then the solution to (iii) whihc remains finite at $z = a, b$ is given by (see [24]):

$$H(z) = \sqrt{(z-a)(z-b)}\int_C \frac{d\lambda}{2\pi i}\frac{\log\lambda - \frac{1}{2}y}{(z-\lambda)\sqrt{(\lambda-a)(\lambda-b)}}. \tag{6.4}$$

The contour C encircles the segment $b \le \lambda \le a$ once. Deforming C to a large contour, picking up poles and discontinuities, we find

$$H(z) = -\frac{y}{2} - 2\log\left\{\frac{(\sqrt{a(z-b)} - \sqrt{b(z-a)})(\sqrt{z-a} - \sqrt{z-b})}{z(a-b)}\right\}. \tag{6.5}$$

The asymptotic behaviour (ii) is satisfied if

$$\sqrt{a} = e^{y/4} + \sqrt{x/2}, \qquad \sqrt{b} = e^{y/4} - \sqrt{x/2}. \tag{6.6}$$

This solution is thus valid for

$$\frac{x}{2} \le \exp\left(\frac{y}{2}\right). \tag{6.7}$$

Using (iv), we obtain the eigenvalue density

$$\rho(\lambda) = \frac{2}{\pi}\arg\left(\lambda + \sqrt{ab} + i\sqrt{(a-\lambda)(\lambda-b)}\right), \qquad b \le \lambda \le a. \tag{6.8}$$

The free energy for this density is given by

$$\begin{aligned} F[\rho] &= \int \{-2\lambda(\log\lambda - 1) + y\}\,\rho(\lambda)d\lambda + \frac{1}{2}\int\int \log(\lambda-\lambda')^2\rho(\lambda)d\lambda\rho(\lambda')d\lambda' \\ &= 2xe^{y/2} - \frac{3}{4}x^2 + \frac{x^2}{2}\log\left(\frac{x}{2}\right) + \frac{x^2 y}{4}. \end{aligned} \tag{6.9}$$

This, however, disagrees with the free energy of the CP^1 model:

$$F_0 = \frac{x^2 y}{2} + e^y. \tag{6.10}$$

(They agree at the boundary $\frac{x}{2} = \exp\left(\frac{y}{2}\right)$ of the admissible domain (6.7).)

Let us next consider a solution whose support is a segment $b_* \leq \lambda \leq a_*$ with $b_* < 0$. There are two ways to analytically continue the solution $H(z)$ in (6.5) from $b > 0$ to $b_* < 0$: $\sqrt{b} \to \pm i\sqrt{|b_*|}$. Taking half the sum of the two, we obtain

$$H_*(z) = -\frac{y}{2} - \log\left(\frac{(\sqrt{z-a_*} - \sqrt{z-b_*})^2}{(a_* - b_*)z}\right). \tag{6.11}$$

The asymptotic behaviour (ii) requires

$$a_* = x + 2e^{y/2}, \qquad b_* = x - 2e^{y/2}. \tag{6.12}$$

Validity domain is hence the same as the previous one (6.7):

$$\frac{x}{2} \leq \exp\left(\frac{y}{2}\right).$$

If we use (iv), we obtain the "eigenvalue density"

$$\rho_*(\lambda) = \begin{cases} \dfrac{2}{\pi}\arg\left(\sqrt{\lambda - b_*} + i\sqrt{a_* - \lambda}\right) & 0 \leq \lambda \leq a_* \\ \dfrac{2}{\pi}\arg\left(\sqrt{\lambda - b_*} + i\sqrt{a_* - \lambda}\right) - 1 & b_* \leq \lambda \leq 0. \end{cases} \tag{6.13}$$

Note that $\rho_*(\lambda)$ is *negative* for $b_* < \lambda < 0$ and therefore we call it a density in the quotation mark. In the ordinary situation we throw away such a solution. However, $\rho_*(\lambda)$ possesses several properties which strongly indicates that it is the correct solution for the CP^1 model.

First when the free energy is calculated with this "density", the result agrees with the CP^1 model:

$$F[\rho_*] = \frac{x^2 y}{2} + e^y. \tag{6.14}$$

The solution ρ_* is preferrable than ρ in that $\exp N^2 F[\rho_*] \geq \exp N^2 F[\rho]$ in the domain $\frac{x}{2} \leq \exp\left(\frac{y}{2}\right)$ of validity. Namely, ρ_* is more stable than ρ.

We also find that ρ_* reproduces the correct one point functions. Let us perform the change of variable:

$$\lambda = p + x + e^y p^{-1}, \qquad p = \exp\left(\frac{y}{2} + i\theta\right). \tag{6.15}$$

Note that

$$\lambda = b_* \Leftrightarrow \theta = \pi, \qquad \lambda = a_* \Leftrightarrow \theta = 0.$$

Then, ρ_* has a simple expression:

$$\rho_*(\lambda) = \begin{cases} \dfrac{\theta}{\pi} & 0 \le \lambda \le a_* \\ \dfrac{\theta}{\pi} - 1 & b_* \le \lambda \le 0. \end{cases} \tag{6.16}$$

Recall the correspondence of topological fields with matrix model potentials:

$$\sigma_n(Q) \leftrightarrow N\mathrm{Tr}\Big(\frac{1}{n+1}M^{n+1}\Big), \qquad \sigma_n(P) \leftrightarrow 2N\mathrm{Tr}\Big(M^n(\log M - c_n)\Big). \tag{6.17}$$

In the large N limit, one point functions of these potentials are given as follows:

$$\begin{aligned} \int \frac{\lambda^{n+1}}{n+1}\rho_*(\lambda)d\lambda &= \frac{1}{(n+1)(n+2)}\int \rho_*(\lambda)d\lambda^{n+2} = -\frac{1}{(n+1)(n+2)}\int_{b_*}^{a_*}\lambda^{n+2}d\rho_*(\lambda) \\ &= \frac{1}{(n+1)(n+2)}\int_0^\pi \lambda^{n+2}\frac{d\theta}{\pi} \\ &= \frac{1}{(n+1)(n+2)}\mathrm{res}\Big(\lambda^{n+2}/p\Big), \end{aligned} \tag{6.18}$$

$$\begin{aligned} 2\int \lambda^n(\log|\lambda| - c_n)\rho_*(\lambda)d\lambda &= -\frac{2}{n+1}\int_{b_*}^{a_*}\lambda^{n+1}(\log|\lambda| - c_{n+1})d\rho_*(\lambda) \\ &= \frac{2}{n+1}\int_0^\pi \lambda^{n+1}(\log|\lambda| - c_{n+1})\frac{d\theta}{\pi} \\ &= \frac{2}{n+1}\mathrm{res}\Big(\lambda^{n+1}(\log|\lambda| - c_{n+1})/p\Big). \end{aligned} \tag{6.19}$$

In the above expressions, "res" means the integration $\oint dp/2\pi i$ along the contour $|p| = \exp(y/2)$. These agree with the genus 0 one point functions $\langle\sigma_n(Q)\rangle_0$, $\langle\sigma_n(P)\rangle_0$ expressed in the Landau-Ginzburg formulation.

7 Generalizations

Finally I want to discuss generalizations of our analysis for other manifolds than CP^1. In the following we consider the example of CP^2 in some detail. We recall that at genus$=0$ CP^2 model has a free energy of the form $F_0 = \frac{1}{2}t_P t_Q^2 + \frac{1}{2}t_P^2 t_R + f(t_Q, t_R)$ where the quantum part f has an instanton expansion $f = \sum \frac{N_d}{(3d-1)!}t_R^{3d-1}\exp(dt_Q)$. Let us introduce a basic set of two-point functions

$$u = \langle PP\rangle,\ v = \langle PQ\rangle,\ w = \langle PR\rangle. \tag{7.1}$$

In the small phase space they are give by $u = t_R$, $v = t_Q$, $w = t_P$. Then we have $\langle QQ\rangle = t_P + \sum \frac{N_d d^2}{(3d-1)!}t_R^{3d-1}e^{dt_Q} = w + \partial^2 f(u,v)/\partial v^2$ where

$$f(u,v) \equiv \sum \frac{N_d}{(3d-1)!}u^{3d-1}e^{dv}. \tag{7.2}$$

Similarly we find $\langle QR\rangle = \partial_u\partial_v f$, $\langle RR\rangle = \partial_u^2 f$ in the small phase space. These are the constitutive relations of the CP^2 model. As in the CP^1 case it is possible to show that these relations actually hold in the large phase space. Thus the formulas

$$\langle QQ\rangle = w + f_{vv}, \quad \langle QR\rangle = f_{uv}, \quad \langle RR\rangle = f_{uu} \tag{7.3}$$

replace (3.3) of CP^1 (we denote $\partial_u f$ as f_u etc.).

Let us derive an import identity obeyed by the function f. We start from the operator-product-expansion (OPE) relations

$$\begin{aligned}
QQ &= \langle QQP\rangle R + \langle QQQ\rangle Q + \langle QQR\rangle P \\
&= R + f_{vvv}Q + f_{uvv} \\
QR &= \langle QRP\rangle R + \langle QRQ\rangle Q + \langle QRR\rangle P \\
&= f_{uvv}Q + f_{uuv} \\
RR &= \langle RRP\rangle R + \langle RRQ\rangle Q + \langle RRR\rangle P \\
&= f_{uuv}Q + f_{uuu}.
\end{aligned} \tag{7.4}$$

By imposing associativity conditions of OPE, $(Q)^2R = Q(QR)$, $Q(RR) = R(QR)$ we find

$$f_{uvv}^2 = f_{uuu} + f_{vvv}f_{uuv}. \tag{7.5}$$

This leads to a recursion relation for the expansion coefficients $N_d, d = 1, 2\cdots$ [4].

Now I want to point out that the above equation of associativity (7.5) in fact guarantees the integrability of CP^2 model. We can, for instance, analyze the flow equations and find that the flows commute due to the associativity equation. We can then integrate the flow equations in t_P and obtain two-point functions

$$\begin{aligned}
\langle\sigma_1(P)P\rangle &= uw + \frac{1}{2}v^2, \\
\langle\sigma_1(P)Q\rangle &= vw - f_v + uf_{uv} + vf_{vv}, \\
\langle\sigma_1(P)R\rangle &= \frac{1}{2}w^2 + uf_{uu} + vf_{uv} - f_u \\
\langle\sigma_1(Q)P\rangle &= vw + f_v, \\
\langle\sigma_1(Q)Q\rangle &= vf_{uv} + \frac{1}{2}(w + f_{vv})^2, \\
\langle\sigma_1(Q)R\rangle &= -f_{uu} + \frac{2}{3}f_{uv}f_{vv} + vf_{uu} + wf_{uv} \\
\langle\sigma_1(R)P\rangle &= \frac{1}{2}w^2 + f_u, \\
\langle\sigma_1(R)Q\rangle &= wf_{uv} + \frac{1}{3}f_{uv}f_{vv} + f_{uu}, \\
\langle\sigma_1(R)R\rangle &= wf_{uu} + \frac{1}{2}(f_{uv})^2, \quad \text{etc..}
\end{aligned} \tag{7.6}$$

We can check, for instance, that the t_P-derivative of $\langle\sigma_1(P)P\rangle$

$$\partial_{t_P}\langle\sigma_1(P)P\rangle = (uw + \frac{1}{2}v^2)' = u'w + uw' + vv' \tag{7.7}$$

agrees with the expression derived from the topological recursion relation

$$\langle \sigma_1(P)PP\rangle = \langle PR\rangle\langle PPP\rangle + \langle PQ\rangle\langle QPP\rangle + \langle PP\rangle\langle RPP\rangle. \tag{7.8}$$

In the case of $\langle \sigma_1(R)Q\rangle$ we have

$$\partial_{t_P}\langle \sigma_1(R)Q\rangle = \left(wf_{uv} + \frac{1}{3}f_{uv}f_{vv} + f_{uv}\right)' \tag{7.9}$$

which agrees with

$$\begin{aligned}&\langle \sigma_1(R)PQ\rangle \\ &= \langle RP\rangle\langle RPQ\rangle + \langle RQ\rangle\langle QPQ\rangle + \langle RR\rangle\langle PPQ\rangle \\ &= wf'_{uv} + f_{uv}(w + f_{vv})' + f_{uu}v'\end{aligned} \tag{7.10}$$

due to an identity

$$-v'f_{uu} + f_{vv}f'_{uv} = (-f_{uu} + \frac{2}{3}f_{uv}f_{vv})'. \tag{7.11}$$

(7.11) follows from the associativity equation (7.5). It is again possible to introduce suitable Hamiltonian structures and the above two-point functions (7.6) are identified as the Hamiltonian densities as in the CP^1 model.

We can repeat a similar construction for a varieties of manifolds (with $c_1(K) > 0$) and convince ourselves that the topological σ-models coupled to 2-dimensional gravity are completely integrable systems. Novelty of these theories is the existence of a function (quantum part of the genus-0 free energy) whose associativity conditions (analogue of (7.5)) guarantee the complete integrability. Much more details will be published elsewhere [16].

Acknowlegments

I would like to thank K. Hori, C. S. Xiong and S. K. Yang for extensive discussions and collaborations. I also would like to P. Di Francesco for discussions on CP^2 model. Research of T.E. is supported in part by Grant-in-Aid for Scientific Research on Priority Area 231 "Infinite Analysis", Japan Ministry of Education.

References

[1] E. Witten, Commun. Math. Phys. 117 (1988) 353.

[2] E. Witten, Commun. Math. Phys. 118 (1988) 411.

[3] M. Bershadsky, S. Cecotti, H. Ooguri and C. Vafa, Nucl. Phys. B405 (1993) 279; Commun. Math. Phys. 165 (1994) 311.

[4] M. Kontsevich and Yu. Manin, Commun. Math. Phys. 164 (1994) 525.

[5] B. Dubrovin, "Geometry of 2D topological field theories", SISSA-89/94/FM, hep-th/9407018.

[6] C. Itzykson, "Counting rational curves on rational surfaces", Saclay T94/001.

[7] P. Di Francesco and C. Itzykson, "Quantum intersection rings", SPhT94/111, hep-th/9412175.

[8] M. Jinzenji and Y. Sun, "Calculation of Gromov-Witten invariants for CP^3, CP^4, and Gr(2,4)", UT-696, hep-th/941226.

[9] T. Eguchi and S.-K. Yang, Mod. Phys. Lett. A9 (1994) 2893.

[10] T. Eguchi, K. Hori and S.-K. Yang, preprint UT-700, March, 1995.

[11] S. Cecotti and C. Vafa, Commun. Math. Phys. 158 (1993) 569.

[12] P. Fendley and K. Intriligator, Nucl. Phys. B380 (1992) 265.

[13] T. Eguchi and S.-K. Yang, Mod. Phys. Lett. A5 (1990) 1693.

[14] E. Witten, Nucl. Phys. B340 (1990) 281.

[15] R. Dijkgraaf and E. Witten, Nucl. Phys. B342 (1990) 486.

[16] T. Eguchi, K. Hori and C.S. Xiong, to appear.

[17] K. Hori, Nucl. Phys. B439 (1995) 395.

[18] C. Itzykson and J.-B. Zuber, Int. J. Mod. Phys. A7 (1992) 5661.

[19] T. Eguchi, Y. Yamada and S.-K. Yang, Rev. Math. Phys. 7 (1995) 279.

[20] E. Witten, Surveys in Diff. Geom. 1 (1991) 243.

[21] R. Dijkgraaf, E. Verlinde and H. Verlinde, Nucl. Phys. B352 (1991) 59.

[22] T. Eguchi, H. Kanno, Y. Yamada and S.-K. Yang, Phys. Lett. B305 (1993) 235.

[23] A. Losev, Theor. Math. Phys. 95 (1993) 595.

[24] N.I. Muskhelishvili, *Singular Integral Equations*, P. Noordhoff N.V., Groningen-Holland, 1953.

CHIRALITY DECONFINEMENT BEYOND THE $C = 1$ BARRIER OF TWO DIMENSIONAL GRAVITY

JEAN-LOUP GERVAIS

LABORATOIRE DE PHYSIQUE THÉORIQUE
DE L'ÉCOLE NORMALE SUPÉRIEURE*
24 RUE LHOMOND, 75231 PARIS CEDEX 05, FRANCE

1 INTRODUCTION

In recent times we have witnessed important developments in our understanding of two dimensional gravity in the weak coupling regime, especially in what concerns the dressing by gravity of matter with central charge smaller than one. Nevertheless, the popular approaches seem to be unable to overcome the so called $c = 1$ barrier, powerful and elegant as they may be. The point of these two lectures is to show, following refs.[1, 2, 3, 4] how the operator approach to Liouville theory which I started to develop long ago with A. Neveu, remains applicable in the strong coupling regime. The principle is as follows. In the weak coupling regime, the Liouville exponentials are expressed in terms of chiral vertex operators, themselves constructed from a free field in two dimensions. This is the quantum version of the well known classical Bäcklund transformation. Locality and closure of their OPA (operator product algebra) uniquely determines this chiral decomposition. In the strong coupling regime, this latter construction looses meaning since the OPA of the Liouville exponentials involves operators and/or highest weight states with complex Virasoro weights. Nevertheless the general operator-family of chiral components may still be used, when truncation theorems[5, 1, 2] apply, that is with central charges $C_L = 7$, 13, 19. Indeed for these values there exist subfamilies of the above chiral operators which form closed operator algebras, only involving real Virasoro weights. In the present operator approach they are used to construct a new set of local fields which replace the

*Unité Propre du Centre National de la Recherche Scientifique, associée à l'École Normale Supérieure et à l'Université de Paris-Sud.

Low Dimensional Applications of Quantum Field Theory
Edited by L. Baulieu *et al.*, Plenum Press, New York, 1997

Liouville exponentials. Since both sets are constructed out of the same free Bäcklund fields, they may be considered as related by a new type of quantum Bäcklund transformation, that connect the weak and strong coupling regimes of two-dimensional gravity. In the present lecture notes, I summarise the basic features (deconfinement of chirality, new expression for the string susceptibility) of these new set of local fields that replace the Liouville exponentials, in the strong coupling regime. Moreover, the basic properties of the new topological models will be recalled, where the gravity part is in the strong coupling regime. The message will be that the derivation of the new features is very close to the previous weak-coupling one, once the new set of local physical fields is established.

2 Basic points about Liouville theory

2.1 The classical case

In order to set the stage, we recall the classical structure for the special case of the Liouville theory. We shall work with Euclidean coordinates σ, τ. As a preparation for the quantum case, the classical action is defined as

$$S = \frac{1}{8\pi}\int d\sigma d\tau \left((\frac{\partial\Phi}{\partial\sigma})^2 + (\frac{\partial\Phi}{\partial\tau})^2 + e^{2\sqrt{\gamma}\Phi} \right). \tag{2.1}$$

We use world sheet variables σ and τ, which are local coordinates such that the metric tensor takes the form $h_{ab} = \delta_{a,b}\, e^{2\Phi\sqrt{\gamma}}$. The complex structure is assumed to be such that the curves with constant σ and τ are everywhere tangent to the local imaginary and real axis respectively. The action 2.1 corresponds to a conformal theory such that $\exp(2\sqrt{\gamma}\Phi) d\sigma d\tau$ is invariant. It is convenient to let

$$x_\pm = \sigma \mp i\tau, \quad \partial_\pm = \frac{1}{2}(\partial_\sigma \pm i\partial_\tau). \tag{2.2}$$

By minimizing the above action, one derives the Liouville equations

$$\frac{\partial^2\Phi}{\partial\sigma^2} + \frac{\partial^2\Phi}{\partial\tau^2} = \sqrt{\gamma}\, e^{2\sqrt{\gamma}\Phi}. \tag{2.3}$$

The chiral modes may be separated very simply using the fact that the field $\Phi(\sigma, \tau)$ satisfies the above equation if and only if[1]

$$e^{-\sqrt{\gamma}\Phi} = \frac{i\sqrt{\gamma}}{2} \sum_{j=1,2} f_j(x_+) g_j(x_-); \quad x_\pm = \sigma \mp i\tau \tag{2.4}$$

where f_j (resp.(g_j)), which are functions of a single variable, are solutions of the same Schrödinger equation (primes mean derivatives)

$$-f_j'' + T(x_+) f_j = 0, \quad (\text{ resp. } - g_j'' + \overline{T}(x_-) g_j\,). \tag{2.5}$$

[1]The factor i means that these solutions should be considered in Minkowsky space-time

The solutions are normalized such that their Wronskians $f_1' f_2 - f_1 f_2'$ and $g_1' g_2 - g_1 g_2'$ are equal to one. The proof goes as follows.

(1) First check that Eq.2.4 is indeed solution. Taking the Laplacian of the logarithm of the right-hand side gives

$$\frac{\partial^2 \Phi}{\partial \sigma^2} + \frac{\partial^2 \Phi}{\partial \tau^2} \equiv 4\partial_+ \partial_- \Phi = -4\sqrt{\gamma}/\Big(\sum_{i=1,2} f_i g_i\Big)^2$$

where $\partial_\pm = (\partial/\partial\sigma \pm i\partial/\partial\tau)/2$. The numerator has been simplified by means of the Wronskian condition. This is equivalent to Eq.2.3.

(2) Conversely check that any solution of Eq.2.3 may be put under the form Eq.2.4. If Eq.2.3 holds one deduces

$$\partial_\mp T^{(\pm)} = 0; \quad \text{with}\, T^{(\pm)} := e^{\sqrt{\gamma}\Phi} \partial_\pm^2 e^{-\sqrt{\gamma}\Phi} \tag{2.6}$$

$T^{(\pm)}$ are thus functions of a single variable. Next the equation involving $T^{(+)}$ may be rewritten as

$$(-\partial_+^2 + T^{(+)}) e^{-\Phi} = 0 \tag{2.7}$$

with solution

$$e^{-\sqrt{\gamma}\Phi} = \frac{i\sqrt{\gamma}}{2} \sum_{j=1,2} f_j(x_+) g_j(x_-); \quad \text{with} - f_j'' + T^{(+)} f_j = 0$$

where the g_j are arbitrary functions of x_-. Using the equation 2.6 that involves $T^{(-)}$, one finally derives the Schrödinger equation $-g_j'' + T^{(-)} g_j = 0$. Thus the theorem holds with $T = T^{(+)}$ and $\overline{T} = T^{(-)}$. One may deduce from Eq.2.6 that the potentials of the two Schrödinger equations coincide with the two chiral components of the stress-energy tensor. Thus these equations are the classical equivalent of the Ward identities that ensure the decoupling of Virasoro null vectors. Next a Bäcklund transformation to free fields is easily obtained as follows. It is easy to verify explicitly that, since f_1 is a solution of Eq.2.5, as second solution is given by $f_2(x) = f_1(x) \int^x dy f_1^{-2}(y)$. So we may introduce a field $q(x)$ by letting

$$f_1(x) = e^{q(x)\sqrt{\gamma}}, \quad f_2(x) = f_1(x) \int^x dx_1 e^{-2q(x_1)\sqrt{\gamma}}, \tag{2.8}$$

It follows from the canonical Poisson brackets of Liouville theory that

$$\Big\{\vartheta(\sigma_1 - i\tau), \vartheta(\sigma_2 - i\tau)\Big\}_{\mathrm{PB}} = 2\pi\delta'(\sigma_1 - \sigma_2). \tag{2.9}$$

where $\vartheta = q'$. Doing the same construction with the functions g, one introduces another field $\bar{\vartheta}$ with opposite chirality. Altogether ϑ and $\bar{\vartheta}$ are the two chiral components of the two dimensional free field derived by the quantum Bäcklund transformation. Clearly the differential equation Eq.2.5 gives $T = p^2 + p'/\sqrt{\gamma}$, or equivalently $T = (\vartheta')^2 + \vartheta''/\sqrt{\gamma}$. The last expression coincides with U_1 Sugawara stress-energy tensor with a linear term. From the viewpoint of differential equations, these relations are simply the well-known Riccati equations associated with the Schrödinger equation Eq.2.5. An easy computation shows that T satisfies the Poisson bracket Virasoro algebra with $C_{\text{class}} = 3/\gamma$. We shall consider the typical situation of a cylinder with $0 \leq \sigma \leq \pi$, and $-\infty \leq \tau \leq \infty$. After appropriate coordinate change, this may describe one handle of a surface with arbitrary genus. Then T is periodic in σ with period 2π, and the standard Virasoro generators are given by $L_n = \int_0^{2\pi} d\sigma e^{-in\sigma} T(\sigma - i\tau)$. Under the corresponding Poisson bracket action, it is easy to see that f_1, and g_1 are primary fields with weights $-1/2, 0$ and $0, -1/2$ respectively, so that f_1^{-2} has weight $1, 0$. Thus it is the classical version of the screening operator in accordance with Eq.2.5. Classically, the Virasoro weights are simply additive, so that the weight of $\exp(-J\sqrt{\gamma}\Phi) = (\exp(-\Phi\sqrt{\gamma}/2))^{2J}$ is $-J, -J$. Therefore, we get $1, 1$ for $J = -1$ in agreement with the fact that the potential term of the action Eq.2.1 Is conformally invariant.

2.2 Quantization (outline)

The parameter γ plays the role of $\hbar$. In this operator method, one starts from a free quantum field with chiral components ϑ (and ϑ) satisfying the quantum version of Eq.2.9, that is, one replaces i times Poisson brackets by commutators. The above free-field expressions are extended to the quantum case by using normal ordering and allowing for finite renormalizations of the classical constants. So we write $T \propto: \vartheta'^2 : +Q\vartheta''$, which gives a central charge $C_L = 1 + 3Q^2$. C_L is the new free paremeter. The chiral vertex operatos are constructed from the quantum version of (powers of) Eq.2.8 with "powers" of f_1 defined by normal ordered exponentials of the type $: \exp(2J\alpha\vartheta) :$, with α specified as follows. One imposes that the choice $J = -1$ again gives an operator of weight $1, 0$, which is the screening operator. There are two choices, the two well known screening charges:

$$\alpha_\pm = \sqrt{\frac{C_L - 1}{12}} \pm \sqrt{\frac{C_L - 25}{12}} \quad ; \quad \alpha_+\alpha_- = 2. \tag{2.10}$$

With them, the fields with $J = 1/2$ satisfy a quantum version of the Schrödinger equation 2.5 which is equivalent to the Virasoro null-field decoupling equation at level 2. Using its monodromy, and the associativity of the operator products —or, more precisely the polynomial equations within the Moore Seiberg formalism[6]—, one deduces[7, 8, 2] the OPA of the chiral vertex operators. A quantum group structure emerges of the type $U_q(sl(2)) \odot U_{\widehat{q}}(sl(2))$, with

$$q = e^{ih}, \quad \widehat{q} = e^{i\widehat{h}}, \qquad h = \pi\frac{\alpha_-^2}{2}, \quad \widehat{h} = \pi\frac{\alpha_+^2}{2}. \tag{2.11}$$

The two quantum group parameters are related by

$$h\widehat{h} = \pi^2, \quad h + \widehat{h} = \frac{C_L - 13}{6}. \tag{2.12}$$

The first relation shows that they are, in a sense, dual pairs. The above symbol $\odot$ has a special meaning which will be clarified as we go on. The Hilbert space of states is of course a direct sum of Virasora Verma modules. They are characterized by the eigenvalue of the free field rescaled zero-mode momentum ϖ defined by writing

$$\vartheta(x) = q_0 + ip_0 x + i \sum_{n \neq 0} e^{-inx} \frac{p_n}{n}, \quad \varpi = \frac{2i}{\alpha_-} p_0. \tag{2.13}$$

A Verma module is charaterized by the highest-weight eigenvalue

$$\Delta(\varpi) = \frac{h}{4\pi}(\varpi_0^2 - \varpi^2), \quad \varpi_0 = 1 + \frac{\pi}{h} \tag{2.14}$$

of L_0. For $\varpi = \pm\varpi_0$, Δ vanishes. This describes the two $Sl(2, C)$ invariant states. In general, the spectrum of ϖ eigenvalues is of the form

$$\varpi_{J,\widehat{J}} = \varpi_0 + 2J^e. \tag{2.15}$$

where J^e which is called the effective spin specifies the representations[2] of $U_q(sl(2)) \odot U_{\widehat{q}}(sl(2))$. If J^e is generic, the corresponding Verma module is simply the bosonic Fock space of the free field ϑ. If J^e may be written as $J^e = J + \frac{\pi}{h}\widehat{J}$, with $2J$ $2\widehat{J}$ non negative integers, are this Fock space involves null states and the irreducible Verma modules must be projected out. We shall assume that this is done if needed. In this latter case J and $\widehat{J}$ are ordinary quantum group spins.

2.3 Chiral operator algebra

For continuous spins, J and $\widehat{J}$ loose meaning in general, and only J^e makes sense. However, in our strong coupling discussion—where $\frac{h}{\pi}$ is not rational—we shall only need to deal with the case where J^e takes the form $J + \widehat{J}\frac{\pi}{h}$, with J $\widehat{J}$ rational numbers (not necessarily positive integers). Then J and $\widehat{J}$ may be defined uniquely from J^e, so that we use them in our discussion. We shall try to display the structure of the result as imply as possible by drawing pictures. There will be a double line for each Verma module, with J and $\widehat{J}$ represented by a solid and a dashed line, respectively, Concerning the three-leg conformal blocks, the matrix elements of the chiral vertex operators (which are the chiral components of the Liouville exponentials) denoted[3] by $\widetilde{V}$ are represented by three leg vertices. The operator product algebra (OPA) is thus pictured by diagrams with basic elements

(2.16)

[2]We refer to the original articles[7, 8, 2] to explain this point which is not central in these lectures. For recent developments in this direction see refs.[9].

[3]The tilde is to distinguish from amother normalisation (see ref.[7]).

which correspond to

$$< J_2^e\{\beta_2\}|\widetilde{V}^{J_1^e,\{\beta_1\}}(z)J_3^e,\{\beta_3\}>, \quad \sum_{\{\beta\}} \frac{|J^e,\{\beta\}><J^e,\{\beta'\}|}{<J^e,\{\beta'\}|J^e,\{\beta\}>}. \tag{2.17}$$

From now on we use the world sheet variable $z = \exp(ix_+)$. The symbol $\{\beta\}$ stands for a multi-index that characterizes the descendent of the Verma module—or the primary field. In order to avoid clumsy drawings, we omit it from the diagrams. It is understood that it is summed over for each intermediate (double) line, as shown on the last equation above. On the contrary, *for the effective spins, the summation over intermediate line should be performed only when indicated.* Operatorially, this reflects the fact that the $V^{(J^e)}$ operators possess an additional index which specifies the shift of J^e between bra and ket. More precisely, for continuous J^e, one defines operators $\widetilde{V}_{m^e}^{(J^e),\{\beta\}}$ with the condition that ($\mathcal{Z}$ represents the set of non negative integers)

$$J^e + m^e = \nu + \frac{\pi}{h}\widehat{\nu}, \quad \nu,\, \widehat{\nu} \in \mathcal{Z}, \tag{2.18}$$

such that

$$< J_2^e\{\beta_2\}|\widetilde{V}_{m_1^e}^{J_1^e,\{\beta_1\}}(z)|J_3^e,\{\beta_3\}>= 0, \text{ unless } m_1^e = J_3^e - J_2^e \tag{2.19}$$

For instance, a four leg conformal block is drawn as

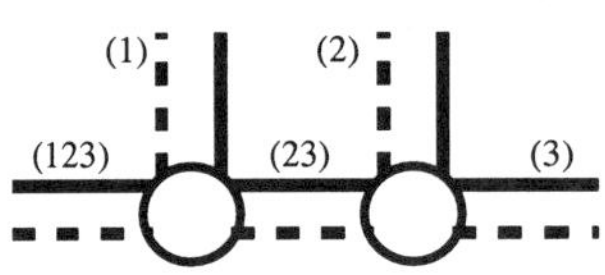

$$\Leftrightarrow < J_{123}^e,\{\beta_{123}\}|\widetilde{V}_{J_{23}^e-J_{123}^e}^{J_1^e,\{\beta_1\}}(z_1)\widetilde{V}_{J_3^e-J_{23}^e}^{J_2^e,\{\beta_2\}}(z_2)|J_3^e,\{\beta_3\}> \tag{2.20}$$

These chiral vertex operator are closed by fusion and braiding. Let us illustrate this in the particular case where all the $\widehat{J}$'s are zero. Using therefore simple lines one has the following representation. For fusion,

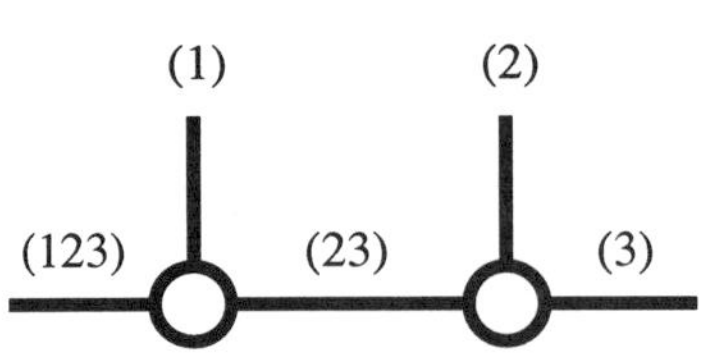

$$= \sum_{J_{12}} F_{J_{23},J_{12}}\begin{bmatrix} J_1 & J_2 \\ J_{123} & J_3 \end{bmatrix} \times$$

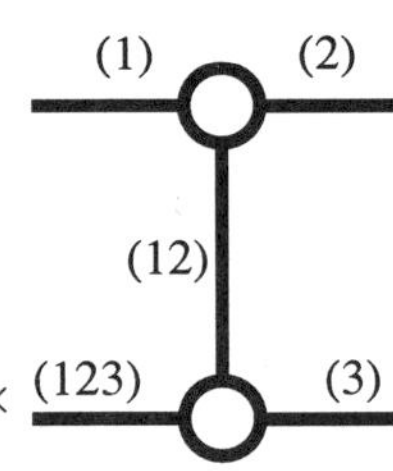

(2.21)

where the fusion matrix is given by

$$F_{J_{23},J_{12}}\begin{bmatrix} J_1 & J_2 \\ J_{123} & J_3 \end{bmatrix} = \left\{ \begin{matrix} J_1 & J_2 \\ J_3 & J_{123} \end{matrix} \middle| \begin{matrix} J_{12} \\ J_{23} \end{matrix} \right\}_q . \tag{2.22}$$

On the right there appears the q 6j symbol. Similarly, the braiding is represented by

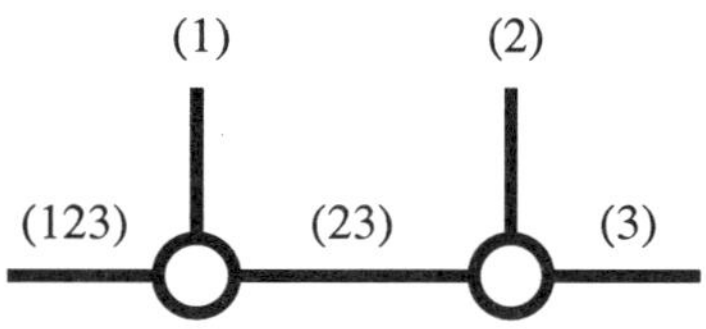

$$= \sum_{J_{13}} B^{\pm}_{J_{23},J_{13}}\begin{bmatrix} J_1 & J_2 \\ J_{123} & J_3 \end{bmatrix}$$

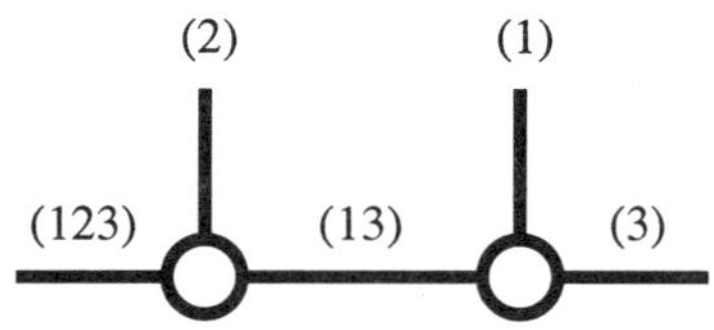

(2.23)

where the braiding matrix is related to the above by

$$B^{\pm}_{J_{23},J_{12}}\begin{bmatrix} J_1 & J_3 \\ J_{123} & J_2 \end{bmatrix} = e^{\pm i\pi(\Delta_{J_{123}} + \Delta_{J_2} - \Delta_{J_{23}} - \Delta_{J_{12}})} F_{J_{23},J_{12}}\begin{bmatrix} J_1 & J_2 \\ J_{123} & J_3 \end{bmatrix}. \tag{2.24}$$

In the general case the structure is similar, with the corresponding extended 6j symbols associated with our $U_q(sl(2)) \odot U_{\widehat{q}}(sl(2))$, which are given by

$$F_{J^e_{23},J^e_{12}}\begin{bmatrix} J^e_1 & J^e_2 \\ J^e_{123} & J^e_3 \end{bmatrix} = \left\{\left\{ \begin{matrix} J^e_1 & J^e_2 \\ J^e_3 & J^e_{123} \end{matrix} \Big| \begin{matrix} J^e_{12} \\ J^e_{23} \end{matrix} \right\}\right\} \left\{\left\{ \begin{matrix} \widehat{J}^e_1 & \widehat{J}^e_2 \\ \widehat{J}^e_3 & \widehat{J}^e_{123} \end{matrix} \widehat{\Big|} \begin{matrix} \widehat{J}^e_{12} \\ \widehat{J}^e_{23} \end{matrix} \right\}\right\}_{\widehat{q}}. \tag{2.25}$$

The double brace means 6j symbols with shifted entries (see ref.[2] for details). The symbol $\widehat{J}^e$ is defined by $\widehat{J}^e = \widehat{J} + \frac{h}{\pi} J \equiv \frac{h}{\pi} J^e$. The normalization of the chiral vertex operators $\widetilde{V}$ is fixed by the condition that their fusion and braiding matrices be exactly equal to q 6j symbols—not simply proportional as they would be with other choices. Then, their highest weight matrix element is given by

$$< J^e_2 | \widetilde{V}^{(J^e_1)}(z) | J^e_3 >= g^{J^e_2}_{J^e_1,J^e_3} \tag{2.26}$$

where the g's, which are called coupling constant, are not simply determined by the quantum group symmetry—they are not trigonometrical functions—but from the monodromy properties of the quantum version of Eq.2.5, the null vector decoupling equation.

2.4 The weak coupling regime

THE LIOUVILLE EXPONENTIALS

We have to put the two chiralities together. The x_- components—which we did not consider so far—are of course similar to the above. Their quantum numbers will be distinguished by a bar, so we have $\overline{\varpi}$, $\overline{J}$, $\widehat{\overline{J}}$ and so on. Graphically, the Verma modules are represented by gray lines in contrast with the black ones of the x_+ components. The Liouville exponentials may be determined as follows: one writes general chiral decompositions and imposes that they be local, with σ and τ being space and time coordinates respectively. Remarkably, this determines them completely, and the result may be consistently restricted to Verma modules such that

$$J = \overline{J}, \quad \widehat{J} = \widehat{\overline{J}} \tag{2.27}$$

To represent this restriction, we introduce springs that link our lines as follows

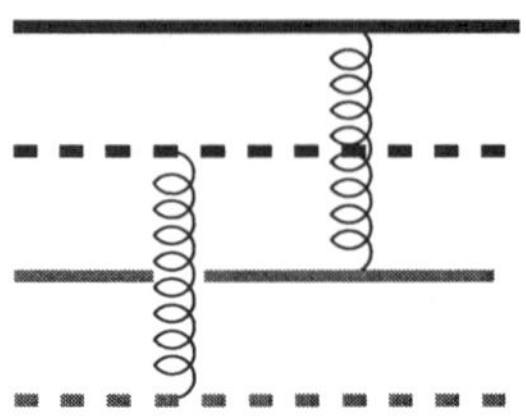

$$J \qquad \widehat{J} \qquad \overline{J} = J \qquad \widehat{\overline{J}} = \widehat{J} \tag{2.28}$$

Then the Liouville three point function is represented by

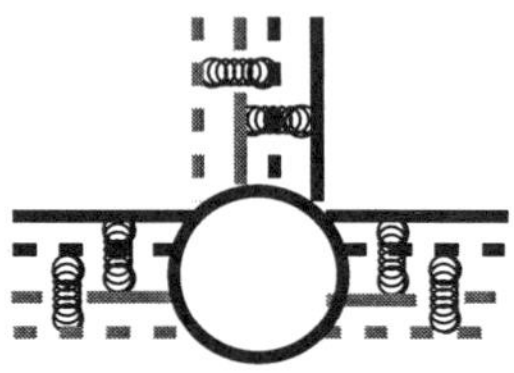

such that

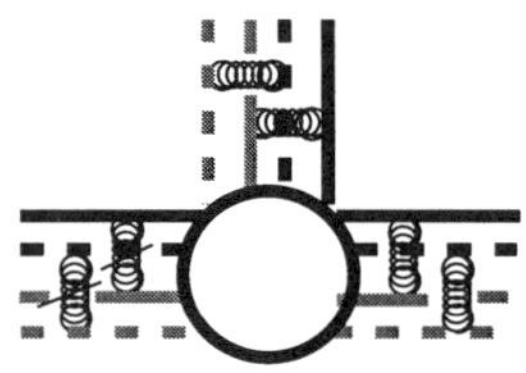

$= 0$ (2.29)

The right-hand side drawing means that the Liouville exponentials applied to a state satisfying condition Eq.2.27 only give states satisfying the same condition. Operatorially, the Liouville exponentials are given by

$$e^{-J^e \alpha_- \Phi(z,\bar{z})} = \sum_{m^e} \widetilde{V}_{m^e}^{(J^e)}(z) \widehat{\overline{V}}_{\overline{m}^e}^{(\overline{J}^e)}(\bar{z}). \tag{2.30}$$

This chiral decomposition only involves $\widetilde{V}$ and $\widehat{\overline{V}}$ operators with the same J^e and m^e, in accordance with the diagrams just written (recall Eq.2.19). It is the quantum version of the classical expression Eq.2.4 and of its higher spin generalizations. The properties of the quantum Liouville exponentials under fusion and braiding are deduced from the chiral ones summarized above, and are consequences of the orthogonality of the q 6j symbols. For the standard ones it reads

$$\sum_{J_{23}} \left\{ \begin{matrix} J_1 & J_2 \\ J_3 & J_{123} \end{matrix} \middle| \begin{matrix} J_{12} \\ J_{23} \end{matrix} \right\} \left\{ \begin{matrix} J_1 & J_2 \\ J_3 & J_{123} \end{matrix} \middle| \begin{matrix} K_{12} \\ J_{23} \end{matrix} \right\} = \delta_{J_{12}-K_{12}}. \tag{2.31}$$

As a result, the braiding matrix of the Liouville exponentials are equal to one, so that they are mutally local, i.e.

$\sum_{J_{23}}$

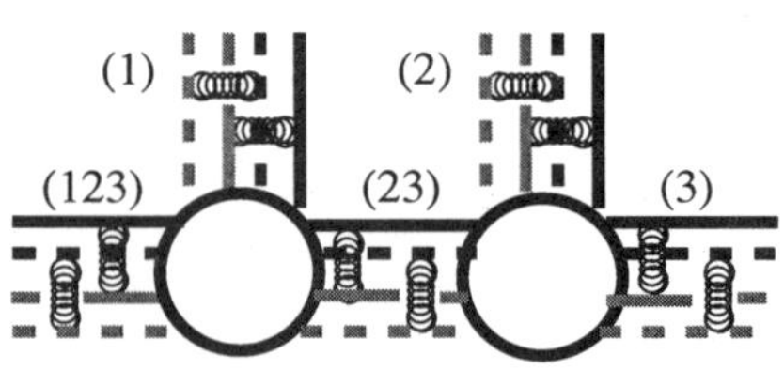

$$= \sum_{J_{13}}$$
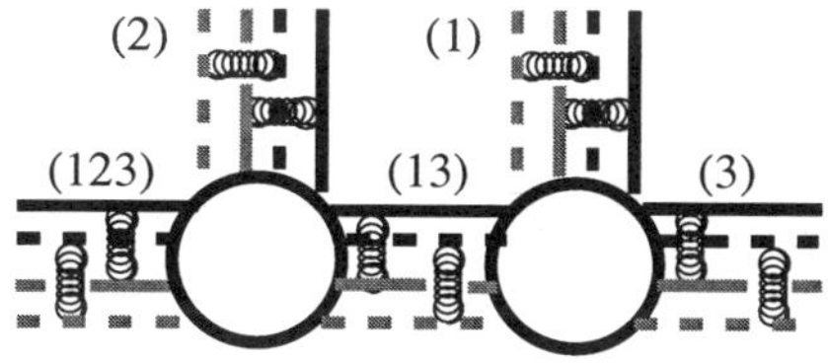

(2.32)

They are moreover closed by fusion, following the diagram

$$\sum_{J_{23}}$$
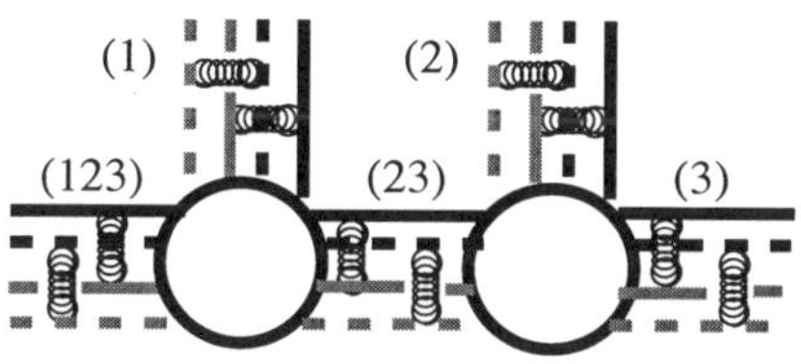

$$= \sum_{J_{12}}$$
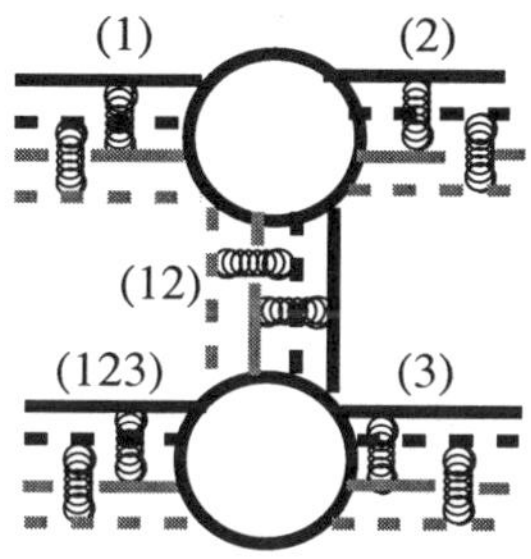

As usual, the upper vertex on the right hand side is a book keeping device for descendents. To leading order, at short distance, the coefficient is determined by its expectation value between highest weight states. According to Eqs.2.262.30 it follows that

$$< J_2^e | e^{-J_1^e \alpha_- \Phi(z,\bar z)} | J_3^e >= |g_{J_1^e, J_3^e}^{J_2^e}|^2. \tag{2.33}$$

and the contribution of $\exp(-J_{12}\alpha_-\Phi)$—and of its descendent—in the operator product of $\exp(-J_1^e\alpha_-\Phi)$ with $\exp(-J_2^e\alpha_-\Phi)$, is proportional to $|g_{J_1^e J_2^e}^{J_{12}^e}|^2$. Thus the coupling constants determine the three point functions of the OPA, as expected. Concerning the braiding diagram, it explicitly means that

$$e^{-J_1^e\alpha_-\Phi(z_1,\bar z_1)} e^{-J_2^e\alpha_-\Phi(z_2,\bar z_2)} = e^{-J_2^e\alpha_-\Phi(z_2,\bar z_2)} e^{-J_1^e\alpha_-\Phi(z_1,\bar z_1)}, \tag{2.34}$$

where we are are equal τ, so that $z_1 = \exp(\tau + i\sigma_1)$, $z_2 = \exp(\tau + i\sigma_2)$.

In general, applying a similar discussions to the higher point functions, one generates a set of diagrams where condition Eq.2.27 is obeyed on every line. Since this has an obvious analogy with quark diagrams, we may say that chirality is confined.

THE COUPLING TO MATTER

One represents[11] matter by another copy of the theory summarized above, now with central charge $c = 26 - C_L$. One constructs local fields in analogy with Eq.2.30:

$$e^{-(J\alpha'_- + \widehat{J}\alpha'_+)\Phi'(z,\bar z)} = \sum_{m,\widehat{m}} \widetilde{V}'^{(J,\widehat{J})}_{m\widehat{m}}(z)\, \widehat{\overline{V}}'^{(J,\widehat{J})}_{m\widehat{m}}(\bar z). \tag{2.35}$$

Symbols pertaining to matter are distinguished by a prime. In particular $\Phi'(z,\bar z)$ is the matter field (it commutes with $\Phi(z,\bar z)$), and $\alpha'_\pm$ are the matter screening charges

$$\alpha'_\pm = \mp i\alpha_\mp. \tag{2.36}$$

The correct dressing of these operators by gravity is achieved by considering the vertex operators

$$\mathcal{W}^{J,\widehat J} \equiv e^{-((-\widehat J-1)\alpha_- + J\alpha_+)\Phi - (J\alpha'_- + \widehat J\alpha'_+)\Phi'} \tag{2.37}$$

which is an operator of weights $\Delta = \bar\Delta = 1$. In particular for $J=\widehat J = 0$, we get the cosmological term $\exp(\alpha_-\Phi)$. The three-point function was computed in refs[11, 2]. The corresponding product of coupling constants gives the correct leg factors after drastic simplifications. Of course, the dressing is such that the Verma modules involved satisfy the appropriate BRST condition. Graphically, they may be represented as

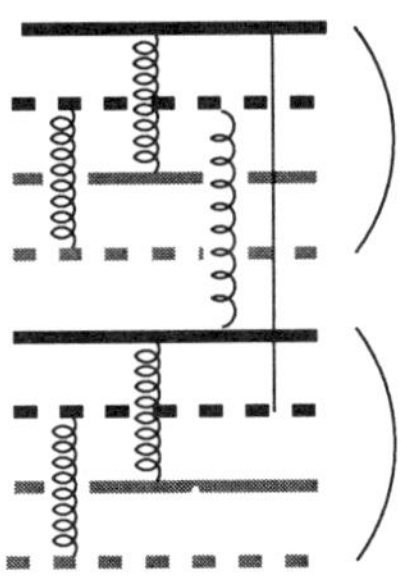

(2.38)

In this drawing the straight line between the first and sixth lines indicates that the J's satisfy the condition $J + \widehat J' + 1 = 0$, as required by the dressing condition.

2.5 The strong coupling regime

THE BARRIER

Eqs.2.10 for the screening charges are real for $C_L > 25$, pure imaginary for $C_L < 1$ and complex otherwise. The weights of the primary fields with spins $J\ \widehat J$ are given by Kac's formula

$$\Delta(J,\widehat J) = \frac{C_L-1}{24} - \frac{1}{24}\left[(J+\widehat J+1)\sqrt{C_L-1} - (J-\widehat J)\sqrt{C_L-25}\right]^2, \tag{2.39}$$

which, for real $J\ \widehat J$, is real for $C_L > 25$, and $C_L < 1$, and complex in general otherwise. In the zero coupling limit $\gamma \to 0$, $C_L \sim 3/\gamma$ blows up. Thus the region $C_L > 25$ is called the weak coupling regime of gravity, which is connected with the classical limit $\gamma \to 0$. There the above formula is real and the construction of the Liouville exponentials just summarized applies.

Can we go beyond $C_L = 25$?. At this moment it seems that one cannot make sense of operators with complex Δ. For $1 < C_L < 25$, Eq.2.39 still gives real weights in two particular cases. First Δ is real and negative if $J = \widehat J$. In the type of picture we are using (see 2.28), this means that a spring is introduced between the two lines, but here they have the same chirality. Second, Δ is real and positive if $J + \widehat{J} + 1 = 0$. This kind of condition (a sort of repulsive "interaction") already appeared on the diagram 2.38. It is represented by a solid line. Thus the subspaces with real Δ's are represented graphically as follows.

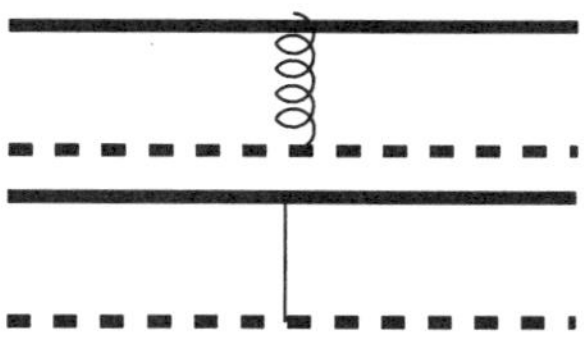

$$\leftrightarrow \quad \widehat{J} = J, \quad \Delta(J, J) \text{ real} < 0,$$

$$\leftrightarrow \quad \widehat{J} = -J - 1, \quad \Delta(J, -J-1) \text{ real} > 0. \quad (2.40)$$

Now we return to the Liouville exponential. Does it respect this reality conditions? The answer is no, since, in Eq.2.30, one sums over m^e and $\overline{m}^e$ independently, so that the shift of J^e and $\overline{J}^e$ are not corrolated. Graphically, this is illustrated as follows

$$\neq 0, \qquad (2.41)$$

and by a similar relation with the other condition. Operiatorially, this means that we may try to use Liouville exponentials with real weights (depicted by the additional springs on the vertical lines) but applied to states with real weights they give states with complex weights so that they loose meaning in the strong coupling regime.

The new set of chiral fields

The solution to this problem is as follows. The chiral operator algebra of the V fields has a natural extension to $C_L < 25$, albeit with complex weights. One constructs different types of local operators, where the conditions $\overline{J} = J$ and $\widehat{\overline{J}} = \widehat{J}$ are not imposed. They are built from the same chiral vertex operators as the Liouville exponentials, but one takes different combinations so that they respect one of the two reality conditions Eq.2.40. Following our graphical rules, their three point function are of the type

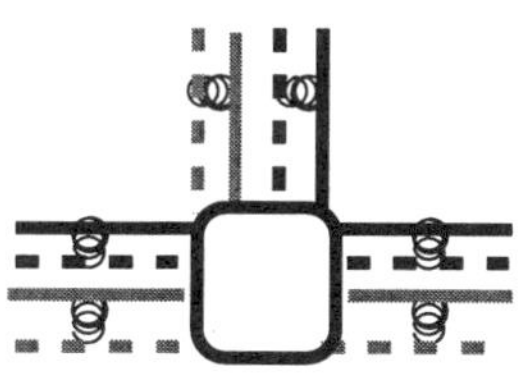

or

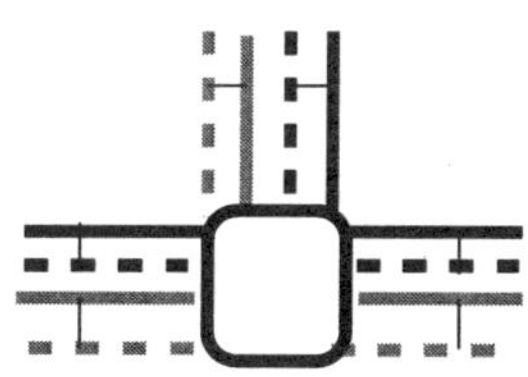

The reality conditions on the Verma modules do not link the two chiralities any longer, so we may say that chirality is deconfined. The possibility of constructing these operators consistently is based on the truncation theorems that hold for special values of C_L, i.e. $C_L = 7,\ 13,\ 19$. One first constructs chiral operators which are closed by fusion and braiding. They are particular linear combinations of the chiral vertex operators drawn on figure 2.16 These so called called physical operators $\chi_\pm^{(J)}$ have three point functions of the type (for the x_+ components)

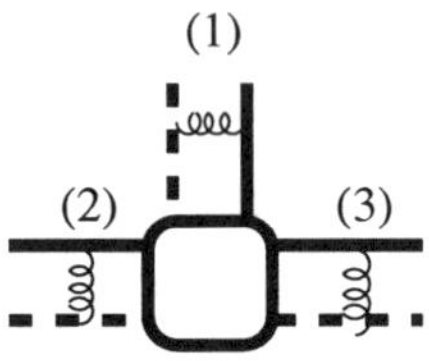

and

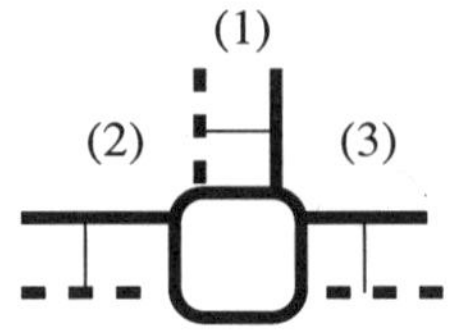

which represent

$$< J_2^{e\pm}\{\beta_2\}|\chi_\pm^{J_1^{e\pm},\{\beta_1\}}(z)|J_3^{e\pm},\{\beta_3\} >, \tag{2.42}$$

where

$$J^{e\pm} = J(1 \mp \frac{\pi}{h}) + \frac{\pi}{2h}(1 \pm 1) \tag{2.43}$$

A similar representation with gray lines holds of course for the x_- components. We do not discuss it explicitly.

Now is a good time to recall the truncation theorems which hold for

$$C = 1 + 6(s+2), \quad s = 0, \pm 1. \tag{2.44}$$

First define the physical Hilbert space

$$\mathcal{H}_{\text{phys}}^{\pm} \equiv \bigoplus_{r=0}^{1\mp s} \bigoplus_{n=-\infty}^{\infty} \mathcal{H}_{r/2(2\mp s)+n/2}^{\pm} \tag{2.45}$$

where $\mathcal{H}_{J^e}^{\pm}$ denotes the Verma modules with highest weights $|J^{e\pm} >$. The tree leg conformal bloch of the χ operators take the form

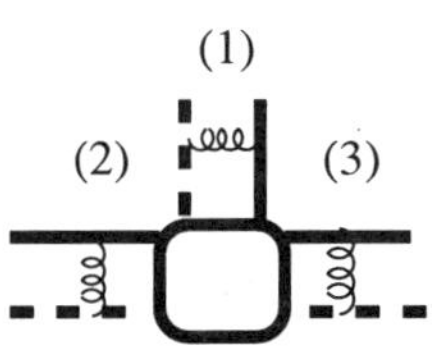

$$= (-1)^{(2-s)(2J_3+\nu(\nu+1)/2)}$$

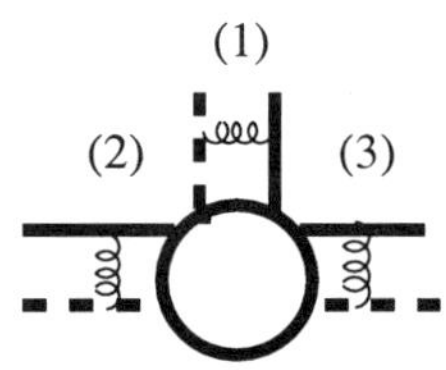

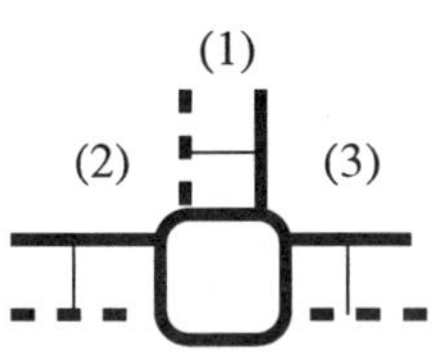

$$= (-1)^{(2+s)(2J_3+\nu(\nu+1)/2)}$$

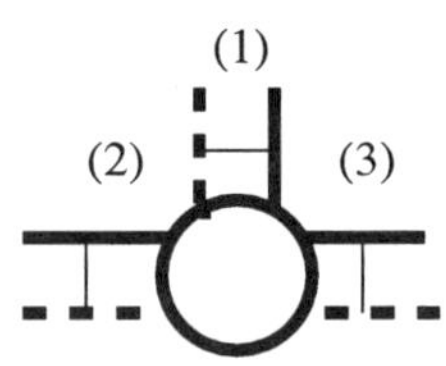

if $\nu = J_1 + J_3 - J_2 \in \mathcal{Z}_+$, and $2J_i \in \mathcal{Z}/(2\mp s)$; and they are taken to vanish otherwise. By construction the physical operators are restricted to $\mathcal{H}_{\text{phys}}^{\pm}$. Denote their ensemble by $\mathcal{A}_{\text{phys}}^{\pm}$. The basic properties of the special values Eq.2.44 is the

TRUNCATION THEOREM:

For $C_L = 1 + 6(s+2)$, $s = 0, \pm 1$, the above set $\mathcal{A}^+_{\mathrm{phys}}$ (resp. $\mathcal{A}^-_{\mathrm{phys}}$) is closed by braiding and fusion.

Let us display, for instance the braiding properties. One has

$\sum_{J_{23}}$ 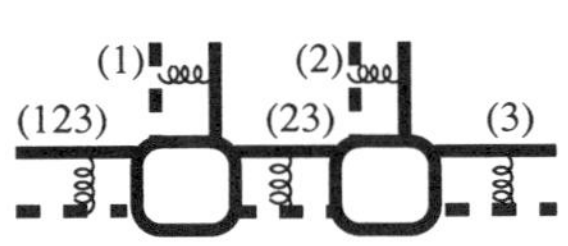

$= \sum_{J_{13}} (-1)^{\pm 2i\pi(2-s)J_1J_2}$ 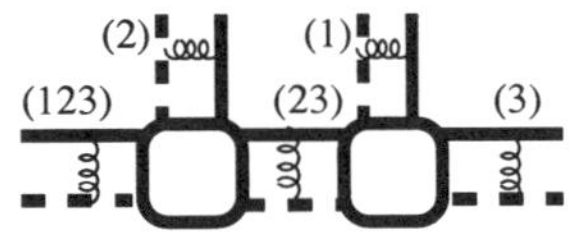

$\sum_{J_{23}}$ 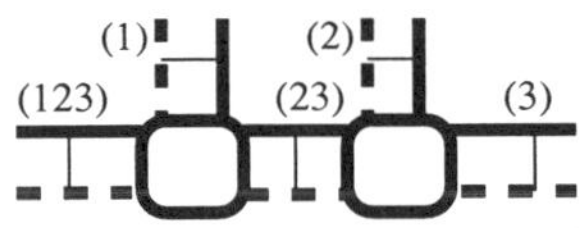

$= \sum_{J_{13}} (-1)^{\pm 2i\pi(2+s)J_1J_2}$ 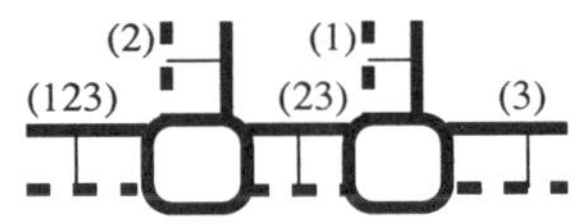

Note an important difference with the braiding properties of the previous chiral components themselves (figure 2.23). On the left hand side of the last two drawings one has to sum over the intermediate J_{23}, contrary to 2.23. Thus the χ operators, contrary to the $\widetilde{V}$ do not have a quantum number specifying the shift of J^e. As a matter of fact, the operatorial relation between χ and $\widetilde{V}$ operators consistent with the three and four leg diagrams just displayed is as follows.

$$\chi_\pm^{(J_1)} \mathcal{P}_{\mathcal{H}^\pm{}_{J_3}} = \sum_{\nu \equiv J_1 + m \in \mathcal{Z}_+} (-1)^{(2\mp s)(2J_3 + \nu(\nu+1)/2)} \widetilde{V}_{m^{e\pm}}^{(J_1^{e\pm})} \mathcal{P}_{\mathcal{H}^\pm_{J_3}}, \tag{2.46}$$

where $\mathcal{P}_{\mathcal{H}^\pm_J}$ is the projector on $\mathcal{H}^\pm_J$, and $m^{e\pm} = m(1 \pm \frac{\pi}{h})$. Note that, on the contrary, the operators $V_{m^e}^{(J^{e\pm})}$ themselves cannot be consistently restricted to either $\mathcal{H}^+_{\mathrm{phys}}$ or $\mathcal{H}^+_{\mathrm{phys}}$. Operatorially, the last two figures correspond to

$$\chi_\pm^{(J_1)} \chi_\pm^{(J_2)} = e^{2i\pi\epsilon(2\mp s)J_1J_2} \chi_\pm^{(J_2)} \chi_\pm^{(J_1)}, \tag{2.47}$$

where $\epsilon = \pm 1$ is fixed by the ordering of the operator on the left-hand side in the usual way. Next we construct local fields out of the chi fields. The braiding of the chi fields is a simple phase. From the spectrum of the J's, it follows that this phase factor is of the form $\exp(i\pi N/2(2 \mp s))$, where $N \in \mathcal{Z}$. Thus, we have parafermions. As shown in ref.[2], simple products of the form $\chi_\pm^{(J)} \overline{\chi}_\pm^{(\overline{J})}$, with $J - \overline{J} \in \mathcal{Z}$ are local. In such a product, the summations over m, and $\overline{m}$ are independent, while the summations over m, $\widehat{m}$, and $\overline{m}$, $\widehat{\overline{m}}$ are correlated. Now we have a complete reversal of the weak coupling situation summarized by the drawing 2.29 :the new fields preserve the reality condition, but DO NOT PRESERVE THE EQUALITY BETWEEN J AND $\overline{J}$ QUANTUM NUMBERS. Thus, as stressed in ref.[2], in the strong coupling regime we observe a sort of deconfinement of chirality.

3 The Liouville string

One may consider two different problems. First, one may build a full-fledged string theory, by coupling, for instance, the above with $26-C_L$ free fields $\vec{X}$. A typical string vertex is of the form $\exp(i\vec{k}.\vec{X})\chi_+^{(J)}\overline{\chi}_+^{(\overline{J})}$, where $\vec{k}$, J, and $\overline{J}$ are related so that this is a $1,1$ operator. Here obviously, the restriction to real weight is instrumental. Moreover, since one wants the representation of Virasoro algebra to be unitary, one only uses the chi+ fields. This line was already persued with noticable success in refs.[10]. However, the N-point functions seem to be beyond reach at present. Second a simpler problem seems to be tractable, namely, we may proceed as in the construction of topological models just recalled. We consider another copy of the present strongly coupled theory, with central charge $c = 26 - C_L$. Since this gives $c = 1+6(-s+2)$, we are also at the special values, and the truncation theorems applies to matter as well. This "string theory" has no transverse degree of freedom, and is thus topological. The complete dressed vertex operator is now

$$\mathcal{V}^{J,,\overline{J},} = \chi_+^{(J)}\overline{\chi}_+^{(\overline{J})}\,\chi'^{(J)}_-\overline{\chi}'^{(\overline{J})}_- \tag{3.1}$$

As in the weak coupling formula, operators relative to matter are distinguished by a prime. The definition of the $\overline{\chi}$ is similar to the above, with an important difference. Clearly, the definition of χ_+ is not symmetric between α_+, and α_-. The truncation theorems also holds if we interchange the two screening charges. We re-establish some symmetry between them by taking the other possible definition for $\overline{\chi}_+$. Our results will then be invariant by complex conjugation provided we exchange J's and $\overline{J}$'s. Thus left and right movers are interchanged, which seems to be a sensible requirement. The spectrum of Verma modules involved may be depicted graphically as follows:

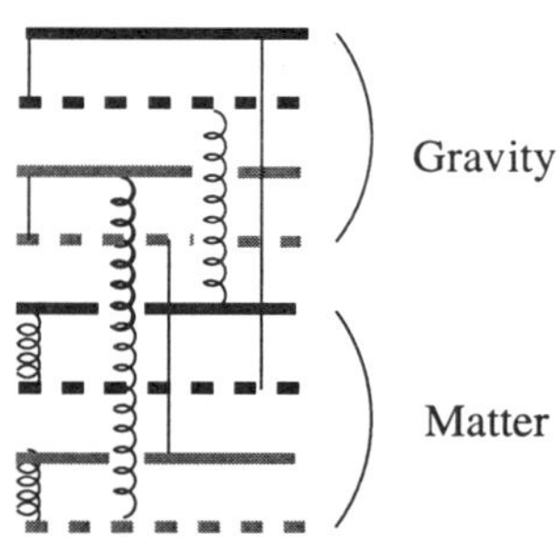

(3.2)

It is suggestive to compare with the corresponding drawing 2.38of the weak coupling regime. The vertical links between the lines are of the same nature, but they are distributed differently. A basic difference here is that the present drawing is not connected since there are no links between the two chiral components. Thus, chirality is deconfined. For $J = \overline{J} = 0$, we get the new cosmological term

$$\mathcal{V}^{0,,0,} = \chi_+^{(0)}(z)\overline{\chi}_+^{(0)}(\bar{z}). \tag{3.3}$$

Thus the area element of the strong coupling regime is $\chi_+^{(0)}(z)\overline{\chi}_+^{(0)}(\bar{z})dzd\bar{z}$. It is factorized into a simple product of a single z component by a $\bar{z}$ component. From this expression one may compute the string susceptibility using the operator version of the DDK argument developed in ref.[11] for the weak coupling regime. We refer to refs.[3, 4] for details. One finds

$$\gamma_{\rm str} = (2 - s)/2. \tag{3.4}$$

The result is real for $c > 1$ ($C_L < 25$), contrary to the continuation of the weak-coupling equation $\gamma_{\rm str} = 2 - Q/\alpha_-$. Explicitly one has

$$\left\{ \begin{array}{cccc} s & c & C_L & \gamma_{\rm str} \\ 1 & 7 & 19 & 1/2 \\ 0 & 13 & 13 & 1 \\ -1 & 19 & 7 & 3/2 \end{array} \right. , \qquad \left\{ \begin{array}{cccc} s & c & C_L & \gamma_{\rm str} \\ 2 & 1 & 25 & 0 \\ -2 & 25 & 1 & 2 \end{array} \right. \tag{3.5}$$

The last two are the extreme points of the strong coupling regime. The values at $c = 1$, and $c = 25$ agree with the weak-coupling formula. The result is always positive, contrary to the weak-coupling regime. At $c = 7$, we find the value $\gamma_{\rm str} = 1/2$ of branched polymers.

Finally, we have computed the N-point functions with one incoming and N-1 outgoing legs, defined as follows. First in general[7] the two-point function of two $\widetilde{V}$ fields with spins J_1, $\overline{J}_1$, and J_2, $\overline{J}_2$ vanishes unless $J_1 + J_2 + 1 = 0$, and $\overline{J} + \overline{J}_2 + 1 = 0$. Thus conjugation involves the transformation $J \to -J - 1$. Taking account of the exchange between J and $\widehat{J}$, due to complex conjugation, yields the following vertex operator for the conjugate representation:

$$\mathcal{V}_{conj}^{J,,}\overline{J}, = \chi_+^{(J)}\,\overline{\chi}_+^{(\overline{J})}\,\chi'{}_-^{(-J-1)}\,\overline{\chi}'{}_-^{(-\overline{J}-1)}. \tag{3.6}$$

Thus we have computed the matrix elements $\left\langle \mathcal{V}_{conj}^{J_1,,\overline{J}_1,} \mathcal{V}^{J_2,,\overline{J}_2,} \ldots \mathcal{V}^{J_N,,\overline{J}_N,} \right\rangle$. The method is similar to the one developed in ref.[12], with a reshuffling of quantum numbers. In the weak coupling regime, left and right quantum numbers are kept equal, while the ones associated with different screening charge are chosen independently. In the strong coupling regime, the situation is reversed: the reality condition ties the quantum numbers which differ by the screening charge, but the quantum numbers with different chiralities become independent. See refs.[3, 4] for details.

4 Concluding remarks

We should probably stress that no total conformal spin has been introduced. Indeed, although, the conformal spin is non zero for the gravity, and matter components of our vertex operators separately, the total weights for the left and right components are kept equal to one.

One may wonder why the present approach succeeds to break through the $c = 1$ barrier, in sharp contrast with the other ones. This may be traced to the fact that we first deal with the chiral components of gravity and matter separately before reconstructing the vertex operators. This is more painful than the matrix model approach which directly constructs the expectation values of the dressed matter operators.

However, in this way we have a handle over the way the gravity quantum numbers are coupled, and so we may build up vertex operators which change the gravity chirality. This seems to be the key to the $c = 1$ problem, since this quantum number plays the role of order parameter.

Let us turn to a final remark. The redefinition of the cosmological term led us to modify the KPZ formula[13]. On the other hand, in standard studies of the matrix models or KP flows, one first derives γ_{str} and deduces the value of the central charge by assuming that the KPZ formula holds. In this way of thinking, one would start from our formula Eq.3.4 and apply KPZ, which would lead to a different value of the central charge, say d. It is easy to see that for $c = 1 + 6(-s + 2)$ one gets $d = 1 - 6(2 - s)^2/2s$. This is the value of a $2, s$ minimal model! What happens is that in terms of d, we have $\gamma_{\text{str}} = (d - 1 + \sqrt{(d-1)(d-25)})/12$, in contrast with the KPZ formula $(d - 1 - \sqrt{(d-1)(d-25)})/12$. Thus the strongly coupled topological theories may be given by another branch of $d < 1$ theories.

References

[1] J.-L. Gervais, *Comm. Math. Phys.*, 138 (1991) 301 .

[2] J.-L. Gervais, J.-F. Roussel, *Nucl. Phys.*, B426 (1994) 140 .

[3] J.-L. Gervais, J.-F. Roussel, *Phys. Lett.*, B338 (1994) 437 .

[4] J.-L. Gervais, J.-F. Roussel, "Solving the strongly coupled 2D gravity III: String susceptibility and topological N-point functions", in preparation.

[5] J.-L. Gervais and A. Neveu, *Phys. Lett.*, B151 (1985) 271 .

[6] G. Moore, N. Seiberg, *Comm. Math. Phys.*, 123 (1989) 77 .

[7] E. Cremmer, J.-L. Gervais, J.-F. Roussel, *Nucl. Phys.*, B413 (1994) 244 ; *Comm. Math. Phys.*, 161 (1994) 597 .

[8] J.-L. Gervais, J. Schnittger, *Phys. Lett.*, B315 (1993) 258 ; *Nucl. Phys.*, B341 (1994) 273 .

[9] E. Cremmer, J.-L. Gervais, J. Schnittger, "*Operator coproduct-realisation of quantum group transformations in two dimensional; gravity* I" LPTENS preprint hep-th/9503198, *ibid II: Bloch waves*, in preparation.

[10] A. Bilal, J.-L. Gervais, *Nucl. Phys.*, B284 (1987) 397 , *Phys. Lett.*, B187 (1987) 39 , *Nucl. Phys.*, B293 (1987) 1 , *Nucl. Phys.*, B295 [FS21] (1988) 277 .

[11] J.-L. Gervais, *Nucl. Phys.*, B391 (1993) 287 .

[12] P. Di Francesco, D. Kutasov, *Nucl. Phys.*, B375 (1992) 119 .

[13] V. Knizhnik, A. Polyakov, A.A. Zamolodchikov, *Mod. Phys. Lett.* A3 (1988) 819.

LECTURES ON SUPERSYMMETRIC GAUGE THEORIES AND ELECTRIC–MAGNETIC DUALITY

K. INTRILIGATOR AND N. SEIBERG

DEPARTMENT OF PHYSICS
RUTGERS UNIVERSITY
PISCATAWAY, NJ 08855-0849, USA
AND
INSTITUTE FOR ADVANCED STUDY
PRINCETON, NJ 08540, USA

1 INTRODUCTION

Recently, it has become clear that certain aspects of four dimensional supersymmetric field theories can be analyzed exactly, providing a laboratory for the analysis of the dynamics of gauge theories (for a recent elementary presentation and a list of references see Ref.1). For example, the phases of gauge theories and the mechanisms for phase transitions can be explored in this context. The dynamical mechanisms explored are standard to gauge theories and thus, at least at a qualitative level, the insights obtained are expected to also be applicable for non-supersymmetric theories. We summarize some of the recent ideas. The discussion is not in historical order and other examples appear in the literature.

1.1 Phases of Gauge Theories

The phases of gauge theories can be characterized by the potential $V(R)$ between electric test charges separated by a large distance R. Up to a non-universal, additive constant, the potential is conjectured to behave as

$$\begin{array}{lcl} \text{Coulomb} & V(R) \sim & \dfrac{1}{R} \\ \text{free electric} & V(R) \sim & \dfrac{1}{R\log(R\Lambda)} \end{array}$$

Low Dimensional Applications of Quantum Field Theory
Edited by L. Baulieu *et al.*, Plenum Press, New York, 1997

$$\begin{array}{rlll} \text{free magnetic} & V(R) & \sim & \dfrac{\log(R\Lambda)}{R} \\ \text{Higgs} & V(R) & \sim & constant \\ \text{confining} & V(R) & \sim & \sigma R. \end{array} \tag{1.1}$$

The first three phases have massless gauge fields and potentials of the form $V(R) \sim e^2(R)/R$. In the Coulomb phase, the electric charge $e^2(R)$ ~constant. In the free electric phase, massless electrically charged fields renormalize the charge to zero at long distances as $e^{-2}(R) \sim \log(R\Lambda)$. Similar behavior occurs when the long distance theory is a non-Abelian theory which is not asymptotically free. The free magnetic phase occurs when there are massless magnetic monopoles, which renormalize the electric coupling constant to infinity at large distance with a conjectured behavior $e^2(R) \sim \log(R\Lambda)$. In the Higgs phase, the condensate of an electrically charged field gives a mass gap to the gauge fields by the Anderson-Higgs mechanism and screens electric charges, leading to a potential which, up to the additive non-universal constant, has an exponential Yukawa decay to zero at long distances. In the confining phase, there is a mass gap with electric flux confined into a thin tube, leading to the linear potential with string tension σ.

All of the above phases can be non-Abelian as well as Abelian. In particular, in addition to the familiar Abelian Coulomb phase, there are theories which have a non-Abelian Coulomb phase with massless interacting quarks and gluons exhibiting the above Coulomb potential. This phase occurs when there is a non-trivial, infrared fixed point of the renormalization group. These are thus non-trivial, interacting four dimensional conformal field theories.

We can also consider the behavior of the potential $V(R)$ for magnetic test charges separated by a large distance R. Up to an additive, non-universal constant, the potential behaves as

$$\begin{array}{rlll} \text{Coulomb} & V(R) & \sim & \dfrac{1}{R} \\ \text{free electric} & V(R) & \sim & \dfrac{\log(R\Lambda)}{R} \\ \text{free magnetic} & V(R) & \sim & \dfrac{1}{R\log(R\Lambda)} \\ \text{Higgs} & V(R) & \sim & \rho R \\ \text{confining} & V(R) & \sim & constant. \end{array} \tag{1.2}$$

The behavior in the first three phases can be written as $V(R) = g^2(R)/R$ where the effective magnetic charge $g^2(R)$ is related to the effective electric charge appearing in (1.2) by the Dirac condition, $e(R)g(R) \sim 1$. The linear potential in the Higgs phase reflects the string tension in the Meissner effect.

The above behavior is modified when there are matter fields in the fundamental representation of the gauge group because virtual pairs can be popped from the vacuum and completely screen the sources. Indeed, in this situation there is no invariant distinction between the Higgs and the confining phases [2]. In particular, there is no phase with a potential behaving as the "confining" potential in (1.2) at large distances – the flux tube can break. For large expectation values of the fields, a Higgs description is most natural while, for small expectation values, it is more natural to interpret the theory as "confining." Because there is really no distinction, it is possible to smoothly interpolate from one interpretation to the other.

Note that under electric-magnetic duality, which exchanges electrically charged fields with magnetically charged fields, the behavior in the free electric phase is exchanged with that of the free magnetic phase. Mandelstam and 't Hooft suggested that, similarly, the Higgs and confining phases are exchanged by duality. Confinement can then be understood as the dual Meissner effect associated with a condensate of monopoles. As we will review, in supersymmetric theories it is possible to show that this picture is indeed correct.

Dualizing a theory in the Coulomb phase, we remain in the same phase (the behavior of the potential is unchanged). For an Abelian Coulomb phase with free massless photons, this follows from a standard duality transformation. What is not obvious is that this is also the case in a non-Abelian Coulomb phase. This was first suggested by Montonen and Olive [3]. The simplest version of their proposal is true only in $N = 4$ supersymmetric field theories [4] supersymmetric theories [5, 6, 7]. The extension of these ideas to asymptotically free $N = 1$ theories first appeared in [8] and will be reviewed here.

1.2 Super Yang-Mills Theories

We briefly review standard lore concerning $N = 1$ supersymmetric Yang-Mills theories – *i.e.* pure super glue with no matter. The lore presented here can be proven via the analysis in the following sections, by adding vector-like matter and then integrating it out.

We consider a theory based on a simple group G; the generalization to semi-simple Yang-Mills theories is obvious. The theory consists of the G vector bosons and gauginos λ_α in the adjoint of G. There is a classical $U(1)_R$ symmetry, gaugino number, which is broken to a discrete Z_{2h} subgroup by instantons, $\langle(\lambda\lambda)^h\rangle = \text{const.}\Lambda^{3h}$, where $h = C_2(A)$ is the Casimir in the adjoint normalized so that, for example, $h = N_c$ for $SU(N_c)$. The lore is that this theory confines, gets a mass gap, and that there are h vacua associated with the spontaneous breaking of the Z_{2h} symmetry to Z_2 by gaugino condensation, $\langle\lambda\lambda\rangle = \text{const.}\Lambda^3$. These vacua each contribute $(-1)^F = 1$ and thus the Witten index [9] is $\text{Tr}\,(-1)^F = C_2(A)$.

1.3 Outline

In sect. 2, we discuss general techniques for analyzing supersymmetric theories. In sect. 3, we discuss the classical moduli spaces of degenerate vacua which supersymmetric gauge theories with matter often have. In particular, we discuss the classical vacuum degeneracy for $N = 1$ supersymmetric QCD. In sect. 4, we discuss supersymmetric QCD for $N_f \leq N_c + 1$ massless flavors of quarks in the fundamental representation of $SU(N_c)$. In sect. 5, we discuss $N_f > N_c + 1$ and duality. Sect. 6 is a brief discussion of the phases and duality of supersymmetric $SO(N_c)$ with matter fields in the N_c dimensional representation. In sect. 7, we discuss general aspects of supersymmetric theories which have a low energy Abelian Coulomb phase. In sect. 8, we consider the examples of $SU(2)$ with $N_f = 1$ and $N_f = 2$ adjoints, which have electric-magnetic-dyonic triality.

2 Holomorphy and Symmetries

2.1 General Considerations

The basic approach will be to consider the low energy effective action for the light fields, integrating out degrees of freedom above some scale. Assuming that we are working above the scale of possible supersymmetry breaking, the effective action will have a linearly realized supersymmetry which can be made manifest by working in terms of superfields. The light matter fields can be combined into chiral superfields $X_r = \phi_r + \theta_\alpha \psi_r^\alpha + \ldots$, where the ϕ_r are scalars and the ψ_r^α are Weyl fermions. In addition, there are the conjugate anti-chiral superfields $X_r^\dagger = \phi_r^\dagger + \bar\theta_{\dot\alpha}\psi_r^{\dagger\dot\alpha} + \ldots$. Similarly, light gauge fields combine into supermultiplets involving a gauge boson A_μ and gauginos λ_α and $\lambda_{\dot\alpha}^\dagger$.

We will focus on a particular contribution to the effective Lagrangian – the superpotential term

$$\int d^2\theta W_{\text{eff}}(X_r, g_I, \Lambda), \tag{2.1}$$

where X_r are the light fields, the g_I are various coupling constants, and Λ is the dimensional transmutation scale associated with the gauge dynamics, $-\frac{8\pi^2}{g^2(\mu)} \sim \log \Lambda/\mu$. Upon doing the θ integral, the superpotential yields a potential for the scalars and a Yukawa type interaction with the scalars and the fermions.

The key fact is that supersymmetry requires W_{eff} to be holomorphic in the chiral superfields X_r, independent of the $X_r^\dagger$.

We will think of all the coupling constants g_I in the tree level superpotential W_{tree} and the scale Λ as background fields [10]. Then, the quantum, effective superpotential, $W_{\text{eff}}(X_r, g_I, \Lambda)$ is constrained by:

1. Symmetries and selection rules: By assigning transformation laws both to the fields and to the coupling constants, the theory has a large symmetry. The effective Lagrangian should be invariant under it.

2. Holomorphy: W_{eff} is independent of $g_I^\dagger$ [10]. This is the key property. Just as the superpotential is holomorphic in the fields, it is also holomorphic in the coupling constants (the background fields). This is unlike the effective Lagrangian in non-supersymmetric theories, which is not subject to any holomorphy restrictions. This use of holomorphy extends considerations of [11, 12, 13]. It is similar in spirit to the proofs of non-renormalization in sigma model perturbation theory [14] and in semi-classical perturbation theory [15] in string theory.

3. Various limits: W_{eff} can be analyzed approximately at weak coupling. The singularities have physical meaning and can be controlled.

Often these conditions completely determine W_{eff}. The point is that a holomorphic function (more precisely, a section) is determined by its asymptotic behavior and singularities. The results can be highly non-trivial, revealing interesting non-perturbative dynamics.

When there is a Coulomb phase, the kinetic terms for the gauge fields are also constrained by the above considerations. The relevant term in the effective Lagrangian is

$$\int d^2\theta \; Im[\tau_{\text{eff}}(X_r, g_I, \Lambda) W_\alpha^2]; \tag{2.2}$$

W_α^2 gives the supersymmetric completion of $F^2 + iF\tilde{F}$ so

$$\tau_{\text{eff}} \sim \frac{\theta_{\text{eff}}}{2\pi} + \frac{4\pi i}{g_{\text{eff}}^2} \tag{2.3}$$

is the effective gauge coupling constant. $\tau_{\text{eff}}(X_r, g_I, \Lambda)$ is holomorphic in its arguments and can often be exactly determined.

2.2 Example: Wess-Zumino Model

In order to demonstrate the above rules, we consider the simplest Wess-Zumino model and rederive the known non-renormalization theorem. Consider the theory based on the tree level superpotential

$$W_{\text{tree}} = m\phi^2 + g\phi^3. \tag{2.4}$$

We will make use of two $U(1)$ symmetries. The charges of the field ϕ and the coupling constants m and g are

$$\begin{array}{ccccc} & U(1) & \times & U(1)_R \\ \phi & 1 & & 1 \\ m & -2 & & 0 \\ g & -3 & & -1 \end{array} \tag{2.5}$$

where $U(1)_R$ is an R symmetry under which $d^2\theta$ has charge -2 and thus, in order for (2.1) to be invariant, the superpotential has charge 2. Note that non-zero values for m and g explicitly break both $U(1)$ symmetries. However, they still lead to selection rules.

The symmetries and holomorphy of the effective superpotential restrict it to be of the form

$$W_{\text{eff}} = m\phi^2 f\left(\frac{g\phi}{m}\right). \tag{2.6}$$

Consider the limit $g \to 0$ and $m \to 0$ with arbitrary g/m. We must have $W_{eff} \to W_{tree}$ and, therefore, $f(t = (g\phi/m)) \to 1 + t$ in this limit of vanishing coupling. Now, because t is arbitrary in this limit, $f(t)$ is thus evaluated to be given by $f(t) = 1 + t$ for all t. Therefore, the exact effective superpotential is found to be

$$W_{\text{eff}} = m\phi^2 + g\phi^3 = W_{\text{tree}}; \tag{2.7}$$

i.e. the superpotential is not renormalized [10].

This argument rederives the standard perturbative non-renormalization theorem [16] and extends it beyond perturbation theory. Strictly speaking, the Wess Zumino model probably does not exist as an interacting quantum field theory in four dimensions except as an effective low energy theory of a more fundamental theory and, therefore, this non-perturbative result is only of limited interest. This non-perturbative proof of the Wess-Zumino non-renormalization can be directly applied to two dimensions, where such models do exist as interacting quantum field theories.

If there are several fields, some light and some heavy, the heavy fields can be integrated out to yield a low energy effective Lagrangian for the light fields. The contribution of tree diagrams with intermediate heavy fields should be then included in the effective action. The above simple rules allow such diagrams to contribute and are thus compatible with the known tree level renormalization of the superpotential.

2.3 The 1PI Effective Action

There are two different objects which are usually called "the effective action:" the 1PI effective action and the Wilsonian one. When there are no interacting massless particles, these two effective actions are identical. This is often the case in the Higgs or confining phases. However, when interacting massless particles are present, the 1PI effective action suffers from IR ambiguities and might suffer from holomorphic anomalies [12]. These are absent in the Wilsonian effective action.

Consider the theory with a tree level superpotential with sources for the gauge invariant polynomials X^r in the matter fields, $W_{\rm tree} = \sum_r g_r X^r$, with the g_r regarded as background chiral superfield sources [10]. The functional integral with the added source terms gives the standard generating function for the correlation functions, $\Gamma(g)$. If supersymmetry is not broken, $\Gamma(g)$ is supersymmetric (otherwise we should include the Goldstino field and supersymmetry will be realized non-linearly) and[1] $\Gamma(g) = \ldots + \int d^2\theta W_L(g)$. Using $W_L(g)$ we can compute the expectation values

$$\frac{\partial W_L(g)}{\partial g_r} = \langle X^r \rangle. \tag{2.8}$$

It is standard to perform a Legendre transform to find the 1PI effective action for the operators X_r:

$$W_{\rm dyn}(X^r) = \left(W_L(g_r) - \sum_r g_r X^r \right)_{\langle g_r \rangle}, \tag{2.9}$$

where the $\langle g_r \rangle$ are the solutions of (2.8). The transformation from $W_L(g_r)$ to $W_{\rm dyn}(X_r)$ can be inverted by the inverse Legendre transform as

$$W_L(g) = \left(W_{\rm dyn}(X^r) + \sum_r g_r X^r \right)_{\langle X^r \rangle}, \tag{2.10}$$

[1] In writing this expression we should think of the coupling constants g_r as background superfields. Otherwise, $W_L(g)$ is a constant superpotential, which has no effect in global supersymmetry. Indeed, the following equation can be interpreted as differentiating the action with respect to the F component of g_r.

where the X^r are evaluated at their expectation values $\langle X^r \rangle$, which solve

$$\frac{\partial W_{\text{dyn}}}{\partial X_r} + g_r = 0. \tag{2.11}$$

The 1PI effective superpotential

$$W_{\text{eff}}(X, g) = W_{\text{dyn}}(X^r) + \sum_r g_r X^r \tag{2.12}$$

has the property that the equations of motion (2.11) for the fields X^r derived from it determine their expectation values. In some cases the superpotential W_{eff} obtained by the above Legendre transform is the same as the Wilsonian superpotential for the light fields. In applying this procedure we should be careful of the following pitfalls:

1. The theory with the sources should have a gap. Otherwise, the 1PI action is ill defined.

2. The theory with the sources might break supersymmetry. In that case W_L is ill defined.

3. As the sources are turned off, some particles become massless. Their interpolating fields should be among the composite fields X^r. If some massless particles cannot be represented by a gauge invariant operator X^r, the effective superpotential derived this way will not include them. This often leads to singularities.

4. The theory might also have other branches which are present only when some sources vanish. In this case there are new massless particles at that point and the 1PI action might miss some of the branches. In other words, then the Legendre transform does not exist.

5. If some composites do not represent massless particles, they should be integrated out. Although we can use the effective superpotential to find their expectation values, we cannot think of them as fields corresponding to massive particles except near a point where they become massless.

There are known examples[2] of each of these situations; in these cases the 1PI effective action is misleading, failing to capture important aspects of the physics.

When we can use this procedure to find the Wilsonian action, the linearity of W_{eff} (2.12) in the sources provides a derivation of the linearity of the Wilsonian effective action in the sources. See [18, 19] for a related discussion.

[2]As a simple example of a situation in which the Legendre transform analysis is incomplete because supersymmetry is dynamically broken by the added source terms, consider $N = 1$ supersymmetric $SU(2)$ with a single field Q in the **4** of $SU(2)$[17]. The theory without added source terms has a one complex dimensional smooth moduli space of vacua labeled by $\langle X \rangle$, where $X = Q^4$ is the basic gauge invariant, with a superpotential $W(X) = 0$. Adding a source $W = gX$ does not lead to a supersymmetric effective superpotential $W(g)$ – rather, it breaks supersymmetry [17]. (As discussed in [17], it is also possible that there is a non-Abelian Coulomb phase at the origin of the moduli space and that supersymmetry is unbroken with the added source term. In that case the 1PI analysis again fails to capture the physics.)

This approach is particularly useful when we know how to compute $W_L(g_r)$ exactly. Then, $W_{\rm dyn}$ and $W_{\rm eff}$ follow simply from the Legendre transform (2.9); this is the "integrating in" discussed in [20, 21]. One situation where $W_L(g_r)$ can be determined is when the X^r are all quadratic in the elementary fields. In that case, the sources g_r are simply mass terms for the matter fields and $W_L(g)$ is the superpotential for the low energy gauge theory with the massive matter integrated out, expressed in terms of the quantities in the high-energy theory.

These issues will be exhibited and further discussed in later sections.

2.4 Extended Supersymmetry

Theories with extended supersymmetry are further constrained. For example, $N = 2$ supersymmetry combines an $N = 1$ vector superfield with a chiral superfield Φ in the adjoint representation of the gauge group into an $N = 2$ vector multiplet. In particular, $N = 2$ supersymmetry includes a global $SU(2)_R$ symmetry which rotates the gluino component of an $N = 1$ vector superfield into the fermion component of the chiral superfield. This symmetry relates the Kahler potential for the chiral superfield to the τ_{eff} in the kinetic term (2.2) for the gauge field. Therefore, the Kahler potential for chiral superfields which are part of $N = 2$ vector multiplets are determined in terms of holomorphic functions and can thus often be obtained exactly.

Another type of $N = 2$ supermultiplet is the hypermultiplet, consisting of two $N = 1$ chiral multiplets Q and $\widetilde{Q}$. The global $SU(2)_R$ symmetry rotates the scalar component of Q into the scalar component of $\widetilde{Q}^\dagger$. This symmetry implies that the Kahler potential for these fields must yield a metric which is hyper-Kahler. $N = 2$ supersymmetry further implies that this metric is not corrected by quantum effects.

Another condition on theories with $N = 2$ supersymmetry is the Bogomol'nyi Prasad, Sommerfeld bound [22, 23], which is related to the central term in the $N = 2$ algebra [24]. The mass of any state satisfies

$$M \geq \sqrt{2}|Z|, \tag{2.13}$$

where Z is the central term of the $N = 2$ algebra, involving the gauge and global quantum numbers of the state. The BPS saturated states, for which the inequality (2.13) is saturated, are some of the stable states in the spectrum. Because Z is a holomorphic object, it can often be exactly determined; (2.13) then yields the exact mass spectrum for some of the stable states.

This review will focus on $N = 1$ supersymmetry. We refer the reader to the literature for further details about extended supersymmetry.

3 Classical Super Gauge Theories – Classical Moduli Spaces of Vacua

The classical Lagrangian of a supersymmetric theory with gauge group G, matter superfields Φ_f in representations $R(f)$ of G, and zero tree level superpotential is

$$\mathcal{L} = \mathcal{L}_o + \sum_{f,a} \phi_f^\dagger \lambda^a T_f^a \psi_f + h.c. + \sum_a (\sum_f \phi_f^\dagger T_f^a \phi_f)^2, \tag{3.1}$$

where $\mathcal{L}_o$ are the obvious gauge invariant kinetic terms for the gauge and matter fields and T_f^a are the G generators in representation $R(f)$. The interactions in (3.1) are related by supersymmetry to the coupling in $\mathcal{L}_o$ of gauge fields to matter.

Classical gauge theories often have "D-flat" directions of non-zero $\langle\phi_f\rangle$ along which the squark potential in (3.1) vanishes. In other words, these theories often have classical moduli spaces of degenerate vacua. As a simple example, consider $U(1)$ gauge theory with a matter superfield Q of charge 1 and $\widetilde{Q}$ of charge -1. The squark potential in (3.1) is $V = (Q^\dagger Q - \widetilde{Q}^\dagger\widetilde{Q})^2$ and thus there is a continuum of degenerate vacua labeled, up to gauge equivalence, by $\langle Q\rangle = \langle\widetilde{Q}\rangle = a$, for any complex a. In vacua with $a \neq 0$ the gauge group is broken by the super Higgs mechanism. The gauge superfield gets mass $|a|$ by "eating" one chiral superfield degree of freedom from the matter fields. Since we started with the two superfields Q and $\widetilde{Q}$, one superfield degree of freedom remains massless. The massless superfield can be given a gauge invariant description as $X = Q\widetilde{Q}$. In the vacuum labeled as above by a, $\langle X\rangle = a^2$. Because a is arbitrary, there is no potential for X, $W_{cl}(X) = 0$ – classically X is a "modulus" field whose expectation value labels a classical moduli space of degenerate vacua. The classical Kahler potential of the microscopic theory is $K_{cl} = Q^\dagger e^V Q + \widetilde{Q}^\dagger e^{-V}\widetilde{Q}$. In terms of the light field X, $K_{cl} = 2\sqrt{X^\dagger X}$, which has a conical ($Z_2$ orbifold) singularity at $X = 0$. A singularity in a low energy effective action generally reflects the presence of additional massless fields which should be included in the effective action. Indeed, the singularity at $X = 0$ corresponds to the fact that the gauge group is unbroken and all of the original microscopic fields are classically massless there.

As in the above simple example, the classical moduli space of vacua is the space of squark expectation values $\langle\phi_f\rangle$, modulo gauge equivalence, along which the potential in (3.1) vanishes. It can always be given a gauge invariant description in terms of the space of expectation values $\langle X_r\rangle$ of gauge invariant polynomials in the fields subject to any classical relations. This is because setting the potential in (3.1) to zero and modding out by the gauge group is equivalent to modding out by the complexified gauge group. The space of chiral superfields modulo the complexified gauge group can be parameterized by the gauge invariant polynomials modulo any classical relations. These results follow from results in geometrical invariant theory [25]; see also [26] for a recent discussion. As in the above example, the fields X_r correspond to the matter fields left massless after the Higgs mechanism and are classical moduli, $W_{cl}(X_r) = 0$.

The vacuum degeneracy of classical moduli spaces of vacua is not protected by any symmetry. In fact, vacua with different expectation values of the fields are physically inequivalent: as in the above example, the masses of the massive vector bosons depend on the expectation values $\langle X_r\rangle$. Therefore, the degeneracy of a classical moduli space of vacua is accidental and can be lifted in the quantum theory by a dynamically generated $W_{\text{eff}}(X_r)$. We will often be able to determine $W_{\text{eff}}(X_r)$ by the considerations discussed above.

Example: SUSY QCD

Consider supersymmetric $SU(N_c)$ gauge theory with N_f quark flavors Q^i in the fundamental representation and $\widetilde{Q}_{\tilde{i}}$ in the anti-fundamental, $i, \tilde{i} = 1 \ldots N_f$. In the absence of mass terms, there is a classical moduli space of vacua given, up to gauge and global symmetry transformations, by:

$$Q = \tilde{Q} = \begin{pmatrix} a_1 & & & \\ & a_2 & & \\ & & \cdot & \\ & & & a_{N_f} \end{pmatrix} \tag{3.2}$$

for $N_f < N_c$, with a_i arbitrary, and by

$$Q = \begin{pmatrix} a_1 & & & \\ & a_2 & & \\ & & \cdot & \\ & & & a_{N_c} \\ & & & \\ & & & \end{pmatrix}, \quad \tilde{Q} = \begin{pmatrix} \tilde{a}_1 & & & \\ & \tilde{a}_2 & & \\ & & \cdot & \\ & & & \tilde{a}_{N_c} \\ & & & \\ & & & \end{pmatrix}, \tag{3.3}$$

$$|a_i|^2 - |\tilde{a}_i|^2 = \text{independent of } i,$$

for $N_f \geq N_c$.

For $N_f < N_c$, the gauge invariant description of the classical moduli space is in terms of arbitrary expectation values of the "mesons" $M^i_{\tilde{j}} = Q^i \tilde{Q}_{\tilde{j}}$. For $N_f \geq N_c$, it is also possible to form "baryons" $B^{i_1 \ldots i_{N_c}} = Q^{i_1} \ldots Q^{i_{N_c}}$ and $\tilde{B}_{\tilde{i}_1 \ldots \tilde{i}_{N_c}} = \tilde{Q}_{\tilde{i}_1} \ldots \tilde{Q}_{\tilde{i}_{N_c}}$. The gauge invariant description of the classical moduli space for $N_f \geq N_c$ is given in terms of the expectation values of M, B and $\tilde{B}$, subject to the following classical constraints. Up to global symmetry transformations, the expectation values are

$$M = \begin{pmatrix} a_1\tilde{a}_1 & & & & \\ & a_2\tilde{a}_2 & & & \\ & & \cdot & & \\ & & & a_{N_c}\tilde{a}_{N_c} & \\ & & & & \\ & & & & \end{pmatrix} \tag{3.4}$$

$$\begin{aligned} B^{1,\ldots,N_c} &= a_1 a_2 \cdots a_{N_c} \\ \tilde{B}_{1,\ldots,N_c} &= \tilde{a}_1 \tilde{a}_2 \cdots \tilde{a}_{N_c} \end{aligned}$$

with all other components of M, B and $\tilde{B}$ vanishing. Therefore, the rank of M is at most N_c. If it is less than N_c, either $B = 0$ with $\tilde{B}$ having rank at most one or $\tilde{B} = 0$ with B having rank at most one. If the rank of M is equal to N_c, both B and $\tilde{B}$ have rank one and the product of their eigenvalues is the same as the product of non-zero eigenvalues of M.

As discussed above, the physical interpretation of the flat directions is that the gauge group is Higgsed. If $B = \tilde{B} = 0$ and M has rank k, $SU(N_c)$ is broken to $SU(N_c - k)$ with $N_f - k$ light flavors.

Already at the classical level, we can integrate out the massive fields and consider an effective Lagrangian for the massless modes. Their expectation values label the particular ground state we expand around, and hence they are coordinates on the classical moduli space. The classical moduli space is not smooth. Its singularities are at the points of enhanced gauge symmetry. For instance, when $a_i = \tilde{a}_i = 0$ for every i the gauge symmetry is totally unbroken. Therefore, the low energy effective theory of the moduli is singular there. This should not surprise us. At these singular points there are new massless particles – gluons. An effective Lagrangian without them is singular. If we include them in the low energy description, the Lagrangian is smooth.

In the next two sections we will see how this picture changes in the quantum theory. At large expectation values of the fields, far from the classical singularities, the gauge symmetry is broken at a high scale, the quantum theory is weakly coupled, and semi- classical techniques are reliable. We expect the quantum corrections to the classical picture to be small there. On the other hand, at small field strength the quantum theory is strongly coupled and the quantum corrections can be large and dramatically modify the classical behavior. In particular, the nature of the classical singularities, which are at strong coupling, is generally totally different in the quantum theory.

4 SUSY QCD FOR $N_f \le N_c + 1$

Because these theories have matter fields in the fundamental representation of the gauge group, as mentioned in the introduction, there is no invariant distinction between the Higgs and the confining phases [2]. It is possible to smoothly interpolate from one interpretation to the other.

4.1 $N_f < N_c$ - No Vacuum

The first question to ask is whether the classical vacuum degeneracy can be lifted quantum mechanically by a dynamically generated superpotential. The form of such a superpotential is constrained by the symmetries. At the classical level, the symmetries are

$$SU(N_f)_L \times SU(N_f)_R \times U(1)_A \times U(1)_B \times U(1)_R \tag{4.1}$$

where the the quarks transform as

$$\begin{aligned} &Q \quad (N_f, 1, 1, 1, \tfrac{N_f - N_c}{N_f}) \\ &\tilde{Q} \quad (1, \overline{N}_f, 1, -1, \tfrac{N_f - N_c}{N_f}). \end{aligned} \tag{4.2}$$

$U(1)_R$ is an R symmetry (the gauginos have charge $+1$, the squark components of Q and $\tilde{Q}$ have the charge $R(Q)$ indicated above, and the charge of the fermion components is $R(\psi) = R(Q) - 1$). The charges were chosen so that only the symmetry $U(1)_A$ is anomalous in the quantum theory. Considering the anomalous $U(1)_A$ as explicitly broken by fermion zero modes in an instanton background, $U(1)_A$ leads to a selection rule. The instanton amplitude is proportional to $e^{-S_{Inst.}} = e^{-8\pi^2 g^{-2}(\mu)+i\theta} = (\Lambda/\mu)^{3N_c - N_f}$, where Λ is the dynamically generated scale of the theory, and we integrated the 1-loop beta function. Therefore, $U(1)_A$ is respected provided we assign

$\Lambda^{3N_c-N_f}$ charge $2N_f$ to account for the charge of the fermion zero modes under $U(1)_A$. The dependence on the scale Λ is thus determined by the $U(1)_A$ selection rule.

There is a unique superpotential which is compatible with these symmetries [27]

$$W_{\text{eff}} = C_{N_c,N_f}\left(\frac{\Lambda^{3N_c-N_f}}{\det \tilde{Q}Q}\right)^{1/(N_c-N_f)}, \tag{4.3}$$

where C_{N_c,N_f} are constants which depend on the subtraction scheme. Therefore, if the vacuum degeneracy is lifted, this particular superpotential must be generated. For $N_f \geq N_c$ this superpotential does not exist (either the exponent diverges or the determinant vanishes) and therefore the vacuum degeneracy cannot be lifted. We will return to $N_f \geq N_c$ in the next subsections.

Note that the superpotential (4.3) is non-perturbative and is thus not in conflict with the *perturbative* non-renormalization theorem. Indeed, the above argument demonstrates that the perturbative non-renormalization and its non-perturbative violation can be understood simply as the need for obtaining a well-defined charge violation of the anomalous $U(1)_A$ as occurs, for example, in an instanton background.

The superpotential (4.3) is further constrained by considering various limits. For example, consider the limit of large $M^{N_f}_{\tilde{N}_f}$, *i.e.* large a_{N_f} in (3.2), which breaks $SU(N_c)$ with N_f flavors to $SU(N_c-1)$ with N_f-1 light flavors by the Higgs mechanism at energy a_{N_f}. Matching the running gauge coupling at energy a_{N_f}, the low energy theory has scale $\Lambda_L^{3(N_c-1)-(N_f-1)} = \Lambda^{3N_c-N_f}/a^2_{N_f}$. The fact that the scales are so matched without any threshold factors reflects a choice of subtraction scheme; this is the correct matching, for example, in the $\overline{DR}$ scheme. Requiring (4.3) to properly reproduce the superpotential of the low energy theory in this limit gives $C_{N_c,N_f} = C_{N_c-N_f}$.

Next consider giving Q_{N_f} and $\tilde{Q}_{\tilde{N}_f}$ a large mass by adding $W_{\text{tree}} = mM^{N_f}_{\tilde{N}_f}$. The low energy theory is $SU(N_c)$ with N_f-1 flavors. Matching the running gauge coupling at the transition scale m, the low energy theory has scale $\Lambda_L^{3N_c-(N_f-1)} = m\Lambda^{3N_c-N_f}$. Again, this equality is up to a scheme dependent threshold factor which is one in the $\overline{DR}$ scheme. Using the symmetries, the exact superpotential with the added mass term is of the form

$$W_{\text{exact}} = \left(\frac{\Lambda^{3N_c-N_f}}{\det M}\right)^{1/(N_c-N_f)} f\left(t = mM^{N_f}_{\tilde{N}_f}\left(\frac{\Lambda^{3N_c-N_f}}{\det M}\right)^{-1/(N_c-N_f)}\right). \tag{4.4}$$

In the limit of small mass and weak coupling, we know that $f(t) = C_{N_c,N_f} + t$. Because all values of t can be obtained in this limit, the function $f(t)$ is evaluated in this understood limit to be $f(t) = C_{N_c,N_f} + t$ for all t. The exact superpotential with the added mass term is thus

$$W_{\text{exact}} = C_{N_c,N_f}\left(\frac{\Lambda^{3N_c-N_f}}{\det M}\right)^{1/(N_c-N_f)} + mM^{N_f}_{\tilde{N}_f}. \tag{4.5}$$

Requiring (4.5) to give the correct superpotential in the low energy theory upon integrating out $M^{N_f}_{\tilde{N}_f}$ relates C_{N_c,N_f} to C_{N_c,N_f-1} which, when combined with $C_{N_c,N_f} = C_{N_c-N_f}$, determine that $C_{N_c,N_f} = (N_c-N_f)C^{1/(N_c-N_f)}$ with C a universal constant.

For $N_f = N_c - 1$, the superpotential (4.3) is proportional to the one-instanton action and, thus, the constant C can be exactly computed via a one-instanton calculation. Because the gauge group is completely broken by the Higgs mechanism for $N_f = N_c - 1$ (for $\det M \neq 0$), the instanton calculation is reliable (there is no infra-red divergence). The universal constant C can be computed by considering the particular case $N_c = 2$, $N_f = 1$. The direct instanton calculation [28] reveals that the constant $C \neq 0$. The more detailed analysis of [29] shows that $C = 1$ in the $\overline{DR}$ scheme. For $N_f < N_c$ there is thus a dynamically generated superpotential

$$W_{\text{eff}} = (N_c - N_f)\left(\frac{\Lambda^{3N_c-N_f}}{\det \widetilde{Q}Q}\right)^{1/(N_c-N_f)}. \tag{4.6}$$

While W_{eff} is generated by instantons for $N_f = N_c - 1$, for $N_f < N_c - 1$ it is associated with gaugino condensation in the unbroken $SU(N_c - N_f)$ gauge group [28]. In particular, the low energy theory has a WZ term[3]

$$\int d^2\theta \log(\det M)(W_\alpha W^\alpha)_{SU(N_c-N_f)} = (M^{-1})_i^{\tilde{j}} F_{M^i_{\tilde{j}}} \lambda_\alpha \lambda^\alpha + \ldots, \tag{4.7}$$

to account for the matching to the high energy theory. If gaugino condensation occurs in the low energy theory, $\langle\lambda\lambda\rangle = c\Lambda^3_{N_c-N_f}$ where c is a constant and $\Lambda_{N_c-N_f}$ is the scale of the low energy $SU(N_c - N_f)$ gauge theory, given in terms of the scale Λ of the high-energy theory by $\Lambda^{3(N_c-N_f)}_{N_c-N_f} = \Lambda^{3N_c-N_f}/\det M$, (4.7) yields a term $c\Lambda^3_{N_c-N_f} M^{-1} F_M$ in the effective Lagrangian. Our superpotential (4.6) indeed gives exactly such a term,
$(\Lambda^{3N_c-N_f}/\det M)^{1/(N_c-N_f)} M^{-1} F_M = \Lambda^3_{N_c-N_f} M^{-1} F_M$, in the effective Lagrangian. Therefore, gaugino condensation occurs in $N = 1$ $SU(N_c - N_f)$ Yang-Mills theory with the normalization

$$\langle\lambda_\alpha\lambda^\alpha\rangle = e^{2\pi i k/(N_c-N_f)}\Lambda^3_{N_c-N_f} \qquad \text{with} \qquad k = 1\ldots(N_c - N_f), \tag{4.8}$$

where we explicitly exhibit the phase. Using (4.6) (whose normalization follows from a well understood instanton calculation), we have *derived* gaugino condensation (4.8), including its normalization in the $\overline{DR}$ scheme, in the low energy $N = 1$ Yang-Mills theory. (For related work on this model see also references [11] and [30].)

The dynamically generated superpotential (4.6) leads to a squark potential which slopes to zero for $\det M \to \infty$. Therefore, the quantum theory does not have a ground state. We started with an infinite set of vacua in the classical theory and ended up in the quantum theory without a vacuum!

Consider adding $W_{\text{tree}} = \operatorname{Tr} mM$, giving masses to the N_f flavors. As in (4.5), symmetries and the weak coupling and small mass limit determine that the exact superpotential is $W_{\text{full}} = W_{\text{eff}} + W_{\text{tree}}$. The vacua are given by $\langle M\rangle$ solving $\frac{\partial W_{\text{full}}}{\partial M^i_{\tilde{j}}}|_{\langle M\rangle} = 0$. This gives N_c vacua

$$\langle M^i_{\tilde{j}}\rangle = (\det m\Lambda^{3N_c-N_f})^{1/N_c}\left(\frac{1}{m}\right)^i_{\tilde{j}}, \tag{4.9}$$

---

[3] We absorb a factor of $1/32\pi^2$ into the definition of W_α^2.

corresponding to the N_c branches of the N_c-th root. For large m, the matter fields are very massive and decouple, leaving a low energy $SU(N_c)$ pure Yang-Mills theory. Indeed, the low energy theory has confinement with a mass gap and N_c vacua.

Evaluating (4.6) with the expectation values (4.9) yields

$$W_L(m) = N_c(\Lambda^{3N_c-N_f} \det m)^{1/N_c} = N_c\Lambda_L^3, \tag{4.10}$$

where Λ_L is the scale of the low energy $SU(N_c)$ Yang-Mills theory. Expressed in terms of Λ_L, the superpotential (4.10) is interpreted as the result of gaugino condensation in the low energy $SU(N_c)$ Yang-Mills theory. Expressed in terms of m and Λ, the superpotential (4.10) is the effective superpotential which yields the expectation values (4.9) via (2.8), $\frac{\partial W_L(m)}{\partial m} = \langle M \rangle$. As in (2.9), $W_L(m)$ leads to an effective superpotential for the operators M:

$$W_{\rm dyn} = (N_c - N_f)\left(\frac{\Lambda^{3N_c-N_f}}{\det M}\right)^{1/(N_c-N_f)}. \tag{4.11}$$

In this case $W_{\rm dyn}$ agrees with the Wilsonian effective superpotential (4.6), as could have been expected because this theory satisfies all of the conditions spelled out in sect. 2.3. The fact that the Wilsonian effective action is here the same as the 1PI effective action provides a simple alternate derivation of the linearity derived in (4.5).

It is also possible to "integrate in" operators which do not correspond to massless particles. Then, the effective action can be used only to compute their expectation values, rather than for studying them as massive particles. An example is the "glueball" field $S = -W_\alpha^2$, whose source is $\log \Lambda^{3N_c-N_f}$. Integrating in S by the Legendre transform of (4.11) with respect to the source $\log \Lambda^{3N_c-N_f}$ yields

$$W(S, M) = S\left[\log\left(\frac{\Lambda^{3N_c-N_f}}{S^{N_c-N_f}\det M}\right) + (N_c - N_f)\right], \tag{4.12}$$

the superpotential obtained in [31]. Adding mass terms $W_{tree} = \mathrm{Tr}\, mM$ and integrating out M yields

$$W(S) = S\left[\log\left(\frac{\Lambda_L^{3N_c}}{S^{N_c}}\right) + N_c\right], \tag{4.13}$$

where Λ_L is the scale of the low energy $SU(N_c)$ Yang-Mills theory,

$$\Lambda_L^{3N_c} = \det m \Lambda^{3N_c-N_f}.$$

This superpotential simply gives the information discussed above: supersymmetric $SU(N_c)$ Yang-Mills theory has N_c supersymmetric vacua with the gaugino condensates (4.8) and superpotential $W = N_c\Lambda_L^3$. Working with such effective superpotentials which include massive fields can be convenient when interesting but complicated dynamics is encoded in the integrating out of these massive fields. Several such examples can be found in [20]. However, as stressed above, we should not think of S as a field describing a massive particle.

4.2 $N_f = N_c$ *– Quantum Moduli Space with Confinement with Chiral Symmetry Breaking*

As we said above, for $N_f \geq N_c$ the vacuum degeneracy can not be lifted. Therefore, the quantum theory also has a continuous space of inequivalent vacua. Since this space can be different than the classical one, we will refer to it as the "quantum moduli space." The most interesting questions about it are associated with the nature of its singularities. Classically, the singularities were associated with massless gluons. Are there singularities in the quantum moduli space? What massless particles are associated with those singularities?

The classical moduli space for $N_f = N_c$ is given in terms of the gauge invariant description (3.4) as the the space of expectation values of mesons $M^i_{\tilde{j}}$, and baryons B and $\widetilde{B}$ subject to the classical constraint

$$\det M - \widetilde{B}B = 0, \tag{4.14}$$

which follows from Bose statistics of Q and $\tilde{Q}$. This space has a singular submanifold $B = \widetilde{B} = 0$ and $\text{rank}(M) \leq N_c - 2$, where $d(\det M - B\widetilde{B}) = 0$. Physically, the classical singularities reflect the fact that there are additional degrees of freedom, $SU(N_c - \text{rank}(M))$ gluons, on this submanifold.

The quantum moduli space is parameterized by the same fields but the constraint is modified [32] to

$$\det M - \widetilde{B}B = \Lambda^{2N_c}. \tag{4.15}$$

This can be seen by adding mass terms $W_{\text{tree}} = \text{Tr}\, mM$ and taking $m \to 0$. It follows from (4.9) that $\det\langle M\rangle = \Lambda^{2N_c}$, independent of m, for $N_f = N_c$. This agrees with (4.15) for $\langle B\widetilde{B}\rangle = 0$, which is the case with only added mass terms. Considering more generally $W_{\text{tree}} = \text{Tr}\, mM + bB + \tilde{b}\widetilde{B}$ with $m, b, \tilde{b} \to 0$, $\langle B\widetilde{B}\rangle$ can be non-zero and the expectation values are found to satisfy (4.15) [21]. Because the right hand side of (4.15) is proportional to the one-instanton action, the quantum modification of the classical constraint is exactly given by a one-instanton contribution.

There are no singularities on the quantum moduli space given by (4.15) – all of the classical singularities have been smoothed out by quantum effects. As a similar but simpler example of the deformed moduli space, consider the space $XY = \mu$ in C^2. For $\mu = 0$ the space is a pair of cones (corresponding to the X plane and the Y plane) touching at their tips; the space is singular at the origin, where the cones touch. In the deformed space $\mu \neq 0$, the two cones are smoothed out to a single sheeted hyperboloid. It asymptotes to the two cones far from the origin but has a smooth hourglass shape where they connect. Similarly, the quantum space (4.15) is smooth, which is dramatically different from the classical space (4.14) near the origin. For large expectation values M, B and $\widetilde{B}$, the difference becomes negligible, as it should be in the weak coupling region.

Since there are no singularities on the quantum moduli space, the only massless particles are the moduli, the fluctuations of M, B, and $\widetilde{B}$ preserving (4.15). In the semi-classical region of large fields it is appropriate to think of the theory as "Higgsed." Near the origin, because the quantum theory is smooth in terms of the mesons and baryons, it is appropriate to think of the theory as being confining. There is a smooth transition from the region where a Higgs description is more appropriate to the region where a confining description is more appropriate. Again, this is possible because of the presence of matter fields in the fundamental representation of the gauge group [2].

Because the origin $M = B = \widetilde{B} = 0$ is not on the quantum moduli space (4.15), the quantum dynamics necessarily break the anomaly free, chiral $SU(N_f)\times SU(N_f)\times U(1)_B \times U(1)_R$ symmetry in (4.1). Different points on the quantum moduli space exhibit different patterns of chiral symmetry breaking. For example at $M^i_{\tilde{j}} = \Lambda^2\delta^i_{\tilde{j}}$, $B = \widetilde{B} = 0$ the symmetry is broken as

$$SU(N_f)_L \times SU(N_f)_R \times U(1)_B \times U(1)_R \to SU(N_f)_V \times U(1)_B \times U(1)_R. \quad (4.16)$$

At $M = 0$, $B = -\widetilde{B} = \Lambda^{N_c}$ the breaking pattern is

$$SU(N_f)_L \times SU(N_f)_R \times U(1)_B \times U(1)_R \to SU(N_f)_L \times SU(N_f)_R \times U(1)_R. \quad (4.17)$$

Some of the moduli are Goldstone bosons of the broken symmetry whereas others take the theory to vacua with different breaking patterns. For no vacuum in the quantum moduli space (4.15) is the full chiral symmetry unbroken. It is straightforward to check that the massless fermion spectrum, consisting of the fermionic components of the chiral superfield moduli, satisfies the 't Hooft anomaly conditions for the unbroken symmetries.

The constraint (4.15) can be implemented with a superpotential $W = A(\det M - B\widetilde{B} - \Lambda^{2N_c})$, with A a Lagrange multiplier. The reader can verify that, upon adding $W_{\text{tree}} = mM^{N_c}_{\tilde{N}_c}$ to give a mass to the N_c-th flavor, the low energy theory with $N_f = N_c - 1$ light flavors has the appropriate superpotential (4.6).

4.3 $N_f = N_c + 1$ - Confinement Without Chiral Symmetry Breaking

We now add another massless flavor to the previous case. The classical moduli space is again described as in (3.4) by the mesons M, baryons $B_i = \epsilon_{ij_1\ldots j_{N_c}} Q^{j_1}\ldots Q^{j_{N_c}}$ and $\widetilde{B}^{\tilde{i}} = \epsilon^{\tilde{i}\tilde{j}_1\ldots\tilde{j}_{N_c}}\widetilde{Q}_{\tilde{j}_1}\ldots\widetilde{Q}_{\tilde{j}_{N_c}}$ subject to the constraints

$$\begin{aligned} \det M \left(\tfrac{1}{M}\right)^{\tilde{j}}_i - B_i\widetilde{B}^{\tilde{j}} &= 0 \\ M^i_{\tilde{j}}B_i = M^i_{\tilde{j}}\widetilde{B}^{\tilde{j}} &= 0. \end{aligned} \quad (4.18)$$

Unlike the previous case, for $N_f = N_c + 1$ the quantum moduli space is the same as the classical one [32]. This can be seen by adding $W_{\text{tree}} = \text{Tr}\, mM$, giving masses to all $N_f + 1$ flavors. The expectation values of M are then given by (4.9). By taking the limit $m^{\tilde{j}}_i \to 0$ in different ratios, it is found that $\langle M^i_{\tilde{j}}\rangle$ can be anywhere on the classical moduli space (4.18) of vacua.

Because the quantum moduli space of vacua is the same as the classical moduli space, it has singularities at strong coupling. The singularities, however, are interpreted differently than in the classical theory. Rather than being associated with massless gluons, the singularities are associated with additional massless mesons and baryons! In particular, at the point $M = \widetilde{B} = B = 0$ the global chiral symmetry

$$SU(N_f)_L \times SU(N_f)_R \times U(1)_B \times U(1)_R \tag{4.19}$$

is unbroken and all the components of M, B and $\widetilde{B}$ are massless and physical [32]. It is a non-trivial consistency check that this massless spectrum at the origin satisfies the 't Hooft anomaly matching conditions for the full global symmetry (4.19).

Away from the origin, all the degrees of freedom in M, B and $\widetilde{B}$ are physical and they couple through the superpotential [32]

$$W_{\text{eff}} = \frac{1}{\Lambda^{2N_c-1}}(M^i_{\bar{j}} B_i \widetilde{B}^{\bar{j}} - \det M). \tag{4.20}$$

The classical constraints (4.18) appear as the equations of motion $\frac{\partial W_{\text{eff}}}{\partial M} = \frac{\partial W_{\text{eff}}}{\partial B} = \frac{\partial W_{\text{eff}}}{\partial \widetilde{B}} = 0$. Far from the origin, in the region of weak coupling, the number of independent massless fields is the same as in the classical theory because the components of M, B, and $\widetilde{B}$ which are classically constrained get a large mass from (4.20).

We conclude that the spectrum at the origin of field space consists of massless composite mesons and baryons and that the chiral symmetry of the theory is unbroken there. This is confinement without chiral symmetry breaking. Again, we see a smooth transition [2] from the semi-classical region where a Higgs description is more appropriate to a strongly coupled region where a confining description is more appropriate.

The reader can verify that adding $W_{\text{tree}} = mM^{N_c+1}_{\widetilde{N_c+1}}$ to (4.20) to give a mass to the $N_c + 1$-th flavor yields the quantum moduli space with constraint (4.15) in the low energy theory with $N_f = N_c$ light flavors.

5 SUSY QCD for $N_f > N_c + 1$

5.1 $N_f \geq 3N_c$

In this range the theory is not asymptotically free. This means that, because of screening, the coupling constant becomes smaller at large distances. Therefore, the spectrum of the theory at large distance can be read off from the Lagrangian – it consists of the elementary quarks and gluons. The long distance behavior of the potential between external electric test charges has the free electric behavior in (1.2). For this range of N_f, the theory is in a non-Abelian free electric phase.

We should add here that, strictly speaking, such a theory is not well defined as an interacting quantum field theory because of the Landau pole at $R \sim \Lambda^{-1}$. However, it can be a consistent description of the low energy limit of another theory.

5.2 $\frac{3}{2}N_c < N_f < 3N_c$*; Interacting Non-Abelian Coulomb Phase*

In this range the theory is asymptotically free. This means that at short distance the coupling constant is small and it becomes larger at longer distances. However, for this range of N_f [32, 8], rather than growing to infinity, it reaches a finite value – a fixed point of the renormalization group.

The exact beta function in supersymmetric QCD satisfies [33, 12]

$$\begin{aligned} \beta(g) &= -\tfrac{g^3}{16\pi^2}\tfrac{3N_c-N_f+N_f\gamma(g^2)}{1-N_c\frac{g^2}{8\pi^2}} \\ \gamma(g^2) &= -\tfrac{g^2}{8\pi^2}\tfrac{N_c^2-1}{N_c}+\mathcal{O}(g^4), \end{aligned} \tag{5.1}$$

where $\gamma(g^2)$ is the anomalous dimension of the mass. Since there are values of N_f and N_c where the one loop beta function is negative but the two loop contribution is positive, there might be a non-trivial fixed point [34]. Indeed, by taking N_c and N_f to infinity holding $N_c g^2$ and $\frac{N_f}{N_c}=3-\epsilon$ fixed, one can establish the existence of a zero of the beta function at $N_c g_*^2 = \frac{8\pi^2}{3}\epsilon + \mathcal{O}(\epsilon^2)$. Therefore, at least for large N_c and $\epsilon = 3-\frac{N_f}{N_c} \ll 1$, there is a non-trivial fixed point. It was argued in Ref.8 that such a fixed point exists for every $\frac{3}{2}N_c < N_f < 3N_c$.

Therefore, for this range of N_f, the infrared theory is a non-trivial four dimensional superconformal field theory. The elementary quarks and gluons are not confined but appear as interacting massless particles. The potential between external electric sources behaves as

$$V \sim \frac{1}{R}$$

and therefore we refer to this phase of the theory as the non-Abelian Coulomb phase.

Given that such a fixed point exists, we can use the superconformal algebra to derive some exact results about the theory. This algebra includes an R symmetry. It follows from the algebra that the dimensions of the operators satisfy

$$D \geq \frac{3}{2}|R|; \tag{5.2}$$

the inequality is saturated for chiral operators, for which $D=\frac{3}{2}R$, and for anti-chiral operators, for which $D=-\frac{3}{2}R$. Exactly as in N=2 theories in two dimensions, this has important consequences. Consider the operator product of two chiral operators, $\mathcal{O}_1(x)\mathcal{O}_2(0)$. All the operators in the resulting expansion have $R=R(\mathcal{O}_1)+R(\mathcal{O}_2)$ and hence $D \geq D(\mathcal{O}_1)+D(\mathcal{O}_2)$. Therefore, there is no singularity in the expansion at $x=0$ and we can define the product of the two operators by simply taking the limit of x to zero. If this limit does not vanish, it leads to a new chiral operator $\mathcal{O}_3$ whose dimension is $D(\mathcal{O}_3)=D(\mathcal{O}_1)+D(\mathcal{O}_2)$. We conclude that the chiral operators form a ring.

The R symmetry of the superconformal fixed point is not anomalous and commutes with the flavor $SU(N_f)\times SU(N_f)\times U(1)_B$ symmetry. Therefore, it must be the anomaly free R symmetry appearing in (4.2). Hence the gauge invariant operators $\widetilde{Q}Q$ have [8]

$$D(\widetilde{Q}Q) = \frac{3}{2}R(\widetilde{Q}Q) = 3\frac{N_f-N_c}{N_f} \tag{5.3}$$

and similarly

$$D(B) = D(\widetilde{B}) = \frac{3N_c(N_f-N_c)}{2N_f}. \tag{5.4}$$

The value of $D(\widetilde{Q}Q)$ also follows from (5.1) – at the zero of the beta function $\gamma = -3\frac{N_c}{N_f} + 1$ and hence $D = \gamma + 2 = 3\frac{N_f - N_c}{N_f}$.

All of the gauge invariant operators at the infrared fixed point should be in unitary representations of the superconformal algebra. The complete list of such representations was given in [35] by extending the analysis [36] of the ordinary conformal algebra. One of the constraints on the representations which follows already from the analysis of [36] is that spinless operators have $D \geq 1$ (except the identity operator with $D = 0$) and the bound is saturated for free fields (satisfying $\partial_\mu \partial^\mu \Phi = 0$). For $D < 1$ ($D \neq 0$) a highest weight representation includes a negative norm state which cannot exist in a unitary theory.

The fixed point coupling g_* gets larger as the number of flavors is reduced. For N_f at or below $\frac{3}{2}N_c$ the theory is very strongly coupled and goes over to a new phase, different from the interacting non-Abelian Coulomb phase. To see that the theory must be in a different phase, note that the value of $D(\widetilde{Q}Q)$ in (5.3) is inconsistent with the unitarity bound $D \geq 1$ for $N_f < \frac{3}{2}N_c$. The new phase will be explained below. A clue is the fact that the dimension of $M = \widetilde{Q}Q$ becomes one for $N_f = \frac{3}{2}N_c$, which shows that M becomes a free field, i.e. $\partial^2 M = 0$. This suggests that in the correct description for $N_f = \frac{3}{2}N_c$ the field M, and perhaps even the whole IR theory, is free.

5.3 *Duality*

The physics of the interacting fixed point obtained for the range $\frac{3}{2}N_c < N_f < 3N_c$ has an equivalent, "magnetic," description [8]. It is based on the gauge group $SU(N_f - N_c)$, with N_f flavors of quarks q_i and $\widetilde{q}^i$ and gauge invariant fields $M^i_{\bar{j}}$ with a superpotential

$$W = \frac{1}{\mu} M^i_{\bar{j}} q_i \widetilde{q}^{\bar{j}}. \tag{5.5}$$

We will refer to this gauge group as the magnetic gauge group and to its quarks as magnetic quarks. Without the superpotential (5.5), the magnetic theory also flows to a non-Abelian Coulomb phase fixed point because $\frac{3}{2}(N_f - N_c) < N_f < 3(N_f - N_c)$ for the above range of N_f. At this fixed point M is a free field of dimension one and, using (5.3), $D(q\widetilde{q}) = 3N_c/N_f$. Because the dimensions of chiral operators add, the superpotential (5.5) has dimension $D = 1 + 3N_c/N_f < 3$ at the fixed point of the magnetic gauge theory and is thus a relevant perturbation, driving the theory to a new fixed point. The surprising fact is that this new fixed point is identical to that of the original, "electric," $SU(N_c)$ theory. Note that the two theories have different gauge groups and different numbers of interacting particles. Nevertheless, they describe the same fixed point. In other words, there is no experimental way to determine whether the $\frac{1}{R}$ potential between external sources is mediated by the interacting electric or the interacting magnetic variables. Such a phenomenon of two different Lagrangians describing the same long distance physics is common in two dimensions and is known there as quantum equivalence. These four dimensional examples generalize the duality [3] in finite $N = 4$ supersymmetric theories [4] and in finite $N = 2$ theories [5] to asymptotically free $N = 1$ theories.

The scale μ in (5.5) is needed for the following reason. In the electric description $M^i_{\bar{j}} = Q^i\widetilde{Q}_{\bar{j}}$ has dimension two at the UV fixed point and acquires anomalous dimension (5.3) at the IR fixed point. In the magnetic description, M_m is an elementary field of dimension one at the UV fixed point which flows to the same operator with dimension (5.3) at the IR fixed point. In order to relate M_m to M of the electric description in the UV, a scale μ must be introduced with the relation $M = \mu M_m$. Below we will write all the expressions in terms of M and μ rather than in terms of M_m.

The magnetic theory has a scale $\widetilde{\Lambda}$ which is related to the scale Λ of the electric theory by

$$\Lambda^{3N_c-N_f}\widetilde{\Lambda}^{3(N_f-N_c)-N_f} = (-1)^{N_f-N_c}\mu^{N_f}, \tag{5.6}$$

where μ is the dimensionful scale explained above. This relation of the scales has several consequences:

1. It is easy to check that it is preserved under mass deformations and along the flat directions (more details will be given below). The phase $(-1)^{N_f-N_c}$ is important in order to ensure that this is the case.

2. It shows that as the electric theory becomes stronger the magnetic theory becomes weaker and vice versa. It is the analog of $g \to 1/g$ for asymptotically free theories.

3. Differentiating the action with respect to $\log\Lambda$ relates the field strengths of the electric and the magnetic theories as $W_\alpha^2 = -\widetilde{W}_\alpha^2$. The minus sign in this expression is common in electric magnetic duality, which maps $E^2 - B^2 = -(\widetilde{E}^2 - \widetilde{B}^2)$. In our case it shows that the gluino bilinear in the electric and the magnetic theories are related by $\lambda\lambda = -\widetilde{\lambda}\widetilde{\lambda}$.

4. Because of the phase $(-1)^{N_f-N_c}$, the relation (5.6) does not look dual – if we perform another duality transformation it becomes $\Lambda^{3N_c-N_f}\widetilde{\Lambda}^{3(N_f-N_c)-N_f} = (-1)^{N_c}\widetilde{\mu}^{N_f}$ and therefore

   $$\widetilde{\mu} = -\mu. \tag{5.7}$$

   This minus sign is important when we dualize again. The dual of the dual magnetic theory is an $SU(N_c)$ theory with scale Λ, quarks d^i and $\widetilde{d}_{\bar{j}}$, and additional singlets $M^i_{\bar{j}}$ and $N_i^{\bar{j}} = q_i\widetilde{q}^{\bar{j}}$, with superpotential

   $$W = \frac{1}{\widetilde{\mu}}N_i^{\bar{j}}d^i\widetilde{d}_{\bar{j}} + \frac{1}{\mu}M^i_{\bar{j}}N_i^{\bar{j}} = \frac{1}{\mu}N_i^{\bar{j}}(-d^i\widetilde{d}_{\bar{j}} + M^i_{\bar{j}}). \tag{5.8}$$

   The first term is our standard superpotential of duality transformations[4] and the second term is simply copied from (5.5). M and N are massive and can be integrated out using their equations of motion: $N = 0$, $M^i_{\bar{j}} = d^i\widetilde{d}_{\bar{j}}$. This last relation shows that the quarks d and $\widetilde{d}$ can be identified with the original electric quarks Q and $\widetilde{Q}$. The dual of the magnetic theory is the original electric theory.

[4]The relative minus sign between it and (5.5), which follows from (5.7), is common in Fourier or Legendre transforms. (Compare with (2.9) and (2.10).)

The electric and magnetic theories have different gauge symmetries. This is possible because gauge symmetries really have to do with a redundant description of the physics rather than with symmetry. There is no problem with having two different redundant descriptions of the same physics. On the other hand, global symmetries are physical and should be the same in the electric and magnetic theories. Indeed, the magnetic theory has the same anomaly free global $SU(N_f)\times SU(N_f)\times U(1)_B\times U(1)_R$ as the electric theory, with the singlet $M^i_{\bar{j}}$ transforming as $Q^i\widetilde{Q}_{\bar{j}}$ and the magnetic quarks transforming as

$$\begin{aligned} q \quad &\text{in} \quad (\overline{N}_f, 1, \tfrac{N_c}{N_f-N_c}, \tfrac{N_c}{N_f}) \\ \widetilde{q} \quad &\text{in} \quad (1, N_f, -\tfrac{N_c}{N_f-N_c}, \tfrac{N_c}{N_f}). \end{aligned} \tag{5.9}$$

This symmetry is anomaly free in the magnetic theory and it is preserved by the superpotential (5.5). Furthermore, the magnetic spectrum with these charges satisfies the 't Hooft anomaly matching conditions.

In order for the dual theory to describe the same physics as the electric theory, there must be a mapping of all gauge invariant primary operators of the electric theory to those of the dual theory. For example, the electric mesons $M^i_{\bar{j}}=Q^i\widetilde{Q}_{\bar{j}}$ and the singlets $M^i_{\bar{j}}$ of the magnetic theory become identical in the infrared. All such mappings must be compatible with the global symmetry charges discussed above. Another set of gauge invariant operators of the electric theory are the baryons $B^{i_1\ldots i_{N_c}}=Q^{i_1}\cdots Q^{i_{N_c}}$ and $\widetilde{B}_{\bar{i}_1\ldots\bar{i}_{N_c}}=\widetilde{Q}_{\bar{i}_1}\cdots\widetilde{Q}_{\bar{i}_{N_c}}$. In the magnetic theory we can similarly form the baryons $b_{i_1\ldots i_{\widetilde{N}_c}}=q_{i_1}\cdots q_{i_{\widetilde{N}_c}}$ and $\widetilde{b}^{\bar{i}_1\ldots\bar{i}_{\widetilde{N}_c}}=\widetilde{q}^{\bar{i}_1}\cdots\widetilde{q}^{\bar{i}_{\widetilde{N}_c}}$, where $\widetilde{N}_c=N_f-N_c$. At the fixed point, these operators are related via

$$\begin{aligned} B^{i_1\ldots i_{N_c}} &= C\epsilon^{i_1\ldots i_{N_c}j_1\ldots j_{\widetilde{N}_c}}b_{j_1\ldots j_{\widetilde{N}_c}}, \\ \widetilde{B}_{\bar{i}_1\ldots\bar{i}_{N_c}} &= C\epsilon_{\bar{i}_1\ldots\bar{i}_{N_c}\bar{j}_1\ldots\bar{j}_{\widetilde{N}_c}}\widetilde{b}^{\bar{j}_1\ldots\bar{j}_{\widetilde{N}_c}}, \end{aligned} \tag{5.10}$$

with $C=\sqrt{-(-\mu)^{N_c-N_f}\Lambda^{3N_c-N_f}}$. Note that these mappings respect the global symmetries discussed above. The normalization constant C was fixed by symmetries and by limits to be discussed below. It follows from (5.6) that (5.10) respects the Z_2 nature of the duality.

5.4 $N_c+2\le N_f\le\frac{3}{2}N_c$

Recall that the electric $SU(N_c)$ theory with N_f flavors was in a non-Abelian free electric phase for $N_f\ge 3N_c$ and in the non-Abelian Coulomb phase for $\frac{3}{2}N_c<N_f<3N_c$, with the fixed point at larger electric coupling for smaller N_f. In the magnetic description of the non-Abelian Coulomb phase fixed point, the magnetic coupling at the fixed point is small for N_f near $3(N_f-N_c)$ and gets larger with $3(N_f-N_c)-N_f$; in the magnetic description the fixed point is at weaker coupling for smaller N_f. It was seen that for $N_f\le\frac{3}{2}N_c$ the theory must be in a different phase. In the magnetic description, the situation is clear: since $3(N_f-N_c)\le N_f$, the magnetic $SU(N_f-N_c)$ gauge theory is not asymptotically free (and the added superpotential (5.5) is irrelevant) and thus weakly coupled at large distances. Therefore, the low energy spectrum of the theory consists of the $SU(N_f - N_c)$ gauge fields and the fields M, q, and $\tilde{q}$ in the dual magnetic Lagrangian [8]. These magnetic massless states are composites of the elementary electric degrees of freedom. The massless composite gauge fields exhibit gauge invariance which is not visible in the underlying electric description. The theory generates new gauge invariance! Because there are massless magnetically charged fields, the theory is in a non-Abelian free magnetic phase.

5.5 *Deformations of the Dual Theories*

We now consider deforming the theories, showing in detail how the electric-magnetic duality exchanges strong coupling with weak coupling and Higgs with confinement.

Consider starting from the electric theory and giving a mass to the N_f-th flavor by adding $W_{\text{tree}} = mM^{N_f}_{\tilde{N}_f}$. The low energy theory has $N_f - 1$ light flavors and a scale Λ_L related to the scale Λ of the high energy theory by $\Lambda_L^{3N_c-(N_f-1)} = m\Lambda^{3N_c-N_f}$. As expected, the low energy electric theory is at stronger coupling; the mass term is a relevant operator, sending the theory to a more strongly coupled fixed point.

In the magnetic theory, adding W_{tree} gives

$$W = \frac{1}{\mu} M^i_{\tilde{i}} q_i \tilde{q}^{\tilde{i}} + mM^{N_f}_{\tilde{N}_f}. \tag{5.11}$$

The equations of motion of $M^{N_f}_{\tilde{N}_f}$, $M^i_{\tilde{N}_f}$ and $M^{N_f}_{\tilde{i}}$ lead to

$$q_{N_f}\tilde{q}^{\tilde{N}_f} = -\mu m, \qquad q_i\tilde{q}^{\tilde{N}_f} = q_{N_f}\tilde{q}^{\tilde{i}} = 0, \tag{5.12}$$

(color indices are suppressed and summed) which show that the magnetic gauge group is broken by the Higgs mechanism to $SU(N_f - N_c - 1)$ with $N_f - 1$ quarks left massless. The equations of motion of the massive quarks lead to

$$M^{N_f}_{\tilde{N}_f} = M^i_{\tilde{N}_f} = M^{N_f}_{\tilde{i}} = 0. \tag{5.13}$$

The low energy superpotential is

$$W = \frac{1}{\mu}\widehat{M}^i_{\tilde{i}}\hat{q}_i\hat{\tilde{q}}^{\tilde{i}}, \qquad i, \tilde{i} = 1, \ldots, N_f - 1, \tag{5.14}$$

where $\widehat{M}$, $\hat{q}$ and $\hat{\tilde{q}}$ are the light fields with $N_f - N_c - 1$ colors and $N_f - 1$ flavors. The scale of the low energy magnetic theory is given by $\widetilde{\Lambda}_L^{3(\widetilde{N}_c-1)-(N_f-1)} = \widetilde{\Lambda}^{3\widetilde{N}_c-N_f}/\langle q_{N_f}\tilde{q}^{\tilde{N}_f}\rangle$. Note that (5.6) and (5.10) are preserved in the low energy theories. The low energy magnetic theory is at weaker coupling and is the dual of the low energy electric theory. The duality is preserved under the mass term deformation and exchanges a more strongly coupled electric description of the new fixed point with a more weakly coupled magnetic description of the new fixed point.

The above discussion of the mass term deformation is incomplete for $N_f = N_c+2$, where the mass term for the $(N_c + 2)$-th flavor triggers complete breaking of the magnetic gauge group. The low energy theory contains the mesons $\widehat{M}^j_{\tilde{i}}$, where the hat refers to the flavors $i, \tilde{i} = 1 \ldots N_c + 1$, and the singlets $\hat{q}_i$ and $\hat{\tilde{q}}^{\tilde{i}}$, which are the components of the magnetic quarks remaining massless after the Higgs mechanism, with a superpotential as in (5.14). The map (5.10) for $N_f = N_c + 1$ implies that the singlets $\hat{q}_i$ and $\hat{\tilde{q}}^{\tilde{i}}$ are, in fact, the baryons B_i and $\widetilde{B}^{\tilde{i}}$ of the low energy electric theory. It has already been suggested, at least for large N_c, that baryons can be thought of as solitons in the pion Lagrangian [37]. Here we see an explicit realization of a related idea – the baryons are magnetic monopoles of the elementary quarks and gluons! Taking the normalization in (5.10) into account, the superpotential (5.14) obtained in the low energy theory from flowing down from that of the magnetic, $N_f = N_c + 2$, $SU(2)$ theory is

$$W_o = \frac{1}{\Lambda_L^{2N_c-1}} \widehat{M}_{\tilde{j}}^{i} B_i \widetilde{B}^{\tilde{j}}, \tag{5.15}$$

where Λ_L is the scale of the low energy electric theory with $N_f = N_c + 1$ light flavors. However, because the magnetic theory is completely Higgsed in the flow down from $N_f = N_c + 2$, the low energy theory also includes instanton contributions in the broken magnetic group. In particular, the superpotential has an additional term which is the magnetic analog of (4.6)

$$W_{inst} = \frac{\widetilde{\Lambda}_L^{6-(N_c+2)} \det(\mu^{-1}\widehat{M})}{q^{N_f+2}\tilde{q}^{N_f+2}} = -\frac{\det \widehat{M}}{\Lambda_L^{3N_c-(N_c+1)}}, \tag{5.16}$$

where use was made of (5.6). Adding this to the superpotential (5.15) and dropping the hats and the L subscript, the superpotential of the low energy theory with $N_f = N_c + 1$ light flavors is

$$W = \frac{1}{\Lambda^{2N_c-1}} (M_{\tilde{j}}^{i} B_i \widetilde{B}^{\tilde{j}} - \det M). \tag{5.17}$$

This is precisely the superpotential (4.20). In the electric description (4.20) is associated with strong coupling effects. In the magnetic description it is thus rederived in a weakly coupled framework.

Another way to analyze the theory with mass terms is to consider the massless theory for generic values of $\langle M \rangle$. The dual quarks acquire mass $\mu^{-1} M$ and the low energy magnetic theory is pure glue $SU(N_f - N_c)$ Yang-Mills theory with scale $\widetilde{\Lambda}_L^{3(N_f-N_c)} = \mu^{-N_f} \widetilde{\Lambda}^{3(N_f-N_c)-N_f} \det M$. As in (4.10), gluino condensation in this theory leads to an effective superpotential

$$W_{\text{eff}} = (N_f - N_c)\widetilde{\Lambda}_L^3 = (N_c - N_f) \left(\frac{\Lambda^{3N_c-N_f}}{\det M} \right)^{1/(N_c-N_f)}, \tag{5.18}$$

where we used (5.6) (the $(-1)^{N_f-N_c}$ sign in that relation plays a crucial role in getting the correct overall sign here), which is the same as the continuation of (4.6) to these values of N_c, N_f. This guarantees that the superpotentials (4.6) and the expectation values of $\langle M_{\tilde{j}}^{i} \rangle$ are reproduced correctly when mass terms are added to the magnetic theory.

The infrared fixed point can also be deformed by giving the fields in the electric theory expectation values along the D-flat directions (3.3). Consider, for example, large $\langle Q^{N_f} \rangle = \langle \widetilde{Q}_{\tilde{N}_f} \rangle$, breaking the electric $SU(N_c)$ theory with N_f flavors to $SU(N_c-1)$ with N_f-1 light flavors. The low energy electric theory has scale $\Lambda_L^{3(N_c-1)-(N_f-1)} = \Lambda^{3N_c-N_f}/\langle Q^{N_f}\widetilde{Q}_{N_f}\rangle$ and the fixed point is at weaker coupling. In the magnetic description, the large $\langle M_{\tilde{N}_f}^{N_f}\rangle$ gives a large mass $\mu^{-1}\langle M_{\tilde{N}_f}^{N_f}\rangle$ to the flavor $q_{N_f}\tilde{q}^{\tilde{N}_f}$. The low energy magnetic theory is $SU(N_f-N_c)$ with N_f-1 light flavors and scale $\widetilde{\Lambda}_L^{3(N_f-N_c)-(N_f-1)} = \mu^{-1}\langle M_{\tilde{N}_f}^{N_f}\rangle\widetilde{\Lambda}^{3(N_c-N_f)-N_f}$. The low energy magnetic theory is at stronger coupling and is the dual of the low energy electric theory. Note that the flow preserves (5.6) and (5.10). Deformations along the flat directions with $\langle B\rangle \neq 0$ directions were analyzed in Ref.38.

Classically, and to all orders of perturbation theory, the electric and magnetic theories have different moduli spaces of vacua – it is only after taking non-perturbative effects into account that they are seen to be identical. For example, in the electric theory there is a classical constraint rank $\langle M\rangle \leq N_c$. In the dual theory, M is an independent field whose expectation value is unconstrained to all orders of perturbation theory – the constraint arises in the dual theory by quantum effects! Defining $N_i^{\tilde{j}} = q_i\tilde{q}^{\tilde{j}}$, the M equations of motion imply that vacua of (5.5) are at $\langle N_i^{\tilde{j}}\rangle = 0$. However, because the magnetic theory has $N_f - \text{rank}\langle M\rangle$ massless flavors, for $N_f - \text{rank}\langle M\rangle < N_f - N_c$ the magnetic theory generates a superpotential analogous to (4.6) and there is no vacuum with $\langle N\rangle = 0$. The vacua of the dual theory thus also satisfy $\text{rank}\langle M\rangle \leq N_c$ but as a result of quantum effects rather than as a classical constraint. Similarly, for $\text{rank}\langle M\rangle = N_c$, the magnetic theory has $N_f - N_c = \widetilde{N}_c$ massless fields and thus develops a constraint analogous to (4.15)

$$\det N - b\tilde{b} = \widetilde{\Lambda}_L^{2\widetilde{N}_c}, \tag{5.19}$$

where $\widetilde{\Lambda}_L^{2\widetilde{N}_c} = \det'\langle\mu^{-1}M\rangle\widetilde{\Lambda}^{3\widetilde{N}_c-N_f}$, with $\det'\langle\mu^{-1}M\rangle$ the product of the N_c non-zero eigenvalues of $\langle\mu^{-1}M\rangle$. Using the M equation of motion, $\langle N\rangle = 0$, and the mapping (5.10) and (5.6), the relation (5.19) yields

$$\langle B\widetilde{B}\rangle = \det{}'\langle M\rangle. \tag{5.20}$$

This derives a constraint which is classical in the electric description (it was discussed following (3.3)) via quantum dynamics in the dual theory. In addition to providing a necessary check on the duality, the fact that classical relations arise via quantum effects in the dual is interesting in that, conversely, quantum effects in one theory can be related to classical identities in the dual.

6 $SO(N_c)$ WITH $N_f \times \mathbf{N_c}$

In the $SU(N_c)$ theories discussed above, with matter in the fundamental representation, there is no invariant distinction between Higgs and confinement [2]. This is not the case in theories based on $SO(N_c)$ with $N_f \times \mathbf{N_c}$ because the matter cannot screen sources in the spinor representation[5]. These theories, therefore, lead to a clearer pic-

[5] We will limit our considerations to the Lie algebra and not distinguish between $SO(N_c)$ and $\text{Spin}(N_c)$.

ture of the dynamics. In particular, here the transition from the Higgs phase to the Confining phase occurs with a well defined phase transition.

Many of the results in these $SO(N_c)$ theories [28, 20, 39, 40, 8, 41] are similar to the results in $SU(N_c)$, showing that some phenomena are generic. These $SO(N_c)$ theories also exhibit many new phenomena, which are not present in the $SU(N_c)$ examples. The most dramatic of them is oblique confinement [42, 43], driven by the condensation of dyons (particles with both electric and magnetic charges). This phenomenon is best described by another equivalent theory – a dyonic theory. Therefore, these theories exhibit electric-magnetic-dyonic triality [41, 44]. The discussion here will be brief; many more details can be found in [41].

6.1 The Phases

For $N_f \geq 3(N_c-2)$ massless quarks Q in the N_c dimensional representation of $SO(N_c)$, the theory is not asymptotically free and the infrared theory is thus in a free electric phase. For $\frac{3}{2}(N_c-2) < N_f < 3(N_c-2)$, the theory is asymptotically free and flows to a non-Abelian Coulomb phase fixed point in the infrared. This phase has a dual magnetic description in terms of an $SO(N_f-N_c+4)$ gauge theory which we review below. For $N_c-2 < N_f \leq \frac{3}{2}(N_c-2)$, the theory is in a free magnetic phase with a dynamically generated composite $SO(N_f-N_c+4)$ gauge invariance. For $N_f = N_c-2$, the theory is in an Abelian Coulomb phase.

6.2 Duality

As discussed in Refs.8 and 41, the infra-red behavior of these theories has a dual, magnetic description in terms of an $SO(N_f-N_c+4)$ gauge theory with N_f flavors of dual quarks q_i and an additional gauge singlet field $M^{ij} = Q^i \cdot Q^j$ with a superpotential

$$W = \frac{1}{2\mu} M^{ij} q_i \cdot q_j \tag{6.1}$$

(an additional term is required for $N_f = N_c-1$). The scale μ is needed for the same reason as in the $SU(N_c)$ duality. For generic N_c and N_f the scale $\widetilde{\Lambda}$ of the magnetic theory is related to the scale Λ of the electric theory by

$$2^8 \Lambda^{3(N_c-2)-N_f} \widetilde{\Lambda}^{3(N_f-N_c+2)-N_f} = (-1)^{N_f-N_c} \mu^{N_f}. \tag{6.2}$$

The constant 2^8 reflects the $\overline{DR}$ conventions[6].

The electric theory has an anomaly free global $SU(N_f) \times U(1)_R$ symmetry with the fields Q transforming as $(N_f)_{(N_f-N_c+2)/N_f}$. The dual theory has the same global symmetry with M transforming as $Q \cdot Q$ and q transforming as $(\overline{N}_f)_{(N_c-2)/N_f}$, which

[6]In these conventions the matching relation between the scale of the high energy theory with N_f flavors and the mass term $W_{tree} = \frac{1}{2} m Q^{N_f} \cdot Q^{N_f}$ and the low energy theory with N_f-1 flavors is $\Lambda_L^{3(N_c-2)-N_f+1} = m \Lambda^{3(N_c-2)-N_f}$ (it is $\Lambda_L^{6-2(N_f-1)} = m^2 \Lambda^{6-2N_f}$ for $N_c = 3$). Similarly, the matching relation associated with breaking $SO(N_c)$ with N_f vectors to $SO(N_c-1)$ with N_f-1 vectors by an expectation value $\langle Q^{N_f} \rangle$ is $\Lambda_L^{3(N_c-2)-N_f-2} = \Lambda^{3(N_c-2)-N_f} (M^{N_f N_f})^{-1}$ (for breaking $SO(4) \cong SU(2)_1 \times SU(2)_2$ to $SO(3)$, it is $\Lambda_L^{6-2(N_f-1)} = 4 \Lambda_1^{6-N_f} \Lambda_2^{6-N_f} (M^{N_f N_f})^{-2}$).

is anomaly free and respected by the superpotential (6.1). Furthermore, the 't Hooft anomalies of this magnetic spectrum match those of the electric spectrum.

For $N_c - 2 < N_f \leq \frac{3}{2}(N_c - 2)$, the magnetic degrees of freedom are free in the infra-red while, for $\frac{3}{2}(N_c - 2) < N_f < 3(N_c - 2)$, the electric and the magnetic theories flow to the same non-trivial fixed point of the renormalization group. Although the two theories are different away from the extreme infra-red, they are completely equivalent at long distance. This means that the two (super) conformal field theories at long distance are identical, having the same correlation functions of all of the operators, including high dimension (irrelevant) operators.

The gauge invariant (primary) chiral operators of the electric theory are

$$\begin{aligned} M^{ij} &= Q^i Q^j \\ B^{[i_1,\ldots,i_{N_c}]} &= Q^{i_1} \ldots Q^{i_{N_c}} \\ b^{[i_1,\ldots,i_{N_c-4}]} &= W_\alpha^2 Q^{i_1} \ldots Q^{i_{N_c-4}} \\ \mathcal{W}_\alpha^{[i_1,\ldots,i_{N_c-2}]} &= W_\alpha Q^{i_1} \ldots Q^{i_{N_c-2}} \end{aligned} \tag{6.3}$$

with the gauge indices implicit and contracted. These operators get mapped to gauge invariant operators of the magnetic theory as

$$\begin{aligned} M^{ij} &\to M^{ij} \\ B^{[i_1\ldots i_{N_c}]} &\to \epsilon^{i_1\ldots i_{N_f}} \tilde{b}_{i_{N_c+1}\ldots i_{N_f}} \\ b^{[i_1\ldots i_{N_c-4}]} &\to \epsilon^{i_1\ldots i_{N_f}} \tilde{B}_{[i_{N_c-3}\ldots i_{N_f}]} \\ \mathcal{W}_\alpha^{[i_1,\ldots,i_{N_c-2}]} &\to \epsilon^{i_1\ldots i_{N_f}} (\widetilde{\mathcal{W}}_\alpha)_{[i_{N_c-1},\ldots,i_{N_f}]}, \end{aligned} \tag{6.4}$$

where $\tilde{B}$, $\tilde{b}$, and $\widetilde{\mathcal{W}}_\alpha$ are the magnetic analogs of the operators in (6.3). These maps are compatible with the global symmetries discussed above.

Deforming the electric theory along a flat direction of non-zero $\langle Q^{N_f} \rangle$ gives $SO(N_c - 1)$ with $N_f - 1$ massless flavors and the new fixed point is at weaker coupling. In the dual description this deformation gives a mass $\mu^{-1}\langle M^{N_f N_f} \rangle$ to q_{N_f} and the low energy dual theory is $SO(N_f - N_c + 4)$ with $N_f - 1$ massless quark flavors, which is the dual of the low energy electric theory. In the magnetic description the new fixed point is at stronger coupling and the relation (6.2) is preserved in the low energy theories.

Some classical identities of the electric theory arise from quantum effects in the dual. For example, it is seen classically in the electric theory that $\text{rank}(M) \leq N_c$. In the magnetic description, this condition arises because, for larger rank, a superpotential is dynamically generated and the magnetic theory has no vacuum. Similarly, the electric theory has a classical relation that, when $\text{rank}(M) = N_c$, the baryon $B^{[i_1,\ldots i_{N_c}]}$ has the non-zero eigenvalue $B = \pm\sqrt{\det{}' M}$. This is mapped under duality to the relation that, for $N_f = N_c - 4$, the operator b defined above is given by $b = \pm\sqrt{\Lambda^{2(N_c-1)}}$, which is related to gaugino condensation.

Deforming the electric theory by giving a mass to Q^{N_f} gives $SO(N_c)$ with $N_f - 1$ massless flavors and the new fixed point is at stronger coupling. In the dual description, adding $W_{\text{tree}} = \frac{1}{2} m M^{N_f N_f}$ to (6.1) and integrating out the massive fields gives $\langle q_{N_f} \cdot q_{N_f} \rangle = -\mu m$ which, along with the D-terms, implies that the dual gauge group is broken by the Higgs mechanism to $SO(N_f - N_c + 3)$ with $N_f - 1$ massless flavors, which is the dual of the low energy electric theory. In the magnetic description the new fixed point is at weaker coupling and the relation (6.2) is preserved in the low energy theories.

For $N_f = N_c - 1$ the dual gauge group is $SO(3)$ and the superpotential (6.1) is modified to

$$W = \frac{1}{2\mu} M^{ij} q_i \cdot q_i - \frac{1}{64\Lambda_{N_c,N_c-1}^{2N_c-5}} \det M. \tag{6.5}$$

Also, the relation (6.2) between the scales is modified in this case to

$$2^{14}(\Lambda_{N_c,N_c-1}^{2N_c-5})^2 \widetilde{\Lambda}_{3,N_c-1}^{6-2(N_c-1)} = \mu^{2(N_c-1)}. \tag{6.6}$$

These modifications arise upon going from $N_f = N_c$ with a mass term added for Q^{N_c} to the low energy theory with $N_f = N_c - 1$ because of peculiarities associated with the breaking of the magnetic $SO(4) \cong SU(2) \times SU(2)$ to the diagonally embedded magnetic $SO(3)$ [8, 41].

7 Abelian Coulomb Phase

7.1 General Features

Consider a general theory with a low energy $N = 1$ Abelian Coulomb phase. For simplicity we consider only the situation with a single photon. As discussed in sect. 2, the effective gauge coupling $\tau_{\text{eff}}(X_r, g_I)$ depends holomorphically on the light fields X_r and the coupling constants, including the scale Λ of the underlying non-Abelian G theory in which the low energy, Abelian theory is embedded. However, τ_{eff} is not a single valued function of X_r and the couplings. This is possible because τ gives a redundant description of the physics: τ is physically identified under $SL(2, Z)$ transformations, generated by $S : \tau \to -1/\tau$, which is associated with the possibility of exchanging electric with magnetic in ordinary Maxwell theory, and $T : \tau \to \tau + 1$, which is a unit shift of the theta angle. In order for physics to be single valued, τ need only be a section of an $SL(2, Z)$ bundle [39].

For simplicity, we will consider the case where τ only depends on a single light field (or a single function of the light fields) U whose expectation value serves as an order parameter for breaking the underlying non-Abelian G gauge theory to a $U(1)$ subgroup. For large U, the G gauge theory is weakly coupled and the one loop beta function in the microscopic theory leads to

$$\tau \approx \frac{ib}{2\pi} \log \frac{U}{\Lambda^p}, \tag{7.1}$$

for some integers b and p, where Λ is the scale of G. As we circle around infinity, $U \to e^{2\pi i} U$, $\tau \to \tau - b$; i.e. τ is transformed by $\mathcal{M}_\infty = T^{-b}$. So even at weak coupling τ is not single valued. The low energy effective gauge coupling $\frac{1}{g_{\text{eff}}^2} \sim \text{Im}\tau$ is invariant under $\mathcal{M}_\infty$. However, if $\text{Im}\tau$ were single valued everywhere in the interior of the moduli space, because it is a harmonic function, it couldn't be everywhere positive definite [39]. There would then be regions in the moduli space where $g_{\rm eff}$ is imaginary. This unphysical conclusion can be avoided if the topology of the moduli space is complicated in the interior or, as found in Ref.39, there are several (at least two) singular values U_i of U with monodromies $\mathcal{M}_i$ around them which do not commute with $\mathcal{M}_\infty = T^{-b}$.

The monodromies $\mathcal{M}_i$ around the U_i must have a physical interpretation. The simplest one is that they are associated with k_i massless particles at the singularity. The low energy superpotential near U_i then has the form

$$W_L{}^{(i)} = (U - U_i)\sum_{l=1}^{k_i} c_l^{(i)} \widetilde{E}_l^{(i)} E_l^{(i)} + \mathcal{O}((U - U_i)^2) \tag{7.2}$$

where $\widetilde{E}_l^{(i)}$ and $E_l^{(i)}$ are the new massless states. If the constants $c_l^{(i)}$ are nonzero, these states acquire a mass of order $\mathcal{O}((U - U_i))$ away from the singularity. Therefore, the one loop beta function in the low energy theory leads to

$$\tau_i \approx -\frac{ik_i}{2\pi}\log(U - U_i) \tag{7.3}$$

(we assume for simplicity that, as in Refs.39 and 5, all the $E_l^{(i)}$ have charge one; the generalization to other cases is straightforward) where τ_i is the coupling to the low energy photon. τ_i is related to τ in the asymptotic region by a duality transformation N_i. It is clear from (7.3) that the monodromy in τ_i is T^{k_i}. Therefore, the monodromy in τ is

$$\mathcal{M}_i = N_i^{-1} T^{k_i} N_i. \tag{7.4}$$

For $\mathcal{M}_i$ to not commute with $\mathcal{M}_\infty = T^{-b}$, the transformation N_i must include S. This means that the massless particles $E_l^{(i)}$ at U_i are magnetically charged.

As discussed in Ref.39, because τ is a section of an $SL(2, \mathbf{Z})$ bundle it is naturally described as the modular parameter τ of a torus. A torus is conveniently described by the one complex dimensional curve in C^2:

$$y^2 = x^3 + Ax^2 + Bx + C, \tag{7.5}$$

where $(x, y) \in C^2$ and A, B and C are parameters. The modular parameter of the torus (7.5) is given by

$$\tau(A, B, C) = \frac{\int_b \frac{dx}{y}}{\int_a \frac{dx}{y}}, \tag{7.6}$$

where a and b refer to a basis of cycles around the branch cuts of (7.5) in the x plane. The problem of finding the section τ is thus reduced to the simpler problem of finding A, B, and C as functions (rather than sections) of U and the various coupling constants and scales.

The τ obtained from (7.5) is singular when the torus is singular, which is when

$$x^3 + Ax^2 + Bx + C = 0 \qquad \text{and} \qquad 3x^2 + 2Ax + B = 0. \tag{7.7}$$

Eliminating x, this is when the discriminant of the cubic equation in (7.7) vanishes: $\Delta(A, B, C) = 0$ where

$$\Delta = 4A^3C - B^2A^2 - 18ABC + 4B^3 + 27C^2. \tag{7.8}$$

As discussed in Refs.39 and 5, the order of the zero can be used to determine the monodromy (7.4) around the singularity and, thus, the charge of the associated massless fields.

Important constraints [5] on the dependence of the coefficients in (7.5) on U and the coupling constants are the following:

1. In the weak coupling limit $\Lambda = 0$ the curve should be singular for every U. Without loss of generality we can then take $y_0^2 = x^2(x - U)$.

2. The parameters A, B, C in (7.5) are holomorphic in U and the various coupling constants. This guarantees that τ is holomorphic in them.

3. The curve (7.5) must be compatible with all the global symmetries of the theory including those which are explicitly broken by the coupling constants or the anomaly.

4. In various limits (e.g. as some mass goes to zero or infinity) we should recover the curves of other models.

5. The curve should have physical monodromies around the singular points.

7.2 $SO(N_c)$ with $N_f = N_c - 2$

For $N_f = N_c - 2$ the gauge group is broken by $\langle Q \rangle$ to $SO(2) \cong U(1)$ and the theory has an Abelian Coulomb phase. Applying the considerations of the previous section, it is found that the Abelian Coulomb phase has an effective gauge coupling $\tau_{\text{eff}}(\det M, \Lambda)$ which is exactly given by the curve [41]

$$y^2 = x^3 + x^2(-\det M + 8\Lambda^{2N_c-4}) + 16\Lambda^{4N_c-8}x. \tag{7.9}$$

For example, at weak coupling (large $\det\langle M \rangle$), (7.9) properly reproduces the one loop beta function of $SO(N_c)$ with N_f fields Q. The curve (7.9) has singularities at the solutions of (7.8), which are $\det M = 0$ and $\det M = 16\Lambda^{2N_c-4}$.

Classically, the submanifold $\det M = 0$ has a singularity associated with a non-Abelian Coulomb phase with some of the $SO(N_c)$ gluons becoming massless. In the quantum theory, the monodromy of τ implied by (7.9) around $\det M = 0$ reveals that this submanifold of the moduli space of vacua is actually in an Abelian free magnetic phase associated with massless monopoles. At the origin there are N_f massless monopoles $q_i^{\pm}$. Away from the origin, they obtain a mass matrix proportional to $\langle M \rangle$ via

$$W \sim M^{ij} q_i^+ q_j^-. \tag{7.10}$$

The monodromy obtained from (7.9) around $\det M = 16\Lambda^{2N_c-4}$ reveals that this submanifold is in a free dyonic phase associated with a single dyonically charged field

E which is massless at $\det M = 16\Lambda^{2N_c-4}$; near the singularity the dyon gets a mass via

$$W \sim (\det M - 16\Lambda^{2N_c-4})E^+E^-. \tag{7.11}$$

The quantum numbers of the dyon E are such that $E \sim q_i Q^i$. For example, if we take the q_i to be monopoles with zero electric charge, E is a dyon with electric charge one. Thus, as a matter of convention, we will refer to the q_i as monopoles and to E as a dyon.

To summarize, we should expect to find two dual descriptions of the original electric theory, one in which the monopoles q_i are taken as fundamental fields and one in which the dyon E is a taken as a fundamental field.

Consider the duality discussed in the previous section. For $N_f = N_c - 2$ the dual magnetic theory is $SO(2) \cong U(1)$ with N_f charged fields $q_i^{\pm}$ and neutral fields M^{ij} with a superpotential

$$W \sim M^{ij} q_i^+ q_j^-. \tag{7.12}$$

To see the relation of this Abelian dual to the non-Abelian duals considered in the previous section, consider flowing from $N_f = N_c - 1$ to $N_f = N_c - 2$ by giving an electric quark a mass. In the magnetic theory this generically Higgses the magnetic $SO(3)$ to a magnetic $SO(2)$ and the low energy superpotential is (7.12). (There are additional contributions to (7.12) from instantons in the broken magnetic $SO(3)$ [41].) The components $q_i^{\pm}$ of the magnetic quarks become magnetic monopoles in the low energy Abelian Coulomb phase. The N_f monopoles $q_i^{\pm}$ with superpotential (7.12) is precisely the situation (7.10), determined above from the curve (7.9).

The dyon (7.11) is seen by a strong coupling analysis in either the electric or the magnetic theories [8, 41]. There is also a dyonic dual description in which the dyons E and $\tilde{E}$ are taken as fundamental fields. Just as the $N_f = N_c - 2$, Abelian magnetic dual can be obtained by flowing down from the non-Abelian, $N_f = N_c - 1$, $SU(2)$ magnetic dual, the Abelian dyonic dual can be obtained by flowing down from a dyonic dual, $N_f = N_c - 1$, $SO(N_c)$ gauge theory. This dyonic dual theory [41] has composite $SO(N_c)$ gauge fields interacting with a theta angle which differs from that of the original electric theory by π and it has composite quarks d^i, satisfying $d^i \cdot d^j = M^{ij}$, with superpotential

$$W = -\frac{1}{32\Lambda^{2N_c-5}} \det d \cdot d. \tag{7.13}$$

Adding $W_{tree} = \frac{1}{2} m Q^{N_c-1} \cdot Q^{N_c-1} = \frac{1}{2} m d^{N_c-1} \cdot d^{N_c-1}$ the low energy electric theory has $N_f = N_c - 2$ and the low energy dyonic dual has an unbroken $SO(2)$ with a charged pair which becomes massless at $\det M = 16\Lambda_L^{2N_c-4}$, coming from d^{N_c-1}, which is seen at weak coupling.

Consider giving a mass to the $N_c - 2$-th flavor by adding $W_{\rm tree} = \frac{1}{2} m M^{N_c-2, N_c-2}$, giving $SO(N_c)$ with $N_f = N_c - 3$ light flavors at low energy. Adding $W_{\rm tree}$ eliminates the Coulomb phase and the low energy theory has two distinct branches. One branch is associated with adding $W_{\rm tree}$ to (7.12) which, upon integrating out the massive fields, gives the monopole condensate $\langle q_{N_f} q_{N_f} \rangle \sim m$, yielding a confining phase by the dual Meissner effect. The low energy theory has a moduli space of vacua labeled by $\langle \widehat{M} \rangle$ with $N_c - 3$ massless fields at the origin, with a superpotential given by

$$W \sim M^{ij} q_i q_j, \tag{7.14}$$

where we dropped the hats and $i, j = 1 \dots N_c - 3$. In the magnetic description, the fields q_i are left-over components of the magnetic quarks. By the map (6.5), the quark component q_i in this case is mapped to the exotic $q_i \sim (Q)^{N_c-4} W_\alpha W^\alpha$. Intuitively, one thinks of such exotics as being large and heavy bound states. Here we see that they become massless at $\langle M \rangle = 0$. This phenomenon is similar to the massless composite mesons and baryons found in $SU(N_c)$ with $N_f = N_c + 1$ [32]. As was the case there, we have confinement without chiral symmetry breaking.

The other branch is associated with adding W_{tree} to (7.11). Integrating out the massive fields gives the dyon condensate $\langle E^+ E^- \rangle \sim m / \det \widehat{M}$, yielding an oblique confining phase. The low energy theory has a superpotential

$$W_{\text{oblique}} = \frac{8\Lambda_L^{2N_c-3}}{\det \widehat{M}}. \tag{7.15}$$

In the electric description, this superpotential is associated with gaugino condensation in the unbroken electric $SO(3)$ along with contributions from instantons in the broken $SO(N_c)/SO(3)$ [41]. In the magnetic description (7.15) arises from dyon condensation. In the dyonic description (7.13) it arises at tree level.

8 $SO(3) \simeq SU(2)$ Examples

The discussion of the previous sections is modified slightly for $SO(3) \cong SU(2)$, where the 3 dimensional representation is the adjoint representation.

8.1 *One Adjoint, Q; an Abelian Coulomb Phase*

This is the $N = 2$ theory discussed in Ref.39. The theory has a quantum moduli space of vacua labeled by the expectation value of the massless meson field $M = Q^2$. The $SU(2)$ gauge symmetry is broken to $U(1)$ on this moduli space, so the theory has a Coulomb phase with a massless photon.

As discussed in Refs.39 and 5, the effective gauge coupling in the Coulomb phase is given by the $\tau_{\text{eff}}(M)$ obtained from the curve

$$y^2 = x^3 - Mx^2 + 4\Lambda^4 x. \tag{8.1}$$

(This curve is expressed using the convention for the normalization of τ discussed in Ref.5, $\tau = \frac{\theta}{\pi} + 8\pi i g^{-2}$.) This gives a $\tau_{\text{eff}}(M)$ which has singularities associated with a massless magnetic monopole field $q_{(+)}$, at $M = 4\Lambda^2$ and a massless dyon $q_{(-)}$ at[7] $M = -4\Lambda^2$. Therefore, these two points are in a free magnetic and a

[7]We use the conventions of Refs.20 and 29 where the normalization of Λ^2 (in the $\overline{DR}$ scheme) differs by a factor of 2 from that of Ref.39; our order parameter M is related to u of Ref.39 as $u = \frac{1}{2}M$.

free dyonic phase, respectively. Here $q_{(+)}$ is a doublet charged under the magnetic $U(1)_M$, which is related to the electric $U(1)_E$ by the electric-magnetic transformation S: $F \to \widetilde{F}$ (modulo $\Gamma(4) \subset SL(2,Z)$). Similarly, $q_{(-)}$ is a doublet charged under a dyonic $U(1)_D$, related to $U(1)_E$ by the $SL(2,Z)$ transformation ST^2 (again, modulo $\Gamma(4) \subset SL(2,Z)$), where T is a rotation of the theta angle by π. Near where these fields are massless, they couple through the effective superpotentials

$$W_{\pm} \sim \left(M \mp 4\Lambda^2\right) q_{(\pm)} \cdot q_{(\pm)}. \tag{8.2}$$

Referring to the underlying $SU(2)$ theory as "electric," we can say that it has two dual theories. One of them, which we can refer to as the "magnetic dual," describes the physics around $M = 4\Lambda^2$ with the superpotential W_+. The other dual, which can be called the "dyonic dual," is valid around $M = -4\Lambda^2$ and is described by W_-.

Consider giving Q a mass by adding a term $W_{\text{tree}} = \frac{1}{2}mM$ in the electric theory. Adding W_{tree} to (8.2), the equations of motion give $\langle q_{(\pm)} \cdot q_{(\pm)}\rangle \sim m$ and lock $\langle M\rangle = \pm 4\Lambda^2$. The condensate of monopoles/dyons Higgses the dual theory and thus gives confinement/oblique confinement of the electric theory by the dual Meissner effect [39].

Consider analyzing, as in sect. 2.3, the 1PI effective action for this theory with sources. Starting from the analog of (4.10) for this theory, $W_L(m) = \pm 2(\Lambda^4 m^2)^{1/2}$, equations (2.8) and (2.9) give $W = 0$ with the constraint $\langle M\rangle = \pm 4\Lambda^2$. Indeed, adding the source m for M drives the theory to the confining or oblique confining phase with $\langle M\rangle = \pm 4\Lambda^2$. The Coulomb phase cannot be explored in the theory with a mass term for Q. As discussed in sect. 2.3, this method of analyzing the theory must fail to capture some of the physics because the theory without sources has massless fields (the monopole or dyon) which cannot be represented by the gauge invariant observables.

Because this theory actually has $N = 2$ supersymmetry, it can be further analyzed using the additional techniques applicable for $N = 2$ theories, yielding the Kahler potential for Q and a BPS mass bound [39].

8.2 $SU(2)$ with Two Adjoints; A Non-Abelian Coulomb Phase

This theory has $N = 1$ (not $N = 2$) supersymmetry. Writing the matter fields as Q^i with $i = 1,2$ a flavor index, there is a 3 complex dimensional moduli space of classical vacua parameterized by the expectation values of the gauge singlet fields $M^{ij} = Q^i \cdot Q^i$. In the generic vacuum $\langle Q^1\rangle$ breaks $SU(2)$ to a $U(1)$ which is then broken by $\langle Q^2\rangle$. For $\det\langle M^{ij}\rangle \neq 0$, the gauge group is completely broken and the theory is in the Higgs phase. On the non-compact two complex dimensional subspace of vacua with $\det M = 0$, there is an unbroken $U(1)$ gauge symmetry and thus a light photon along with a pair of massless electrically charged fields. At the point $\langle M\rangle = 0$ the $SU(2)$ gauge group is unbroken.

We now turn to the quantum theory. The theory has the global symmetry group $SU(2) \times U(1)_R$, with Q transforming as $\mathbf{2}_{\frac{1}{2}}$, which determines that any dynamically generated superpotential must be of the form

$$W = \frac{c}{\Lambda} \det M, \tag{8.3}$$

with c a dimensionless constant. Its behavior at $M \to \infty$ is incompatible with asymptotic freedom, as signaled by the presence of the scale Λ in the denominator. Therefore, no superpotential can be generated and the classical vacuum degeneracy outlined above is not lifted quantum mechanically.

The generic ground state with generic M is in the Higgs phase. Consider now the subspace of the moduli space with $\det M = 0$. The low energy degrees of freedom there are a single photon, a pair of massless electrically charged fields and some neutral fields. This theory cannot become strong in the infrared. In fact, the loops of the massless charged fields renormalize the electric charge to zero. Therefore, this subspace of the moduli space is in a free electric phase.

Now consider adding a tree level superpotential $W_{\text{tree}} = \frac{1}{2}\text{Tr } mM$. Taking $m = \begin{pmatrix} 0 & 0 \\ 0 & m_2 \end{pmatrix}$, Q^2 gets a mass and can be integrated out. The low energy theory is $SU(2)$ with a single massless adjoint matter field, which is the example of the previous subsection. This low energy theory has a scale Λ_L, which is related to that of the original theory by $\Lambda_L^4 = m_2^2\Lambda^2$, and a massless monopole or dyon at $\langle M^{11}\rangle = \pm 4\Lambda_L^2 = \pm 4m_2\Lambda$. Note that as $m_2 \to 0$ the point $\langle M\rangle = 0$ has both massless monopoles and dyons. These are mutually non-local[8] and signal another phase at this point in the theory with $m_2 = 0$. We interpret this as a non-Abelian Coulomb phase [40].

Starting from the theory with $m_2 \neq 0$, turning on $m_1 \neq 0$ drives the monopole or dyon to condense and the vacuum is locked at $\langle M^{11}\rangle = \pm 4m_2\Lambda$. The $+$ sign is a vacuum with monopole condensation and thus confinement. The $-$ sign is a vacuum with dyon condensation and thus oblique confinement. More generally, these vacua are at $\langle M^{ij}\rangle = \pm 4\Lambda \det m (m^{-1})^{ij}$. These expectation values can be obtained from

$$W_e = \frac{e}{8\Lambda}\det M + \frac{1}{2}\text{Tr } mM, \tag{8.4}$$

with $e = \mp 1$ for confinement and oblique confinement, respectively.

The theory has various phase branches. For mass $m = 0$ there is a Higgs phase which, in terms of W_e, corresponds to $e = 0$. There is a subspace $\det M = 0$ in the free electric phase and the point $M = 0$ in a non-Abelian Coulomb phase. For $m \neq 0$ but with $\det m = 0$ the theory is in the Coulomb phase with a free magnetic point and a free dyonic point. For $\det m \neq 0$ the theory is either confining and described by the superpotential (8.4) with $e = -1$ or it is oblique confining and described by the superpotential (8.4) with $e = 1$.

If we consider the 1PI effective action, the analog of (4.10) is

$$W_L(m) = \pm 2(\Lambda^2 \det m)^{1/2}.$$

Integrating in gives the confining or oblique confining phase branches of the superpotential (8.4), with $e = \pm 1$, missing the $e = 0$ Higgs phase branch. Again, as discussed in sect. 2.3, the 1PI superpotential necessarily fails to capture some of the physics because the theory without the sources has massless particles, the quarks and the gluons, which cannot be represented by the gauge invariant observables.

[8]A similar situation was found in $N = 2$ $SU(3)$ Yang Mills theory [45].

The analysis of Ref.41 reveals that this electric theory has two dual descriptions, similar to the magnetic and dyonic duals discussed in the previous section, labeled by $\epsilon = \pm 1$. The two dual theories are based on an $SU(2)$ gauge group with two fields q_i in its adjoint representation and three gauge singlet fields M^{ij}. The difference between the two theories is in the superpotential

$$W_\epsilon = \frac{1}{12\sqrt{\Lambda\widetilde{\Lambda}}} M^{ij} q_i \cdot q_j + \epsilon \left(\frac{1}{24\Lambda} \det M + \frac{1}{24\widetilde{\Lambda}} \det q_i \cdot q_j \right), \tag{8.5}$$

where $\widetilde{\Lambda}$ is the scale of the dual $SU(2)$ (we expressed μ in terms of Λ and $\widetilde{\Lambda}$). The theory with $\epsilon = 1$ is a "magnetic" dual and that with $\epsilon = -1$ a "dyonic" dual.

We now analyze the dynamics of these dual theories. Since they are similar to the original electric theory, we proceed as we did there. These theories have three phases: Higgs, confining and oblique confining. We study them using the gauge invariant order parameters $N_{ij} \equiv q_i \cdot q_j$. Its effective superpotential is obtained by writing the tree level superpotential (8.5) in terms of N and adding to it $\frac{\tilde{e}}{8\widetilde{\Lambda}} \det N$ where, in the Higgs, confining and oblique confinement branches, $\tilde{e} = 0, -1, 1$, respectively

$$W_{\epsilon,\tilde{e}} = \frac{1}{12\sqrt{\Lambda\widetilde{\Lambda}}} \mathrm{Tr}\; MN + \epsilon \left(\frac{1}{24\Lambda} \det M + \frac{1}{24\widetilde{\Lambda}} \det N \right) + \frac{\tilde{e}}{8\widetilde{\Lambda}} \det N. \tag{8.6}$$

Now we can integrate out the massive field N to find

$$W_{\text{eff}} = \frac{1}{8\Lambda} \frac{\tilde{e} - \epsilon}{1 + 3\tilde{e}\epsilon} \det M. \tag{8.7}$$

This is the same as the effective superpotential (8.4) of the electric theory with

$$e = \frac{\tilde{e} - \epsilon}{1 + 3\tilde{e}\epsilon}. \tag{8.8}$$

We see that the various phases are permuted in the different descriptions as:

THEORY		PHASES	
electric	Higgs ($e = 0$)	conf. ($e = -1$)	obl. conf. ($e = 1$)
magnetic ($\epsilon = 1$)	obl. conf. ($\tilde{e} = 1$)	Higgs ($\tilde{e} = 0$)	conf. ($\tilde{e} = -1$)
dyonic ($\epsilon = -1$)	conf. ($\tilde{e} = -1$)	obl. conf. ($\tilde{e} = 1$)	Higgs ($\tilde{e} = 0$)

It is a simple exercise to check that by dualizing the magnetic and dyonic theories as we above dualized the electric theory (two duals of each), we find permutations of the same three theories. The S_3 triality permuting the phases and branches is associated with a quotient of the $SL(2, Z)$ electric-magnetic duality symmetry group: the theories are preserved under $\Gamma(2) \subset SL(2, Z)$, leaving the quotient $S_3 = SL(2, Z)/\Gamma(2)$ with a non-trivial action.

This discussion leads to a new interpretation of the first term in (8.4). In the electric theory this term appears as a consequence of complicated strong coupling dynamics in the confining and the oblique confinement branches of the theory. In the dual descriptions it is already present at tree level.

Consider the theory with a mass m_2 for Q^2. As discussed above, the low energy electric theory has a Coulomb phase with massless monopoles or dyons at the strong coupling singularities $\langle M^{11}\rangle = \pm 4m_2\Lambda$. We now derive this result in the dual theories. Adding $W_{\rm tree} = \frac{1}{2}m_2 M^{22}$ to the superpotential (8.5) of the dual theory, the equations of motion give

$$\begin{aligned} \frac{1}{12\sqrt{\Lambda\tilde\Lambda}}q_2\cdot q_2 + \frac{8\epsilon}{24\Lambda}M^{11} + \tfrac{1}{2}m_2 = 0 \quad & M^{22} = -\tfrac{1}{2}\epsilon\sqrt{\tfrac{\tilde\Lambda}{\Lambda}}q_1\cdot q_1 \\ q_1\cdot q_2 = 0 \quad & \\ & M^{12} = 0. \end{aligned} \tag{8.9}$$

For $q_2^2 \neq 0$, $\langle q_2\rangle$ breaks the gauge group to $U(1)$ and the remaining charged fields $q_1^{\pm}$ couple through the low energy superpotential

$$\frac{1}{16\sqrt{\Lambda\tilde\Lambda}}(M^{11} - 4\epsilon m_2\Lambda)q_1^+ q_1^- . \tag{8.10}$$

(This superpotential is corrected by contributions from instantons in the broken magnetic $SU(2)$ theory. However, these are negligible near $M^{11} = 4\epsilon m_2\Lambda$.) We see that the theory has a charged doublet of massless fields $q_1^{\pm}$ at $M^{11} = 4\epsilon m_2\Lambda$, exactly as expected from the analysis of the electric theory. There these states appeared as a result of strong coupling effects. Here we see them as weakly coupled states in the dual theories. This is in accord with the interpretation of the $\epsilon = 1$ ($\epsilon = -1$) theory as magnetic (dyonic).

The other monopole point on the moduli space of the theory with $m_1 = 0$ but $m_2 \neq 0$ is at $M^{11} = -4\epsilon m_2\Lambda$. It arises from strong coupling dynamics in the dual theories. To see that, note that the above analysis is not valid when the expectation value of q_2 is on the order of or smaller than the mass of q_1. In that case, q_1 should be integrated out first. The equations of motion in the low energy theory yield a single massless monopole point at $M^{11} = -4\epsilon m_2\Lambda$ [41].

An analysis similar to the one above leads to a strongly coupled state in the dual theories along the flat directions with $\det M = 0$ in the $m = 0$ case. This state can be interpreted as the massless quark of the electric theory in that free electric phase.

To conclude, this theory has three branches which are in three different phases: Higgs, confining and oblique confinement (various submanifolds of these branches are in Coulomb, free electric, free magnetic and free dyonic phases). They touch each other at a point in a non-Abelian Coulomb phase. Corresponding to the three branches there are three different Lagrangian descriptions of the theory: electric, magnetic and dyonic. Each of them describes the physics of one of the branches, where it is Higgsed, in weak coupling and the other two in strong coupling.

In both examples of this section, the theory has a discrete symmetry which relates the confining and the oblique confinement phases[9]. Therefore, in these cases the effects of confinement are indistinguishable from the effects of oblique confinement. Correspondingly, the magnetic and the dyonic descriptions are similar – they differ

[9]This symmetry is manifest only in the electric description. In the dual descriptions it is realized as a quantum symmetry [41].

only in the sign of ϵ. In the other $SO(N_c)$ examples discussed in the previous section, these two phases are not related by a symmetry and the two dual descriptions look totally different.

9 Conclusions

To conclude, supersymmetric field theories are tractable and many of their observables can be computed exactly[10]. Our analysis led us to find new phases of non-Abelian gauge theories, like the non-Abelian Coulomb phase with its quantum equivalence and the free magnetic phase with its massless composite gauge fields. The main dynamical lesson is the role of electric-magnetic duality in non-Abelian gauge theories in four dimensions. To be clear we should distinguish several different notions of duality:

1. The exact $\tau \to -1/\tau$ duality of Maxwell theory and its generalization to the Montonen-Olive [3] duality of finite, interacting, non-Abelian theories.
2. The $\tau_{eff} \to -1/\tau_{eff}$ duality of low energy theories with an Abelian Coulomb phase. This is not a symmetry but, rather, an ambiguity in the description of the low energy physics.
3. The duality of two asymptotically free theories which flow to the same non-Abelian Coulomb phase fixed point in the infrared.
4. The duality of the free magnetic phase, which provides a relation between the UV and the IR behavior of a theory which is free in the IR.

There are relations between many of the phenomena discussed above [41, 50] but, at a deeper level, they remain to be really understood.

Acknowledgments

We would like to thank T. Banks, D. Kutasov, R. Leigh, M.R. Plesser, P. Pouliot, S. Shenker, M. Strassler and especially E. Witten for many helpful discussions. This work was supported in part by DOE grant #DE-FG05-90ER40559.

References

[1] N. Seiberg, hep-th/9506077, RU-95-37, IASSNS-HEP-95/46, to appear in the Proc. of PASCOS 95, the Proc. of the Oskar Klein lectures, and in the Proc. of the Yukawa International Seminar '95.

[2] L. Susskind, unpublished; T. Banks, E. Rabinovici, *Nucl. Phys.* B160 (1979) 349; E. Fradkin and S. Shenker, *Phys. Rev.* D19 (1979) 3682.

[10] Although we did not discuss them here, we would like to point out that many other examples were studied [28, 11, 46, 20, 5, 40, 8, 17, 65] exhibiting many new interesting phenomena.

[3] C. Montonen and D. Olive, *Phys. Lett.* 72B (1977) 117; P. Goddard, J. Nuyts and D. Olive, *Nucl. Phys.* B125 (1977) 1.

[4] H. Osborn, *Phys. Lett.* 83B (1979) 321; A. Sen, hep-th/9402032, *Phys. Lett.* 329B (1994) 217; C. Vafa and E. Witten, hep-th/9408074, *Nucl. Phys.* B432 (1994) 3.

[5] N. Seiberg and E. Witten, hep-th/9408099, *Nucl. Phys.* B431 (1994) 484.

[6] A. Hanany and Y. Oz, TAUP-2248-95, WIS-95/19, hep-th/9505075.

[7] P.C. Argyres, M.R. Plesser and A. Shapere, IASSNS-HEP-95/32, UK- HEP/95-06, hep-th/9505100.

[8] N. Seiberg, hep-th/9411149 , *Nucl. Phys.* B435 (1995) 129.

[9] E. Witten, *Nucl. Phys.* B202 (1982) 253.

[10] N. Seiberg, hep-ph/9309335, *Phys. Lett.* 318B (1993) 469.

[11] D. Amati, K. Konishi, Y. Meurice, G.C. Rossi and G. Veneziano, *Phys. Rep.* 162 (1988) 169 and references therein.

[12] M.A. Shifman and A. I. Vainshtein, *Nucl. Phys.* B277 (1986) 456; *Nucl. Phys.* B359 (1991) 571.

[13] J. Polchinski and N. Seiberg, (1988) unpublished.

[14] E. Witten, *Nucl. Phys.* B268 (1986) 79.

[15] M. Dine and N. Seiberg, *Phys. Rev. Lett.* 57 (1986) 2625.

[16] M.T. Grisaru, W. Siegel and M. Rocek, *Nucl. Phys.* B159 (1979) 429.

[17] K. Intriligator, N. Seiberg, and S. Shenker, hep-ph/9410203, *Phys. Lett.* 342B (1995) 152.

[18] V. Kaplunovsky and J. Louis, *Nucl. Phys.* B422 (1994) 57.

[19] C.P. Burgess, J.-P. Derendinger, F. Quevedo, and M. Quiros, hep-th/9595171, CERN-Th/95-111.

[20] K. Intriligator, R.G. Leigh and N. Seiberg, hep-th/9403198, *Phys. Rev.* D50 (1994) 1092.

[21] K. Intriligator, hep-th/9407106, *Phys. Lett.* 336B (1994) 409.

[22] E. B. Bogomol'nyi, Sov. J. Nucl. Phys. 24 (1976) 449.

[23] M. K. Prasad and C. M. Somerfeld, *Phys. Rev. Lett.* 35 (1975) 760.

[24] E. Witten and D. Olive, *Phys. Lett.* 78B (1978) 97.

[25] D. Mumford and J. Fogarty, *Geometric Invariant Theory* (Springer, 1982).

[26] M. A. Luty and W. Taylor IV, hep-th/9506098, MIT-CTP-2440.

[27] A.C. Davis, M. Dine and N. Seiberg, *Phys. Lett.* 125B (1983) 487.

[28] I. Affleck, M. Dine and N. Seiberg, *Nucl. Phys.* B241 (1984) 493; *Nucl. Phys.* B256 (1985) 557.

[29] D. Finnell and P. Pouliot, RU-95-14, SLAC-PUB-95-6768, hep-th/9503115.

[30] V.A. Novikov, M.A. Shifman, A.I. Vainshtein and V.I. Zakharov, *Nucl. Phys.* B260 (1985) 157; A.I. Vainshtein, V.I. Zakharov, and M.A. Shifman, Sov. Phys. Usp. 28 (1985) 709.

[31] G. Veneziano and S. Yankielowicz *Phys. Lett.* 113B (1982) 231; T. Taylor, G. Veneziano and Yankielowicz, *Nucl. Phys.* B218 (1983) 439.

[32] N. Seiberg, hep-th/9402044, *Phys. Rev.* D49 (1994) 6857.

[33] V. Novikov, M. Shifman, A. Vainshtein and V. Zakharov, *Nucl. Phys.* B229 (1983) 381.

[34] T. Banks and A. Zaks, *Nucl. Phys.* B196 (1982) 189.

[35] M. Flato and C. Fronsdal, Lett. Math. Phys. 8 (1984) 159; V.K. Dobrev and V.B. Petkova, *Phys. Lett.* 162B (1985) 127.

[36] G. Mack, *Comm. Math. Phys.* 55 (1977) 1.

[37] T.H.R. Skyrme, Proc.Roy.Soc. A260 (1961) 127; E. Witten, *Nucl. Phys.*, 160 (1979) 57 ; *Nucl. Phys.* B223 (1983) 422; *Nucl. Phys.* B223 (1983) 433.

[38] O. Aharony, hep-th/9502013, *Phys. Lett.* 351B (1995) 220.

[39] N. Seiberg and E. Witten, hep-th/9407087, *Nucl. Phys.* B426 (1994) 19.

[40] K. Intriligator and N. Seiberg, hep-th/9408155, *Nucl. Phys.* B431 (1994) 551.

[41] K. Intriligator and N. Seiberg, hep-th/9503179,*Nucl. Phys.* B444 (1995) 125.

[42] G. 't Hooft, *Nucl. Phys.* B190 (1981) 455.

[43] J. Cardy and E. Rabinovici, *Nucl. Phys.* B205 (1982) 1; J. Cardy, *Nucl. Phys.* B205 (1982) 17.

[44] K. Intriligator and N. Seiberg, hep-th/9506084, to appear in the Proceedings *Strings '95.*

[45] P. Argyres and M. Douglas, hep-th/9505062, IASSNS-HEP-95-31.

[46] A.Yu. Morozov, M.A. Olshansetsky and M.A. Shifman, *Nucl. Phys.* B304 (1988) 291.

[47] A. Klemm, W. Lerche, S. Theisen and S. Yankielowicz, hep-th/9411048, *Phys. Lett.* 344B (1995) 169; hep-th/9412158.

[48] P. Argyres and A. Faraggi, hep-th/9411057, *Phys. Rev. Lett.* 74 (1995) 3931.

[49] D. Kutasov, hep-th/9503086, *Phys. Lett.* 351B (1995) 230.

[50] R. Leigh and M. Strassler, hep-th/9503121, *Nucl. Phys.* B447 (95) 1995.

[51] U. Danielsson and B. Sundborg, USITP-95-06, UUITP-4/95, hep-th/9504102.

[52] M.R. Douglas and S.H. Shenker, RU-95-12, RU-95-12, hep-th/9503163.

[53] S. Elitzur, A Forge, A. Giveon and E. Rabinovici, RI-4-95 hep-th/9504080.

[54] O. Aharony, J. Sonnenschein and S. Yankielowicz, TAUP–2246–95, CERN-TH/95–91, hep-th/9504113.

[55] D. Kutasov and A. Schwimmer, EFI–95–20, WIS/4/95, hep-th/9505004.

[56] K. Intriligator and P. Pouliot, hep-th/9505006, *Phys. Lett.* 353B (1995) 471.

[57] K. Intriligator, hep-th/9505051, *Nucl. Phys.* B448 (1995) 187.

[58] P.C. Argyres and M.R. Douglas, RU-95-31, hep-th/9505062.

[59] M. Berkooz, RU-95-29, hep-th/9505067.

[60] R. Leigh and M. Strassler, hep-th/9505088, RU-95-30.

[61] K. Intriligator, R. Leigh and M. Strassler, hep-th/9506148, RU-95-38.

[62] P. Pouliot, hep-th/9507018, RU-95-46.

[63] I. Pesando, hep-th/9506139, NORDITA-95/42 P.

[64] S.B. Giddings and J. M Pierre, hep-th/9506196, UCSBTH-95-14.

[65] E. Poppitz and S. P. Trivedi, hep-th/9507169, EFI-95-44, Fermilab-Pub-95/258-T.

BILINEAR FUNCTIONAL EQUATIONS OF HIROTA TYPE IN STATISTICAL MODELS ON RANDOM SURFACES

IVAN K. KOSTOV

SERVICE DE PHYSIQUE THÉORIQUE DE SACLAY
CEA-SACLAY, F-91191 GIF-SUR-YVETTE, FRANCE

1 INTRODUCTION

The most transparent microscopical construction of $2D$ quantum gravity is given by the random-lattice versions of the $sl(2)$-related statistical models: the Ising model [1], the $O(\mathrm{n})$ [2] and the q-state Potts [3] models, the SOS and RSOS models and their ADE and $\hat{A}\hat{D}\hat{E}$ generalizations [4]-[5]. The corresponding matrix models have been solved perturbatively (i.e., order by order in the genus expansion) in the continuum limit, by means of a collective field theory in which the effective potential is given by the free energy of pure gravity[6]. The perturbative solution is formulated as a system of loop-space Feynman rules, where the tadpole is the partition function of the disc, etc. Some recent results in the $O(\mathrm{n})$ model [7] indicate that the exact perturbative solution may be found also at the microscopical level.

An interesting issue, which had not been studied until recently, is whether the above construction contains nonperturbative integrable structures. The nonperturbative integrability is one of the most appealing features of the matrix-model formulation of $2D$ quantum gravity [8]. The difficulty here is that the formalism of orthogonal polynomials, which can link the "standard" matrix chains to integrable hierarchies of KdV type [9], is not applicable, with the exception of the Ising model.

Recently, the partition functions of the ADE and $\hat{A}\hat{D}\hat{E}$ matrix models have been identified as τ-functions of the KP hierarchy [10]. In particular, the (canonical) partition function of the $O(2)$ model is a τ-function for the KdV hierarchy. We expect that, more generally, the partition function of the $\hat{A}_n$ model (whose target space is a circle with $n+1$ points) is a τ-function of the $(n+1)$-reduced KP hierarchy. The relation of the D and E series to the Drinfeld-Sokolov reductions of KP is still to be clarified.

Low Dimensional Applications of Quantum Field Theory
Edited by L. Baulieu *et al.*, Plenum Press, New York, 1997

In this talk we address another aspect of the relation with the KP hierarchy. We will derive bilinear functional equations of Hirota type for the nonperturbative partition functions of the $sl(2)$-related matrix models. These equations can be viewed as deformations of the Hirota equations for the KP hierarchy characterizing the hermitian one-matrix model. Unlike the traditional loop equations, which are written for boson-like quantities (traces), the Hirota-like equations relate correlation functions of fermion-like quantities (determinants). At the heart of the construction, which is in a sense a generalization of the method of orthogonal polynomials, lies the fact that for each of the considered models there exists a differential operator of second order $\mathbf{H}$ such that the partition function of the model is obtained by acting with the operator $e^{\mathbf{H}}$ on the partition function of one or several decoupled one-matrix models. The bilinear functional equations are then obtained as a deformation of the Hirota equations for the one-matrix model. Namely, the operator $e^{\mathbf{H}}$ transforms the vertex operators from $\mathbf{V}_\pm$ to $e^{\mathbf{H}}\mathbf{V}_\pm e^{-\mathbf{H}}$. In a similar way the "string equations" needed to determine completely the partition function are obtained from the string equations for the one-matrix model. Here we restrict ourselves to the "discrete" version of the bilinear equations. A similar construction exists in the continuum limit. The bilinear equations in the continuum limit are deformations of the KdV hierarchy and their perturbative solution is given by the string field diagrammatic rules formulated in [6]. A detailed derivation of these results will appear in ref. [11].

2 The Hirota equations for the hermitean matrix model

In order to make the presentation self-consistent we first derive the Hirota equations for the hermitean one-matrix model using the formalism of orthogonal polynomials. The partition function of the ensemble of all two-dimensional random lattices is given by the hermitian $N \times N$ matrix integral

$$Z_N[t] \sim \int dM \exp\Big(\mathrm{tr} \sum_{n=0}^{\infty} t_n M^n\Big). \tag{2.1}$$

The integrand depends on the matrix variable M only through its eigenvalues λ_i, $i = 1, \ldots, N$. Retaining, therefore, only the radial part of the integration measure we can write the integral (2.1) as the partition function for a system of N Coulomb particles in a common potential

$$Z_N[t] = \int \prod_{i=1}^{N} d\lambda_i \, \exp\Big(\sum_{n=0}^{\infty} t_n \lambda_i^n\Big) \prod_{1 \le i < j \le N} (\lambda_i - \lambda_j)^2. \tag{2.2}$$

In the following we will denote by $\langle \quad \rangle_{[N,t]}$ the mean value on the ensemble with partition function (2.2):

$$\langle f(\lambda_1, \ldots, \lambda_N) \rangle_{[N,t]} = \frac{1}{Z_N[t]} \int \prod_{i=1}^{N} d\lambda_i f(\lambda_1, \ldots, \lambda_N) \exp\Big(\sum_{n=0}^{\infty} t_n \lambda_i^n\Big) \prod_{1 \le i < j \le N} (\lambda_i - \lambda_j)^2. \tag{2.3}$$

In order to make all integrals finite, we assume that the integration with respect to the eigenvalues is restricted to the finite interval $[\lambda_L, \lambda_R]$ on the real axis. The choice of the integration interval does not affect the form of the bilinear equations.

It is well known [12] that the partition function (2.1) is a τ-function of the KP hierarchy of soliton equations. The global form of this hierarchy is given by the Hirota's bilinear equations [13], to be derived below (for a review on the theory of the τ-functions see, for example, [14]).

Let us first recall that the polynomials

$$P_{[N,t]}(z) = \langle \det(z - M) \rangle_{[N,t]} \equiv \Big\langle \prod_{i=1}^{N} (z - \lambda_i) \Big\rangle_{[N,t]}, \qquad N = 0, 1, 2, \ldots \tag{2.4}$$

satisfy the orthogonality relations [15], [16]

$$\int_{\lambda_L}^{\lambda_R} d\lambda e^{\sum_n t_n \lambda^n} P_{[N,t]}(\lambda) P_{[k,t]}(\lambda) = \delta_{N,k} \frac{Z_{N+1}[t]}{(N+1) Z_N[t]}. \tag{2.5}$$

These relations can be written in the form of a contour integral, namely,

$$\frac{1}{2\pi i} \oint_C dz \left\langle \det(z - M) \right\rangle_{[N,t]} \left\langle \frac{1}{\det(z - M)} \right\rangle_{[k+1,t]} = \delta_{N,k}, \tag{2.6}$$

where the integration contour encloses the interval $[\lambda_L, \lambda_R]$. Indeed, after applying the Cauchy formula, the residue of each of the n poles is equal to the left hand side of (2.5) multiplied by $Z_k[t]/Z_{k+1}[t]$.

A set of more powerful identities follow from the fact that the polynomial $P_N[z,t]$ is orthogonal to *any* polynomial of degree less than N and in particular to the polynomials $P_{N'-1}[z,t'], N' = 1, 2, \ldots, N$, where $t' = \{t'_n, n = 1, 2, \ldots\}$ is another set of coupling constants. Written in the form of contour integrals, these orthogonality relations state

$$\oint_C dz \, \exp\Big(\sum_{n=1}^{\infty} (t_n - t'_n) z^n \Big) \left\langle \det(z - M) \right\rangle_{[N,t]} \left\langle \frac{1}{\det(z - M)} \right\rangle_{[N',t']} = 0 \qquad (N' \leq N). \tag{2.7}$$

The Hirota equations for the KP hierarchy are obtained from eq. (2.7) after expressing the mean values of $\langle \det^{\pm 1} \rangle$ in terms of the vertex operators

$$\mathbf{V}_{\pm}(z) = \exp\Big(\pm \sum_{n=0}^{\infty} t_n z^n \Big) \exp\Big(\mp \ln \frac{1}{z} \frac{\partial}{\partial t_0} \mp \sum_{n=1}^{\infty} \frac{z^{-n}}{n} \frac{\partial}{\partial t_n} \Big). \tag{2.8}$$

Namely, from the definition of the partition function (2.2) we find for $|z| > \max\{\lambda_L, \lambda_R\}$

$$\mathbf{V}_{\pm}(z) \cdot Z_N[t] = e^{\pm \sum_{n=0}^{\infty} t_n z^n} \Big\langle \det(z - M)^{\pm 1} \Big\rangle_{[N,t]} Z_N[t] \tag{2.9}$$

and eq. (2.7) can be written as

$$\oint_C dz \, \Big(\mathbf{V}_+(z) \cdot Z_N[t] \Big) \Big(\mathbf{V}_-(z) \cdot Z_{N'}[t'] \Big) = 0 \qquad (N' \leq N), \tag{2.10}$$

which is one of the forms of the Hirota equation for KP [14]. Since the integrand has its only singularity at infinity, the contour integral can be evaluated by taking the residue at infinity. After a change of variables

$$x_n = \frac{t_n + t'_n}{2}, \quad y_n = \frac{t_n - t'_n}{2}, \tag{2.11}$$

the Hirota equations (2.10) take their canonical form

$$\oint_\infty dz\, z^{N-N'} \exp\Big(\sum_{n=1}^{\infty} 2y_n z^n\Big) \exp\Big(-\sum_{n=1}^{\infty} \frac{z^{-n}}{n}\frac{\partial}{\partial y_n}\Big) Z_N[x+y] Z_{N'}[x-y] = 0 \quad (N' \leq N). \tag{2.12}$$

The differential equations of the KP hierarchy are obtained by expanding (2.12) in y_n. For example, for $N' = N$, the coefficient in front of y_1^4 is

$$\Big(\frac{\partial^4}{\partial y_1^4} + 3\frac{\partial^2}{\partial y_2^2} - 4\frac{\partial}{\partial y_1}\frac{\partial}{\partial y_3}\Big) Z_N[t+y] Z_{N'}[t-y]\Big|_{y=0} = 0 \tag{2.13}$$

and one finds for the "free energy" $u[t] = 2\frac{\partial^2}{\partial t_1^2}\log Z_N$ the KP equation

$$3\frac{\partial^2 u}{\partial t_2^2} + \frac{\partial}{\partial t_1}\Big[-4\frac{\partial u}{\partial t_3} + 6u\frac{\partial u}{\partial t_1} + \frac{\partial^3 u}{\partial t_1^3}\Big] = 0. \tag{2.14}$$

The simplest equation for $N' = N - 1$ is the so called Miura transformation relating the functions $u = 2\partial_1^2 \log Z_N[t]$ and $v = \log(Z_{N+1}[t]/Z_N[t])$

$$-\partial_2 v + \partial_1^2 v + (\partial_1 v)^2 + u = 0, \tag{2.15}$$

etc. The Hirota equations themselves are not sufficient to determine uniquely the partition functions $Z_N[t]$. The additional constraints to be imposed (the so called "string equations") are[1] [16]

$$\int_{-\infty}^{\infty} d\lambda P_{[N-1,t]}(\lambda)\frac{d}{d\lambda}\Big[P_{[N,t]}(\lambda)\exp\Big(\sum_n t_n\lambda^n\Big)\Big] = -\frac{Z_N[t]}{Z_{N-1}[t]} \tag{2.16}$$

$$\int_{-\infty}^{\infty} d\lambda P_{[N,t]}(\lambda)\frac{d}{d\lambda}\Big[P_{[N,t]}(\lambda)\exp\Big(\sum_n t_n\lambda^n\Big)\Big] = 0. \tag{2.17}$$

These relations are expressed in terms of vertex operators as

$$\oint_C dz\, \frac{\mathbf{V}_-(z)\cdot Z_N[t]}{Z_N[t]}\, \frac{d}{dz}\Big[\frac{\mathbf{V}_+(z)\cdot Z_N[t]}{Z_N[t]}\Big] = N \tag{2.18}$$

$$\oint_C dz\, \frac{\mathbf{V}_-(z)\cdot Z_{N+1}[t]}{Z_{N+1}[t]}\, \frac{d}{dz}\Big[\frac{\mathbf{V}_+(z)\cdot Z_N[t]}{Z_N[t]}\Big] = 0. \tag{2.19}$$

[1]Strictly speaking, these equations are valid only in the limit $|\lambda_{L,R}| \to \infty$ where the integration by parts does not produce boundary terms.

3 BILINEAR EQUATIONS FOR THE $O(\mathrm{n})$ MODEL

The partition function of the $O(\mathrm{n})$ model[2] on a random lattice is defined by the matrix integral [2]

$$Z_N^{O(\mathrm{n})}[t] \sim \int dM \exp\Big[\sum_{n=0}^{\infty} t_n \;\mathrm{tr}\; M^n + \frac{\mathrm{n}}{2} \sum_{n+m\geq 1} \frac{T^{-n-m}}{n+m} \frac{(n+m)!}{n!\; m!} \mathrm{tr}\; M^n \mathrm{tr}\; M^m\Big]. \quad (3.1)$$

(the sum over n and m runs the set of nonnegative integers) and describes the ensemble of nonintersecting loops on a random graph. A loop of length n is weighed by a factor $\mathrm{n}T^{-n}$ where T is the temperature of the loop gas. For $\mathrm{n} \in [-2,2]$, the model exhibits critical behaviour with spectrum of the central charge $C = 1 - 6(g-1)^2/g$, $g = \frac{1}{\pi}\arccos(-\mathrm{n}/2)$; the critical behavior for $|\mathrm{n}| > 2$ is this of a branched polymer [7].

The matrix integral (3.1) can be again interpreted as the partition function of a Coulomb gas with more complicated but still pairwise interaction between particles:

$$Z_N^{O(\mathrm{n})}[t] = T^{-\frac{1}{2}\mathrm{n}N^2} \int \prod_{i=1}^{N} d\lambda_i \; \exp\Big(\sum_{n=0}^{\infty} t_n \lambda_i^n\Big) \frac{\prod_{1\leq i<j\leq N}(\lambda_i - \lambda_j)^2}{\prod_{i,j}(T-\lambda_i-\lambda_j)^{\mathrm{n}/2}}. \quad (3.2)$$

The integration goes along some interval $[\lambda_L, \lambda_R]$ on the real half-axis $\lambda \leq T/2$. With the last restriction the denominator in the integrand never vanishes[3].

The $O(\mathrm{n})$ model reduces to the hermitean matrix model in the limit $\mathrm{n} \to 0$ and/or $T \to \infty$, and can be considered as a perturbation of the latter in the following sense. Introduce the second-order differential operator

$$\mathbf{H} = \frac{\mathrm{n}}{2}\Big[-\ln T \frac{\partial^2}{\partial t_0^2} + \sum_{n+m\geq 1} \frac{T^{-n-m}}{n+m} \frac{(n+m)!}{n!\; m!} \frac{\partial}{\partial t_n} \frac{\partial}{\partial t_m}\Big]. \quad (3.3)$$

It is evident that the partition function (3.2) is obtained from the partition function of the one-matrix model by acting with the operator $e^{\mathbf{H}}$:

$$Z_N^{O(\mathrm{n})}[t] = e^{\mathbf{H}} \cdot Z_N[t], \quad Z_N[t] = e^{-\mathbf{H}} \cdot Z_N^{O(\mathrm{n})}[t]. \quad (3.4)$$

As a consequence, the Hirota equations (2.10) produce, after being dressed by the operator $\mathbf{H}$, bilinear relations for the partition function of the $O(\mathrm{n})$ model

$$\oint_C dz \Big(\mathbf{V}_+^{O(\mathrm{n})}(z) \cdot Z_N^{O(\mathrm{n})}[t]\Big) \Big(\mathbf{V}_-^{O(\mathrm{n})}(z) \cdot Z_{N'}^{O(\mathrm{n})}[t']\Big) = 0 \qquad (N' \leq N) \quad (3.5)$$

where $\mathbf{V}_\pm^{O(\mathrm{n})}(z)$ are the transformed vertex operators

$$\mathbf{V}_\pm^{O(\mathrm{n})}(z) = e^{\mathbf{H}} \mathbf{V}_\pm(z) e^{-\mathbf{H}} \quad (3.6)$$

[2]We use a Roman letter for the parameter $\mathrm{n} \in [-2,2]$ to avoid confusion with the summation index n running the set of natural numbers.

[3]In the perturbative solution of the $O(\mathrm{n})$ model the last restriction is not really needed because the support of the classical eigenvalue distribution lies automatically in the interval $\lambda \leq T/2$.

and the integration contour $\mathcal{C}$ encloses the interval $[\lambda_L, \lambda_R]$ but leaves outside the interval $[T-\lambda_R, T-\lambda_L]$.

From eqs (3.6) and (3.3) one finds $\mathbf{V}_{\pm}^{O(\mathrm{n})}(z)$:

$$\mathbf{V}_{\pm}^{O(\mathrm{n})}(z) = (T-2z)^{-\mathrm{n}/2} \exp\Big(\pm \sum_{n=1}^{\infty} t_n z^n \Big)$$

$$\exp\Big(\mp \ln\Big[\frac{(T-z)^{\mathrm{n}}}{z}\Big] \frac{\partial}{\partial t_0} \mp \sum_{n=1}^{\infty} \frac{z^{-n} - \mathrm{n}(T-z)^{-n}}{n} \frac{\partial}{\partial t_n} \Big). \tag{3.7}$$

After being expanded in $t_n - t'_n$, eq. (3.5) provides a hierarchy of differential equations, each of them involving derivatives with respect to an infinite number of "times" t_n.

The functional relations (3.5) can be also proved directly using the definition of the partition function of the $O(\mathrm{n})$ model. They read, in terms correlation functions of determinants,

$$\oint_{\mathcal{C}} \frac{dz}{(T-2z)^{\mathrm{n}}} \exp\Big(\sum_{n=1}^{\infty} (t_n - t'_n) z^n \Big)$$

$$\Big\langle \frac{\det(z-M)}{\det(T-z-M)^{\mathrm{n}}} \Big\rangle_{[N,t]} \Big\langle \frac{\det(T-z-M)^{\mathrm{n}}}{\det(z-M)} \Big\rangle_{[N',t']} = 0 \;\; (N' \le N) \tag{3.8}$$

where $\langle \;\; \rangle_{[N,t]}$ denotes the average in the ensemble described by the partition function (3.2). Let us sketch the proof. The contour of integration $\mathcal{C}$ encloses the the integration interval $[\lambda_L, \lambda_R]$ but not the cuts starting at the branchpoints $z = T - \lambda_i$ and $z = T/2$. Therefore the integral in z can be evaluated using the residue theorem. The result is

$$\int_{\lambda_L}^{\lambda_R} \frac{d\lambda \; e^{\sum_n t_n \lambda^n}}{(T-2\lambda)^{\frac{\mathrm{n}}{2}}} \Big\langle \frac{\det\big(\lambda - M\big)}{\det(T-\lambda-M)^{\mathrm{n}}} \Big\rangle_{[N,t]} \Big\langle \det(\lambda - M) \Big\rangle_{[N'-1,t']} = 0 \;\; (N \le N').$$

To prove the last identity, it is sufficient to show that the functions

$$Q_{[N,t]}(\lambda) = \langle \det(\lambda - M) \det(T - \lambda - M)^{-\mathrm{n}} \rangle_{[N,t]} \tag{3.9}$$

are ortogonal, with respect to the measure $d\lambda (T-2\lambda)^{-\mathrm{n}/2} e^{\sum_n t_n \lambda^n}$, to all polynomials of degree less than N. This follows from the identity

$$\int_{\lambda_L}^{\lambda_R} d\lambda \frac{e^{\sum_n t_n \lambda^n}}{(T-2\lambda)^{\frac{\mathrm{n}}{2}}} Q_{[N,t]}(\lambda) \; \lambda^k = \delta_{N,k} \; \frac{Z_{N+1}^{O(\mathrm{n})}[t]}{(N+1) \; Z_N^{O(\mathrm{n})}} \tag{3.10}$$

which is proved in the same way as the relations (2.5) in the one-matrix model [16]. Namely, denoting $\lambda = \lambda_{N+1}$ we can rewrite the left hand side of (3.10) as the integral

$$\frac{1}{Z_N^{O(\mathrm{n})}} \int_{\lambda_L}^{\lambda_R} \prod_{i=1}^{N+1} d\lambda_i \; e^{\sum_n t_n \lambda_i^n} \; \frac{\prod_{i<j}^{N+1} (\lambda_i - \lambda_j)}{\prod_{i,j=1}^{N+1} (T - \lambda_i - \lambda_j)^{\frac{\mathrm{n}}{2}}} [\lambda_{N+1}^k \prod_{i<j}^{N} (\lambda_i - \lambda_j)] \tag{3.11}$$

in which, except for the term in square brackets, the integrand is antisymmetric with respect to permutations of the λ_i, $i = 1, \ldots, N+1$. After antisymmetization, the term in the square bracket vanishes if $k < N$, since it is equal to a determinant with two identical columns. When $k = N$, the result of the integration is equal to $Z_{N+1}^{O(n)}[t]/(N+1)Z_N^{O(n)}$.

The "string equations", needed to determine completely the partition function, are obtained from (2.18) and (2.19) by replacing $\mathbf{V}_{\pm}(z) \to \mathbf{V}_{\pm}^{O(n)}(z)$. However, one has to take care of the boundary terms because of the restriction $\lambda_R < T/2$ or, alternatively, to introduce an infinite wall at $\lambda = T/2$ by adding small term $\epsilon(T/2 - \lambda)^{-1}$ to the Coulomb gas potential. If one is interested only in the perturbative solution, the boundary terms can be neglected.

4 Bilinear equations for the ADE and $\hat{A}\hat{D}\hat{E}$ matrix models

The ADE and $\hat{A}\hat{D}\hat{E}$ matrix models [5] give a nonperturbative microscopic realization of the rational string theories with $C \leq 1$. Within the $1/N$ expansion these models can be interpreted as height models on random lattices with different topologies [4]. Each one of these models is associated with a rank r classical simply laced Lie algebra (that is, of type $A_r, D_r, E_{6,7,8}$) or its affine extension, and represents a system of r coupled random matrices. The matrices $M_{(a)}$ of size $N_a \times N_a$ ($a = 1, 2, \ldots, r$) are associated with the nodes of the Dynkin diagram of the simply laced Lie algebra, the latter being defined by its adjacency matrix G with elements

$$G^{ab} = \begin{cases} 1 & \text{if the two nodes are the extremities of a link} < ab > \\ 0 & \text{otherwise.} \end{cases} \tag{4.1}$$

The partition function $Z_{\vec{N}}^G[\vec{t}]$ depends on r sets of coupling constants

$$\vec{t} = \{t_n^{(a)} \mid a = 1, \ldots, r;\ n = 0, 1, 2, \ldots\}.$$

The interaction is of nearest-neighbor type and the measure is a product of factors associated with the nodes a and the (nonoriented) links $< ab > = < ba >$ of the Dynkin diagram

$$Z_{\vec{N}}^G[\vec{t}] \sim \int \prod_{a=1}^{r} dM_{(a)} e^{\sum_n t_n^{(a)} \mathrm{tr}\, M_{(a)}^n} \prod_{<ab>} \exp\left(\sum_{m+n \geq 1}^{\infty} \frac{T^{-m-n}}{m+n} \frac{(m+n)!}{m!\, n!} \mathrm{tr}\, M_{(a)}^m \mathrm{tr}\, M_{(b)}^n \right). \tag{4.2}$$

The only nontrivial integration is with respect to the eigenvalues $\lambda_{(a)i}$ ($i = 1, \ldots, N_a$) of the matrices $M_{(a)}$:

$$Z_{\vec{N}}^G[\vec{t}] = \prod_{a=1}^{r} \prod_{i=1}^{N_a} d\lambda_{(a)i} e^{\sum_n t_n^{(a)} \lambda_{(a)i}^n} \frac{\prod_a \prod_{i<j} (\lambda_{(a)i} - \lambda_{(a)j})^2}{\prod_{<ab>} \prod_{i,j} (T - \lambda_{(a)i} - \lambda_{(b)j})}. \tag{4.3}$$

As in the previous section we will restrict the intergation to the interval $[\lambda_L, \lambda_R]$ with $\lambda_R \leq T/2$.

The partition function (4.3) can be obtained by acting with an operator $e^{\mathbf{H}}$ on the product of r one-matrix partition functions

$$Z_{\vec{N}}^G[\vec{t}] = e^{\mathbf{H}} \cdot \prod_{a=1}^{r} Z_{N_a}[t^{(a)}]. \tag{4.4}$$

The operator $\mathbf{H}$ is a sum of terms associated with the links $< ab >$ of the Dynkin diagram

$$\mathbf{H} = \sum_{<ab>} \left[\ln T^{-1} \frac{\partial}{\partial t_a^{(0)}} \frac{\partial}{\partial t_b^{(0)}} + \sum_{n+m\geq 1} \frac{T^{-n-m}}{n+m} \frac{(n+m)!}{n!\ m!} \frac{\partial}{\partial t_n^{(a)}} \frac{\partial}{\partial t_m^{(b)}} \right]. \tag{4.5}$$

Each of the one-matrix partition functions on the right hand side of (4.4) satisfies the Hirota equations (2.10). As a consequence, the left hand side satisfies a set of r functional equations associated with the nodes $a = 1, \ldots, r$ of the Dynkin diagram. Introducing the transformed vertex operators

$$\mathbf{V}_{\pm}^{(a)}(z) = e^{\mathbf{H}} e^{\pm \sum_{n=0}^{\infty} t_n^{(a)} z^n} e^{\mp \ln z^{-1} \frac{\partial}{\partial t_0^{(a)}} \mp \sum_{n=1}^{\infty} \frac{z^{-n}}{n} \frac{\partial}{\partial t_n^{(a)}}} e^{-\mathbf{H}} \tag{4.6}$$

we have for each node a

$$\oint_{\mathcal{C}} dz \left(V_{+}^{(a)}(z) \cdot Z_{\vec{N}}^{G}[\vec{t}] \right) \left(V_{-}^{(a)}(z) \cdot Z_{\vec{N}'}^{G}[\vec{t}'] \right) = 0 \qquad (N_a' \leq N_a) \tag{4.7}$$

where the integration contour $\mathcal{C}$ in (4.7) encloses the interval $[\lambda_L, \lambda_R]$ but leaves outside the interval $[T - \lambda_R, T - \lambda_L]$.

Inserting (4.5) in the definition (4.6) we find the explicit form of the operator $\mathbf{V}_{\pm}^{a}(z)$

$$\mathbf{V}_{\pm}^{(a)}(z) = \prod_b (T - 2z)^{-\frac{1}{2}G^{ab}} e^{\pm \sum_{n=0}^{\infty} t_n^{(a)} z^n} \exp\left(\mp \left[\ln z^{-1} \frac{\partial}{\partial t_0^{(a)}} + \sum_{n=1}^{\infty} \frac{z^{-n}}{n} \frac{\partial}{\partial t_n^{(a)}} \right] \right)$$
$$\exp\left(\pm \sum_b G^{ab} \left[\ln(T-z)^{-1} \frac{\partial}{\partial t_0^{(b)}} + \sum_{n=1}^{\infty} \frac{(T-z)^{-n}}{n} \frac{\partial}{\partial t_n^{(b)}} \right] \right) \tag{4.8}$$

and write the generalized Hirota equations (4.7) in terms of correlation functions of determinants

$$\oint_{\mathcal{C}} dz \frac{e^{\sum_{n=1}^{\infty} [t_n^{(a)} - t_n'^{(a)}] z^n}}{\prod_b (T-2z)^{G^{ab}}} \left\langle \frac{\det(z - M_{(a)})}{\prod_b \det(T - z - M_{(b)})^{G^{ab}}} \right\rangle_{[\vec{N}, \vec{t}]}$$
$$\left\langle \frac{\prod_b \det(T - z - M_b)^{G^{ab}}}{\det(z - M_{(a)})} \right\rangle_{[\vec{N}', \vec{t}']} = 0, \qquad (N_a' \leq N_a). \tag{4.9}$$

Here $\langle\ \rangle_{[\vec{N},\vec{t}]}$ denotes the average in the ensemble described by the partition function (4.3). One can prove eq.(4.9) also directly by performing the contour integration. It is essential for the proof that the interval $[T - \lambda_R, T - \lambda_L]$ is outside the contour of integration $\mathcal{C}$.

Finally, let us remark that the the Virasoro constraints satisfied by each of the one-matrix partition functions on the right hand side of eq.(4.4) imply r sets of constraints $L_n^{(a)} Z^G = 0; n \geq -1,\ a = 1, \ldots, r$, where the operators $L_n^{(a)}$ defined by

$$L_n^{(a)} = \sum_{k=0}^{n} \frac{\partial^2}{\partial t_k^{(a)} \partial t_{n-k}^{(a)}} + \sum_k k t_k^{(a)} \frac{\partial}{\partial t_{n+k}^{(a)}} + \sum_b G^{ab} \sum_{k,m} T^{-k-m-1} \frac{(k+m)!}{k!\ m!} \frac{\partial^2}{\partial t_{n+k+1}^{(a)} \partial t_m^{(b)}}$$

and satisfying commuting Virasoro algebras.

References

[1] V.Kazakov, *Phys. Lett.*119 A (1986) 140; D. Bulatov and V.Kazakov, *Phys. Lett.*186 B (1987) 379.

[2] I. Kostov, *Mod. Phys. Lett.* A4 (1989) 217; M. Gaudin and I. Kostov, *Phys. Lett.*B 220 (1989) 200; I. Kostov and M. Staudacher, *Nucl. Phys.* B 384 (1992) 459; B. Eynard and J. Zinn-Justin, *Nucl. Phys.* B 386 (1992) 55.

[3] V. Kazakov, *Nucl. Phys.* B (Proc. Supp.) 4 (1988) 93; J.-M. Daul, Q-states Potts model on a random planar lattice, hep-th/9502014.

[4] I. Kostov, *Nucl. Phys.* B 326 (1989)583; *Nucl. Phys.* B 376 (1992) 539.

[5] I. Kostov, *Phys. Lett.*297 (1992) 74.

[6] S. Higuchi and I. Kostov, *Phys. Lett.*B 357 (1995) 62.

[7] B. Eynard and C. Kristjansen, SPhT/95-068 (hep-th/9506193), SPhT/95/133(hep-th/9512052).

[8] M. Douglas and S. Shenker, *Nucl. Phys.* B 335 (1990) 635; E. Brézin and V. Kazakov, *Phys. Lett.*B 236 (1990); D. Gross and A. Migdal, *Phys. Rev. Lett.*64 (1990) 127 ; *Nucl. Phys.* B 340 (1990) 333.

[9] M. Douglas, *Phys. Lett.*B 238 (1990) 176; M. Fukuma, H. Kawai and R. Nakayama, *Int. J. Mod. Phys.* A 6 (1991) 1385; R. Dijkgraaf, H. Verlinde and E. Verlinde, *Nucl. Phys.* B 348 (1991) 435.

[10] I. Kostov, Solvable Statistical Models on Random Lattices, Proceedings of the *Conference on recent developments in statistical mechanics and quantum field theory* (10–12 April 1995), Trieste, Italy.

[11] J. Alfaro and I. Kostov, to be published.

[12] S. Kharchev, A. Marshakov, A. Mironov and A. Morozov, *Nucl. Phys.* B 397 (1993) 339 and references therein.

[13] R. Hirota, *Phys. Rev. Lett.*27 (1971) 1192.

[14] M. Jimbo and T. Miwa, "Solitons and infinite dimensional Lie algebras", *RIMS* Vol. 19 (1983) 943-1001.

[15] M. L. Mehta, *Random Matrices* (second edition), Academic Press, New York, 1990.

[16] C. Itzykson and J.-B. Zuber, J. Math. Phys. 21 (1980) 411.

CALOGERO-SUTHERLAND MODEL AND SINGULAR VECTORS OF W ALGEBRAS

Y. Matsuo

YITP, Kyoto University

Japan

1 Introduction

Calogero-Sutherland model (CSM) describes a system of non-relativistic particles on a circle under the inverse square potential, whose Hamiltonian is given by

$$H_{CS} = \sum_{j=1}^{N_0} \frac{1}{2}\left(\frac{1}{i}\frac{\partial}{\partial q_j}\right)^2 + \left(\frac{\pi}{L}\right)^2 \sum_{\substack{i,j=1 \\ i<j}}^{N_0} \frac{\beta(\beta-1)}{\sin^2 \frac{\pi}{L}(q_i - q_j)}, \tag{1.1}$$

where β is a coupling constant. This model was introduced by Sutherland several years ago and has been known to describe a system with the generalized exclusion principle in $1+1$ dimension.

The purpose of this paper is to explain the intimate relation of CSM with the representation of Virasoro, or more generally, W algebras. Our discussion is based on papers, [AMOS1] [AMOS2] [AMOS3], which is done in collaboration with H. Awata, S.Odake, and J. Shiraishi. See also the related work, [MY].

Before we give the detailed explanation, let us sketch briefly some of the features of CSM.

1.1 Vacuum wave function

Since the Hamiltonian (1.1) can be decomposed as,

$$H_{CS} = \sum_{j=1}^{N} h_j^\dagger h_j + E_0, \qquad h_j(\beta) = \frac{1}{i}\frac{\partial}{\partial q_j} + \beta \sum_{i(\neq j)} \cot\left(\frac{q_j - q_i}{2}\right), \tag{1.2}$$

Low Dimensional Applications of Quantum Field Theory
Edited by L. Baulieu *et al.*, Plenum Press, New York, 1997

with vacuum energy $E_0 = \frac{1}{6}(\frac{\pi}{L})^2\beta^2(N_0^3 - N_0)$, the ground state wave function ψ_0 is easily found by solving the first order differential equation, $h_j\psi_0 = 0$,

$$\psi_0 \equiv \Delta^\beta = \Big(\frac{L}{\pi}\prod_{\substack{i,j=1\\ i<j}}^{N_0} \sin\frac{\pi}{L}(q_i - q_j)\Big)^\beta . \tag{1.3}$$

When $\beta = 1$, this wave function is identified with the free fermion vacuum (Vandermonde determinant). For general β, it shows the generalized statistics, namely when we put two particles very close, the wave function approaches zero as $(q - q')^\beta$.

1.2 Spectrum

To describe the excited states, we decompose the wave function as $\psi = J\psi_0$ and rewrite the Hamiltonian that directly acts on the new wave function J. J becomes symmetric function of the coordinates. After we make a coordinate transformation to the "plane wave coordinates", $x_j = e^{2\pi i q_j/L}$, the Hamiltonian becomes,

$$H_\beta = \sum_{i=1}^N D_i^2 + \beta \sum_{\substack{i,j=1\\ i<j}}^N \frac{x_i + x_j}{x_i - x_j}(D_i - D_j), \tag{1.4}$$

where $D_i = x_i\frac{\partial}{\partial x_i}$. It is known that the eigenstates of this Hamiltonian is indexed by Young diagrams which is parametrized by the decreasing series of positive integers. We denote them, for example, $\lambda = (\lambda_1, \cdots, \lambda_M)$ with $\lambda_1 \geq \lambda_2 \geq \cdots \geq \lambda_M > 0$, where each number λ_i is the number of boxes at ith row. It is clear that we can express this Young diagram in alternative way, namely by the number of boxes at each columns. We denote such parametrization as, $\lambda = {}^t(\lambda_1', \lambda_2', \cdots, \lambda_{M'})$.

The energy eigenvalue for the state with index λ can be given by two different fashions,

$$\epsilon_{\beta,\lambda} = \sum_{i=1}^M\Big(\lambda_i^2 + \beta(N_0 + 1 - 2i)\lambda_i\Big) = \sum_{i=1}^{M'}\Big(-\beta\lambda_i'^2 + (\beta N_0 + 2i - 1)\lambda_i'\Big). \tag{1.5}$$

Let us introduce the "pseudo-momentum" of particles, $k_i = \frac{2\pi}{L}\Big(\lambda_i + \frac{\beta}{2}(N_0 + 1 - 2i)\Big)$ and holes, $\tilde{k}_i = \frac{2\pi}{L}\Big(\lambda_i' - \frac{1}{2\beta}(\beta N_0 - 1 + 2i)\Big)$. The energy is then expressed by the sum of square of them,

$$\epsilon_{\beta,\lambda} = \sum_{i=1}^N k_i^2 + \text{const.} = -\beta\sum_{j=1} \tilde{k}_i^2 + \text{const.} \tag{1.6}$$

The Calogero-Sutherland system is "almost" free in this sense. The non-triviality comes from the fact that the momentum of two particles should be at least separated by β, which is another manifestation of fractional statistics. It is also interesting that there are alternative description by "holes" where the minimum separation is separated by $1/\beta$. This *deformed duality* (or charge conjugation) is one of the most interesting features of CSM.

2 Collective Coordinate Representation

To find a correspondence between CSM and Virasoro symmetry, we introduce a collective field that describe the system in the large N limit.

We introduce a standard free boson,

$$\phi(z) = \hat{q} + a_0 \log z - \sum_{n\neq 0} \frac{1}{n} a_n z^{-n}, \quad \phi_-(z) = \sum_{n>0} \frac{1}{n} a_{-n} z^n,$$
$$[a_n, a_m] = n\delta_{n+m,0}, \quad [a_0, \hat{q}] = 1. \tag{2.1}$$

The boson Fock space $\mathcal{F}_\alpha$ is generated over oscillators of negative mode by the state $|\alpha\rangle$ such that

$$a_n|0\rangle = 0 \quad (n \geq 0), \qquad |\alpha\rangle = e^{\alpha\hat{q}}|0\rangle. \tag{2.2}$$

$\langle\alpha|$ is similarly defined, with the normalization $\langle\alpha|\alpha'\rangle = \delta_{\alpha,\alpha'}$.

We consider the following map from a state $|f\rangle$ into $\mathcal{F}_\alpha$ to a symmetric function $f(x)$,

$$|f\rangle \mapsto f(x) = \langle\alpha|C|f\rangle,$$
$$C = \exp\Big(\sqrt{\beta}\sum_{n>0}\frac{1}{n}a_n p_n\Big), \quad p_n = \sum_i x_i^n, \tag{2.3}$$

where $\sqrt{\beta}$ is a parameter. Under this correspondence, a_{-n} and a_n are interpreted as the creation and annihilation operator of power-sum, $\sqrt{\beta}p_n$ and $\frac{n}{\sqrt{\beta}}\frac{\partial}{\partial p_n}$, respectively, because $\langle\alpha|Ca_{-n} = \sqrt{\beta}p_n\langle\alpha|C$ and $\langle\alpha|Ca_n = \frac{n}{\sqrt{\beta}}\frac{\partial}{\partial p_n}\langle\alpha|C$.

The correspondence between the Hilbert spaces is one to one in the large N limit. Therefore we can translate any operator which act on one Hilbert space to another which is acomplished by aplying the rule, $\mathcal{O}\langle\alpha|C = \langle\alpha|C\hat{\mathcal{O}}$. In particular, the Hamiltonian is translated as,

$$H_\beta\langle\alpha|C = \langle\alpha|C\hat{H}_\beta, \tag{2.4}$$

where

$$\hat{H}_\beta = \sum_{n,m>0}\sqrt{\beta}\Big(a_{-n-m}a_n a_m + a_{-n}a_{-m}a_{n+m}\Big) + \sum_{n>0} a_{-n}a_n\Big((1-\beta)n + N_0\beta\Big). \tag{2.5}$$

This is a typical ϕ^3 Hamiltonian which appears in some string theories. To find the eigenstates of CSM now reduces to the diagonalization of this Hamiltonian. This is of course nontrivial task but the knowledge of CFT gives some crucial hints as we shal see below.

3 Virasoro Singular Vectors

A critical hint to find the Jack polynomials in the collective field language is to express the Hamiltonian (2.4) in terms of the Virasoro generators. Let us define them as,

$$T(z) = \sum_n L_n z^{-n-2} = \frac{1}{2} :\partial\phi(z)\partial\phi(z): + \alpha_0\partial^2\phi(z),$$

$$c = 1 - 12\alpha_0^2, \quad 2\alpha_0 = \alpha_+ + \alpha_-,$$
$$\alpha_+ = \sqrt{\beta} \qquad \alpha_- = -1/\sqrt{\beta}. \tag{3.1}$$

We note that these generators satisfy the Virasoro algebra with central extension $c = 1 - \frac{6(1-\beta)^2}{\beta}$. If we further put $\beta = p/q$, it is identical with the central charge of minimal model.

With this definition, the Hamiltonian can be rewritten as,

$$\hat{H}_\beta = \sqrt{2\beta}\sum_{n>0} a_{-n}L_n + \sum_{n>0} a_{-n}a_n(N_0\beta + \beta - 1 - \sqrt{2\beta}a_0). \tag{3.2}$$

Our main idea to diagonalize the Hamiltonian is following. *The cubic term which was the main obstrucle to diagonalize the Hamiltonian is now replaced by the Virasoro charges. If we apply this operator to the singular vectors of the Virasoro algebra, this term vanishes and the remaining bilinear part is already diagonalized.*

The explicit form of the null state is well-known. We introduce $\alpha_{r,s}$ as

$$\alpha_{r,s} = \frac{1}{2}(1+r)\alpha_+ + \frac{1}{2}(1+s)\alpha_-. \tag{3.3}$$

The singular vector indexed by $\{r, s\}$ has two representations, which we write,

$$\begin{aligned} |\chi^+_{r,s}\rangle &= \oint \prod_{j=1}^{r} \frac{dz_j}{2\pi i} \cdot \prod_{i=1}^{r} : e^{\alpha_+\phi(z_i)} : |\alpha_{-r,s}\rangle \\ &= \oint \prod_{j=1}^{r} \frac{dz_j}{2\pi i z_j} \cdot \prod_{\substack{i,j=1\\i<j}}^{r} (z_i - z_j)^{2\beta} \cdot \prod_{i=1}^{r} z_i^{(1-r)\beta - s} \cdot \prod_{j=1}^{r} e^{\alpha_+\phi_-(z_j)}|\alpha_{r,s}\rangle, \end{aligned} \tag{3.4}$$

$$|\chi^-_{r,s}\rangle = \oint \prod_{j=1}^{s} \frac{dz_j}{2\pi i} \cdot \prod_{i=1}^{s} : e^{\alpha_-\phi(z_i)} : |\alpha_{r,-s}\rangle, \tag{3.5}$$

Applying the Hamiltonian to these states, we find,

$$\hat{H}_\beta|\chi^\pm_{r,s}\rangle = rs\Big((N_0 - r)\beta + s\Big)|\chi^\pm_{r,s}\rangle \tag{3.6}$$

By comparing the eigenvalues, we find that the singular vector $|\chi^+_{r,s}\rangle$ (resp. $|\chi^-_{r,s}\rangle$) may be identified with the Jack polynomial whose Young diagram is given by a rectangle s^r.

By using the dictionary we introduced, we can get the integral representation of the Jack polynomial in two different ways,

$$\begin{aligned} &J_{(s^r)}(x) \\ &\propto \oint \prod_{j=1}^{r} \frac{dz_j}{2\pi i z_j} \cdot \prod_{\substack{i,j=1\\i<j}}^{r} (z_i - z_j)^{2\beta} \cdot \prod_{i=1}^{r} z_i^{(1-r)\beta - s} \cdot \prod_i \prod_{j=1}^{r} (1 - x_i z_j)^{-\beta} \\ &\propto \oint \prod_{j=1}^{s} \frac{dz_j}{2\pi i z_j} \cdot \prod_{\substack{i,j=1\\i<j}}^{s} (z_i - z_j)^{2/\beta} \cdot \prod_{i=1}^{s} z_i^{(1-s)/\beta - r} \cdot \prod_i \prod_{j=1}^{s} (1 - x_i z_j). \end{aligned} \tag{3.7}$$

4 W_N SINGULAR VECTORS AND JACK POLYNOMIAL

The identification of the Jack polynomials with the singular vectors of the Virasoro algebra is simple and exact. However, the story is not over here because the Young diagrams thus obtained are only the special type, the rectangular. To get more general ones, we need to extend the algebra. In this section, we give such a realization in terms of W algebras.

As shown by Fateev and Lukyanov, the W_N algebra can be realized by $N-1$ free bosons defined by OPE,

$$\phi^a(z)\phi^b(w) \sim A^{ab}\log(z-w), \tag{4.1}$$

where A^{ab} is the standard A_{N-1} Cartan matrix. Each generator is generated through the Miura transformation,

$$\begin{aligned} &:\prod_{i=1}^{N}\Big(\alpha_0\partial+\vec{h}_i\cdot\partial\vec{\phi}(z)\Big):=\sum_{k=0}^{N}W^k(z)(\alpha_0\partial)^{N-k}, \\ &\alpha_0=\alpha_++\alpha_-, \quad \alpha_+=\sqrt{\beta}, \quad \alpha_-=\frac{-1}{\sqrt{\beta}}. \end{aligned} \tag{4.2}$$

In particular, the spin two (Virasoro) and the spin three (W^3) generators are given by

$$\begin{aligned} T(z) &= -W^2(z)=\frac{1}{2}:\partial\vec{\phi}(z)\cdot\partial\vec{\phi}(z):+\alpha_0\vec{\rho}\cdot\partial^2\vec{\phi}(z), \\ W(z) &= W^3(z)=\sum_n W_n z^{-n-3} \\ &= \sum_{a=1}^{N-1}:\partial\phi_a(z)\partial\phi_a(z)\Big(\partial\phi_{a+1}(z)-\partial\phi_{a-1}(z)\Big): \\ &\quad +\alpha_0\sum_{a,b=1}^{N-1}(1-a)A^{ab}:\partial\phi_a(z)\partial^2\phi_b(z):+\alpha_0^2\sum_{a=1}^{N-1}(1-a)\partial^3\phi_a(z), \end{aligned} \tag{4.3}$$

with the central charge,

$$c=N-1-12\alpha_0^2\vec{\rho}^2 \qquad \vec{\rho}^2=\frac{1}{12}N(N^2-1) \tag{4.4}$$

The singular vectors of the W_N algebra are realized in two different ways. At level $\sum_{a=1}^{N-1}r^a s^a$, $|\vec{\lambda}^{\pm}_{\vec{r},\vec{s}}\rangle$ is expressed by using screening currents $:e^{\alpha_\pm\phi^a(z)}:$,

$$\begin{aligned} |\chi^+_{\vec{r},\vec{s}}\rangle &= \oint\prod_{a=1}^{N-1}\prod_{j=1}^{r^a}\frac{dz_j^a}{2\pi i}\cdot\prod_{a=1}^{N-1}\prod_{j=1}^{r^a}:e^{\alpha_+\phi^a(z_j^a)}:|\vec{\lambda}^+_{\vec{r},\vec{s}}\rangle-\alpha_+\sum_{a=1}^{N-1}r^a\vec{\alpha}^a \\ &= \oint\prod_{a=1}^{N-1}\prod_{j=1}^{r^a}\frac{dz_j^a}{2\pi i z_j^a}\cdot\prod_{a=1}^{N-1}\prod_{\substack{i,j=1\\ i<j}}^{r^a}(z_i^a-z_j^a)^{2\beta}\cdot\prod_{a=1}^{N-2}\prod_{i=1}^{r^a}\prod_{j=1}^{r^{a+1}}(z_i^a-z_j^{a+1})^{-\beta} \\ &\quad\times\prod_{a=1}^{N-1}\prod_{j=1}^{r^a}(z_j^a)^{(1-r^a+r^{a+1})\beta-s^a}\cdot\prod_{a=1}^{N-1}\prod_{j=1}^{r^a}e^{\alpha_+\phi_-^a(z_j^a)}|\vec{\lambda}^+_{\vec{r},\vec{s}}\rangle \\ |\chi^-_{\vec{r},\vec{s}}\rangle &= \oint\prod_{a=1}^{N-1}\prod_{j=1}^{s^a}\frac{dz_j^a}{2\pi i}\cdot\prod_{a=1}^{N-1}\prod_{j=1}^{s^a}:e^{\alpha_-\phi^a(z_j^a)}:|\vec{\lambda}^-_{\vec{r},\vec{s}}\rangle-\alpha_-\sum_{a=1}^{N-1}s^a\vec{\alpha}^a. \end{aligned} \tag{4.5}$$

In the following we assume $r^1 > \cdots > r^{N-1}$.

Unlike the Virasoro case, the correspondence between the Hilbert spaces is not one-to-one even in the large N limit. In some sense, in defining the dictionary to translate states, we need to project out some of the degree of freedom in the W_N module. We define such a machinary by the vertex operator,

$$\begin{aligned} f(x) &= \langle\vec{\lambda}|C|f\rangle, \\ C &= \exp\Big(\sqrt{\beta}\vec{\Lambda}_1\cdot\sum_{n>0}\frac{1}{n}\vec{a}_n p_n\Big), \quad p_n = \sum_i x_i^n, \end{aligned} \tag{4.6}$$

This project out states that can be written as the linear combination of $a^{\nu}_{-n}(\cdots)|\vec{\lambda}\rangle$ $(\nu = 2,\ldots,N-1)$.

We define the bosonized Hamiltonian as before, $H_\beta\langle\vec{\lambda}|C = \langle\vec{\lambda}|C\hat{H}_\beta$. We note, however, that $\hat{H}_\beta$ can not be determined uniquely this time.

$$\begin{aligned} \hat{H}_\beta &= \frac{2}{N}\sqrt{\beta}\sum_{n>0}a^1_{-n}L_n + \sum_{n>0}\vec{a}_{-n}\cdot\vec{a}_n(N_0\beta+\beta-1-2\sqrt{\beta}a_{1,0}) \\ &\quad +\sqrt{\beta}(W_0 - W_{0,zero}) + \sum_{n>0}\sum_{a>1}a^a_{-n}\times(\cdots). \end{aligned} \tag{4.7}$$

Here $W_{0,zero}$ is the zero mode part of W_0,

$$\begin{aligned} W_{0,zero} &= \sum_{a=1}^{N-1}\Big(a_{a,0}a_{a,0}(a_{a+1,0}-a_{a-1,0}) \\ &\quad +\alpha_0\big(2(a-1)a_{a,0}a_{a,0} + (1-2a)a_{a,0}a_{a+1,0}\big) + \alpha_0^2 2(1-a)a_{a,0}\Big). \end{aligned} \tag{4.8}$$

In this generalized situation, the cubic part of the Hamiltonian is again given by the Virasoro generators and zero mode of the W^3 charges. It becomes again vanishing or trivially diagonalized when it is acted on the null state of the W_N algebra. Indeed, we get,

$$\begin{aligned} \hat{H}_\beta|\chi^+_{\vec{r},\vec{s}}\rangle &= \Big(\sum_{a=1}^{N-1}r^a s^a\times\big(N_0\beta+\beta-1-2\sqrt{\beta}(\alpha_+ r_1+\alpha_- s_1+\alpha_0\rho_1)\big) \\ &\quad +\sqrt{\beta}\big(w(\vec{\lambda}^+_{\vec{r},\vec{s}} - \alpha_+\sum_{a=1}^{N-1}r^a\vec{\alpha}^a) - w(\vec{\lambda}^+_{\vec{r},\vec{s}})\big)\Big)|\chi^+_{\vec{r},\vec{s}}\rangle \\ &\quad +\sum_{n>0}\sum_{a>1}a^a_{-n}\times(\cdots)|\chi^+_{\vec{r},\vec{s}}\rangle \\ &= \Big(\sum_{a=1}^{N-1}r^a s^a s^a + 2\sum_{\substack{a,b=1\\a>b}}^{N-1}r^a s^a s^b + \beta\sum_{a=1}^{N-1}r^a s^a(N_0 - r^a)\Big)|\chi^+_{\vec{r},\vec{s}}\rangle \\ &\quad +\sum_{n>0}\sum_{a>1}a^a_{-n}\times(\cdots)|\chi^+_{\vec{r},\vec{s}}\rangle. \end{aligned} \tag{4.9}$$

Therefore $|\chi^+_{\vec{r},\vec{s}}\rangle$ is the eigenstate of $\hat{H}_\beta$ up to the last term, which will vanish after multiplying by $\langle\vec{\lambda}^+_{\vec{r},\vec{s}}|C$. This eigenvalue is just $\epsilon_{\beta,\lambda}$ with $\lambda' = ((r^1)^{s^1},(r^2)^{s^2},\cdots,(r^{N-1})^{s^{N-1}})$, namely, $\lambda = ((s^1+\cdots+s^{N-1})^{r^{N-1}},(s^1+\cdots+s^{N-2})^{r^{N-2}-r^{N-1}},\cdots,(s^1)^{r^1-r^2})$,

$\lambda =$ 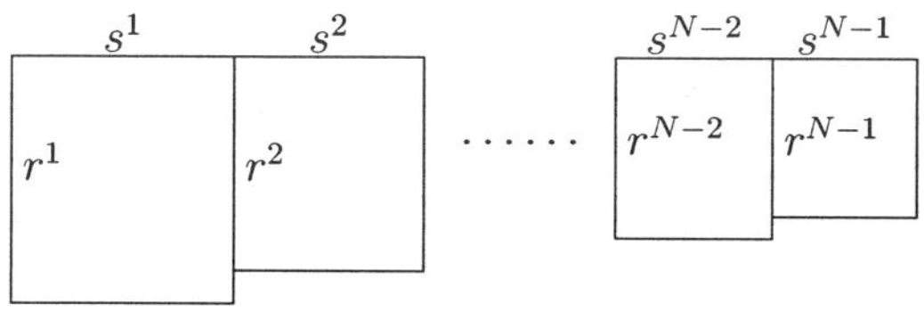

Using the state-function correspondence (4.6), we obtain an integral representation of the Jack polynomial with the Young diagram λ,

$$
\begin{aligned}
J_\lambda(x) &\propto \langle \vec{\lambda}^{+}_{\vec{r},\vec{s}} | C | \chi^{+}_{\vec{r},\vec{s}} \rangle \\
&= \oint \prod_{a=1}^{N-1} \prod_{j=1}^{r^a} \frac{dz_j^a}{2\pi i z_j^a} \cdot \prod_{a=1}^{N-1} \prod_{\substack{i,j=1 \\ i<j}}^{r^a} (z_i^a - z_j^a)^{2\beta} \cdot \prod_{a=1}^{N-2} \prod_{i=1}^{r^a} \prod_{j=1}^{r^{a+1}} (z_i^a - z_j^{a+1})^{-\beta} \\
&\qquad \times \prod_{a=1}^{N-1} \prod_{j=1}^{r^a} (z_j^a)^{(1-r^a+r^{a+1})\beta - s^a} \cdot \prod_{i} \prod_{j=1}^{r^1} (1 - x_i z_j^1)^{-\beta}. \qquad (4.10)
\end{aligned}
$$

5 Conclusion and discussion

In this talk, we show that some of the essential materials of the Virasoro and the W_N algebras are encoded in CSM. In particular, the null state for (p, q) minimal models for these algebras are neatly expressed by the eigenstates of the CSM.

The q-deformation of such an approach is pursued in [SKAO] and it gives the definition of q deformation of Virasor and W_N algebra.

Virasoro algebra and W_N algebra are known to be deeply related to the KdV or KP system. The connection that we found here may be related to the fact that (quantum) KdV system is in some sense related to CSM through the Miura transformation.

The study of the similar correspondence for the Calogero-Sutherland model with internal degree of freedoms is another interesting direction. We will report on this aspect in our future communication.

References

[AMOS1] H. Awata, Y.Matsuo, S.Odake, J.Shiraishi, *Phys. Lett.* B347 (1995) 49–55.

[AMOS2] H. Awata, Y.Matsuo, S.Odake, J.Shiraishi, *Nucl. Phys.* B449 (1995) 347–374.

[AMOS3] H. Awata, Y.Matsuo, S.Odake, J.Shiraishi, hep-th/950321.

[MY] K. Mimachi and Y. Yamada, preprint (November 1994).

[SKAO] J. Shiraishi, H. Kubo, H. Awata, S. Odake, q-alg/950734.

MATRIX MODELS, QUANTUM PENNER ACTION AND TWO-DIMENSIONAL STRING THEORY[1]

CAMILLO IMBIMBO

INFN, SEZIONE DI GENOVA
I-16146 GENOVA, ITALY

SUNIL MUKHI

TATA INSTITUTE OF FUNDAMENTAL RESEARCH
BOMBAY 400 005, INDIA

1 INTRODUCTION

Random-matrix models have been intensively studied in the last few years[1]. Although the bulk of the attention has been paid to models describing dynamically triangulated random surfaces[2], and their double-scaling limits[3], it has turned out that many of the results obtained can be equivalently deduced from a far simpler class of matrix models, which moreover require no double-scaling limit. These are sometimes called "topological matrix models" because the random-matrix integral can be thought of as a generating function for certain topological invariants.

The first such model was the one constructed by Penner[4][5] to study the virtual Euler characteristic of the moduli space of Riemann surfaces. Subsequently, Kontsevich[6] obtained a model which generates intersection numbers on this moduli space (see also Ref.[7]), and generalizations of the Kontsevich model[8] were soon found which describe more complicated topological problems associated to vector bundles over moduli space.

The simplest of this class of models will be the subject of this article. We will show that a simple (almost trivial) model of hermitian *positive* random matrices coupled to an external matrix gives rise to a fascinating theory[9] which can eventually be seen to generalize both the Penner and Kontsevich models. Moreover, our theory

[1]Talk given by the second author

Low Dimensional Applications of Quantum Field Theory
Edited by L. Baulieu *et al.*, Plenum Press, New York, 1997

will be shown to satisfy W_∞ identities which uniquely fix its perturbative expansion to be that of noncritical $c = 1$ string theory at self-dual radius[10][11].

The two-dimensional string at this radius has been recently argued[12] to determine universal properties of type II superstrings compactified on Calabi-Yau manifolds with conifold singularities. This suggests the exciting possibility that the matrix model described here could have a direct bearing on physical properties of type II superstrings. We will make some speculations in this and other directions below.

2 A Topological Matrix Model

Consider a single $N \times N$ hermitian matrix M whose eigenvalues are constrained to be positive-semidefinite. We will choose a *linear* action, describing the coupling of M to a fixed external matrix A. Remarkably, this almost trivial choice leads us to a matrix model which describes a well-known string theory, as we will see below.

The random-matrix integral is

$$Z(A) = \int dM e^{-\nu \mathrm{tr}\ MA} \tag{2.1}$$

where ν is a coupling constant. The eigenvalue part of the integral is taken over the range $[0, \infty)$. Thus this integral is convergent.

The redefinition $M \to MA^{-1}$ transfers the A-dependence to a prefactor:

$$Z(A) = (\det A)^{-N} \int dM e^{-\nu \mathrm{tr}\ M} \tag{2.2}$$

We would like to think of this as a toy version of a quantum field theory. Treating A as a fixed source (a background), we add an extra source term and compute the Legendre transform of the free energy to obtain the effective action. Thus we start now with

$$Z_A(J) = \int dM e^{-\nu \mathrm{tr}\ MA - \mathrm{tr}\ JM} = (\det(A + \frac{J}{\nu}))^{-N} \int dM e^{-\nu \mathrm{tr}\ M} \tag{2.3}$$

The free energy as a function of the background A and the source J is minus the log of this integral, which is (dropping additive constants):

$$F_A(J) = N \mathrm{tr}\ \log (A + \frac{J}{\nu}) \tag{2.4}$$

To find the effective action, define the "classical field" $\hat{M}$ by

$$\hat{M} \equiv \frac{\partial F_A(J)}{\partial J} = \frac{N}{\nu}(A + \frac{J}{\nu})^{-1} \tag{2.5}$$

Then make the Legendre transformation

$$\Gamma(\hat{M}) = F_A(J) - \mathrm{tr}\ \hat{M} J \tag{2.6}$$

and eliminate J from Eq.(2.5) to get (again dropping an additive constant)

$$\Gamma(\hat{M}) = \nu \mathrm{tr}\ \hat{M} A - N \mathrm{tr}\ \log \hat{M} \tag{2.7}$$

It is perhaps surprising that in such a trivial theory, the effective action is not identically equal to the classical one! The difference is an additive logarithmic term, which is generated dynamically — essentially by the boundary of the integration region at 0. The coefficient N in front of the log term is undesirable since we ultimately intend to take the limit of large N with the coupling μ fixed. It can be easily removed by including in the "bare" action a term of the same logarithmic type, with coefficient $(\nu - N)$.

This finally leads us to consider the model with matrix integral

$$Z(A) = \int dM e^{-\nu \mathrm{tr}\ MA + (\nu - N)\mathrm{tr}\ \log M} \tag{2.8}$$

whose effective action is

$$\Gamma(\hat{M}) = \nu(\mathrm{tr}\ \hat{M}A - \mathrm{tr}\ \log \hat{M}) \tag{2.9}$$

Continuing to treat this as a toy field theory, we will show right away that this action gives rise to a rich set of "amplitudes". Later we will see that these turn out to be precisely the *tachyon scattering amplitudes of* $c = 1$ *string theory* at the self-dual radius, with $\frac{1}{\nu^2}$ as the genus-expansion parameter!

To evaluate amplitudes, we need to first solve the equations of motion coming from the effective action $\Gamma(\hat{M})$:

$$0 = \frac{\partial \Gamma}{\partial \hat{M}} = A - \hat{M}^{-1} \tag{2.10}$$

and then shift the "field" $\hat{M}$ about the classical solution $\hat{M} = A^{-1}$ by $\hat{M} = A^{-1} + \hat{m}$. Then $\hat{m}$ is the "quantum field", and its amplitudes are determined by the expansion of the effective action $\Gamma(\hat{M})$ about the classical solution, which leads to the "quantum Penner action":

$$\Gamma(\hat{m}) = \nu \mathrm{tr}\ \log A + \nu \sum_{k=2}^{\infty} \frac{(-1)^k}{k} \mathrm{tr}\ (A\hat{m})^k \tag{2.11}$$

For the special case $A = 1$ this coincides with the classical action of the Penner model, whose partition function computes the virtual Euler characteristic of the moduli space of Riemann surfaces. Here we see that a general background A gives rise to a full-fledged theory of amplitudes. To compute these, we read off the 1PI vertices, for example the two- and three-point vertices are

$$\Gamma^{(2)}_{i_1 j_1; i_2 j_2} = \nu\, A_{i_2 j_1} A_{i_1 j_2} \tag{2.12}$$

$$\Gamma^{(3)}_{i_1 j_1; i_2 j_2; i_3 j_3} = -\nu\, [A_{i_3 j_1} A_{i_1 j_2} A_{i_2 j_3} + A_{i_2 j_1} A_{i_3 j_2} A_{i_1 j_3}] \tag{2.13}$$

while the propagator is the inverse of the two-point vertex:

$$G^{(2)}_{i_1 j_1; i_2 j_2} = \langle \hat{m}_{i_1 j_1} \hat{m}_{i_2 j_2} \rangle = \frac{1}{\nu} A^{-1}_{i_1 j_2} A^{-1}_{i_2 j_1} \tag{2.14}$$

Now we can compute n-point functions in terms of the background A. At this point we take the limit of large N, and parametrize the background in terms of an infinite number of independent parameters t_n via the Kontsevich-Miwa transform

$$t_n = \frac{1}{\nu} \mathrm{tr}\ \frac{A^{-n}}{n} \tag{2.15}$$

Then the two- and three-point functions are easily computed to be

$$\langle \mathrm{tr}\ \hat{M}^2 \rangle = \nu(2t_2 + (t_1)^2) \tag{2.16}$$

$$\langle \mathrm{tr}\ \hat{M}^3 \rangle = \nu(3t_3 + 6t_1t_2 + (t_1)^3 + \frac{1}{\nu^2}3t_3) \tag{2.17}$$

Higher point functions follow using the higher vertices calculated from the action, by computing all connected and disconnected tree diagrams.

The structure of the amplitudes already begins to look interesting. Because of obvious homogeneity properties of the quantum Penner action, the n-point function will be a quasi-homogeneous function of the t_i where each term has $\sum i = n$. Moreover, apart from an overall factor of ν, the amplitudes have the form of a power series expansion in $\frac{1}{\nu^2}$ which *terminates* at order $\left(\frac{1}{\nu^2}\right)^{[\frac{n-1}{2}]}$ where [ ] denotes the integer part. In the next section we show that the amplitudes that we obtain in this way represent the complete perturbative solution of $c = 1$ string theory at the self-dual radius, with the operators $tr\hat{M}^n$ representing the tachyons of momentum $-n$ in that theory, and the parameters t_n representing the couplings to tachyons of positive momentum.

3 RELATION TO $c = 1$ STRING

The $c = 1$ string is a background of bosonic string theory where the spacetime is two-dimensional. One of these dimensions may be compact, and we focus on the case where this compact direction has radius unity in suitable units, the self-dual value under T-duality. The perturbative solution of this theory takes the form of a generating function for "tachyon" scattering amplitudes, satisfying a set of recursive W_∞ ward identities which completely determine it (at the self-dual radius, these have been obtained from matrix models in Ref.[10] and from the topological Landau-Ginzburg model in Ref.[11]). Denote the partition function of this theory $Z_{W_\infty}(t, \bar{t})$ where $t_n, \bar{t}_n$ $(n = 1, 2, \ldots)$ are respectively the couplings to tachyons of positive and negative momentum $|n|$. The W_∞ identities can be written

$$\frac{1}{\mu^2}\frac{\partial Z_{W_\infty}}{\partial \bar{t}_n}(t, \bar{t}) = \frac{1}{(n+1)(i\mu)^{n+1}} \oint dz\, W_{n+1}(z, t, \frac{\partial}{\partial t}) Z_{W_\infty} \tag{3.1}$$

Here, μ is the cosmological constant of the theory, and the W-generators are differential operators defined in terms of certain free-fermion operators by

$$W_{n+1} = \ :\bar{\Psi}\, \partial_z^{n+1}\, \Psi: \tag{3.2}$$

The free-fermion operators in turn are defined through bosonization,

$$\Psi(z, t, \frac{\partial}{\partial t}) = e^{i\mu\phi(z,t,\frac{\partial}{\partial t})} \tag{3.3}$$

$$\bar{\Psi}(z, t, \frac{\partial}{\partial t}) = e^{-i\mu\phi(z,t,\frac{\partial}{\partial t})} \tag{3.4}$$

$$\partial\phi(z, t, \frac{\partial}{\partial t}) = \frac{1}{z} + \sum_{n>0} n t_n z^{n-1} - \frac{1}{\mu^2}\sum_{n>0}\frac{\partial}{\partial t_n} z^{-n-1} \tag{3.5}$$

To make contact with the matrix model of the previous section, we replace the couplings t_n by an external hermitian matrix A defined through Eq.(2.15) above. The limit $N \to \infty$ on this matrix is implicit, since this is required for the parameters t_n to be all independent. A lengthy but straightforward computation[9] shows that the W_∞ constraint can be rewritten

$$\frac{1}{i\mu}\frac{\partial Z_{W_\infty}}{\partial \bar{t}_n}(t,\bar{t}) = \frac{1}{(i\mu)^n}(\det A)^{-i\mu}\,\mathrm{tr}\left(\frac{\partial}{\partial A}\right)^n (\det A)^{i\mu}\, Z_{W_\infty}(t,\bar{t}) \tag{3.6}$$

It is easy to check explicitly that this constraint is solved by the random-matrix model with partition function

$$Z_(t,\bar{t}) = (\det A)^\nu \int dM \mathrm{e}^{-\nu \mathrm{tr}\; MA + (\nu - N)\mathrm{tr}\;\log M - \nu\sum_{k>0}\bar{t}_k \mathrm{tr}\; M^k} \tag{3.7}$$

where $\nu = -i\mu$. Differentiating in $\bar{t}_n$ and setting $\bar{t} = 0$, and dividing by the partition function, we find the normalized correlator

$$\nu\langle \mathrm{tr}\; M^n\rangle \equiv \frac{1}{Z(A)}\int dM\; \nu \mathrm{tr}\; M^n\, \mathrm{e}^{-\nu \mathrm{tr}\; MA + (\nu - N)\mathrm{tr}\;\log M} \tag{3.8}$$

where now Z is the partition function of the matrix model defined in Eq.(2.8) above. This shows that the expectation values of tr M^n computed via the quantum Penner action are (after multiplying by ν) just the expectation values of tachyons of negative momentum, $\langle T_{-n}\rangle$, in the $c = 1$ string at self-dual radius. The same can be done for higher derivatives in $\bar{t}$, so we have proved that the simple matrix model defined in the previous section is (at least perturbatively) equivalent to this string theory.

4 Discussion and Conclusions

A more detailed discussion of the model above, and in particular its relation to previous attempts at finding a topological matrix model for the $c = 1$ string, can be found in Ref.[9]. Here we point out some of the interesting open questions that follow from this work.

Although the equivalence of our matrix model to the $c = 1$ string has been demonstrated only perturbatively, the matrix model seems perfectly well-defined outside perturbation theory (it is really no more than the matrix analogue of the Gamma-function). Thus one may ask whether this provides new insight into the tricky question of giving a nonperturbative definition to the $c = 1$ string.

Recently Ghoshal and Vafa[12] have argued that the $c = 1$ string at self-dual radius describes universal properties of type II superstrings compactified on Calabi-Yau manifolds which are developing conifold singularities. Their identification makes use only of the $c = 1$ partition function. It would be interesting to understand whether the amplitudes of the $c = 1$ string likewise tell us something about conifold singularities and related issues concerning Ramond-Ramond states. It might, for example, be possible to address some recent speculations due to Shenker[13] using our topological matrix model.

Finally, the simple way in which the background matrix A enters into our model suggests that one might understand something more about background-independent string field theory from this point of view. In the absence of a background of tachyons, the potential for our matrix model is 0, rather analogous to the situation in topological field theories, which are believed to contain background-independent information about quantum gravity. It has been argued by us in Ref.[9] that this model contains other "vacua" corresponding to string theory in the background of $c < 1$ minimal models, so that one could at least hope to find a background independent picture for all $c \leq 1$ string backgrounds.

Acknowledgements

One of us (S.M.) is grateful to the organisers of the Cargèse School for their kind invitation to lecture on this work, and for their generous hospitality at Cargèse.

References

[1] P. Ginsparg and G. Moore, *Lectures on 2-D Gravity and 2-D String Theory*, hep-th/9304011.

[2] D.V. Boulatov, V.A. Kazakov, I.K. Kostov and A.A. Migdal, Nucl. Phys. B275 (1986) 641;
V.A. Kazakov, Mod. Phys. Lett. A4 (1989) 2125.

[3] E. Brézin and V.A. Kazakov, Phys. Lett. B236 (1990) 144;
M.R. Douglas and S.H. Shenker, Nucl. Phys. B335 (1990) 635;
D.J. Gross and A.A. Migdal, Nucl. Phys. B340 (1990) 333.

[4] R. Penner, Comm. Math. Phys. 113 (1987) 299.

[5] J. Distler and C. Vafa, Mod. Phys. Lett. A6 (1991) 259.

[6] M. Kontsevich, Comm. Math. Phys. 147 (1992) 1.

[7] C. Itzykson and J.B. Zuber, hep-th/9201001, Int. Jour. Mod. Phys. A7 (1992) 5661;
P. Di Francesco, C. Itzykson and J.B. Zuber, hep-th/9206090, Comm. Math. Phys. 151 (1993) 193.

[8] M. Adler and P. van Moerbeke, Comm. Math. Phys. 147 (1992) 24;
S. Kharchev, A. Marshakov, A. Mironov, A. Morozov and A. Zabrodin, hep-th/9111037, Phys. Lett. B275 (1991) 311.

[9] C. Imbimbo and S. Mukhi, hep-th/9505127, Nucl. Phys. B449 (1995) 553.

[10] R. Dijkgraaf, G. Moore and R. Plesser, hep-th/9208031, Nucl. Phys. B394 (1993) 356.

[11] D. Ghoshal, C. Imbimbo and S. Mukhi, hep-th/9410034, Nucl. Phys. B440 (1995) 355.

[12] D. Ghoshal and C. Vafa, hep-th/9506122, Nucl. Phys. B453 (1995) 121.

[13] S. Shenker, "Another Length Scale in String Theory?", hep-th/9509132.

ISING MODEL IN A MAGNETIC FIELD: THE QUANTUM FIELD THEORY APPROACH TO THE CALCULATION OF THE SPIN-SPIN CORRELATION FUNCTION *

G. DELFINO

THEORETICAL PHYSICS, UNIVERSITY OF OXFORD
1 KEBLE ROAD, OXFORD OX1 3NP, UNITED KINGDOM

G. MUSSARDO

INTERNATIONAL SCHOOL FOR ADVANCED STUDIES
AND
ISTITUTO NAZIONALE DI FISICA NUCLEARE
34014 TRIESTE, ITALY

1 INTRODUCTION

The calculation of the spin-spin correlation function $G(x) =< \sigma(x)\sigma(0) >$ of the two-dimensional Ising Model in a magnetic field has been a long-standing problem of statistical mechanics. Solution of this problem has been recently obtained by using a quantum field theory approach based on the spectral representation method for correlation functions [1]. The aim of this talk is to illustrate in a plain language the main features of the solution and to comment on some interesting aspects of the method we used.

Correlation functions of local fields have a quite distinct behaviour at and away from criticality. While at the critical points the correlation functions of the statistical models fall into a scale-invariant regime and their computation may be achieved by solving the linear differential equations obtained by the representation theory of the infinite-dimensional conformal symmetry [2], away from criticality their determination may present in general quite a remarkable degree of complexity. This is

*Talk presented by G. Mussardo

Low Dimensional Applications of Quantum Field Theory
Edited by L. Baulieu *et al.*, Plenum Press, New York, 1997

physically due to the presence of mass scales which destroy the conformal invariance of the critical point and its universal properties.

Several obstacles must be overcomed in order to determine the properties of the statistical models away from criticality. Let us preliminarly clarify that we are interested in applying methods of Quantum Field Theory to those models. This means that we will consider statistical models in the phase space near the critical points, *i.e.* in the so-called *scaling region*[1]. In this region, the physical properties of the statistical models may be expressed in terms of universal scaling functions which depend only on scaling variables, as for instance the combination Mr, where M is the lowest mass scale and r a length variable. There is however the very first problem of determining the exact spectrum of the massive (massless) excitations which rule the dynamics of the model away from criticality. Since this question is often without an answer, in many case the only available tool of investigation of off-critical theories remains just perturbation theory around the conformal points and correlation functions will be then computed in a series expansion based on the conformal correlators. This expansion is particularly appropriate when the fields are located very close each other but, in general, is not particularly efficient for large distance separations.

A remarkable progress in the determination of off-critical correlation functions has been obtained over the past few years by combining concepts and methods of scattering theory for integrable QFT [3] together with spectral representation based on the determination of exact matrix elements of local operators [5, 6]. Several QFTs have been solved through this approach and some examples may be found in the references [7, 8, 9, 10, 11, 12]. A general property of the spectral series of massive integrable QFT, confirmed so far by all models solved in this way, is their very fast rate of convergence for *all* distance scales. This important quality of the spectral representation series may be regarded as the key point to the success of the Form Factor approach, since it allows us to extract the exact value of quantities related to the large distance properties as well as to control their ultraviolet limit.

It is worth to notice that one of the first exact computation of correlation functions, the spin-spin correlator in the thermal deformed Ising model ($T \neq T_c, h = 0$) obtained by a tour-de-force calculation in the classical papers [14, 15, 16, 17], can be elegantly recovered by a Form Factor approach, as shown in the references [5, 8, 18, 19]. In this talk we are concerned with the other deformation of the Ising model, namely with the Ising model at $T = T_c$ but at $h \neq 0$. The corresponding QFT has the important property of being integrable and the scattering theory has been determined by Zamolodchikov [3]. Let us briefly remind the main results of his analysis which are an essential step in our calculations, referring for all details to the original papers.

[1]This region is identified by the relevant directions near the fixed point and has the property that, although the correlation length is finite nevertheless we can still discard the microscopic details of the models and set up a continuum formulation of the properties of the model.

2 Scattering theory of Ising Model in a Magnetic Field

For small values of the magnetic field h, the system is still at $T = T_c$ although the coupling to the magnetic field induces a mass scale $M(h)$ in the problem and destroys the long-range fluctuations of the critical point. The system is integrable though and its properties are ruled by the root system of the E_8 Dynkin diagram [3, 4]. In fact, the spin of its conserved charges are $s = 1, 7, 11, 13, 17, 19, 23, 29 \pmod{30}$, which are nothing else but the Coxeter exponents of E_8. Moreover, the spectrum is given by eight different species of self-conjugated particles A_a, $a = 1, \ldots, 8$ whose mass values are proportional to the components of the Perron-Frobenius eigenvector of the E_8 algebra. The scattering processes in which the eight particles are involved are completely elastic (the final state contains exactly the same particles as the initial one with unchanged momenta) and, due to the factorization of multiparticle scattering amplitudes induced by integrability, they are entirely characterised by the two-particle amplitudes S_{ab}. These are functions of the relativistic invariant Mandelstam variable $s = (p_a + p_b)^2$ or, equivalently, of $u = (p_a - p_b)^2$. S_{ab} has a branch root singularity in the variable s at the threshold $s = (m_a + m_b)^2$. By crossing symmetry, an analogous branch point also appears at the threshold of the u-channel, namely at $s = (m_a - m_b)^2$. Those are the only branch cuts of the S-matrix, due to the elastic nature of the scattering processes. The other possible singularities of the scattering amplitudes S_{ab} are simple and higher-order poles in the interval $(m_a - m_b)^2 < s < (m_a + m_b)^2$ which are related to the bound state structure. An important simplification in the analysis of the analytic structure of the S-matrix comes from the parameterization of the external momenta in terms of the rapidity variable θ, i.e. $p_a^0 = m_a \cosh\theta_a$, $p_a^1 = m_a \sinh\theta_a$. The mapping $s(\theta_{ab}) = m_a^2 + m_b^2 + 2m_a m_b \cosh\theta_{ab}$, where $\theta_{ab} = \theta_a - \theta_b$ transforms the amplitudes S_{ab} into meromorphic functions $S_{ab}(\theta_{ab})$, which satisfy the equations

$$S_{ab}(\theta) S_{ab}(-\theta) = 1 \;; \tag{2.1}$$

$$S_{ab}(\theta) = S_{ab}(i\pi - \theta) \;, \tag{2.2}$$

expressing the unitarity and the crossing symmetry of the theory, respectively. The simple poles of $S_{ab}(\theta)$ with positive residues are related to bound state propagation in the s-channel. Since the bootstrap principle gives the possibility to consider the bound states on the same footing as the asymptotic states, the amplitudes S_{ab} are related each other by the functional equations $S_{il}(\theta) = S_{ij}(\theta + i\overline{u}_{jl}^k)\, S_{ik}(\theta - i\overline{u}_{lk}^j)$, $(\overline{u}_{ab}^c \equiv \pi - u_{ab}^c)$. For the IMMF, the expression of the elastic S-matrix that satisfies all the above constraints can be cast in the form

$$S_{ab}(\theta) = \prod_{\alpha \in \mathcal{A}_{ab}} (f_\alpha(\theta))^{p_\alpha} \tag{2.3}$$

where $f_\alpha(\theta) \equiv \frac{\tanh\frac{1}{2}(\theta + i\pi\alpha)}{\tanh\frac{1}{2}(\theta - i\pi\alpha)}$. The sets of numbers $\mathcal{A}_{ab}$ and their multiplicities p_α, specifying the amplitudes (2.3), can be found, for instance, in [1]. The functions $f_\alpha(\theta)$ have two poles located at the crossing symmetrical positions $\theta = i\pi\alpha$ and $\theta = i\pi(1 - \alpha)$. In addition to simple poles, the S-matrix of the IMMF presents higher-order poles due to multi-scattering processes (for their detailed interpretation see [20, 21, 22]). The odd order poles correspond to bound state poles while those of even order do not. Their appearance is an unavoidable consequence of the iterative application of the bootstrap equations.

3 Correlation Functions and Form Factors

In order to appreciate the significant progress achieved by means of the bootstrap methods in the computation of the correlation functions, let us discuss the most common difficulties which arise in the usual perturbative method. Let

$$\mathcal{A} = \mathcal{A}_0 + \lambda\, \mathcal{A}_{\rm int} \tag{3.1}$$

be the action of the theory, where $\mathcal{A}_0$ corresponds to a solvable QFT (*e.g.* a free theory, CFT, etc.) whereas $\mathcal{A}_{\rm int}$ defines the interactive part. For the perturbative definition of the Green functions we have the formal expressions

$$G^{(N)}(x_1,\ldots,x_N) = \sum_{k=0} \lambda^k\, G_k^{(N)}(x_1,\ldots,x_N) \ , \tag{3.2}$$

where $G_k^{(N)}(x_1,\ldots,x_N)$ is the k-th perturbative term. The above expression usually suffers of several drawbacks: (a) We may face, for instance, the renormalization problem, *i.e.* the presence of infinities which should correctly be handled. In this case, we need to express our final computation in terms of physical quantities by means of the renormalization procedure. This is usually a painful task. The same can be also said in the case of finite renormalization; (b) Assuming we were able to pass through the renormalization program, the resulting expression may present a low rate of convergence, if any! In the light of the above considerations, a more efficient way to compute correlation functions is needed. Once the spectrum and the S-matrix are known, this is provided by the spectral representation methods, *i.e.* by the possibility to express the correlation functions as an infinite series over multi-particle intermediate states

$$\begin{aligned} < \Phi(x)\Phi(0) >= \sum_{n=0}^{\infty} \int_{\theta_1>\theta_2\ldots>\theta_n} \frac{d\theta_1}{2\pi}\cdots\frac{d\theta_n}{2\pi} \\ | < 0|\Phi(0)|A_{a_1}(\theta_1)\cdots A_{a_n}(\theta_n) > |^2 e^{-|x|\sum_{k=1}^n m_k \cosh\theta_k} \ . \end{aligned} \tag{3.3}$$

Basic quantities of this approach are the Form Factors (FF), *i.e.* the matrix elements of the local operators Φ on the asymptotic states defined as

$$F^{\Phi}_{a_1,\ldots,a_n}(\theta_1,\ldots,\theta_n) =< 0|\Phi(0)|A_{a_1}(\theta_1),\ldots,A_{a_n}(\theta_n) > \ . \tag{3.4}$$

Since the spectral representations are based only on the completeness of the asymptotic states, they are general expressions for any QFT. However, for integrable models, they become quite effective because the exact computation of the form factors reduces to finding a solution of a finite set of functional equations. There are other advantages in using the spectral representation method. In fact, since it deals with physical quantities, there is no need of renormalization and the coupling constant dependance is taken into account at all orders by a closed expression of the Form Factors. Secondly, as we mentioned in the introduction, the above representation (3.3) present a very fast rate of convergence for all values of the scaling variable (mr). This is of course expected for large values of (mr) which are dominated by the lowest massive state but the surprising and pleasant result is that in many cases the series is saturated by the lowest terms also for small values of (mr). This may be interpreted as a threshold suppression phenomenon discussed in [7] but we should stress that the same fast convergent behaviour is also observed in massless theories [13]. Finally, let us mention that the two-point correlation functions (3.3) have the form of a Grand-Canonical Partition Function of a one-dimensional fictitious gas (with coordinate position β_i) $\Xi(mr) = \sum_{N=0}^{\infty} z^N Z_N(mr)$ although with a coordinate-dependent activity $z_i(m_i r, \beta_i) = \frac{1}{2\pi} e^{-m_i r \cosh \beta_i}$. This observation is extremely useful to recover ultraviolet data of the theory, as the anomalous dimensions of the fields, in terms of massive quantities [8, 9].

Let us discuss now the main properties of the Form Factors for 2-D Integrable Massive Field Theories which are the crucial quantities entering the spectral representation of correlation functions. For a scalar operator $\Phi(x)$, relativistic invariance requires that its form factors depend only on the rapidity differences $\theta_i - \theta_j$. The elasticity of the scattering processes, together with the crossing symmetry and the completeness relation of the asymptotical states, permit to derive the following monodromy equations satisfied by the FF

$$\begin{aligned} F^{\Phi}_{a_1,\ldots,a_i,a_{i+1},..a_n}(\theta_1,\ldots,\theta_i,\theta_{i+1},\ldots,\theta_n) &= S_{a_i a_{i+1}}(\theta_i - \theta_{i+1}) \\ &\quad \times F^{\Phi}_{a_1,..a_{i+1},a_i,..a_n}(\theta_1,\ldots,\theta_{i+1},\theta_i,\ldots,\theta_n) \\ F^{\Phi}_{a_1,a_2,\ldots a_n}(\theta_1 + 2\pi i,\ldots,\theta_{n-1},\theta_n) &= F^{\Phi}_{a_2,a_3,\ldots,a_n,a_1}(\theta_2,\ldots,\theta_n,\theta_1). \end{aligned} \tag{3.5}$$

Apart from these monodromy properties, the FF are expected to have poles induced by the singularities of the S-matrix, among them the simple poles associated to the kinematic singularities with residue given by [6]

$$-i \lim_{\tilde{\theta}\to\theta}(\tilde{\theta}-\theta) F^{\Phi}_{a,\bar{a},a_1,\cdots,a_n}(\tilde{\theta}+i\pi,\theta,\theta_1,\ldots,\theta_n) = \left(1 - \prod_{j=1}^{n} S_{a,a_j}(\theta-\theta_j)\right) F^{\Phi}_{a_1,\cdots,a_n}(\theta_1,\cdots,\theta_n). \tag{3.6}$$

and those due to the bound state singularities with residue [5, 6]

$$-i \lim_{\theta_{ab}\to i u^c_{ab}} (\theta_{ab} - i u^c_{ab})\, F^{\Phi}_{a,b,a_1,\ldots,a_n}(\theta_a,\theta_b,\theta_1,\ldots,\theta_n) = \Gamma^c_{ab}\, F^{\Phi}_{c,a_1,\ldots,a_n}(\theta_c,\theta_1,\ldots,\theta_n) \tag{3.7}$$

where $\theta_c = (\theta_a \overline{u}^a_{bc} + \theta_b \overline{u}^b_{ac})/u^c_{ab}$. In general, the FF may also present simple poles which do not fall into the two classes above. In addition, they may also have higher-order poles induced by the multi-scattering processes and indeed, their analytic structure may be quite complicated. Their analysis has been pursued in [1] and we refer the reader to this paper for a detailed account on this aspect.

4 Form Factors of the Magnetization Operator $\sigma(x)$ in the Ising Model in a Magnetic Field

In the Form Factor approach to integrable theories, the two-particle FF play a particularly important role, both from a theoretical and from a practical point of view. From a theoretical point of view, they provide the initial conditions which are needed for solving the recursive equations relative to the residue conditions (3.6) and (3.7). Moreover, they also encode all the basic properties that the matrix elements with higher number of particles inherit by factorization, namely the asymptotic behaviour and the analytic structure. In other words, once the two-particle FF of the considered operator have been given, the determination of all other matrix elements is simply reduced to solve a well-defined mathematical problem. From a practical point of view, the truncation of the spectral series at the two-particle level usually provides a very accurate approximation of the correlation function, which goes even further than the crossover region. This section is mainly devoted to the discussion of the basic features of the two-particle FF in the IMMF.

The FF $F_{ab}^{\Phi}(\theta)$ of a local field $\Phi(x)$ must be a meromorphic function of the rapidity difference defined in the strip $Im\theta \in (0, 2\pi)$. Its monodromy properties are dictated by the general equations (3.5), once specialized to the case $n = 2$ Thus, denoting by $F_{ab}^{min}(\theta)$ a solution of those equations free of poles and zeros in the physical strip and also requiring asymptotic power boundness in momenta, we conclude that $F_{ab}^{\Phi}(\theta)$ must be equal to $F_{ab}^{min}(\theta)$ times a rational function of $\cosh\theta$. The poles of this extra function are determined by the singularity structure of the scattering amplitude $S_{ab}(\theta)$.

According to the analysis of the paper [1], a convenient parametrization of the two-particle FF is given by

$$F_{ab}^{\Phi}(\theta) = \frac{Q_{ab}^{\Phi}(\theta)}{D_{ab}(\theta)} F_{ab}^{min}(\theta) \,, \tag{4.1}$$

where $D_{ab}(\theta)$ and $Q_{ab}^{\Phi}(\theta)$ are polynomials in $\cosh\theta$: the former is fixed by the singularity structure of $S_{ab}(\theta)$ while the latter carries the whole information about the operator $\Phi(x)$.

To further constrain the order of the polynomial $Q_{ab}^{\Phi}(\theta)$, a bound extremely useful has been derived in [1]. Briefly stated, the argument consists in looking at the large energy limit of the FF and relating it to the conformal properties of the corresponding operator $\Phi(x)$. Denoted by Δ_{Φ} the conformal weight of the operator $\Phi(x)$ and by y_{Φ} the real quantity defined by

$$\lim_{|\theta_i|\to\infty} F_{a_1,\ldots,a_n}^{\Phi}(\theta_1,\ldots,\theta_n) \sim e^{y_{\Phi}|\theta_i|}$$

we have the bound

$$y_{\Phi} \leq \Delta_{\Phi} \,. \tag{4.2}$$

Taking into account the degree of the polynomial $D_{ab}(\theta)$ in the denominator of the two-particle FF (4.1), it is easy to translate this inequality into an upper bound on the degree of the polynomial $Q_{ab}^{\Phi}(\theta)$.

Let us now apply the aforementioned considerations to the case of the IMMF, considering in order the different terms entering (4.1). For the minimal FF, we have

$$F_{ab}^{min}(\theta) = \left(-i\sinh\frac{\theta}{2}\right)^{\delta_{ab}} \prod_{\alpha\in\mathcal{A}_{ab}} \left(G_\alpha(\theta)\right)^{p_\alpha} , \tag{4.3}$$

where

$$G_\alpha(\theta) = \exp\left\{2\int_0^\infty \frac{dt}{t}\frac{\cosh\left(\alpha-\frac{1}{2}\right)t}{\cosh\frac{t}{2}\sinh t}\sin^2\frac{(i\pi-\theta)t}{2\pi}\right\} . \tag{4.4}$$

For large values of the rapidity $G_\alpha(\theta) \sim \exp(|\theta|/2)$, $|\theta| \to \infty$, independent of the index α. Concerning the pole terms in eq. (4.1) this could be written as

$$D_{ab}(\theta) = \prod_{\alpha\in\mathcal{A}_{ab}} \left(\mathcal{P}_\alpha(\theta)\right)^{i_\alpha} \left(\mathcal{P}_{1-\alpha}(\theta)\right)^{j_\alpha} , \tag{4.5}$$

where

$$\begin{array}{llll} i_\alpha = n+1 , & j_\alpha = n , & \text{if} & p_\alpha = 2n+1 ; \\ i_\alpha = n , & j_\alpha = n , & \text{if} & p_\alpha = 2n , \end{array} \tag{4.6}$$

and we have introduced the notation

$$\mathcal{P}_\alpha(\theta) \equiv \frac{\cos\pi\alpha - \cosh\theta}{2\cos^2\frac{\pi\alpha}{2}} . \tag{4.7}$$

Both $F_{ab}^{min}(\theta)$ and $D_{ab}(\theta)$ have been normalized to 1 in $\theta = i\pi$.

Finally, let us turn our attention to the determination of the polynomials $Q_{ab}^{\Phi}(\theta)$ for the specific operator we are interested in, namely the magnetization field $\sigma(x)$. In view of the relation

$$\Theta(x) = 2\pi\, h(2-2\Delta_\sigma)\,\sigma(x) , \tag{4.8}$$

which relates $\sigma(x)$ to the trace $\Theta(x)$ of the stress-energy tensor, we can concentrate our attention on the latter operator. This is particularly useful. In fact, the conservation equation $\partial_\mu T^{\mu\nu} = 0$ implies the following relations among the FF of the different components of the energy-momentum tensor

$$F_{a_1,\ldots,a_n}^{T^{++}}(\theta_1,\ldots,\theta_n) \sim \frac{P^+}{P_-} F_{a_1,\ldots,a_n}^{\Theta}(\theta_1,\ldots,\theta_n) ; \tag{4.9}$$

$$F_{a_1,\ldots,a_n}^{T^{--}}(\theta_1,\ldots,\theta_n) \sim \frac{P^-}{P_+} F_{a_1,\ldots,a_n}^{\Theta}(\theta_1,\ldots,\theta_n) , \tag{4.10}$$

where $x^\pm = x^0 \pm x^1$ are the light-cone coordinates and $P^\pm \equiv \sum_{i=1}^n p_{a_i}^\pm$. The requirement that all the components of the energy-momentum tensor must exhibit the same singularity structure, leads to conclude that the FF of $\Theta(x)$ must contain a factor P^+P^-. However, the case $n = 2$ is special because, if the two particles have equal masses, the mismatch of the singularities disappears in eqs. (4.10) and no factorisation takes place. From this analysis, we conclude that for our model we can write

$$Q_{ab}^{\Theta}(\theta) = \left(\cosh\theta + \frac{m_a^2+m_b^2}{2m_a m_b}\right)^{1-\delta_{ab}} P_{ab}(\theta) , \tag{4.11}$$

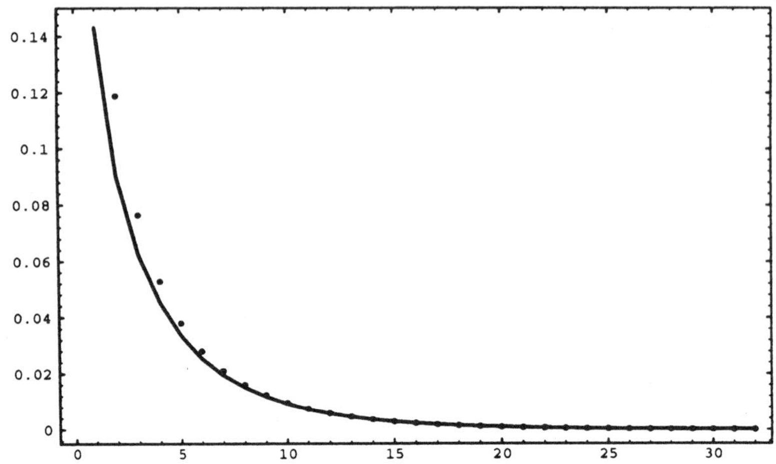

Figure 1. Correlation function $< \sigma(x)\sigma(0) >_c$ versus lattice space distances. The points on the graph represent the numerical data, as extracted from ref. [23], while the continuum curve is the theoretical estimate with only the first three Form Factors.

where

$$P_{ab}(\theta) \equiv \sum_{k=0}^{N_{ab}} a_{ab}^k \cosh^k \theta \ . \tag{4.12}$$

The degree N_{ab} of the polynomials $P_{ab}(\theta)$ can be severely constrained by using eq. (4.2). Additional conditions for these polynomials are provided by the normalization of the operator $\Theta(x)$, that for the diagonal elements F_{aa}^{Θ}, reads

$$F_{aa}^{\Theta}(i\pi) =< A_a(\theta_a)|\Theta(0)|A_a(\theta_a) >= 2\pi m_a^2 \ . \tag{4.13}$$

Using all the information above, in [1] we have computed the first FF of the trace $\Theta(x)$. Their explicit expressions may be found in our original paper. Rather than providing them here, we would like instead to present the final result of our computation, namely the graphs relative to the two-point function $G(x) =< \sigma(x)\sigma(0) >$ compared with a high-precision numerical determination of this quantity obtained by Montecarlo simulations [23]. In Fig. 1 we have only included the first three terms of $G(x)$ (those relative to the form factors of the one-particle states A_1, A_2, A_3). As shown in this figure, they can reproduce correctly the behaviour of the correlation function on the whole infrared and crossover regions. A slight deviation of the theoretical curve from the numerical values is only observed for the first points of the ultraviolet region, where a better approximation can be obtained by including more terms in the form factor series. This is shown in Figure 2, where five more contributions (those relative to form factors up to state $A_1 A_3$) have been added to the series. As far as the Form Factor approach is concerned, this comparison constitutes one of the most remarkable success obtained by means of these techniques.

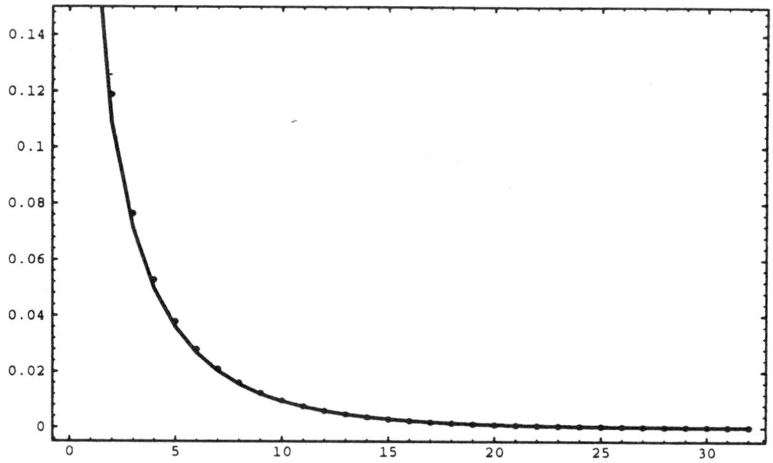

Figure 2. Correlation function $< \sigma(x)\sigma(0) >_c$ versus lattice space distances. The points on the graph represent the numerical data, as extracted from ref. [23], while the continuum curve is the theoretical estimate obtained with the first eight terms of the spectral series.

References

[1] G. Delfino and G. Mussardo, *Nucl. Phys.* B 455 (1995), 724.

[2] A.A. Belavin, A.M. Polyakov and A.B. Zamolodchikov, *Nucl. Phys.* B 241 (1984), 333.

[3] A.B. Zamolodchikov, in *Advanced Studies in Pure Mathematics* 19 (1989), 641; *Int. J. Mod. Phys.* A 3 (1988), 743.

[4] V.A. Fateev and A.B. Zamolodchikov, *Int. J. Mod. Phys.* A 5 (1990), 1025.

[5] B. Berg, M. Karowski, P. Weisz, *Phys. Rev.* D 19 (1979), 2477; M. Karowski, P. Weisz, *Nucl. Phys.* B 139 (1978), 445; M. Karowski, *Phys. Rep.* 49 (1979), 229;

[6] F.A. Smirnov, *Form Factors in Completely Integrable Models of Quantum Field Theory* (World Scientific) 1992, and references therein.

[7] J.L. Cardy and G. Mussardo, *Nucl. Phys.* B 410 [FS] (1993), 451.

[8] V.P. Yurov and Al.B. Zamolodchikov, *Int. J. Mod. Phys.* A 6 (1991), 3419.

[9] J.L. Cardy and G. Mussardo, *Nucl. Phys.* B 340 (1990), 387.

[10] Al.B. Zamolodchikov, *Nucl. Phys.* B 348 (1991), 619.

[11] G. Delfino and G. Mussardo, *Phys. Lett.* B 324 (1994), 40; J. Balog, *Phys. Lett.* B 300 (1993), 145.

[12] A. Fring, G. Mussardo and P. Simonetti, *Nucl. Phys.* B 393 (1993), 413; G. Mussardo and P. Simonetti, *Int. J. Mod. Phys.* A 9 (1994), 3307.

[13] G. Delfino, G. Mussardo and P. Simonetti, *Phys. Rev.* D 51 (1995), 6620.

[14] T.T. Wu, B.M. McCoy, C.A. Tracy and E. Barouch, *Phys. Rev.* B 13 (1978), 316.

[15] B.M. McCoy and T.T. Wu, *The Two-Dimensional Ising Model* (Harvard University Press, Cambridge, 1973).

[16] B.M. McCoy, C.A. Tracy and T.T. Wu, *Jour. Math. Phys.* 18 (1977), 1058.

[17] M. Sato, T. Miwa and M. Jimbo, *Publ. RIMS, Kyoto Univ.* 14 (1978), 223.

[18] J. Palmer and C.A. Tracy, *Adv. in Applied Math.* 2 (1981) 329;

[19] O. Babelon and D. Bernard, *Phys. Lett.* B 288 (1992), 113.

[20] S. Coleman and H.J. Thun, *Commun. Math. Phys.* 61 (1978), 31; C.J. Goebel, *Prog. Theor. Phys. Supplement* 86 (1986), 261.

[21] H.W. Braden, E. Corrigan, P.E. Dorey and R. Sasaki, *Nucl. Phys.* B 338 (1990), 689; *Nucl. Phys.* B 356 (1991), 469.

[22] P. Christe and G. Mussardo, *Nucl. Phys.* B 330 (1990), 465; *Int. J. Mod. Phys.* A 5 (1990), 4581.

[23] P.G. Lauwers and V. Rittenberg, *Numerical Estimates of the Spin-Spin Correlation Function for the Critical 2-D Ising Model in a Magnetic Field*, Bonn-HE-89-11.

REVIEW OF RECENT RESULTS ON N= 1 SUPERSYMMETRIC GAUGE THEORIES

ELIEZER RABINOVICI

RACAH INSTITUTE OF PHYSICS
HEBREW UNIVERSITY OF JERUSALEM
JERUSALEM 91904, ISRAEL

1 INTRODUCTION

As the last speaker in this school it is my pleasure to thank the organizers on behalf of all lecturers and participants for this outstanding school. The tradition of the Cargese schools has set the highest standard of scientific excellence, coupled with a very exceptional interactive environment, thus serving as an inspiration for scientists all over the globe. I also wish to express my deep sorrow on the passing of Claude Iytzkson, on a more personal note I would mention that he was generous to be the first visitor to the particle physics group in Jerusalem. We miss him dearly.

The school has covered a wide range of topics. I will concentrate on those related to the very significant progress in understanding the structure of four dimensional supersymmetric gauge theories. These were discussed in the lectures of M. Douglas, K. Intriligator and E. Verlinde. The references can be found in their talks.

I will discuss three main topics:

1. The phase structure of four-dimensional supersymmetric QCD as studied for regions ranging from weak to strong coupling.

2. The emergence of various surprising "dualities" in the process of analyzing the above phase structure.

3. The discovery of new four-dimensional conformal field theories.

Each of these topics can be discussed separately from the others yet they are strongly interrelated. The features describing the new results are threefold, older ideas have been confirmed, unexpected novel results were obtained and a new style has been created.

Low Dimensional Applications of Quantum Field Theory
Edited by L. Baulieu *et al.*, Plenum Press, New York, 1997

2 Phase structure of supersymmetric gauge theories.

2.1 Ideas

The electromagnetic properties of matter manifest themselves in the existence of various phases. Matter can be found as a gas, a liquid and a solid. It can be ferromagnetic and it can emerge as a superfluid and a superconductor. Systems with a local gauge symmetry can appear in Coulomb, Higgs and confinement phases as well in the more general oblique confinement phase. Systems of elementary particles, as known today, seem to have chosen the first three options, the electromagnetic, weak and colored forces respectively are a realization of those possible phases. In order to fully characterize the different phases one needs to probe the systems with both electric and magnetic charges.

- The Higgs phase: The vacuum is an electric condensate ($V(r)$ is the appropriate potential) electric charge is screened: $V(r) \sim \exp(-mr)$ magnetic charge is confined: $V(r) \sim r$
- The confinement phase: The vacuum is a magnetic condensate electric charge is confined: $V(r) \sim r$ magnetic charge is screened: $V(r) \sim \exp(-mr)$
- The coulomb phase (with massive charged particles) – both the electric and the magnetic charges have coulomb interactions: $-V(r) \sim 1/r$.
- The coulomb phase (with massless charged particles) The electric potential is $V(r) \sim \frac{1}{r \ln r}$ the magnetic potential is $V(r) \sim \frac{\ln r}{r}$.

In this picture the Higgs and confinement phases are of a dual nature. A serious question remained however, where, in QCD, are the monopoles which eventually condense? In the early seventies physicists studied asymptotically free 2 dimensional systems with global symmetries and 2 and 3 dimensional QCD like theories in an attempt to grasp the idea of confinement. $D = 2$ systems were studied mostly for different reasons than those for which they are very actively studied today. Non-linear sigma models on compact manifolds are asymptotically free, in particular the $O(N)/O(N-1)$ non-linear sigma model is one such system. Classically it may seem that the expectation value of the fields is non-zero, the system seems ordered leading to a spontaneous symmetry breaking and its attendant massless Goldstone Bosons. This however is not a good description of the system, actually disorder wins, large quantum fluctuations essentially wipe out the order, the system develops a dynamically generated mass gap and the physical spectrum offers a linear realization of the $O(N)$ symmetry. It was suggested that similar symmetry restoration occurs in QCD leading to confinement. Systems with $Z(N)$ and compact U(1) global symmetries were studied on the lattice and the formation of condensates of various topological configurations was identified as the dynamical source of phase transitions. Analogies were drawn to gauge systems.

In 2 dimensional QED and QCD it was demonstrated that the confinement phase is consistent with quantum field theory however the dynamics resided already in the electric Born term. In an SO(3) three dimensional systems with scalar particles in the adjoint ($I = 1$) representation, instantons (which were the 4-dimensional monopole solutions, monopoles in 4 dimensions may serve as instantons in three) upgraded a logaritmic born term to a linear potential. Researchers studied also lattice regulated versions of QCD, those confined in any dimensions and for any compact symmetry group in the strong coupling limit. Monopoles and their condensation was observed in systems with local $Z(N)$ and compact U(1) symmetry and were eagerly sought for in non abelian systems (whose center was $Z(N)$).

Additional phases were discovered in the presence of a $\mathcal{L} = \Theta F\tilde{F}$ term. It was also shown that for systems with matter fields carrying quantum numbers of the center of the group of the local symmetries (such as the weak interactions) there is no boundary in the phase diagram between the Higgs and confinement phase.

I have just touched upon only a few of the many physical properties of gauge systems. However it is sufficient to realize the amount of detail, extremely important detail, which would be involved once a solution is obtained for such systems. This indeed is the situation since exact results pertaining to the dynamics of supersymmetric gauge theories have become available. To give a feeling as to the part of the new results which serve to confirm ideas about the dynamics of confinement I wish to draw the following analogy ; Dirac has realized the possibility of having magnetic monopoles in a U(1) gauge theory. These configurations involve the presence of the Dirac string. By adding scalar fields and increasing the gauge symmetry it had become possible to dispose of the strings and obtain non-singular classical configurations carrying magnetic charge. In a similar manner the "QCD monopoles" were given life by adding scalar fields and a supersymmetry. It was somewhat of a surprise that the mythical QCD monopoles merged into real monopoles, it was a total surprise that it was possible to follow these configurations all the way from weak coupling, where they are classical and heavy, to strong coupling, where it turns out they sometimes become massless, they are at the verge of condensing. This is true for $N = 2$ supersymmetric systems, the monopoles can be pushed to condense in $N = 1$ supersymmetric systems. The idea of monopole condensation as the reason for confinement has thus been demonstrated. The methods used in the proof and the style of the proof are novel. Dynamical issues were not settled in the past by searching for the appropriate representations of the topological properties.

We have been presented with lots and lots of $N = 1$ and $N = 2$ data in the course of the school, each the purpose of learning about different dynamical properties of the systems, properties which have been before outside our reach.

A feature of many of the supersymmetric systems is the emergence of flat potentials. For any value of the expectation value of a field(called modulus) along the flat potential one obtains a different asymptotically free theory. The dependence of the various properties of the theory on its moduli were obtained as well as the dependance on the matter content of the theory.

As an illustrative example let us consider the $N_c = 2, N(I = 1/2) = 4, N(I = 1) = 1$ supersymmetric system. The microscopic degrees of freedom are

$$Q_i^a(a = 1, \cdots, N_c, \quad i = 1, \cdots, 2N_f), \Phi_\alpha^{ab} \quad (\alpha = 1, \cdots, N(I = 1); N_c = 2; N_f = 2)$$

The macroscopic degrees of freedom are

$$X_{ij} = \varepsilon_{ab} Q_i^a Q_j^b$$

$$M_{\alpha\beta} = \varepsilon_{aa'} \varepsilon_{bb'} \Phi_\alpha^{ab} \Phi_\beta^{a'b'}$$

$$Z_{ij\alpha} = \varepsilon_{aa'} \varepsilon_{bb'} Q_i^a \Phi_\alpha^{a'b'} Q_j^b$$

The quantum effective potential is

$$\Gamma = -2\Lambda^{-1}(PfX)^{1/2}\Gamma + \tilde{m}M + \frac{1}{2} TrMX + \frac{1}{\sqrt{2}} Tr\lambda Z$$

$$\Gamma = M + tr(ZX^{-1})^2$$

The classical moduli space is recaptured in the limit $\Lambda \to 0$. Not only is the phase diagram obtained but the idea that the Higgs and Confinement phases are connected is confirmed as well. One can follow the electric condensate which is a good picture of large values of the appropriate moduli "turn into" a magnetic condensate which is a good picture near the coulomb phase.

2.2 Surprises

Integrating out the $I = 1/2$ particles by allowing them to have a very large mass one returns to the $N = 2$ supersymmetric theory. Classically one expects the coulomb phase at all non zero values for M. For zero values of the field one has classically massless non abelian spin one particles. Following the $D = 2$ knowledge one may have expected that disorder will prevail for $M = O(\Lambda^2)$ leading to confinement inside the disk of radius Λ^2, while order, the Coulomb phase, may have been expected outside the disk. Surprisingly it seems to turn out that analyticity allows 2 points to be on the verge of confinement (and separated form each other by Λ^2) while the rest are in the coulomb phase. Λ is the dynamical scale. Eventually this is less surprising considering that these two points are the vacua of the system in the presence of a mass term for one of the adjoint fields. For that $N = 1$ system two vacua are indeed what was expected. The number of vacua for all $N_c = 2, N = 1$ supersymmetric gauge theories with non IR free one loop beta function were found. All known phases of gauge theories are encountered in the survey (including oblique confinement).

There were more surprises in store for us.

1. For certain values of the representations ($3N_c > N_f > 3N_c/2$ for SU(N_c)) the origin of the classical moduli space turns quantum mechanically into some conformal field theory in a phase termed the non-abelian coulomb phase for obvious reasons.

2. Different points in moduli space describe theories with massive and massless charge carriers. Special points were discovered were both massless electric and magnetic charge carriers are present simultaneously. More generally massless charge carriers, which are not mutually local are present together. The

gauge theory (such as $N = 2$ SU(3)) for which it appears is thus, in the infrared, a Lorentz invariant local description of a theory of massless electrons and monopoles. It is to be seen if this description can be made even more concise.

3. Already more than ten years ago it was discovered that in certain cases ($N_f < N_c$ for SU(N_c) theories) the classical moduli space disappears quantum mechanically, the system seems not to have any ground state.

4. For certain cases ($N_f = N_c+1$ for SU(N_c)) it was found that a phase exists were chiral symmetry is unbroken although the theory confines (a prejudice was that the emergence of a bound state would force chiral symmetry breaking as the constituents would need to turn on their heels in a bound state). Candidates of theories in which massless quarks and leptons are composite objects were searched for in such a phase.

2.3 More suprises — duality

An additional surprise was the emergence of certain unexpected duality properties. Dualities are used in physics in quite a few contexts, in this case two theory are called "dual" when they have the same infrared limit. For example the common wisdom is that QCD is dual in this sense to a theory of massless pions as long as the number of matter multiplets allows the coupling to grow towards the infrared. In the extreme infrared such a theory of massless scalars is a free theory. In supersymmetric $N = 1$ systems dualities were uncovered for various gauge groups(such as SU(N), SP(N), SO(N) and some of the exceptional groups) and various types of matter representations. The first one was found for SU(N_c) gauge group and N_f number of flavors. It is dual to another color group SU($N_f - N_c$) with the same number, N_f, of flavors and some more color singlets. The claims are based on many independent consistency checks(such as matching anomalies à-la 't-Hooft, adding mass terms for the different particles, comparing the moduli spaces and more). It all fits together so well it is very unlikely not to be true. The most striking result obtained so far from this duality is for $3N_c/2 > N_f > N_c+1$ (and an appropriate range for different gauge groups). In that case the SU(N_c) gauge group has as its infrared limit a free theory of SU($N_f - N_c$) color; that is the infrared limit contains massless spin one particles of a different gauge group. This is the first example where it was shown that massless spin one particles emerge out of a strong coupling limit of a theory. Moreover, the emerging color group is different than the input one. In general the infrared limit of the above SU(N_c) theories span gauge theories with the number of colors between 1 (i.e no group) and $N_c/2$. For $3Nc > Nf > 3Nc/2$ and at the origin of moduli space both the theory and its dual flow to some yet to be discovered conformal theory. As N_f increases the input theory becomes more weakly coupled while its dual becomes more strongly coupled. One would expect when the differences in coupling are very pronounced the more strongly coupled theory spend some time in the vicinity of the weaker coupled theory before the crossover to the conformal theory controlling their infrared behavior. Deeper reasons for these dualities were not yet uncovered, one may hope to find it in pure field theory context where it exists in its own right, but one may also search for a simpler structure in the framework of string theory and its own web of dualities. In any case these dualities have brought back the doubt that in gauge theories the role of the gauge symmetry as fundamental is over appreciated. Similar thoughts pertain to the role of general coordinate invariance. Also the issue of the precise impact of chirality is challenged as chiral and non-chiral theories turn out to be in some cases dual to each-other.

2.4 New 4-d conformal field theories

The discovery of new 4 dimensional conformal field theories prompts us to rexamine their classification. One may attempt to learn something about the subject by employing the methods which were so successful in studying two dimensional conformal field theories. Those methods included the identification of the central charge of the Virasoro algebra, c, the search for the marginal operators spanning the moduli space of conformal field theories and the use of various methods to discover a new conformal field theory once one is known. Examples are Landau-Ginsburg analysis around such a theory or the appropriate gauging of conformal theories.

Known four dimensional conformal theories include free theories, $N = 4$ supersymmetric theories, $N = 2$ supersymmetric theories with $N = 2N_c$. New types of conformal four dimensional systems could be searched for in the non-abelian coulomb phases as well as in those conformal theories including various mutually non-local massless charge carriers.

Researchers have tried to find the four dimensional analog of the c-theorem in four dimensions. Although this issue is still unresolved, naive counting arguments of ultraviolet and infrared degrees of freedom allow for the expectation that the number of degrees of freedom decreases towards the infrared. Marginal deformations of $N = 4$ theories were found, for example a $N = 1$ supersymmetric system containing 3 particles in the adjoint representation will flow to an $N = 4$ fixed line even if at the ultraviolet the gauge coupling and the Yukawa coupling differed. This remains true if one adds relevant deformations. The last point in the list of emulating the two dimensional analysis is the construction of new conformal theories out of known ones. The addition of the Landau Ginsburg potential has already played a role in identifying new theories. In two dimensions one can shown that an appropriate gauging of WZW models leads to a new (called co-set) conformal theory. In two dimensions the gauge coupling always flows to stronger coupling in the infrared, in 4 dimensions the above duality considerations indicate that there is more structure involved, however it will still be true that gauging some flavor symmetry group present in a conformal theory (in particular just free fermions, but it could be any one on the list described above of known conformal four dimensional theories) will lead in the infrared to some, hopefully new, conformal field theory. Questions such as the search for a co-set like structure and the determination of the degrees of freedom of the new theory in terms of those of the old theory (such as the construction of parafermions out of gauged WZW variables) are under study.

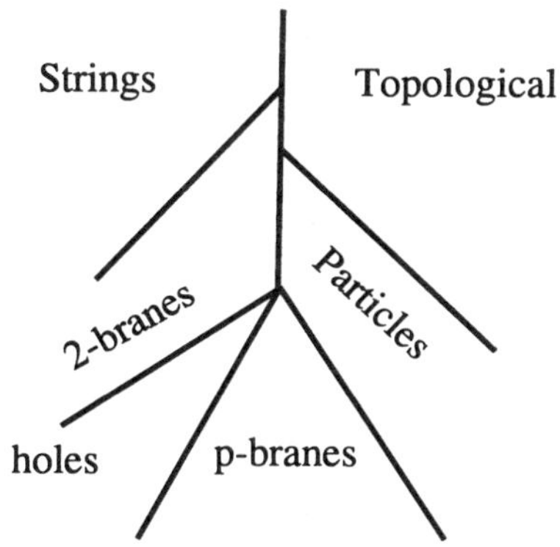

Figure 1. Phases of gravity

The above picture of the phases of gravity seems to have dropped out the year 2000 proceedings of Cargese. It shows phases in which particles, strings, p-branes are the relevant excitations as well as topological phases.

TUNNELING BETWEEN EDGE STATES IN THE FRACTIONAL QUANTUM HALL EFFECT: A PHYSICAL APPLICATION FOR INTEGRABLE 1 + 1 QUANTUM FIELD THEORIES

P. FENDLEY AND H. SALEUR

PHYSICS DEPARTMENT, UNIVERSITY OF SOUTHERN CALIFORNIA
LOS ANGELES, CALIFORNIA 90089-0484, USA

1 INTRODUCTION

The purpose of these lectures is to discuss a physical problem which is experimentally relevant and is a playground for the techniques developed in the last few years in the study of integrable $1 + 1$ quantum field theories: the transport through impurities in a Luttinger liquid.

It has been known for a long time that some experimental systems are described by integrable field theories in one spatial dimension. The classic example is the Kondo model, which consists of electrons interacting with dilute impurities in a metal. This system is actually three-dimensional, but it can be described by a one-dimensional model because, with dilute enough impurities, the interesting physics occurs in s waves around each impurity, and one can restrict attention to the radial coordinate. The example we discuss here is of a related nature, but it leads to a different set of experiments, and new interesting theoretical developments — in fact, it actually allows the solution of aspects of the Kondo problem which had remained untackled so far.

The experiments to be discussed here result from a remarkable property of the fractional quantum Hall effect. A fractional quantum Hall device is made by putting an electron gas trapped in two dimensions into a strong transverse magnetic field. However, when the Hall conductivity is locked to its plateau value, the current flows only along the spatial boundaries ("edges") of the device. This is the reason why the transport properties can be described effectively by a one-dimensional theory. It was shown in [1] that the appropriate field theory is the chiral Luttinger model,

Low Dimensional Applications of Quantum Field Theory
Edited by L. Baulieu *et al.*, Plenum Press, New York, 1997

which is equivalent to a free chiral boson. To probe the transport behavior, the experimentalists can make a point contact connecting two edges which carry edge currents moving in opposite directions, and measure the tunneling current. This situation is described by the Luttinger model with an impurity [2, 3, 4], and this is the theory we will treat here.

The field theories involved in such one-dimensional problems are often simple to formulate: the Kondo problem consists of free electrons interacting with a single impurity spin at the boundary, while the tunneling between Hall edges can be described by a free boson with a boundary interaction. Despite their mild appearance, these models exhibit most if not all phenomena of an interacting field theory, but often in a more tractable form. This is not to say the techniques involved are even close to trivial. Both problems mentioned possess two renormalization group (RG) fixed points, and much interesting physics is related to the crossover between these fixed points. To understand this crossover, non-perturbative techniques are required. For example, understanding the Kondo problem utilized scaling theory [5], the numerical renormalization group [6], and exact Bethe Ansatz calculations [7]. More recent generalizations involve non-Abelian bosonization, conformal field theory [8], (non)Fermi-liquid behavior, kinks [9], and so on. Even though Fermi-liquid theory [10] and conformal field theory [8] allow one to compute the exact temperature dependence of transport properties like the resistivity asymptotically at low temperatures, there is still no analytic exact calculation for the resistivity at arbitrary temperatures.

Like the Kondo model, tunneling in a Luttinger liquid involves also a rich variety of transport behavior. In this problem the two fixed points describe the completely insulating and the completely conducting fixed points, the latter being the unstable one. Critical exponents associated with either one of these two fixed points can be obtained from (abelian) bosonization, and perturbation theory [2]. We have extended these results much further. A very interesting feature of this problem is that the (experimentally-measurable) current-voltage [$I(V)$] curve is characterized by a *universal* scaling function. Universality means that this function is *independent* of the sample-specific details of the point contact of the quantum Hall device, which naturally occur in the fabrication process. Thus it makes sense to compare analytic results for this function, obtained by field theory techniques, with experimentally-measured data. We have computed the exact current and conductance curves [3, 4]. These are not simply properties of a single one of the two fixed points, but rather depend on the entire crossover RG trajectory, connecting the completely-insulating critical point with the with completely-conducting critical point. This is one of the first calculation of exact non-equilibrium transport properties in a model with interacting electrons. Moreover, it gives a calculable universal signature of the non-Fermi liquid state in the edges of the Hall device, which is a consequence of the of the highly correlated quantum Hall state at fractional filling fraction. Since we have already presented our results in brief [3] and in detail [4], in these lecture notes we will attempt to give an introduction only. We also discuss how to solve this problem to all orders perturbatively [11], and how to perform the similar analysis for the Kondo problem [12, 13].

2 The DC conductance

There is now much known experimentally and theoretically about the fractional quantum Hall effect, but we only discuss the few properties necessary to understand the edge modes. The setup is a two-dimensional electron gas, which is confined to within a "bar" with $-l < x < l$ and $h < y < h$, with l/h very large. The Hall state is created by putting a two-dimensional electron gas in a strong transverse magnetic field H in the z-direction. The Hall conductance G is defined as $\partial I/\partial V$, where I is the current in the xth direction induced by an external electric field E_y in the yth direction. Classically, the conductance $G \propto H$, but at high field, low temperatures and in the presence of disorder and electron interactions, the conductance reaches a universal constant over a range of H, developing what are called plateaus with $G = \nu e^2/h$. When ν is a fraction (often called the filling fraction, because it is the portion of the Landau level which is occupied), this is called the fractional quantum Hall effect. It is by itself a remarkable phenomenon, since disorder (dirt) and complex electron interactions give rise to *universal* physics.

A crucial property of the system on a plateau is that the current flows only along the edge of the sample. In this situation, the bulk of the material is in an insulating state. Thus even though the Hall effect is two-dimensional physics, transport properties can be described by a one-dimensional model. In fact, it turns out that for filling fractions $\nu = 1/t$, with t an odd integer, the edge modes can be described by a single chiral free boson [1]. The direction in which these chiral modes move depends on the direction of the magnetic field; let us say that the modes are left moving on the top edge of the bar and right moving on the bottom. In this state, there are two conserved $U(1)$ currents. In terms of the bosons ϕ_L and ϕ_R the currents are $j_L = -\frac{1}{4\pi}(\partial_t + \partial_x)\phi_L$ and $j_R = \frac{1}{4\pi}(\partial_t - \partial_x)\phi_R$. These currents are the charge densities in the two edges: for example, $eQ_L = e\int_{-l}^{l} dx j_L$ is the total charge of the left-moving edge. The fact that the modes are chiral makes the situation much nicer; normally in one-dimensional systems random impurities backscatter particles between different points on the Fermi surface. In one dimension this always leads to localization, i.e. absence of any extended states that would be able to transport electrical current from one end of the sample to the other. Hence one-dimensional metals are usually insulators. In the quantum Hall situation this localization is avoided, since right and left moving charge carriers are spatially separated.

The dynamics of these free massless bosons are generated by a Tomonaga Hamiltonian

$$H_0 = \frac{\pi}{\nu} \int_{-l}^{l} dx \left[j_L^2 + j_R^2 \right], \tag{2.1}$$

quadratic in the two individually-conserved $U(1)$ currents, This model of a free massless boson is equivalent to a model of interacting massless fermions, often known as the Luttinger liquid in condensed-matter circles [14], and the massless Thirring model in particle physics [15]. The fermions are a nonlinear change of basis: $\psi_L \sim \exp(i\phi_L)$ and $\psi_R \sim \exp(i\phi_R)$. In the Hall effect, these fermions correspond to the Laughlin quasiparticles moving along the edge.

To make a current flow, we put a voltage drop of V in the y-direction, which has the effect of placing the charge carriers injected into the left- and right-moving edges at different chemical potentials. This is done by adding a term

$$H_V = -\Delta Q e V/2 \tag{2.2}$$

to the Hamiltonian, where $e\Delta Q \equiv e(Q_L - Q_R)$ is the difference between right- and left-moving charges. This charge difference is conserved ($\partial_t \Delta Q = 0$) when the edges do not interact, and the resulting current in the x-direction is

$$I_0(V) = e v_F \frac{\langle \Delta Q \rangle}{2l}. \tag{2.3}$$

One indeed finds from first-order perturbation theory [2] that the conductance $G = \nu e^2/h$.

The experimentally-measured Hall conductance on the plateaus agrees with the prediction of the Luttinger-liquid edge theory, but a much more detailed probe of the nature of the current-carrying excitations on the edges can be made by including a point-contact interaction coupling right and left edges at (say) the origin $x = 0$. Such devices have been made in the lab [16], and the conductance curve measured. In the presence of the point contact, the Laughlin quasiparticles can backscatter from one edge to another. Thus no longer are the currents Q_L and Q_R conserved individually; only the sum $Q_L + Q_R$ is. The backscattering contribution to the Hamiltonian [1, 2, 17] is

$$\begin{aligned} H_B &\sim \psi_L^\dagger(0)\psi_R(0) + \psi_R^\dagger(0)\psi_L(0) \\ &= \lambda\{e^{i\phi_L(0)}e^{-i\phi_R(0)} + e^{-i\phi_L(0)}e^{i\phi_R(0)}\}. \end{aligned} \tag{2.4}$$

Additional allowed terms (simultaneous backscattering of more than one particle) are irrelevant when $\nu = 1/3$ [2]. In the presence of this term (i.e. for $\lambda \neq 0$) there will be an additional (backscattering) contribution $I_B(V)$ to the current, so that the total current is $I(V) = I_0(V) + I_B(V)$. The conductance curve then runs from the value $G = \nu e^2/h$ with no point contact to the value $G = 0$ where the point contact splits the sample in two and backscatters all particles. Experimentally, λ is controlled by first tuning the gate voltage $\mathcal{V}^*$ across the point contact to a "resonance" value where the conductance is $G = \nu e^2/h$. (For a detailed explanation of why this resonance should arise, see [2].) Then the gate voltage $\mathcal{V}_g$ is taken slightly off the resonance value, with the difference $\mathcal{V}^* - \mathcal{V}_g \propto \lambda$. Near this resonance, the differential conductance is given by a *universal curve* $G(T_B/T, V/T)$ where the "backscattering temperature" T_B is is a scale set by λ. Roughly speaking, T_B is the crossover scale between the two fixed points. The well-known renormalization-group equations for this model (see e.g. [2]) give

$$T_B \propto \lambda^{1/(2(1-\nu))}.$$

The proportionality constant is given in [4]. The experimental curve [16] does indeed show this universal scaling behavior quite nicely. As long as a resonance is found *no fine tuning is necessary to compute* G because H_B is the only relevant operator at $\nu = 1/3$. At other values of ν, the other operators are less relevant, so with generic values of the couplings, the experimental curve should be close to the theoretical curve with point-contact interaction H_B.

In our papers [3, 4] we derived an exact expression for $I(V)$. This derivation relies on the fact that the model with Hamiltonian $H + H_B$ is integrable [18]. We do this by finding the exact quasiparticle spectrum of the model, the transmission rate through the point contact, and the exact "non-Fermi" distribution function of the quasiparticles. What "non-Fermi" means is that the quasiparticles are neither free fermions nor perturbations of them. Thus for example, the distribution function $f(p)$ is not the Fermi one $1/(1 + \exp[(p - p_F)/T])$ or even close. For example, the charge-carrying quasiparticle excitations have $f(p_F) = \nu$, not the free-fermion value 1/2. A striking manifestation of this non-Fermi-liquid behavior seen experimentally [16] is that on the $\nu = 1/3$ plateau, the conductance at low temperature goes as T^4, not the Fermi-liquid dependence T^2.

We first map this impurity problem with left and right movers on to a problem solely of left movers. We see from (2.4) that the interaction involves only the combination $\phi_L(x = 0) - \phi_R(x = 0)$. It is therefore convenient to define a new basis by

$$\phi^{e/o}(x + t) \equiv \frac{1}{\sqrt{2}}[\phi_L(x, t) \pm \phi_R(-x, t)],$$

where the *even* boson ϕ^e has the plus sign and the *odd* boson ϕ^o the minus sign. Both bosons are left-moving, and defined on $-l < x < l$. This leads to replacing j_L, j_R with j^e, j^o in the Hamiltonian (2.1), where $j^{e/o}(x + t) = (1/\sqrt{2})[j_L(x, t) \pm j_R(-x, t)]$ are the charge densities of even and odd bosons. The even and odd charges are thus related to the charges of the original left- and right-moving edges by

$$\Delta Q = Q_L - Q_R = \sqrt{2}Q^o, \qquad Q_L + Q_R = \sqrt{2}Q^e.$$

Two major simplifications arise from this change of basis. First, the interaction involves only the odd boson ϕ^o, while the even boson ϕ^e remains free [19]. Additionally, Q^e measures the total charge on both edges and is conserved even in the presence of the interaction. Thus the backscattering current I_B is related to ΔQ and can be entirely expressed in terms of the odd boson theory. Therefore, the even boson will play essentially no role in the following.

The odd boson theory is integrable [18, 20], and its quasiparticles are known [21]. Integrability means that there is an infinite number of conserved quantities which commute with the Hamiltonian, even in the presence of the point interaction (2.4) [18]. There are many bases for the quasiparticles of the odd-boson theory, which are related by not-necessarily-local mappings. The most obvious basis for a massless scalar field (the odd boson in our case) consists of plane waves, but they are not eigenstates of $H_0 + H_B$ [20]. A more useful basis is obtained [18, 20] by adding an auxiliary bulk mass term to the Hamiltonian, which is judiciously chosen so that bulk and point contact are both integrable. This defines a basis of massive bulk excitations in the odd-boson theory (a massive sine-Gordon model) which scatter off the point contact without particle production. By letting the auxiliary bulk mass tend to zero, a basis of massless particles is obtained. In this basis the conserved quantities are $\sum_i p_i^n$, where p_i are the momenta of the individual particles, and where n runs over an infinite subset of the positive integers [18]. These conservation laws have important consequences for scattering. The 'quasiparticles' of this basis must

scatter off the point contact *without* particle production, *i.e.* one-by-one. Away from the point contact, the quasiparticles scatter off of each other with a completely elastic and factorizable scattering matrix S^{bulk}. When $1/\nu$ is an integer, the bulk scattering is diagonal, so that the only allowed processes are of the form $|j(p_1)\rangle\otimes |k(p_2)\rangle \to |k(p_2)\rangle\otimes |j(p_1)\rangle$. Such a process is described by the S matrix element $S^{bulk}_{jk}(p_1/p_2)$.
The exact quasiparticles and their S matrix are already known for the sine-Gordon model and its massless limit [21, 18]. The Hamiltonian (2.1) and (2.4) written in terms of the odd boson corresponds to the case $\beta^2 = 8\pi\nu$ of [20]. At any value of ν, there are a kink (+) and an antikink (−). These carry (odd) charges $Q^o = 1/\sqrt{2}$ and $-1/\sqrt{2}$, respectively. Moreover, for $j-1 < 1/\nu \leq j$, there are $j-2$ additional states, the breathers, which have no charge. These particles span the Hilbert space of the left-moving odd boson; we label them by indices $j, k, \ldots$ running over the kink (+), antikink (−) and breathers (b).

This sounds a bit formal, but can be seen explicitly classically. Away from the point contact, the bulk equations are simply the Klein-Gordon equation, and any plane wave is a solution. However, the "quasiparticle" solutions discussed above maintain their form after scattering through the point-contact interaction; the effect of the interaction is just a phase shift [20]. This phase shift can be computed in the full quantum theory [18]. Scattering of a single kink by the point contact is described by a one-particle S matrix with elements $S_{++}(p/T_B) = S_{--}(p/T_B)$ for kink → kink, and antikink → antikink, as well as $S_{+-}(p/T_B) = S_{-+}(p/T_B)$ for kink → antikink, and vice versa. These were derived exactly in [18]:

$$|S_{+-}(p/T_B)|^2 = \frac{1}{1+(p/T_B)^{(2/\nu)-2}} \tag{2.5}$$

$$|S_{++}(p/T_B)|^2 = 1 - |S_{+-}(p/T_B)|^2 . \tag{2.6}$$

Without the backscattering, the left and right charges (or equivalently, the even and odd charges) are conserved individually. The backscattering allows processes where a charge carrier of the left-moving edge hops to the right-moving edge or vice versa. In the original basis, the current I_B is the rate at which the charge of the left-moving edge is depleted. By symmetry, $\partial_t Q_L = -\partial_t Q_R$ in each such hopping event, so

$$I_B = \partial_t(\frac{e}{2}\Delta Q) = \partial_t(\frac{e}{\sqrt{2}}Q^o), \tag{2.7}$$

and we see that in the even/odd basis, the tunneling corresponds to the violation of odd charge conservation at the contact. In the quasiparticle basis, a kink scattering to an antikink changes $Q_L - Q_R$ by -2; these correspond to the tunneling events. When a kink hits the point contact, the probability of such an tunneling event occuring is given by $|S_{+-}|^2$. As a check on this picture, we see that at $T_B = 0$, $S_{+-} = 0$, so there are no tunneling events. In the Hall edge this means there is complete transmission and no backscattering. At $T_B \to \infty$, $|S_{+-}| = 1$; so all of the Laughlin quasiparticles are backscattered; effectively the system has been cut in two. These two limits are the fixed points of the problem; they correspond respectively to Neumann and Dirichlet boundary conditions on the boson.

To calculate the conductance, we start with a gas of quasiparticles with a chemical potential difference for kinks and antikinks corresponding to the voltage V. A positive voltage means that there are more kinks. When there are more kinks than antikinks, the backscattering will turn more kinks to antikinks than it turns antikinks to kinks. When a kink of momentum p is scattered into an antikink (the conservation laws require that it have the same momentum p) this changes ΔQ by -2. Since kink and antikink quasiparticles scatter off the point contact one-by-one, we may describe the rate at which this charge transport occurs in terms of two quantities: the probabilities of finding a kink or antikink of momentum p at the contact, and the transition probability $|S_{+-}(p/T_B)|^2$. We therefore study the density of states $n_+(p,V) = n_-(p,V)$ and the distribution functions $f_\pm(p,V)$ in the thermodynamic limit ($l \to \infty$) and in the presence of an applied voltage V. The number of allowed kink or antikink states per unit length with momentum between p and $p+dp$ is given by $n_+(p,V)dp$, while the number of states actually occupied by kinks or antikinks in this momentum range is $n_+(p,V)f_+(p,V)dp$ and $n_+(p,V)f_-(p,V)dp$, respectively. Because at most one of these quasiparticles is allowed per level [20] (a consequence of their being the massless limit of the sine-Gordon kinks), we have $0 \leq f_\pm \leq 1$. Thus

$$\frac{\langle \Delta Q \rangle}{2l} = \int_0^\infty dp\, n_+(p,V)\ [f_+(p,V) - f_-(p,V)]. \tag{2.8}$$

These thermodynamic functions $n_+(p)$ and $f_\pm(p,V)$ are different from the free-fermion functions when the odd-boson kink theory is an interacting Luttinger liquid ($\nu \neq 1/2$), but they will be derived exactly from the TBA below.

Having all these definitions in place, it is now easy to compute the backscattering current from a kinetic (Boltzmann) equation. The number of kinks of momentum p which scatter into antikinks per unit time is given by $|S_{+-}(p/T_B)|^2 n_+(p,V)\rho_{+-}$, where ρ_{+-} is the probability that the initial state $+$ is filled and the final state $-$ is open. We have set the Fermi velocity to 1. In a fermi liquid, we would have simply $\rho_{+-} = f_+[1-f_-]$. However, in our interacting theory, correlations between the particles mean that this does not necessarily factorize in this manner. We can however write $\rho_{+-} = f_+ - f_{+-}$, where f_{+-} is the (unknown) probability that the kink and antikink state are *both occupied.* The rate at which antikinks scatter to kinks is likewise proportional to $\rho_{-+} = f_- - f_{+-}$. Thus the charge $Q_L - Q_R$ changes at a rate proportional to $\rho_{+-} - \rho_{-+} = f_+ - f_-$, independent of the unknown factor. Using (2.7) and (2.8) we have

$$\begin{aligned} I_B = & -e \textstyle\int_0^\infty dp\ n_+(p,V)\ |S_{+-}(p/T_B)|^2 \\ & \times [f_+(p,V) - f_-(p,V)]\,. \end{aligned} \tag{2.9}$$

From (2.9) we obtain the desired backscattering contribution to the linear-response conductance $G_B(T_B/T,0) = \lim_{V\to 0} I_B(V)/V$:

$$\begin{aligned} G_B(T_B/T,0) = -2e \int_0^\infty dp\ n_+(p,0) \\ \times |S_{+-}(p/T_B)|^2 \partial_V f_+(p,V)|_{V=0}. \end{aligned} \tag{2.10}$$

The density of states $n_+(p,0)$ has been evaluated at $V=0$ because $f_+(p,V) - f_-(p,V)$ is already proportional to V. The total conductance is thus $G = \nu e^2/h + G_B$. Note that (2.9) may also be used to compute the differential conductance $G(T_B/T, V/T)$ at *finite* voltage V, i.e. far out of equilibrium. This is done in [4].

The densities of states $n_j(p,V)$ and the distribution functions $f_j(p,V)$ of the massless odd boson theory can be computed exactly using a technique called the thermodynamic Bethe Ansatz [22]. As explained above, these define the occupation numbers of a quasiparticle of type j, which are scattered without particle production by the point contact. In models that are not integrable, one generally uses the free-particle density of states and then uses perturbation theory in the interaction parameters. However, the fact that the scattering is completely elastic and factorizable in an integrable model gives an exact functional relation between the n_j and f_j, known as the Bethe equation. We then write down the free energy in terms of n_j and f_j. The distribution functions f_j of the system at temperature T are then found by the free energy, using the relation of n_j to f_j.

As a simple first example we perform the calculation for a single type of left-moving massless particle, with no impurity interaction. The scattering is described by a single S-matrix element S^{bulk}. Quantizing a gas of such particles a circle of length l requires the momentum of the ith particle to obey

$$\exp(i2pl)\prod_{j\neq i} S^{bulk}(p_i/p_j) = 1. \tag{2.11}$$

One can think of this intuitively as bringing the particle around the world through the other particles; one obtains a product of two-particle S-matrix elements because the scattering is factorizable. This is the renormalized equivalent of the bare Bethe Ansatz relations. We see that the renormalized equations are much easier to derive, at least when the S matrix is diagonal. Going to the $l \to \infty$ limit allows us to introduce the densities. Taking the derivative of the log of (2.11) yields the Bethe equation for a single species of particle:

$$n(p,V) = \frac{1}{h} + \frac{1}{p}\int_0^\infty \frac{dp'}{2\pi}\Phi(p/p')n(p',V)f_k(p',V),$$

where $\Phi(p) = -i(d/d\ln p)\ln S^{bulk}(p)$. As long as the scattering is diagonal, generalizing this relation to more than one species of particle is simple:

$$n_j(p,V) = \frac{1}{h} + \frac{1}{p}\sum_k\int_0^\infty \frac{dp'}{2\pi}\Phi_{jk}(p/p')n_k(p',V)f_k(p',V), \tag{2.12}$$

where [22]

$$\Phi_{jk}(p) = -i\frac{d}{d\ln p}\ln S^{bulk}_{jk}(p).$$

For $\nu = 1/3$ there is only one breather (b) and we have [21, 4]

$$\Phi_{bb}(x) = 2\Phi_{++}(x) = 2\Phi_{+-}(x) = -\frac{4x}{x^2+1}$$

$$\Phi_{b+}(x) = \Phi_{+b}(1/x) = -\frac{4x^3+8x}{x^4+4},$$

where the others follow from the symmetry $+ \leftrightarrow -$.

To determine which fraction of the levels is occupied we do the thermodynamics [22]. The energy is

$$\mathcal{E} = \sum_j \int_0^\infty dp\ p\ n_j(p,V) f_j(p,V),$$

the entropy is

$$\mathcal{S} = \sum_j \int_0^\infty dp \left[n_j \ln n_j - n_j f_j \ln(n_j f_j) - n_j(1-f_j)\ln(n_j(1-f_j)) \right].$$

Including the voltage in H_V from (2.2) is equivalent to adding chemical potentials $\mu_+ = -\mu_- = eV/2$ coupling to the charge difference (2.8). The free energy per unit length $\mathcal{F} = (\mathcal{E} - T\mathcal{S} - eV\Delta Q/2)/L$ is found by minimizing it. Taking its variations, it is convenient to parametrize

$$f_j(p) \equiv \frac{1}{1 + e^{\epsilon_j - \mu_j/T}}.$$

The minimization yields an equation for ϵ_j in terms of the (known) bulk S matrix elements:

$$\epsilon_j(p/T, V/T) = \frac{p}{T} - \sum_k \int_0^\infty \frac{dp'}{2\pi p'} \Phi_{jk}(p/p') \ln[1 + e^{\mu_k/T} e^{-\epsilon_k(p'/T, V/T)}]. \tag{2.13}$$

Solving this equation for ϵ_j gives the functions f_j. Even though the breather does not appear directly in the kinetic equations, it interacts with the kink and antikink via (2.13) and affects the calculation of $f_\pm$. One can include the effects of the backscattering on n_j and f_j, but this gives only contributions to n_j and f_j vanishing with the system size $2l$. These corrections are responsible for the point-contact free energy [20], but they are subleading corrections to the conductance.

We can now evaluate the conductance explicitly. Using (2.13) and (2.12), it is easy to see that $\partial_V n_j(p,V)|_{V=0} = 0$, and that $\partial_V f_\pm(p,V)|_{V=0} = \pm\frac{e}{2T} f[1-f]$, where $f(p) \equiv f_\pm(p, V=0)$. Thus using (2.8) gives us the conductance without a contact:

$$G_0 = \frac{e^2}{T} \int_0^\infty n[1-f] f\ dp.$$

We will verify below that this indeed gives the correct answer $G_0 = \nu e^2/h$. Moreover, this and the unitarity of the S matrix allows us to write the full G as

$$G(T_B/T, 0) = \frac{e^2}{T} \int_0^\infty n\ |S_{++}|^2 [1-f] f\ dp. \tag{2.14}$$

By definition of ϵ, $f(1-f) = -\partial_\epsilon f$. The relations (2.13) and (2.12) imply that $n_j = \frac{T}{h} \partial_p \epsilon_j(p/T, V=0)$. Inserting the explicit expression (2.6) into (2.14) gives our result for the conductance at $\nu = 1/3$:

$$\begin{aligned} G &= -\frac{e^2}{h} \int_0^\infty dp \frac{p^4}{p^4 + T_B^4} \frac{\partial \epsilon_+(p/T,0)}{\partial p} \frac{\partial f(p)}{\partial \epsilon_+} \\ &= \frac{e^2}{h} \int_0^\infty dx\ x^3 \frac{4(T_B/T)^4}{(x^4 + (T_B/T)^4)^2} \frac{1}{1 + e^{\epsilon_+(x,0)}}, \end{aligned}$$

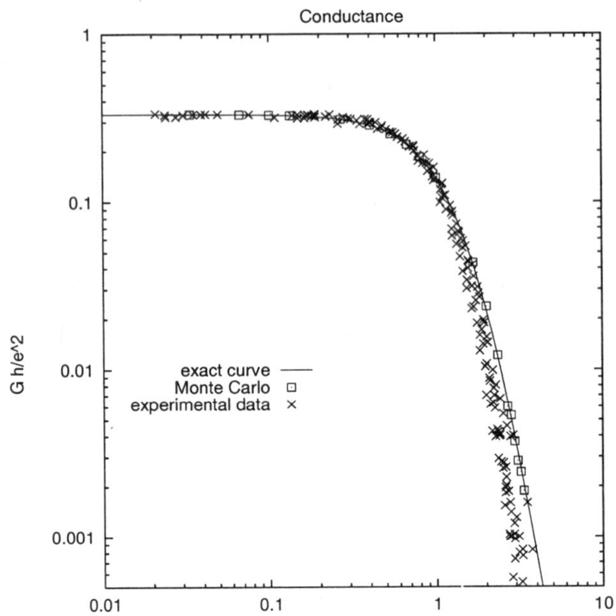

Figure 1. Comparison of experimental and theoretical curves for linear-response conductance for the tunneling through a point contact in a fractional quantum Hall device.

where we integrated by parts and defined the variable $x = p/T$.

As $T_B/T \to \infty$, we have $G \propto (T/T_B)^4$. Thus it goes to zero with the correct exponent, as in [2]. As $T_B/T \to 0$, we can also evaluate the conductance explicitly, becoming $[f(0) - f(\infty)]e^2/h$. The relation (2.13) gives $f(0) = 1/3$ and $f(\infty) = 0$, and we indeed recover $G_0 = e^2/3h$. The agreement with the experimental data [16] is good; see Fig. 1. As another check on our results, the conductance curve agrees very well with imaginary-time [17] and real-time [24] Monte Carlo simulations. The former is also plotted in Fig. 1. The slight discrepancies at low temperature presumably arise from the fact that the experiments do not see a perfect resonance; the data is scaled so that $G(T_B = 0) = 1/3$, but the actual value is somewhat lower. This is because two parameters are needed to tune to resonance in the absence of parity symmetry [2] (relevant for the experiment) whereas only one parameter, $\mathcal{V}_g$, has been varied experimentally. This could be remedied by varying the magnetic field on the Hall plateau, as well as $\mathcal{V}_g$, in a future experiment. Moreover, at low temperature the experiments are far away from the resonance, and it seems likely that the system is leaving the scaling regime (i.e. corrections to scaling are becoming large), another reason for the discrepancy. Indeed, the experimental curve resembles the Monte Carlo simulations [24] with a finite cutoff; the curve gradually goes up to our scaling curve as the cutoff is removed. Even including this slight discrepancy, the exact curve agrees well with the experimental data. In particular, the data still clearly show the predicted $G \propto T^4/(V_g - V_g^*)^6$ behavior in the tail.

These results all apply at non-zero voltage; the details are presented in [4]. They apply at general temperature, but a number of simplifications happen at zero temperature. In fact, one finds a simple duality between the weak backscattering and strong backscattering cases. These are apparent in the power series expressions for $I(V,T_B) = I_0(V) + I_B(V,T_B)$:

$$I(V,T_B) = \frac{e^2V}{h}\left[\nu - \nu^2 \sum_{n=1}^{\infty} a_n(\nu)\left(\frac{eV}{T_B'}\right)^{2n(\nu-1)}\right] \quad (2.15)$$

$$I(V,T_B) = \frac{e^2V}{h}\sum_{n=1}^{\infty} a_n(1/\nu)\left(\frac{eV}{T_B'}\right)^{2n(\frac{1}{\nu}-1)} \quad (2.16)$$

where

$$\begin{aligned} a_n(\nu) &= (-1)^{n+1}\frac{\sqrt{\pi}\,\Gamma(n\nu)}{2\Gamma(n)\Gamma(\frac{3}{2}+n(\nu-1))} \\ T_B' &\equiv T_B\frac{2\sqrt{\pi}\Gamma(1/[2(1-\nu)])}{\nu\Gamma(\nu/[2(1-\nu)])} \end{aligned}$$

The expansion (2.15) is appropriate in the weak-barrier case, where the tunneling occurs by individual Laughlin quasiparticles tunneling. In the strong-barrier limit (near the stable fixed point), the system is almost split in two; therefore only electrons can tunnel. This situation is described by the series (2.16). The identical form of the two expansions makes precise the statement that Laughlin quasiparticles are dual to the electrons under the interchange $\nu \leftrightarrow 1/\nu$. The fact that the exponents had this behavior was already known [1, 2], but this result shows that the duality is much more extensive.

3 The DC noise

Another detailed probe of Luttinger behavior is given by the shot noise. As discussed in [25], the shot noise in the weak-backscattering limit is proportional to the charge of the Laughlin quasiparticle, giving a way of measuring it directly. In [26], we found the DC shot noise at $T = 0$ exactly by showing that arguments given in [27] for the free-electron case can be directly generalized to our interacting quasiparticles, due to the constraints of integrability. We find a stunningly simple relation of the noise $\langle I^2\rangle$ to the current in (2.15) and (2.16):

$$\langle I^2\rangle = \frac{\nu e}{2(1-\nu)}V^2\partial_V\frac{I}{V} = \frac{\nu e}{2(1-\nu)}(VG - I).$$

This expression does indeed agree with the results of [25] in the weak- and strong-backscattering limits.

Recently, we have also obtained an expression for the DC noise at arbitrary temperature [28]. To handle this problem is more difficult, because in addition to the fluctuations in tunneling (the shot noise), there are fluctuations in the numbers of particles. The TBA relies on minimizing the free energy, which is equivalent to finding the saddle point of the path integral for the partition function. Understanding the fluctuations around this therefore requires new techniques to be developed.

4 Perturbative approach and relations with Kondo

There is another way of approaching the problem, completely independent of the above method using quasiparticle scattering. This is to expand the partition function in powers of λ [11]. The coefficients of the power series are then multiple-dimensional integrals with the same integrand (but with different limits of integration) as those which appear in the "Anderson-Yuval" expansion of the Kondo problem [5]. Moreover, by expanding the integrands in terms of Jack polynomials, the integrals (in the regime $\nu < 1/2$ where they converge) can be expressed in terms of a closed-form series [11]. Even though these coefficients are implicit in the Bethe Ansatz expression for the free energy, there is no known way to extact them analytically from the Bethe Ansatz. Thus this alternative method gives us valuable new information, as well as a check on the earlier results. Moreover, the partition function of the Kondo problem can be related to these coefficients, thus solving the original Anderson-Yuval problem [12, 13]. While these perturbative expansion gives rigorously only the partition function, a conjecture in [11, 13] relates the conductance simply to derivatives of the free energy. This conjecture has been checked extensively numerically by comparison with the above results, so it seems that transport properties are also known to all orders perturbatively.

We use an Euclidean formalism where time denoted by τ is a circle of periodicity $1/T$. The bulk action is

$$S = -\frac{1}{4\pi\nu}\int_0^\infty d\sigma \int_0^{1/T} d\tau \ [(\partial_\tau\phi)^2 + (\partial_\sigma\phi)^2] \tag{4.1}$$

while the boundary action is

$$S_B = 2\lambda \int_0^{1/T} d\tau \ \cos\left[\phi(\sigma = 0, \tau)\right]. \tag{4.2}$$

Defining as usual the partition function via the path integral as $Z = \int [d\phi] \ e^{S+S_B}$ we introduce $\mathcal{Z} \equiv Z(\lambda)/Z(\lambda = 0)$. We can rewrite this as a series by expanding out $\exp(S_B)$ in powers of λ. The coefficients of this series are then integrals of correlators in the free-boson theory. Using the bosonic propagator on the edge of a half-cylinder with Neumann boundary conditions,

$$\langle\phi(0,\tau)\phi(0,\tau')\rangle = -2\nu \ln\left|\frac{\kappa}{\pi T}\sin\pi T(\tau-\tau')\right|,$$

(κ is the inverse cutoff), one finds for example

$$\langle e^{i\phi(0,\tau)}e^{-i\phi(0,\tau')}\rangle = \left|\frac{\kappa}{\pi T}\sin\pi T(\tau-\tau')\right|^{-2\nu}$$

Using Wick's theorem one finds the general correlators as ratios of various $\mathcal{C}(\tau,\tau') \equiv |2\sin\pi T(\tau-\tau')|^{2\nu}$. The partition function is the series

$$\mathcal{Z} = \sum_{n=0}^{\infty} x^{2n} Z_{2n} \qquad x \equiv \frac{\lambda}{T}\left(\frac{2\pi T}{\kappa}\right)^{\nu} \tag{4.3}$$

$$Z_{2n} = \frac{1}{(n!)^2}\int_0^{1/T} \prod_i [d\tau_i d\tau_i'] \frac{\prod_{j<l}[\mathcal{C}(\tau_j,\tau_l)\mathcal{C}(\tau_j',\tau_l')]}{\prod_{i,k}\mathcal{C}(\tau_i,\tau_k')} \tag{4.4}$$

where all indices run from 1 to $2n$, and all integrals run from 0 to $1/T$ independently. This expression requires regularization for $\nu \geq 1/2$, but is well-defined otherwise. The sum (4.3) is the grand canonical partition function for a classical two-dimensional Coulomb gas of charged particles restricted to lie on a (one-dimensional) circle of circumference $1/T$. The charges are $\pm\sqrt{2\nu}$, and the whole gas is electrically neutral; x is the fugacity. Scaling arguments [2] indicate that the renormalized coupling x is proportional to $T_B^{(t-1)/t}$. In fact, the exact constant was determined in [4], and is

$$x \equiv \frac{v}{T}\left(\frac{2\pi T}{\kappa}\right)^{\nu} = \Gamma(\nu)\left(\frac{T_B}{T}\frac{\Gamma(\frac{1}{2(1-\nu)})}{2\sqrt{\pi}\Gamma(\frac{\nu}{2(1-\nu)})}\right)^{1-\nu}, \tag{4.5}$$

for any value of ν.

Explicit series expressions were found for the Z_{2n} [11] by expanding the denominator of (4.4) in terms of Jack polynomials [29], and using orthogonality relations of these polynomials. Jack polynomials are indexed by a partition of an integer, so we define $\mathbf{m} \equiv (m_1, m_2, \cdots, m_{(\mathbf{m})})$, where the m_i are integers with $m_1 \geq m_2 \ldots \geq m_{(\mathbf{m})} \geq 0$. One finds [11]

$$Z_{2n} = \left(\frac{1}{\Gamma(\nu)}\right)^{2n} \sum_{\mathbf{m}} \prod_{i=1}^{n} \left[\frac{\Gamma(\nu(n+1-i)+m_i)}{\Gamma(\nu(n-i)+1+m_i)}\right]^2 \tag{4.6}$$

where the sum is over all $\mathbf{m}$ with $(\mathbf{m}) \leq n$. The sum converges only for $\nu < 1/2$, which is where the integrals in (4.4) are well-defined.

In order to extend this approach to $\nu \geq 1/2$, there are a variety of ways to regulate the integrals. In a numerical approach, this would be done using a cut-off. However, the most natural approach here is analytic continuation. This approach, which is very analogous to dimensional regularization, means we *define* the regularized integrals as the analytic continuation of their values for $\nu < 1/2$. Instead of treating the Z_{2n} directly, it is easier to discuss the coefficients of the free-energy expansion, defined by $\sum f_{2n}x^{2n} \equiv -T\ln\mathcal{Z}$ This continuation is illustrated by examing the first coefficient, $f_2 = -Z_2$. For $n = 1$, the sum in (4.6) can be done (or the integral can be done without Jack functions), giving $f_2 = -\Gamma(1-2\nu)/\Gamma^2(1-\nu)$ At $\nu = 1/2$, f_2 has a simple pole. There are no branch points anywhere, and since it is finite for all other $\nu \leq 1$, the analytic continuation is perfectly well-defined. By examining the large-x behavior, one finds that there is a simple pole in $f_{2n}x^{2n}$ at $\nu = 1 - 1/(2n)$ whose residue is $(-1)^{n+1}T_B/(4\pi n^2 T)$.[13] These poles, once properly regulated, result in a logarithmic term $(-1)^n T_B/(2\pi n T)\log T_B/T$ in the perturbative expansion at $\nu = 1 - 1/(2n)$. [13].

We can see these poles explicitly by studying the series expansion (4.6) for the Z_{2n} in the boundary sine-Gordon model. Initially the series looks useless, because it diverges for $\nu \geq 1/2$, and we only know how to resum it for Z_2. However, a first interesting observation is that the series expressions for the f_{2n} converge even where those for the individual Z_{2n} do not. This is because some of the divergences in the Coulomb integrals are cancelled when taking the connected part. More precisely, we define the truncated series $Z_{2n}(\Lambda)$ as the expression (4.6) with all $m_i \leq \Lambda$ and

$$f_4(\Lambda) \equiv \left[\frac{(Z_2(\Lambda))^2}{2} - Z_4(\Lambda)\right].$$

Then one finds that

$$f_4(\Lambda) - f_4(\Lambda - 1) \simeq \left(\frac{2\nu(1-2\nu)+1}{2(1-2\nu)}\right)\Gamma(\nu)^{-4}\Lambda^{4\nu-4}$$

for Λ large. This expression converges as $\Lambda \to \infty$ for $\nu < 3/4$. Moreover, the pole at $\nu = 3/4$ is clearly identified, and its residue can easily be computed, because the divergence is proportional to that of the zeta function. One confirms the earlier result that near $\nu = 3/4$

$$f_4 x^4 \approx -\frac{1}{16\pi(\nu - 3/4)}\left(\frac{T_B}{T}\right)^{4(1-\nu)},$$

where we used the relation (4.5) to relate x and T_B.

Since we know how the series diverges, the continuation around the pole can be constructed by adding and subtracting a zeta function. More precisely, we define the continuation to be

$$f_4(\infty) = f_4^{reg}(\infty) + \frac{2\nu(1-2\nu)+1}{2(1-2\nu)\Gamma(\nu)^4}\zeta(4-4\nu).$$

where

$$f_4^{reg}(\Lambda) \equiv f_4(\Lambda) - \frac{2\nu(1-2\nu)+1}{2(1-2\nu)\Gamma(\nu)^4}\sum_{n=1}^{\Lambda} n^{4\nu-4}.$$

The "regular" part f_4^{reg} is f_4 with the diverging part of the sum subtracted off. This series converges when $\Lambda \to \infty$ for $\nu < 1$. One can extend this result past $\nu = 1$ in the same manner.

The perturbative expansion for the Kondo problem found in [5] is closely related. The Kondo problem is a three-dimensional non-relativistic problem, with free electrons antiferromagnetically coupled to a single fixed impurity. By looking at s-waves only, we restrict to the radial coordinate and this becomes a one-dimensional quantum problem where massless fermions $\psi_i(\sigma, t)$ ($i = 1, 2$ are the spin indices) move on the half-line $\sigma \geq 0$ with a quantum-mechanical spin S_a at $\sigma = 0$. The interaction parameters are $I_z, I_+ = I_-$; in the isotropic case $I_z = I_+$. The impurity action is

$$S_B = \sum_{i,j,a} I_a \int_0^{1/T} d\tau\ \psi_i^\dagger(0,\tau) S_a \sigma_{ij}^a \psi_j(0,\tau) \tag{4.7}$$

where the σ^a are the Pauli matrices. By a well-known bosonization procedure [5], this can be rewritten in terms of a free boson $\phi(x,t)$ with bulk action (4.1) and

$$S_B = I_+ \int_0^{1/T} d\tau \left(S_+ e^{i\phi(0,\tau)} + h.c.\right),$$

where we have absorbed I_z in a redefinition of the coupling ν. The anisotropy is thus parametrized by ν, with $\nu = 1$ the isotropic case and $\nu = 1/2$ the Toulouse (free-fermion) limit.

In the perturbative expansion, we get correlators of $\exp(\pm i\phi)$ like before. The crucial difference arises from the $S_\pm$. When the impurity has spin 1/2, we have $S_+^2 = S_-^2 = 0$, and only terms of the form $S_+S_-S_+S_- \ldots$ survive in the perturbative expansion. In the one-dimensional gas, this is the requirement that charges alternate in sign. We define the partition function of the spin-1/2 Kondo problem as $\mathcal{Z}_K = 2Z_K(I_+)/Z_K(0)$, where the factor of 2 ensures that the entropy is $\ln 2$ in the non-interacting limit $I_+ = 0$. Thus

$$\mathcal{Z}_K(x_K) = 2 + \sum_{n=1}^{\infty} (x_K)^{2n} Q_{2n} \qquad x_K \equiv \frac{I_+}{T}\left(\frac{2\pi T}{\kappa}\right)^g.$$

The Q_{2n} are the integrals in (4.4) times $2(n!)^2$, but where the region of integration is restricted to be $0 \le \tau_1 \le \tau_1' \le \tau_2 \le \ldots \tau_n' \le 1/T$. The periodicity of the integrand means that $Z_2 = Q_2$, but the others are different.

Our central result [12] relates the partition functions of the Kondo model and the model discussed earlier, a boson with a cosine boundary potential. Defining $q = \exp(i\pi g)$, we find that

$$\mathcal{Z}_K[(q - q^{-1})x]\mathcal{Z}(x) = \mathcal{Z}(qx) + \mathcal{Z}(q^{-1}x). \tag{4.8}$$

One simple check is that the equation is consistent with the fact that $\mathcal{Z}(x) = 1$ in the isotropic case $q = -1$. Plugging in the perturbative expansions gives the Kondo coefficients Q_{2n} in terms of the coefficients Z_{2n}. For example

$$4\sin^2(\pi\nu)Q_4 = -4\cos^2(\pi\nu)Z_4 + (Z_2)^2 \tag{4.9}$$

$$16\sin^6(\pi\nu)Q_6 = \sin^2(3\pi\nu)Z_6 + (Z_2)^3\sin^2(\pi\nu) - \left(\sin^2(2\pi\nu) + \sin^2(\pi\nu)\right) Z_2 Z_4. \tag{4.10}$$

Using the series (4.6), we can easily find values for the Z_{2n} and hence the Q_{2n} for $\nu < 1/2$ by truncating the series and summing it numerically. For example, we have $Q_2 = Z_2 = \Gamma(1-2\nu)/(\Gamma(1-\nu))^2$, and

VALUES OF COEFFICENTS			
	$\nu = 2/5$	$\nu = 1/3$	$\nu = 1/4$
Z_4	1.910750624	0.837804224	0.4644013099
Z_6	1.088518710	0.276783311	0.0968299150
Z_8	0.439166887	0.061847648	0.0129159832
Z_{10}	0.135465650	0.010210054	0.0012208002
Q_4	0.982706435	0.432237451	0.2322006549
Q_6	0.291860092	0.074496009	0.0242074788
Q_8	0.061852434	0.008729195	0.0016144979
Q_{10}	0.010067801	0.000757798	0.0000763000

As a check, some of the integrals were explicitly evaluated using Monte Carlo methods; the agreement is good. As discussed in [11], for g rational there are relations between the numbers beyond those in (4.8); for example for $\nu = 1/4$, $Q_{2n} = Z_{2n}/2^{n-1}$. The knowledge of the first few coefficients Q_{2n} provides a very good numerical solution of the problem. The series can be extrapolated using Pade approximants and for instance the flow of the boundary entropy is reproduced fairly accurately [11].

The proof of (4.8) requires first of all the introduction of the "quantum-group" algebra $SU(2)_q$, which is a one-parameter deformation of the $SU(2)$ algebra [30]. There are three generators S_+, S_- and S_z, with commutation relations

$$q^{S_z}S_+q^{-S_z} = qS_+, \qquad q^{-S_z}S_-q^{S_z} = qS_-$$
$$[S_+, S_-] = \frac{q^{2S_z} - q^{-2S_z}}{q - q^{-1}}.$$

Like $SU(2)$, the quantum-group algebra has representations of any spin. A careful analysis shows that the general anisotropic Kondo problem of arbitrary spin is integrable only if the spins in (4.7) obey the $SU(2)_q$ algebra. For the isotropic problem $q = -1$, the distinction is irrelevant because $SU(2)_{-1}$ is identical to $SU(2)$. The distinction is also irrelevant for any q for spin 1/2 or 1, because the the two-dimensional representation of $SU(2)_q$ is given by the Pauli matrices as for $SU(2)$, and the spin-1 representation is also the same up to a rescaling of S_+ and S_-. Other representations differ, however, so one must be careful when discussing the anisotropic Kondo problem at higher spin.

Following [31], let us now introduce the "quantum monodromy operator" $L_j(x)$, which is defined as

$$L_j(x) = \mathcal{T}\ \exp\left[I_+q^{-1}\int_0^{1/T} d\tau \left(e^{2i\phi_L(0,\tau)}q^{S_z}S_+ + e^{-2i\phi_L(0,\tau)}q^{-S_z}S_-\right)\right] \tag{4.11}$$

Here the $\mathcal{T}$ indicates time-ordering, and the $SU(2)_q$ generators are in the spin-j representation. The field ϕ_L is the left-moving component of the field ϕ; with Neumann boundary conditions we have $2\phi_L(0,\tau) = \phi(0,\tau)$. The "quantum transfer matrix" $T_j \equiv \text{Tr}\ L_j$ is a fundamental part of the Kondo problem: $\langle T_j(x)\rangle$ is identical to the spin-j anisotropic Kondo perturbative expansion. In particular,

$$\mathcal{Z}_K(x) = \langle T_{1/2}(x)\rangle.$$

This follows from expanding (4.11) in powers of x (the factors $q^{\pm S_z}$ cancel), and using the fact that the vacuum is an eigenstate of T_j.

As shown in [31], the L_j satisfy the Yang-Baxter equation. This results in a number of remarkable properties. In particular, one finds that all of the T_j can be generated from spin-1/2 via the relation [31]

$$T_j(q^{1/2}x)T_j(q^{-1/2}x) = 1 + T_{j-\frac{1}{2}}(x)T_{j+\frac{1}{2}}(x) \tag{4.12}$$

or via [12]

$$T_{\frac{1}{2}}(q^{j+1/2}x)T_j(x) = T_{j+\frac{1}{2}}(q^{\frac{1}{2}}x) + T_{j-\frac{1}{2}}(q^{-\frac{1}{2}}x) \tag{4.13}$$

This gives enough information to relate the two partition functions $\mathcal{Z}$ and $\mathcal{Z}_K$ (and do much more). When one expands $T_j(x)$ for higher j in powers of x, one obtains integrals with all sorts of charge orderings, with weights depending on q (because S_+ and S_- depend on q for representations other than spin 1/2, and because of the monodromy of the chiral fields). For example, $S_\pm^3 = 0$ for spin 1, so $++--+-$ appears at order x^6 in T_1, but $+++---$ does not. Therefore, the unordered integrals Z_{2n} in (4.4) can be constructed by summing over appropriate combinations of terms from the $T_j(x)$. Since the relations (4.12) or (4.13) give all of the higher $T_j(x)$ in terms of $T_{1/2}$, this means that any Z_{2n} can in principle be expressed in terms of the Q_{2n}. The perturbative results (4.9) and (4.10) follow directly from (4.12).

The quickest way to derive (4.8) is to introduce cyclic representations of $SU(2)_q$. These occur when q is any root of unity, $q^t = \pm 1$, and have no analog in ordinary $SU(2)$. They are indexed by an arbitrary complex parameter δ and have t states $|m\rangle$, $m = 0, \ldots, t-1$ with generators acting as

$$S_\pm|m\rangle = \frac{q^{\delta \mp m} - q^{-\delta \pm m}}{q - q^{-1}}|m \pm 1\rangle \qquad S_z|m\rangle = m|m\rangle$$

where states $|m\rangle$ and $|m+t\rangle$ are identified. Notice that in such a representation one can consider arbitrary powers of the generators. Referring to cyclic representations as spin-δ, one finds that all previous results for spin-j apply. In particular, relation (4.13) holds with $j \to \delta$. In the formal limit where $\delta = -i\Delta$ with $\Delta >> 1$ we have

$$S_\pm \approx \frac{e^{\pi\nu\Delta} q^{\mp m}}{q - q^{-1}}.$$

In this limit the commutator $[S_+, S_-]$ becomes negligible, so the traces of all monomials with n generators S_+ and n generators S_- become identical, e.g.

$$\mathrm{Tr}(S_+S_-)^n = t\left(\frac{q^{1/2}e^{\pi\nu\Delta}}{q - q^{-1}}\right)^{2n}.$$

(One might have been tempted to study $\lim_{j\to\infty} T_j$ where arbitrary powers of the generators also occur, but this is not $\mathcal{Z}$ because the traces do not behave in the right way.) Hence one way of expressing $\mathcal{Z}$ is

$$\mathcal{Z}(x) = \frac{1}{t}\left\langle T_{-i\Delta}\left[(q - q^{-1})q^{-1/2}e^{-\pi\nu\Delta}x\right]\right\rangle_{\Delta>>1}. \tag{4.14}$$

We substitute $j \to \delta = -i\Delta$ and $x \to (q - q^{-1})q^{-\delta - 1/2}x$ into (4.13). Letting $\Delta >> 1$ proves (4.8) for q any root of unity (rational g). The result follows for any $|q| = 1$ (real ν) by continuity. The renormalization of the Kondo coupling in (4.8) can be checked through the identity $Z_2 = Q_2$. This renormalization ensures that (4.8) makes sense in the isotropic limit $q \to -1$, where the Kondo partition function is a non-trivial object but $\mathcal{Z}(x) = 1$.

Having obtained the fundamental relation (4.8), higher-spin Kondo partition functions follow from fusion, using the relations (4.12) or (4.13). For instance one has

$$\begin{gathered}\langle T_1\left[(q - q^{-1})x\right]\rangle \mathcal{Z}(q^{1/2}x)\mathcal{Z}(q^{-1/2}x) = \\ \mathcal{Z}(q^{3/2}x)\mathcal{Z}(q^{1/2}x) + h.c. + \mathcal{Z}(q^{3/2}x)\mathcal{Z}(q^{-3/2}x).\end{gathered} \tag{4.15}$$

5 Conclusion

The next step is to compute time-dependent properties, in particular the AC conductance and the frequency-dependent noise [32]. A priori, this is a much more difficult problem, because the simple Landauer approach is not applicable to interacting particles. Still, the matrix elements of operators on the quasiparticle basis are known, the so-called form-factors. In general, the form-factor-approach to properties of a massless theories is disappointing; only a few results are available [33]. The present case is however especially favorable, mostly because the current has no anomalous dimension. Preliminary analysis shows that the integrals defining the massless form-factors converge, and moreover the form-factors series converges very fast, a limited number of terms giving reliable results all the way from the weak-backscattering limit to the strong-backscattering limit. We hope to report on this soon [34].

The edge of a fractional quantum Hall device presents a clean way of actually observing properties of the one-dimensional Luttinger liquid. Tackling these properties analytically requires the whole armada of integrable techniques, from Jack polynomials and TBA to form-factors. This is a very fortunate situation where beautiful experiments and fundamental theory meet, and we hope there will be much more progress on both fronts in the years to come.

Acknoledgment

We thank our collaborators F. Lesage, A. Ludwig, S. Skorik and N.P. Warner, who participated in the works described. This work was supported by the Packard Foundation, the National Young Investigator program (NSF-PHY-9357207) and the DOE (DE-FG03-84ER40168). The lectures were presented by H. Saleur who thanks the organizers for a very exciting and friendly school, and the participants for many useful questions and discussions.

References

[1] X.G. Wen, Phys. Rev. B41 (1990) 12838; Phys. Rev. B43 (1991) 11025.

[2] C.L. Kane and M.P.A. Fisher, Phys. Rev. B46 (1992) 15233.

[3] P. Fendley, A.W.W. Ludwig and H. Saleur, Phys. Rev. Lett. 74 (1995) 3005, cond-mat/9408068.

[4] P. Fendley, A.W.W. Ludwig and H. Saleur, Phys. Rev. B (1995) in press, cond-mat/9503172.

[5] P.W. Anderson and G. Yuval, Phys. Rev. Lett. 23 (1969) 89; G. Yuval and P.W. Anderson, Phys. Rev. B1 (1970) 1522; K.-D. Schotte, Z. Physik 230 (1970) 99; P.W. Anderson, G. Yuval and D.R. Hamman, Phys. Rev. B1 (1970) 4464

[6] K. Wilson, Rev. Mod. Phys. 47 (1975) 773.

[7] N. Andrei, K. Furuya and J. Lowenstein, Rev. Mod. Phys. 55 (1983) 331; A.M. Tsvelick and P.B. Wiegmann, Adv. Phys. 32 (1983) 453.

[8] I. Affleck and A. Ludwig, Nucl. Phys. B360 (1991) 641; Phys. Rev. B 48 (1993) 7297.

[9] P. Fendley, Phys. Rev. Lett. 71 (1993) 2485, cond-mat/9304031.

[10] P. Nozières, J. Low Temp. Phys. 17 (1974) 31.

[11] P. Fendley, F. Lesage and H. Saleur, J. Stat. Phys. 79 (1995) 799, hep-th/9409176.

[12] P. Fendley and H. Saleur, cond-mat/9506104, to appear in Phys. Rev. Lett.

[13] P. Fendley, F. Lesage and H. Saleur, "A unified framework for the Kondo problem and for an impurity in a Luttinger liquid", cond-mat/9510055.

[14] A. Luther and I. Peschel, Phys. Rev. B12 (1975) 3908; L.P. Kadanoff and A.C. Brown, Ann. Phys. 121 (1979) 318; F.D.M. Haldane, J. Phys. C14 (1981) 2585.

[15] S. Coleman, Phys. Rev. D11 (1975) 2088.

[16] F.P. Milliken, C.P. Umbach and R.A. Webb, "Indications of a Luttinger Liquid in the Fractional Quantum Hall Regime", to appear in Solid State Communications.

[17] K. Moon, H. Yi, C.L. Kane, S.M. Girvin and M.P.A. Fisher, Phys. Rev. Lett. 71 (1993) 4381.

[18] S. Ghoshal and A.B. Zamolodchikov, Int. J. Mod. Phys. A9 (1994) 3841, hep-th/9306002.

[19] E. Wong and I. Affleck, Nucl. Phys. B417 (1994) 403, cond-mat/9311040; I. Affleck and A.W.W. Ludwig, J. Phys. A27 (1994) 5375.

[20] P. Fendley, H. Saleur and N.P. Warner, Nucl. Phys. B430 (1994) 577, hep-th/9406125.

[21] A.B. Zamolodchikov and Al.B. Zamolodchikov, Ann. Phys. 120 (1979) 253.

[22] C.N. Yang and C.P. Yang, J. Math. Phys. 10 (1969) 1115; Al.B. Zamolodchikov, Nucl. Phys. B342 (1991) 695.

[23] T.R. Klassen and E. Melzer, Nucl. Phys. B338 (1990) 485.

[24] K. Leung, R. Egger and C. Mak, "Dynamical simulation of transport in one-dimensional quantum wires", preprint.

[25] C.L. Kane and M.P.A. Fisher, Phys. Rev. Lett. 72 (1994) 724.

[26] P. Fendley, A.W.W. Ludwig and H. Saleur, Phys. Rev. Lett. 75 (1995) 2196, cond-mat/9505031.

[27] R. Landauer, Physica D38 (1989) 226; R. Landauer and Th. Martin, Physica B175 (1991) 167; G.B. Lesovik, JETP Lett. 49 (1989) 594; M. Buttiker, Phys. Rev. Lett. 65 (1990) 2901.

[28] P. Fendley and H. Saleur, in preparation.

[29] I.G. Macdonald, Symmetric functions and Hall polynomials, Clarendon Press (1979); R.P. Stanley, Adv. in Math. 77 (1989) 76.

[30] see e.g. E.K. Sklyanin, Funct. Anal. Appl. 16 (1984) 273.

[31] V. Bazhanov, S. Lukyanov and A.B. Zamolodchikov, "Integrable Structure of Conformal Field Theory, Quantum KdV Theory, and Thermodynamic Bethe Ansatz," hep-th/9412229.

[32] C.de C. Chamon, D.E. Freed and X.G. Wen, Phys. Rev. B51 (1995) 2363, cond-mat/9408064.

[33] G. Delfino, G. Mussardo and P. Simonetti, Phys. Rev. D51 (1995) R6620.

[34] F. Lesage, H. Saleur and S. Skorik, in preparation.

ADVANCES IN LARGE N GROUP THEORY AND THE SOLUTION OF TWO-DIMENSIONAL R^2 GRAVITY

V.A. Kazakov

Laboratoire de physique théorique
de l'École normale supérieure
24 rue Lhomond, 75005 Paris, France

M. Staudacher

Cern, Theory Division
Genève, 1211, Switzerland

The large N expansion of matrix-valued field theories was invented more than twenty years ago by 't Hooft as a way to treat four-dimensional QCD with gauge group $SU(N)$. At $N = \infty$ planar diagrams dominate and it was hoped that this fact would lead either to analytic results or to a reformulation of QCD as a string theory. While the original ideas remain attractive, the program has not yet been successful. It was slowly understood that $N = \infty$ field theories retain much of the complexity of the generic N case.

Some of this complexity remains even in zero and one-dimensional matrix "field theories". In the famous paper [1] it was demonstrated that these so-called matrix models are non-trivial – but still solvable – systems. E.g. the zero dimensional model

$$Z = \int \mathcal{D}M e^{-N\ \mathrm{Tr}[\frac{1}{2}M^2 - \Sigma_{q=1}^{\infty} t_q M^q]}, \tag{1}$$

where M is a $N \times N$ Hermitian matrix and the t_q's are coupling constants parametrizing a general potential, was solved in [1] by changing variables to the eigenvalues of the matrix M and thereby reducing the number of degrees of freedom from N^2 to N. The latter proves possible due to the invariance of the above action and measure under the group $U(N)$. The model's free energy describes at $N = \infty$ a sum over all planar diagrams G of spherical topology, where vertices v_q of order q (there are $\#v_q$

Low Dimensional Applications of Quantum Field Theory
Edited by L. Baulieu *et al.*, Plenum Press, New York, 1997

of them in G) are weighted with a factor t_q:

$$\log Z \sim \mathcal{Z} = \sum_G \prod_{v_q \in G} t_q^{\# v_q} \tag{2}$$

The rather non-trivial combinatorial sum (2) is elegantly calculated by solving (1). At the time of [1] the model was considered as a kind of "toy", zero-dimensional QCD (retaining the diagrammatic structure but nothing else) to test a new technique. A few years later, however, it became clear that the results could be used to obtain the solution of a rather interesting physical problem: two-dimensional quantum gravity [2],[3],[4]. Indeed, it was argued that by tuning the couplings t_q in an appropriate way a continuum limit could be reached at which the planar graphs condense to give a continuum path integral over two-dimensional metrics g_{ab} of spherical topology:

$$\mathcal{Z}_{\text{cont}} = \int \mathcal{D} g_{ab}\, e^{-\int d^2 z \sqrt{\det g}\, (\mu + \frac{1}{\alpha} R_g)}. \tag{3}$$

The distance from the critical point in the space of the t_q's turns into a continuum cosmological constant μ controlling the area of the surfaces. We also wrote the Einstein term, which is however known to be a constant in two dimensions. This approach was subsequently worked out and justified by a large number of researchers.

Thus already the simplest matrix model (1) contains non-trivial physics. Various generalizations were solved and shown to give new physical information; e.g. certain simple multi matrix models describe the coupling of $c \leq 1$ conformal matter to 2D gravity (3). Such generalizations usually required the development of new techniques in order to succeed. Turning this around, each technical advance in large N theory usually allows to address a previously inaccessible physical problem. Keeping in mind the final goals – like large N QCD or string theories in physical dimensions – a valuable strategy is, then, to continue to enlarge our tools and methods.

An encouraging example for this strategy was given recently in a series of papers [5],[6],[7]. We studied a generalization of (1) consisting in the inclusion of an external matrix field A in the potential:

$$Z = \int \mathcal{D}M\, e^{-N \, \mathrm{Tr}[\frac{1}{2} M^2 - \Sigma_{q=1}^{\infty}\, t_q (MA)^q]}. \tag{4}$$

On a technical level, this model seemed for a long time unsolvable: The external field destroys the $U(N)$ invariance of the action. Thus none of the usual methods, whose essence consists in reducing the number of degrees of freedom from N^2 to N, appears applicable to (4). We succeeded in nevertheless finding a reduction of the number of degrees of freedom by developing the new method of large N character expansions . This allows us to address new physical questions. Indeed, it is easy to prove that the perturbative expansion in planar graphs of (4) is given by

$$\log Z \sim \mathcal{Z} = \sum_G \prod_{v_q^*, v_q \in G} t_q^{* \# v_q^*}\, t_q^{\# v_q}, \tag{5}$$

where

$$t_q^* = \frac{1}{q} \frac{1}{N} \,\mathrm{Tr}\, A^q. \tag{6}$$

The sum is, as in eq.(2), taken over all planar graphs G of spherical topology. But now we have an extra set of coupling constants t_q^* at our disposal: They assign weights to the vertices v_q^* of the *dual* lattice (there are $\#v_q^*$ of them in G). Thus the t_q and t_q^* control the coordination numbers of the vertices and the faces of G, respectively. In particular, if we set $t_q = t_q^* = \delta_{q,4}$ the only surviving graphs in the ensemble $\{G\}$ are regular square lattices[1]. Thus, the model (4),(5) is capable of interpolating between fluctuating random lattices and flat, regular lattices! In the continuum formulation, this property would be achieved by adding higher curvature counterterms to the action of (3):

$$\mathcal{Z}_{\rm cont} = \int \mathcal{D}g_{ab}\, e^{-\int d^2z\sqrt{\det g}\,(\mu+\frac{1}{\alpha}R_g+\frac{1}{\beta_0}R_g^2+\ldots)}. \tag{7}$$

Tuning the bare coupling β_0 to zero clearly suppresses any non-flat metric. Our model therefore furnishes a precise invariantly regularized definition of 2D higher curvature gravity Before discussing the physics of the latter, let us first explain the steps that lead to a solution of the apparently intractable matrix model (4). We will only sketch the derivation (and even the results!); a much more detailed discussion can be found in [5],[6],[7].

The idea is to expand the potential in (4) in terms of the characters of the product matrix MA. It is clear that this is possible since the potential is a class function on the group (*i.e.* it only depends on the eigenvalues of the matrix MA). What is less obvious (and proven in [5]) is that the expansion coefficients can themselves be written as the characters of an auxiliary external matrix B generating the couplings t_q, just as A generates the couplings t_q^* (see eq.(6)):

$$t_q = \frac{1}{q}\frac{1}{N}\,\mathrm{Tr}\,B^q. \tag{8}$$

One then has the expansion

$$e^{N\mathrm{Tr}\,\sum_{q=1}^{\infty} t_q (MA)^q} = c\sum_R \chi_R(B)\,\chi_R(MA), \tag{9}$$

with c a numerical constant. The characters are defined by the Weyl formula

$$\chi_{\{h\}}(B) = \frac{\det_{(k,l)}(b_k^{h_l})}{\Delta(b)}, \tag{10}$$

where the b_i are the eigenvalues of the matrix B, $\Delta(b)$ is the Vandermonde determinant of the eigenvalues, the set of $\{h\}$ are a set of ordered, increasing, non-negative integers, and the sum over R is the sum over all such sets. The R's label representations of the group $U(N)$ and the sets of integers $\{h\}$ are the usual Young tableau weights defined by $h_i = i - 1 + \#$boxes in row i (the index i labels the rows in the Young tableau, $i = 1$ corresponding to the lowest row). Note that the restriction on the allowed Young tableaux that any row must have at least as many boxes as the row below implies that the $\{h_i\}$ are a set of increasing integers: $h_{i+1} > h_i$. Substituting

[1] Of course there is no square lattice of spherical topology; one needs to also add some positive curvature defects to be able to close the graph into a sphere; see below.

equation (9) into the integral in equation (4), we can now do the angular integration using the key identity

$$\int (\mathcal{D}\Omega)_H \; \chi_R(\Omega M \Omega^\dagger A) = d_R^{-1} \; \chi_R(M) \; \chi_R(A), \tag{11}$$

where d_R is the dimension of the representation given, up to a constant, by $d_R \sim \Delta(h)$), and arrive, after performing a Gaussian integral over the eigenvalue degrees of freedom, at the expression

$$Z = c \sum_{\{h^e, h^o\}} \frac{\prod_i (h_i^e - 1)!! h_i^o!!}{\prod_{i,j}(h_i^e - h_j^o)} \; \chi_{\{h\}}(A) \; \chi_{\{h\}}(B), \tag{12}$$

where c is an immaterial constant. The sum is taken over a subclass of so-called even representations. These are defined as possessing an equal number of even weights h_i^e and odd weights h_i^o (since the mentioned Gaussian integration vanishes if the latter condition is not satisfied). The formula (12) was originally discovered by Itzykson and Di Francesco [8] by summing up "fatgraphs", using purely combinatoric and group theoretic arguments. We observe that the matrix model (4) is thus reformulated as a sort of "statistical mechanics model" in Young weight space. The important fact is that there are only N weights h_i; therefore the reduction to N degrees of freedom is achieved.

The expansion (12) can be further generalized. Consider the matrix model

$$Z = \int \mathcal{D}\phi \; e^{-N \; \mathrm{Tr}[\frac{1}{2}\phi\phi^+ - \sum_{k=1}^{\infty} g_k (\phi A \phi^+ B)^k]}, \tag{13}$$

where ϕ is a *complex* $N \times N$ matrix. Introducing a third external matrix field C and defining, in analogy with (6),(8),

$$g_k = \frac{1}{k}\frac{1}{N} \; \mathrm{Tr} \; C^k, \tag{14}$$

one finds by the method outlined above the character expansion

$$Z = c \sum_{\{h\}} \frac{\prod_i h_i!}{\Delta(h)} \; \chi_{\{h\}}(A) \; \chi_{\{h\}}(B) \; \chi_{\{h\}}(C), \tag{15}$$

where this time the sum extends over *all* representations. In the special case $g_k = \delta_{k,2}$ the earlier expansion (12) is recovered[2].

The reduction of the number of degrees of freedom is only a *conditio sine qua non.* One next has to take the large N limit of the expansion. The basic idea is the same as for the original model (1) (see [1]), with Young weights replacing eigenvalues: The weights $\frac{1}{N}h_i$ are assumed to freeze into a smooth, stationary distribution $dh \; \rho(h)$, where $\rho(h)$ is a probability density normalized to one. The details, however, turn out to be much more involved. Some simple first examples were worked out in [5]. An unpleasant feature is that, while the saddlepoint always exists, the support of

[2]For this special case the correspondence between the hermitian model (4) and the complex model (13) was already noted in [8]. For a graphical explanation of this correspondence see [6].

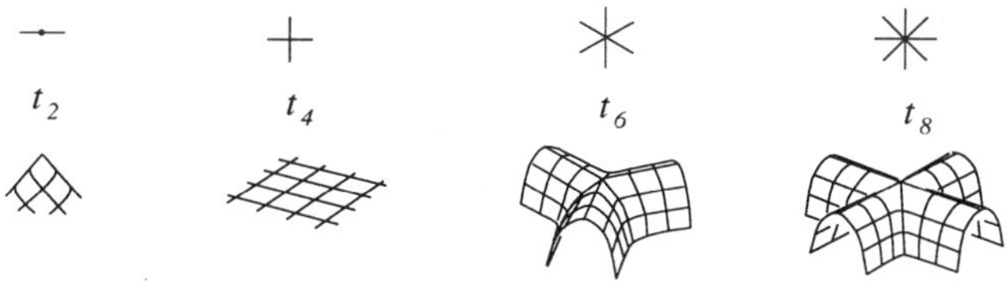

Figure 1. Flat space and curvature defects.

the density $\rho(h)$ does not necessarily remain on the real axis for completely arbitrary couplings t_q,t_q^*, complicating the general analysis in a significant way. However, if we restrict our attention to models in which the matrices A and B are such that traces of all odd powers of A and B are zero the problem does not arise: $t_{2q+1} = t_{2q+1}^* = 0$. This means that our random surfaces are made from vertices and faces with even coordination numbers. Thus it is easiest to consider surfaces made up from squares (as opposed to, say, triangles): $t_q^* = \delta_{q,4}$. A weight $t_{2q} = \frac{1}{2q}\frac{1}{N}\text{Tr}B^{2q}$ is assigned whenever $2q$ squares meet at a vertex (see Fig. 1).

This model is clearly capable of describing the transition from flat to random graphs. The sum (5) over spherical lattices G_4 built from square plaquettes becomes

$$\mathcal{Z} = \sum_{G_4} \prod_{v_{2q} \in G} t_{2q}{}^{\#v_{2q}}, \tag{16}$$

where v_{2q} are the vertices where $2q$ plaquettes meet and $\#v_{2q}$ are the numbers of such vertices in the given graph G_4. In order to investigate (12) in the large N limit, one attempts to locate the stationary point. This leads to the saddlepoint equation

$$2F(h) + \int_0^a dh' \, \frac{\rho(h')}{h - h'} = -\ln h. \tag{17}$$

Here $F(h)$ denotes the large N limit $F(h_k) \to F(h)$ of the variation of the character $\chi_{\{h\}}(B)$ in eq.(12):

$$F(h_k) = 2\frac{\partial}{\partial h_k^e} \ln \frac{\chi_{\{\frac{h^e}{2}\}}(\bar{b})}{\Delta(h^e)}. \tag{18}$$

where we have also used that the matrix B will satisfy $\text{Tr}B^{2q+1} = 0$ if we introduce a $\frac{N}{2} \times \frac{N}{2}$ matrix $\bar{b}^{\frac{1}{2}}$ in terms of which B and the character $\chi_{\{h\}}(B)$ are given by

$$B = \begin{bmatrix} \bar{b}^{\frac{1}{2}} & 0 \\ 0 & -\bar{b}^{\frac{1}{2}} \end{bmatrix} \quad \text{and} \quad \chi_{\{h\}}(B) = \chi_{\{\frac{h^e}{2}\}}(\bar{b})\chi_{\{\frac{h^o-1}{2}\}}(\bar{b}) \ \text{sgn}[\prod_{i,j}(h_i^e - h_j^o)]. \tag{19}$$

The variation of the character of the matrix A is easily computed directly because of the simple choice $t_q^* = \delta_{q,4}$. As has been discussed in detail in [5], the saddlepoint equation (17) actually does not hold on the entire interval $[0, a]$, but only on an interval $[b, a]$ with $0 \leq b \leq 1 \leq a$: Assuming the equation to hold on $[0, a]$ would violate the implicit constraint $\rho(h) \leq 1$ following from the restriction $h_{i+1} > h_i$. The density is in fact exactly saturated at its maximum value $\rho(h) = 1$ on the interval $[0, a]$. It is useful to introduce in addition the weight resolvent $H(h)$ as follows:

$$H(h) = \int_0^a dh' \, \frac{\rho(h')}{h - h'}. \tag{20}$$

We found the weight resolvent $H(h)$ to be very closely related to the standard matrix model *eigenvalue* resolvent (see [1]). It provides a direct link between the statistical distribution of Young weights and the correlators of the model:

$$\frac{1}{N} \, \mathrm{Tr} M^{2q} = \frac{1}{q} \oint \frac{dh}{2\pi i} \, h^q \, e^{qH(h)}. \tag{21}$$

Here the contour encircles the cut of $e^{H(h)}$.

The solution of (17) evidently requires knowing the function $F(h)$. A rather general method for its determination has been one of the main technical achievements of our work. The method of functionally determining an $N = \infty$ character might prove very useful for other applications. We found the following simple result: Further introduce the function $G(h)$ as

$$G(h) = e^{H(h)+F(h)}, \tag{22}$$

where, again, the contour encircles the cut of $e^{H(h)}$. Its importance stems from the fact that it relates in a simple way the introduced functions $F(h)$,$H(h)$ and the coupling constants t_{2q} [5]:

$$t_{2q} = \frac{1}{q} \oint \frac{dh}{2\pi i} \, G(h)^q. \tag{23}$$

It is then easy to deduce, by changing variables from h to G, the expansion

$$h - 1 = \sum_{q=1}^{Q} \frac{t_{2q}}{G^q} + \sum_{q=1}^{\infty} a_q \, G^q, \tag{24}$$

By considering the alternative representation of the Weyl character as a determinant of Schur polynomials one derives[3] [6] that the coefficients a_q of the positive powers of G in (24) are directly related to correlators of the matrix model dual to (4), *i.e.* the model with $t_q \leftrightarrow t_q^*$:

$$a_q = \langle \frac{1}{N} \mathrm{Tr} \, (\tilde{M} B)^{2q} \rangle. \tag{25}$$

We have also assumed for the moment that only a finite number Q of couplings are non-zero (*i.e.* $t_{2q} = 0$ for $q > Q$). Furthermore, we were able to show in our second work [6] that (24) implies the functional equation

$$e^{H(h)} = \frac{(-1)^{(Q-1)} h}{t_Q} \prod_{q=1}^{Q} G_q(h), \tag{26}$$

[3] One also finds the relation $a_q = \frac{2q}{N} \frac{\partial}{\partial t_{2q}} \ln\left(\chi_{\{\frac{h^e}{2}\}}(\bar{b})\right)$. As mentioned in [7], rewriting eq.(24) with this expression for a_q suggests a possible relationship to integrable hierarchies of differential equations. This conjecture was developed in several conversations with I. Kostov.

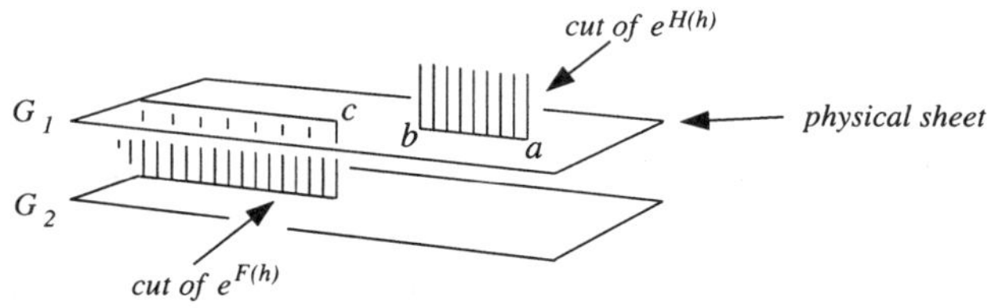

Figure 2. Sheet and cut structure of $G(h)$

where the $G_q(h)$ are the first Q branches of the multivalued function $G(h)$ defined through (24) which map the point $h = \infty$ to $G = 0$. The resulting picture of the analytic structure of $G(h)$ is as follows: the couplings t_{2q} determine the number of sheets attached to the physical sheet by the cut of $e^{F(h)}$; this number is Q for a finite number of non-zero couplings. In turn, the parameters a_q determine the sheets attached to the physical sheet by the cut of $e^{H(h)}$. This picture is easily verified (and was in fact discovered in this way) for the rather trivial cases where the potential in eq.(4) is at most quadratic. The case $Q = 2$ is shown in Fig. 2.

The saddlepoint equation (17), together with (26), defines a well-posed Riemann-Hilbert problem. It was solved exactly and in explicit detail in [6] for the case $Q = 2$, where the Riemann-Hilbert problem is succinctly written in the form

$$\begin{aligned} 2F(h) + \not{H}(h) &= \quad -\ln h \\ 2\not{F}(h) + H(h) &= \ln(-\tfrac{t_4}{h}), \end{aligned} \tag{27}$$

The first equation is the saddlepoint equation (17) with $\not{H}(h)$ denoting the real part of $H(h)$ on the cut $[b, a]$ (the righthand cut in Fig. 2). The second equation is (26) with $\not{F}(h)$ denoting the real part of $F(h)$ on a cut $[-\infty, c]$ with $c < b$ (the lefthand cut in Fig. 2). This case corresponds to an ensemble of squares being able to meet in groups of four (*i.e.* flat points with weight t_4) or two (*i.e.* positive curvature points with weight t_2) (see Fig. 1). We termed the resulting surfaces "almost flat". It turned out that all the introduced functions could be found explicitly in terms of elliptic functions. E.g. the density satisfying (17) is given by

$$\rho(h) = \frac{1}{K}\mathrm{sn}^{-1}\Big(\sqrt{\frac{a-h}{a-b}}, k\Big) \quad \text{with} \quad k = \sqrt{\frac{a-b}{a-c}}, \tag{28}$$

where sn^{-1} is an inverse Jacobi elliptic function and K is the complete elliptic integral of the first kind (depending on the modulus k). The expressions for the cutpoints a, b, c as well as the expressions for $H(h), F(h), G(h)$ and the physical correlators a_q (see (25)) were found as well.

The resulting surfaces are very different from pure gravity. The spherical partition function (16) consists of flat cylinders pinched shut at the two ends, resulting in precisely four defects. Very short, highly twisted cylinders dominate the continuum limit. By exactly calculating the correlators a_q (see eq.(25)) we also analytically solved the combinatorial problem of surfaces with a single negative curvature insertion of arbitrary degree balanced by a gas of positive defects.

To analyse the problem of the transition from flat to random lattices, we merely need to perturb our almost flat lattices by *any* operator containing negative curvature. This physical observation allows us to extend the $Q = 2$ solution in a simple way. Choosing the weights to be

$$t_2 = \sqrt{\lambda}\, t,\ t_4 = \lambda,\ t_6 = \lambda^{\frac{3}{2}} \frac{\beta^2}{t}, \dots t_{2q} = \lambda^{\frac{q}{2}} (\frac{\beta^2}{t})^{(q-2)}, \tag{29}$$

the expansion (24) becomes

$$h - 1 = \frac{t_2}{G} + \frac{t_4}{G\,(G-\epsilon)} + \text{ positive powers of } G, \tag{30}$$

where $\epsilon = \frac{t_4}{t_2}\beta^2$. The derivation of the functional equation (26) is easily modified (note that we now have an infinite number $Q = \infty$ of weights) to give

$$e^{H(h)} = \frac{h}{\epsilon t_2 - t_4}\, G_1(h)\, G_2(h). \tag{31}$$

The essential point is that one keeps the property that only one other sheet $G_2(h)$ is attached to the physical sheet $G_1(h)$ by the cut of $e^{F(h)}$ (see again Fig. 2). The only difference is that the semi-infinite cut $[-\infty, c]$ becomes a finite cut $[d, c]$. This results in modifying the Riemann-Hilbert problem (27) to

$$\begin{aligned} 2F(h) + \not{H}(h) &= \quad -\ln h \\ 2\not{F}(h) + H(h) &= \ \ln(\tfrac{\epsilon t_2 - t_4}{h}), \end{aligned} \tag{32}$$

It may still be explicitly solved in terms of elliptic functions; e.g. the density (28) is generalized to

$$\rho(h) = \frac{u}{K} - \frac{i}{\pi} \ln \left[\frac{\theta_4(\frac{\pi}{2K}(u - iv), q)}{\theta_4(\frac{\pi}{2K}(u + iv), q)} \right] \tag{33}$$

with

$$q = e^{-\pi \frac{K'}{K}}, \quad \text{and} \quad k = \sqrt{\frac{(a-b)(c-d)}{(a-c)(b-d)}}, \tag{34}$$

where K and K' are the complete elliptic integrals of the first kind with respective moduli k and $k' = \sqrt{1-k^2}$ and u and v are given by

$$u = \mathrm{sn}^{-1}(\sqrt{\frac{(a-h)(b-d)}{(a-b)(h-d)}}, k) \quad \text{and} \quad v = \mathrm{sn}^{-1}(\sqrt{\frac{a-c}{a-d}}, k'). \tag{35}$$

Again, the cutpoints a, b, c, d can be explicitly obtained as functions of the couplings λ, β, t.

Let us also mention that the weights (29) can be further generalized while keeping the quadratic (two-sheeted) structure of the function $G(h)$. The idea is to shift away the simple pole at $G = 0$ in the expansion (30), *i.e.*

$$h - 1 = \frac{c_1}{G - \epsilon_1} + \frac{c_2}{G - \epsilon_2} + \text{ positive powers of } G. \tag{36}$$

Here c_1 and c_2 are two constants given by $c_1 = (t_2\epsilon_2 - t_4)/(\epsilon_2 - \epsilon_1)$ and $c_2 = (t_4 - t_2\epsilon_1)/(\epsilon_2 - \epsilon_1)$. It leads to the Riemann-Hilbert problem

$$\begin{aligned} 2F(h) + \not{H}(h) &= -\ln h \\ 2\not{F}(h) + H(h) &= \ln(\frac{(\epsilon_1 + \epsilon_2)t_2 - t_4}{h} + \epsilon_1\epsilon_2\frac{h-1}{h}), \end{aligned} \tag{37}$$

The weights generalizing (29) by one extra parameter can be found explicitly by expanding (36) in inverse powers of G. We have not worked out this further explicitly solvable case in detail. Analyzing it would furnish an interesting universality check of our result.

Having at hand the explicit solution of the model for the weights (29), we are in a position to analyse the problem of discrete 2D R^2 gravity. With these weights it is easy to prove, using Euler's theorem, that the partition sum (16) becomes

$$\mathcal{Z}(t, \lambda, \beta) = t^4 \sum_{G_4} \lambda^A \, \beta^{2(\#v_2 - 4)}, \tag{38}$$

where A is the number of plaquettes of the graph G and $\#v_2$ the number of positive curvature defects. Note that the latter are balanced by a gas of negative curvature defects, whose individual probabilities are given in (29).

We expect this model to describe pure gravity in a sufficiently large interval of β, after tuning the bare cosmological constant λ (controlling the number of plaquettes) to some critical value $\lambda_c(\beta)$. On the other hand, for λ fixed and $\beta = 0$ we entirely suppress curvature defects except for the four positive defects needed to close the regular lattice into a sphere. It is thus clear that β is the precise lattice analog of the bare curvature coupling β_0 in the continuum path integral (7). The phase $\beta = 0$ of "almost flat" lattices – very different from pure gravity – studied in detail in [6] was discussed above.

Let us now summarize the main physical results following from the exact solution (for general λ and β) of this model:

1. A long debated question was whether models of the present type undergo a "flattening" phase transition at a finite, non-zero critical value of $\beta = \beta_c$. The weak coupling region $\beta > \beta_c$ would then correspond to the standard phase of pure gravity while a putative novel "smooth" phase of gravity might exist either at $\beta = \beta_c$ or in the entire interval $0 \leq \beta \leq \beta_c$. This would constitute an existence proof of *continuum* 2D R^2 gravity. We found analyzing the exact solution, to the contrary, that *there is no "flattening" phase transition at non-zero* β. For any given β we find the powerlike scaling of standard pure gravity on large scales. This means that no matter how flat the system is on small scales (of the order of $\beta^{-\frac{1}{2}}$), it destabilizes in the infrared into the familiar ensemble of highly fractal "baby-universes".

2. The dependence of the partition sum (38) on β and the lattice cosmological constant λ in the vicinity of the flat phase $\beta \sim 0$ and close to $\lambda \sim \lambda_c$ is given by a simple, (presumably) universal scaling function $f(x)$ (defined through $\mathcal{Z}(t,\lambda,\beta) = \frac{4t^4}{15\beta^2} f(x)$) reflecting the transition from flat space to pure gravity:

$$f(x) = x^6 - \frac{5}{2}\,x^4 + \frac{15}{8}\,x^2 - \frac{5}{16} - x\,(x^2-1)^{\frac{5}{2}}, \tag{39}$$

where the scaling variable x is given, to leading order, by

$$x = \frac{\sqrt{2}}{\pi}\,\frac{1-\lambda}{\beta}. \tag{40}$$

We can distinguish the following features:

(a) There is a degree $\frac{5}{2}$ singularity at $x = 1$, correctly reproducing the universal string susceptibility exponent $\gamma_s = -\frac{1}{2}$ of pure gravity [2],[3]. In view of eq.(40), the critical value of the lattice cosmological constant λ is therefore given to leading order by $\lambda_c = 1 - \frac{\pi}{\sqrt{2}}\beta + \mathcal{O}(\beta^2)$. Therefore (see (38)), the characteristic growth of the random surfaces as a function of the lattice area A (= number of plaquettes) is given by

$$\mathcal{Z}(t,A,\beta) \;\sim\; \frac{t^4}{\beta^{\frac{9}{2}}}\; e^{\frac{\pi}{\sqrt{2}}\,\beta\,A}\; A^{-\frac{7}{2}}. \tag{41}$$

For any non-zero β we do have exponential growth of the number of surfaces, but one has to go to larger and larger scales (*i.e.* use more and more plaquettes) to be able to take the continuum limit. If β is exactly zero there is no longer any exponential growth and no pure gravity continuum limit is possible. The prefactor $\beta^{-\frac{9}{2}}$ is found in the exact calculation in section 5; we are not sure whether it is universal.

(b) We further see that taking $\beta \to 0$ *before* the limit $\lambda \to \lambda_c$ corresponds to the limit $x \to \infty$. In this limit one finds $f(x) \sim \frac{5}{128}\,\frac{1}{x^2} + \mathcal{O}(\frac{1}{x^3})$, that is, the characteristic critical behavior of 2D gravity "silently" disappears and we recover a power series in $\frac{1}{x}$ corresponding, in view of (40), to a perturbative expansion in lattice defects β. In this limit the characteristic growth of surfaces as a function of area A is

$$\mathcal{Z}(t,A,\beta) \;\sim\; t^4\;(\;A + \mathcal{O}(\beta^2 A^3)\;). \tag{42}$$

The leading order corresponds precisely to the almost flat lattices (with exactly four positive defects) of [6]. The corrections are interpreted as insertions of negative defects, balanced by further positive defects. The typical shape of the surfaces in this limit is a generalization of the one we found for "almost flat" graphs: Long, filamentary cylinders growing out from every negative curvature defect.

(c) It is easy to prove that the scaling function (39) is the simplest possible function with the limiting properties discussed in (a) and (b).

The above results might be interpreted in terms of a continuum model of quantized curvature defects, in which the localized defects move around like particles in a gas on a flat background space. The deficit angle, $\Delta\theta$, of a defect can take on the values $\Delta\theta = \pi, 0$ and $-\pi$. A positive curvature defect is surrounded by a conical geometry, whereas a negative curvature defect corresponds to a saddle-type insertion (see Fig. 1). The higher order negative curvature defects (-2π and higher) would not be expected to play a role in this limit (the entropy from moving two low order defects around would completely dominate that from a single higher order defect). The coupling β can be interpreted as a fugacity controlling the number of defects. The flat space limit $\beta \to 0$ consists of four defects of degree π moving around with respect to one another. Varying the fugacity, β, allows one to smoothly interpolate between flat space, (42) (with four defects), and pure gravity (41) (with an infinite number of defects).

One might also attempt to develop this picture directly in the continuum. One could start with the conformal metric of a flat surface with localized curvature defects. It can be represented locally as $g_{ab} = \delta_{ab}\, e^{\varphi(z)}$ with

$$\varphi(z) = \sum_{j=1}^{M} R_j \ \ln(z - z_j)^2, \tag{43}$$

where $R_j = -1, 1$. Symbolically, the partition function might be written as

$$Z(\mu, \beta) = \sum_{M} \beta^M \int d[z_1, \ldots, z_M]\ e^{-\mu \int d^2z\ \sqrt{\det g(z)}}. \tag{44}$$

Here we introduced the fugacity of curvature defects β instead of the explicit R^2-term in the action. It serves the same purpose: for $\beta \to 0$ we arrive at the completely flat metric, whereas for $\beta \sim 1$ the system should show the behavior of pure quantum gravity, at least in the infrared domain. We retained the notation β to denote the parameter playing a role similar to the R^2 coupling in the above discrete model.

This formulation resembles a little bit the two-dimensional Coulomb gas problem. However, the measure of integration $d[z_1, \ldots, z_M]$ of the positions of the curvature defects is a complicated object: it should take into account the topology of the surface and the existence of zero modes (the action does not depend on some directions in the space of the z_i). It would be very interesting to make this direct continuum formulation more precise.

Another interesting issue is the role of exponential corrections appearing due to the structure of elliptic functions. In fact all physical quantities, such as $\mathcal{Z}(t, \lambda, \beta)$ (see eq.(38)), contain exponentially small terms in the limit $\lambda \to \lambda_c$ and $\beta \to 0$, thus leading to an essential singularity at $\beta = 0$, $\lambda = 1$. One can obtain the first correction of this type in e.g. the free energy $\bar{f}(\beta)$ per unit area in the thermodynamical limit $\lambda = \lambda_c$:

$$\mathcal{Z}(t, A, \beta) \ \sim \ \frac{t^4}{\beta^{\frac{9}{2}}} e^{\bar{f}(\beta)A} A^{-\frac{7}{2}} \quad \text{with} \quad \bar{f}(\beta) = \frac{\pi}{\sqrt{2}}\beta\Big[(1 + \ldots) + e^{-\frac{\pi\sqrt{2}}{\beta}}(4 + \ldots)\Big] \tag{45}$$

where $\bar{f}(\beta) = \lim_{A\to\infty} \frac{1}{A} \ln \mathcal{Z}(t, A, \beta) = \lambda_c$ and the dots denote terms of order β^3 and higher.

These exponential terms are likely to be lattice artifacts. They emerge even in the simplest calculation for the flat closed quadrangulation with four positive curvature defects, where they appear as discrete corrections to the approximation of elliptic sums by integrals close to the continuum limit.

On the other hand, formula (45) corresponds to the critical free energy as a function of the curvature fugacity β (*i.e.* we have already taken the continuum limit). It is possible that the exponential terms might be corrections relevant for the statistical mechanics of random lattices at long distances (of order $\frac{1}{\beta}$) rather than for continuous 2D-gravity.

In conclusion, we have tried to emphasize two points in this presentation:

1. *Technical advances in large N group theory are possible.* Apparently hopelessly difficult problems like the model (4) become tractable when one changes the technique.

Our approach raises a number of interesting questions. First of all, the method we have employed seems rather indirect and is definitely very involved. Is there a hidden simplicity we have missed so far? Are there alternative ways to obtain our results? Secondly, one should ask whether the method could be adapted to other theories describing new interesting physics; in particular gauge theories and strings in physical dimensions.

2. *Each such advance leads to the capability to address new physics questions. The present advance enables us to treat the problem of two-dimensional higher curvature gravity.*

We have presented for the first time an exactly solvable model interpolating between flat space and 2D quantum gravity. This addresses a long-standing open problem. We would like to stress, however, that we have by no means demonstrated yet the *universality* of our result. Could it be that finetuning the weights t_q leads to new phases of gravity? Here we certainly do not mean to rederive the usual multicritical phases of the one-matrix model (1). We rather envision tuning the weights in a subtle way so as to reach a phase of *smooth* 2D gravity. Certainly the analytic complexity of the result could be a hint that much more waits to be discovered. (Compare e.g. the densities (28),(33) of the model (4) under study with the rather trivial algebraic densities obtained in the absence of the external field.) Before addressing this issue, we repeat, one first has to simplify the method enough to allow for a deeper insight into the analytic structure of the solution for general couplings t_q.

References

[1] E. Brézin, C. Itzykson, G. Parisi and J.B. Zuber, Commun. Math. Phys. 59 (1978), 35.

[2] F. David, Nucl. Phys. B257 (1985) 45.

[3] V.A. Kazakov, Phys. Lett. B150 (1985) 282.

[4] J. Fröhlich, in: Lecture Notes in Physics, Vol. 216, Springer, Berlin, 1985; J. Ambjørn, B. Durhuus and J. Fröhlich, Nucl. Phys. B257[FS14](1985) 433.

[5] V.A. Kazakov, M. Staudacher and T. Wynter, École Normale preprint LPTENS-95/9, accepted for publication in Commun. Math. Phys.

[6] V.A. Kazakov, M. Staudacher and T. Wynter, École Normale preprint LPTENS-95/24, accepted for publication in Commun. Math. Phys.

[7] V.A. Kazakov, M. Staudacher and T. Wynter, École Normale preprint LPTENS-95/56, CERN preprint CERN-TH/95-352, submitted for publication to Nuclear Physics B.

[8] P. Di Francesco and C. Itzykson, Ann. Inst. Henri. Poincaré Vol. 59, no. 2 (1993) 117.

QUANTUM INTEGRABLE SYSTEMS AND ELLIPTIC SOLUTIONS OF CLASSICAL DISCRETE NONLINEAR EQUATIONS

I. KRICHEVER

DEPARTMENT OF MATHEMATICS OF COLUMBIA UNIVERSITY
AND
LANDAU INSTITUTE FOR THEORETICAL PHYSICS
KOSYGINA STR. 2, 117940 MOSCOW, RUSSIA

O. LIPAN

JAMES FRANCK INSTITUTE OF THE UNIVERSITY OF CHICAGO
5640 S.ELLIS AVENUE, CHICAGO, IL 60637, USA

P. WIEGMANN

JAMES FRANCK AND ENRICO FERMI INSTITUTES
THE UNIVERSITY OF CHICAGO,
5640 S.ELLIS AVENUE, CHICAGO, IL 60637, USA
AND
LANDAU INSTITUTE FOR THEORETICAL PHYSICS

A. ZABRODIN

JOINT INSTITUTE OF CHEMICAL PHYSICS
KOSYGINA STR. 4, 117334, MOSCOW
AND
ITEP, 117259, MOSCOW, RUSSIA

Low Dimensional Applications of Quantum Field Theory
Edited by L. Baulieu *et al.*, Plenum Press, New York, 1997

1 Introduction

In spite of the diversity of solvable models of quantum field theory and the vast variety of methods, the final results display dramatic unification: the spectrum of an integrable theory with a local interaction is given by a sum of elementary energies

$$E = \sum_i \varepsilon(u_i)\,, \tag{1.1}$$

where u_i obey a system of algebraic or transcendental equations known as *Bethe equations*[1], [2]. The major ingredients of Bethe equations are determined by the algebraic structure of the problem. A typical example of a system of Bethe equations (related to A_1-type models with elliptic R-matrix) is

$$e^{-4\eta\nu}\frac{\phi(u_j)}{\phi(u_j-2)} = -\prod_k \frac{\sigma(\eta(u_j-u_k+2))}{\sigma(\eta(u_j-u_k-2))}\,, \tag{1.2}$$

where $\sigma(x)$ is the Weierstrass σ-function and

$$\phi(u) = \prod_{k=1}^{N} \sigma(\eta(u-y_k))\,. \tag{1.3}$$

Entries of these equations which encode information of the model are the function $\varepsilon(u)$ (entering through $\phi(u)$), quasiperiods ω_1, ω_2 of the σ-function, parameters η, ν, y_k and size of the system N.

Different solutions of the Bethe equations correspond to different quantum states of the model.

In this paper we show that these equations, which are usually considered as a tool inherent to the quantum integrability, arise naturally as a result of the solution of entirely *classical* non-linear *discrete time* integrable equations. This suggests an intriguing interrelation (if not equivalence) between *integrable quantum field theories* and *classical soliton equations in discrete time.* In forthcoming papers we will show that the Bethe equations themselves may be considered as a discrete integrable dynamical system.

In 1981 Hirota [3] proposed a difference equation which unifies the majority of known continuous soliton equations, including their hierarchies [4].

A particular case of the Hirota equation is a bilinear difference equation for a function $\tau(n,l,m)$ of three discrete variables:

$$\begin{aligned}\alpha\tau(n,l+1,m)\tau(n,l,m+1) + \beta\tau(n,l,m)\tau(n,l+1,m+1)&\\ +\gamma\tau(n+1,l+1,m)\tau(n-1,l,m+1) &= 0\end{aligned} \tag{1.4}$$

where it is assumed that $\alpha+\beta+\gamma=0$. Different continuum limits at different boundary conditions then reproduce continuous soliton equations (KP, Toda lattice, etc). On the other hand, $\tau(n,l,m)$ can be identified [4] with the τ-function of a continuous hierarchy expressed through special independent variables.

The same equation (with a particular boundary condition) has quite unexpectedly appeared in the theory of *quantum* integrable systems as a fusion relation for the transfer matrix (trace of the quantum monodromy matrix).

The transfer matrix is one of the key objects in the theory of quantum integrable systems [5]. Transfer matrices form a commutative family of operators acting in the Hilbert space of a quantum problem. Let $R_{i,\mathcal{A}}(u)$ be the R-matrix acting in the tensor product of Hilbert spaces $V_i \otimes V_{\mathcal{A}}$. Then the transfer matrix is a trace over the auxiliary space $V_{\mathcal{A}}$ of the monodromy matrix. The latter being, the matrix product of N R-matrices with a common auxiliary space:

$$\begin{aligned} \hat{T}_{\mathcal{A}}(u|y_i) &= R_{N,\mathcal{A}}(u-y_N)\ldots R_{2,\mathcal{A}}(u-y_2)R_{1,\mathcal{A}}(u-y_1)\,, \\ T_{\mathcal{A}}(u) &= \mathrm{tr}_{\mathcal{A}}\hat{T}_{\mathcal{A}}(u|y_i)\,. \end{aligned} \tag{1.5}$$

The transfer matrices commute for all values of the spectral parameter u and different auxiliary spaces:

$$[T_{\mathcal{A}}(u),\, T_{\mathcal{A}'}(u')] = 0. \tag{1.6}$$

They can be diagonalized simultaneously. The family of eigenvalues of the transfer matrix is an object of primary interest in an integrable system, since the spectrum of the quantum problem can be expressed in terms of eigenvalues of the transfer matrix.

The transfer matrix corresponding to a given representation in the auxiliary space can be constructed out of transfer matrices for some elementary space by means of the *fusion procedure* [6], [7], [8]. The fusion procedure is based on the fact that at certain values of the spectral parameter u the R-matrix becomes essentially a projector onto an irreducible representation space. The fusion rules are especially simple in the A_1-case. For example, the $R_{1,1}(u)$-matrix for two spin-1/2 representations in a certain normalization of the spectral parameter is proportional to the projector onto the singlet (spin-0 state) at $u = +2$ and onto the triplet (spin-1 subspace) at $u = -2$, in accordance with the decomposition $[1/2] + [1/2] = [0] + [1]$. Then the transfer matrix $T_2^1(u)$ with spin-1 auxiliary space is obtained from the product of two spin-1/2 monodromy matrices $\hat{T}_1^1(u)$ with arguments shifted by 2:

$$T_2^1(u) = \mathrm{tr}_{[1]}(R_{1,1}(-2)\hat{T}_1^1(u+1)\hat{T}_1^1(u-1)R_{1,1}(-2))\,.$$

A combination of the fusion procedure and the Yang-Baxter equation results in numerous functional relations (fusion rules) for the transfer matrix [6], [9]. They were recently combined into a universal bilinear form [10], [11]. The bilinear functional relations have the most simple closed form for the models of the A_{k-1}-series and representations corresponding to *rectangular* Young diagrams.

Let $T_s^a(u)$ be the transfer matrix for the rectangular Young diagram of length a and height s. If η can not be represented in the form $\eta = r_1\omega_1 + r_2\omega_2$ with rational r_1, r_2 (below we always assume that this is the case; for models with trigonometric R-matrices this means that the quantum deformation parameter q would not be a root of unity), they obey the following bilinear functional relation:

$$T_s^a(u+1)T_s^a(u-1) - T_{s+1}^a(u)T_{s-1}^a(u) = T_s^{a+1}(u)T_s^{a-1}(u)\,. \tag{1.7}$$

Since $T_s^a(u)$ commute at different u, a, s,, the same equation holds for eigenvalues of the transfer matrices, so we can (and will) treat $T_s^a(u)$ in eq. (1.7) as number-valued functions. The bilinear fusion relations for models related to other Dynkin graphs were suggested in ref. [11].

Remarkably, the bilinear fusion relations (1.7) appear to be identical to the Hirota equation (1.4). Indeed, one can eliminate the constants α, β, γ by the transformation

$$\tau(n,l,m) = \frac{(-\alpha/\gamma)^{n^2/2}}{(1+\gamma/\alpha)^{lm}}\tau_n(l,m),$$

so that

$$\tau_n(l+1,m)\tau_n(l,m+1) - \tau_n(l,m)\tau_n(l+1,m+1) - \tau_{n+1}(l+1,m)\tau_{n-1}(l,m+1) = 0 \quad (1.8)$$

and then change variables from *light-cone* coordinates n, l, m to the *"direct"* variables

$$a = n, \qquad s = l+m, \qquad u = l-m-n,$$
$$\tau_n(l,m) \equiv T^a_{l+m}(l-m-n). \quad (1.9)$$

At least at a formal level, this transformation provides the equivalence between eqs. (1.7), (1.4) and (1.8). In what follows we call eq. (1.8) (or (1.7)) Hirota's bilinear difference equation (HBDE).

Leaving aside more fundamental aspects of this "coincidence", we exploit, as a first step, some technical advantages it offers. Specifically, we treat the functional relation (1.7) not as an identity but as a *fundamental equation* which (together with particular boundary and analytical conditions) completely determines all the eigenvalues of the transfer matrix. The solution to HBDE then appears in the form of the Bethe equations. We anticipate that this approach makes it possible to use some specific tools of classical integrability and, in particular, the finite gap integration technique.

The origin of $T^a_s(u)$ as an eigenvalue of the transfer matrix (1.5) imposes specific boundary conditions and, what is perhaps even more important, requires certain analytic properties of the solutions. As a general consequence of the Yang-Baxter equation, the transfer matrices may always be normalized to be *elliptic polynomials* in the spectral parameter, i.e. finite products of Weierstrass σ-functions (as in (1.3)). The problem therefore is stated as of finding elliptic solutions of HBDE.

A similar problem appeared in the theory of continuous soliton equations since the works [12], [13], wherein a remarkable connection between motion of poles of the elliptic solutions to the KdV equation and the Calogero-Moser dynamical system was revealed. Elliptic solutions to Kadomtsev-Petviashvili (KP), matrix KP equations and the matrix 2D Toda lattice (2DTL) were analyzed in the Refs. [14], [15], [16], respectively. It was shown, in particular, that poles of elliptic solutions to the abelian 2DTL (i.e. zeros of corresponding τ-functions and Baker-Akhiezer functions) move according to the equations of motion for the Ruijsenaars-Schneider (RS) system of particles [17].

Analytic properties of solutions to HBDE relevant to the Bethe Ansatz suggest a similar interpretation of Bethe Ansatz equations. We will show that the nested Bethe Ansatz for A_{k-1}-type models is equivalent to a chain of Bäcklund transformations of HBDE. The nested Bethe Ansatz equations arise as equations of motion for zeros of the Baker-Akhiezer functions in discrete time (discrete time RS system[1].) The discrete time variable is identified with level of the nested Bethe Ansatz.

[1] It should be noted that equations of motion for the discrete time RS system were already written down in the paper [18]. However, the relation to elliptic solutions of discrete soliton equations and their nested Bethe Ansatz interpretation were not discussed there.

The paper is organized as follows. In Sect. 2 we review general properties and boundary conditions of solutions to HBDE that yield eigenvalues of quantum transfer matrices. In Sect. 3 the zero curvature representation of HBDE and the auxiliary linear problems are presented. We also discuss the duality relation between "wave functions" and "potentials" and define Bäcklund flows on the set of wave functions. These flows are important ingredients of the nested Bethe Ansatz scheme. For illustrative purposes, in Sect. 4, we give a self-contained treatment of the A_1-case, where major part of the construction contains familiar objects from the usual Bethe Ansatz. Sect. 5 is devoted to the general A_{k-1}-case. We give a general solution to HBDE with the required boundary conditions. This leads to a new type of determinant formulas for eigenvalues of quantum transfer matrices. The explicit form of generalized Baxter's relations (difference equations for $Q_t(u)$) is presented. They are used for examining the equivalence to the standard Bethe Ansatz results. In Sect. 6 a part of the general theory of elliptic solutions to HBDE is given. Sect. 7 contains a discussion of the results.

2 General properties of solutions to Hirota's equation relevant to Bethe Ansatz

2.1 *Boundary conditions and analytic properties*

HBDE has many different solutions. Not all of them give eigenvalues of the transfer matrix (1.5). There are certain boundary and analytic conditions imposed on the transfer matrix (1.5).

(i) It is known that $T_s^k(u)$, the transfer matrix in the most antisymmetrical representation in the auxiliary space, is a scalar, i.e. it has only one eigenvalue (sometimes called quantum determinant $\det_q \hat{T}_s(u)$ of the monodromy matrix). It depends on the representation in the quantum space of the model and is known explicitly. In the simplest case of the vector representation (one-box Young diagram) in the quantum space it is [19]:

$$T_s^k(u) = \phi(u-s-k)\prod_{l=0}^{k-1}\prod_{p=1}^{s-1}\phi(u+s+k-2l-2p-2)\prod_{l=1}^{k-1}\phi(u+s+k-2l), \quad (2.1)$$

$$T_s^0(u) = 1. \quad (2.2)$$

These values of $T_s^0(u)$ and $T_s^k(u)$ should be considered as boundary conditions. Let us note that they obey the discrete Laplace equation:

$$T_s^k(u+1)T_s^k(u-1) = T_{s+1}^k(u)T_{s-1}^k(u). \quad (2.3)$$

This leads to the boundary condition (b.c.)

$$T_s^a(u) = 0 \quad \text{as } a < 0 \text{ and } a > k \quad (2.4)$$

(with this b.c. eq. (1.8) is known as the discrete two-dimensional Toda molecule equation [20], an integrable discretization of the conformal Toda field theory [21]).

(ii) The second important condition (which follows, eventually, from the Yang-Baxter equation) is that $T_s^a(u)$ has to be an elliptic polynomial in the spectral parameter u. By elliptic polynomial we mean essentially a finite product of Weierstrass σ-functions. For models with rational R-matrix it degenerates to a usual polynomial in u.

To give a more precise formulation of this property, let us note that eq. (1.7) has the gauge invariance under a transformation parametrized by four arbitrary functions χ_i of one variable:

$$T_s^a(u) \to \chi_1(a+u+s)\chi_2(a-u+s)\chi_3(a+u-s)\chi_4(a-u-s)T_s^a(u) \tag{2.5}$$

These transformations can remove all zeros from the characteristics $a \pm s \pm u = \text{const}$. We require that the remaining part of all $T_s^a(u)$ should be an elliptic (trigonometric, rational) polynomial of one and the same degree N, where N is the number of sites on the lattice (see (1.3)).

One can formulate this condition in a gauge invariant form by introducing the gauge invariant combination

$$Y_s^a(u) = \frac{T_{s+1}^a(u)T_{s-1}^a(u)}{T_s^{a+1}(u)T_s^{a-1}(u)} . \tag{2.6}$$

We require $Y_s^a(u)$ to be an elliptic function having $2N$ zeros and $2N$ poles in the fundamental domain. This implies that $T_s^a(u)$ has the general form[2]

$$T_s^a(u) = A_s^a e^{\mu(a,s)u} \prod_{j=1}^{N} \sigma(\eta(u - z_j^{(a,s)})) , \tag{2.7}$$

where $z_j^{(a,s)}$, A_s^a, $\mu(a,s)$ do not depend on u and the following constraints hold:

$$\sum_{j=1}^{N} (z_j^{(a,s+1)} + z_j^{(a,s-1)}) = \sum_{j=1}^{N} (z_j^{(a+1,s)} + z_j^{(a-1,s)}) , \tag{2.8}$$

$$\mu(a,s+1) + \mu(a,s-1) = \mu(a+1,s) + \mu(a-1,s) . \tag{2.9}$$

Another gauge invariant combination,

$$X_s^a(u) = -\frac{T_s^a(u+1)T_s^a(u-1)}{T_s^{a+1}(u)T_s^{a-1}(u)} = -1 - Y_s^a(u) , \tag{2.10}$$

is also convenient.

As a reference, we point out gauge invariant forms of HBDE [11]:

$$Y_s^a(u+1)Y_s^a(u-1) = \frac{(1+Y_{s+1}^a(u))(1+Y_{s-1}^a(u))}{(1+(Y_s^{a+1}(u))^{-1})(1+(Y_s^{a-1}(u))^{-1})} , \tag{2.11}$$

$$X_{s+1}^a(u)X_{s-1}^a(u) = \frac{(1+X_s^a(u+1))(1+X_s^a(u-1))}{(1+(X_s^{a+1}(u))^{-1})(1+(X_s^{a-1}(u))^{-1})} . \tag{2.12}$$

[2]This differs from a more traditional expression in terms of Jacobi θ-functions by a simple normalization factor.

It can be shown that the minimal polynomial appears in the gauge

$$T_s^a(u) \to T_s^a(u) \left(\prod_{l=0}^{a-1} \prod_{p=1}^{s-1} \phi(u+s+a-2l-2p-2) \prod_{l=1}^{a-1} \phi(u+s+a-2l) \right)^{-1}, \tag{2.13}$$

where all the "trivial" zeros (common for all the eigenvalues) of the transfer matrix are removed (see e.g. [23]). The boundary values at $a = 0, k$ then become:

$$\begin{aligned} T_s^0(u) &= \phi(u+s), \\ T_s^k(u) &= \phi(u-s-k). \end{aligned} \tag{2.14}$$

From now on we adopt this normalization.

(iii) The analyticity conditions and b.c. (2.14) lead to a particular "initial condition" in s. It is convenient, however, to take advantage of it before the actual derivation. The condition reads

$$T_s^a(u) = 0 \quad \text{for any} \quad -k < s < 0, \ \ 0 < a < k. \tag{2.15}$$

This is consistent with (1.7), (2.14) and implies

$$T_0^a(u) = \phi(u-a) \tag{2.16}$$

for $0 \le a \le k$.

Under the analyticity conditions (i) and the b.c. (2.14) (and their consequences (2.15), (2.16)) each solution to HBDE (1.7) corresponds to an eigenstate of the A_{k-1}-transfer matrix.

The same conditions are valid for higher representations of the quantum space. However, in that case there are certain constraints on zeros of $\phi(u)$ (they should form "strings"), whence $T_s^a(u)$ acquires extra "trivial" zeros. Here we do not address this question.

2.2 Plücker relations and determinant representations of solutions

Classical integrable equations in Hirota's bilinear form are known to be naturally connected [24], [25], [26], with geometry of Grassmann's manifolds (grassmannians) (see [27] [28], [29]), in general of an infinite dimension. Type of the grassmannian is specified by boundary conditions. Remarkably, the b.c. (2.4) required for Bethe Ansatz solutions corresponds to *finite dimensional* grassmannians. This connection suggests a simple way to write down a general solution in terms of determinants and to transmit the problem to the boundary conditions. Numerous determinant formulas may be obtained in this way.

The grassmannian $\mathbf{G}_{n+1}^{r+1}$ is a collection of all $(r+1)$-dimensional linear subspaces of the complex $(n+1)$-dimensional vector space $\mathbf{C}^{n+1}$. In particular, $\mathbf{G}_{n+1}^{1}$ is the complex projective space $\mathbf{P}^n$. Let $X \in \mathbf{G}_{n+1}^{r+1}$ be such a $(r+1)$-dimensional subspace spanned by vectors $\mathbf{x}^{(j)} = \sum_{i=0}^{n} x_i^{(j)} \mathbf{e}^i$, $j = 1, \ldots, r+1$, where $\mathbf{e}^i$ are basis vectors in

$\mathbf{C}^{n+1}$. The collection of their coordinates form a rectangular $(n+1)\times(r+1)$-matrix $x_i^{(j)}$. Let us consider its $(r+1)\times(r+1)$ minors

$$\det_{pq}(x_{i_p}^{(q)}) \equiv (i_0, i_1, \ldots, i_r), \qquad p, q = 0, 1, \ldots, r\,, \tag{2.17}$$

obtained by choosing $r+1$ lines $i_0, i_1, \ldots, i_r$. These C_{n+1}^{r+1} minors are called *Plücker coordinates* of X. They are defined up to a common scalar factor and provide the *Plücker embedding* of the grassmannian $\mathbf{G}_{n+1}^{r+1}$ into the projective space $\mathbf{P}^d$, where $d = C_{n+1}^{r+1} - 1$ (C_{n+1e}^{r+1} is the bimomial coefficient).

The image of $\mathbf{G}_{n+1}^{r+1}$ in $\mathbf{P}^d$ is realized as an intersection of quadrics. This means that the coordinates $(i_0, i_1, \ldots, i_r)$ are not independent but obey the *Plücker relations* [28], [29]:

$$(i_0, i_1, \ldots, i_r)(j_0, j_1, \ldots, j_r) = \sum_{p=0}^{r} (j_p, i_1, \ldots, i_r)(j_0, \ldots j_{p-1}, i_0, j_{p+1} \ldots, j_r) \tag{2.18}$$

for all i_p, j_p, $p = 0, 1, \ldots, r$. Here it is implied that the symbol $(i_0, i_1, \ldots, i_r)$ is antysymmetric in all the indices, i.e. $(i_0, \ldots, i_{p-1}, i_p, \ldots, i_r) = -(i_0, \ldots, i_p, i_{p-1}, \ldots, i_r)$ and it equals zero if any two indices coincide. If one treats these relations as equations rather than identities, then determinants (2.17) would give a solution to Hirota's equations.

The Plucker relations in their general form (2.18) most likely describe fusion rules for transfer matrices corresponding to arbitrary Young diagrams. In order to reduce them to the 3-term HBDE, one should take $i_p = j_p$ for $p \neq 0, 1$. Then all terms but the first two in the r.h.s. of (2.18) vanish and one is left with the 3-term relation

$$\begin{aligned}(i_0, i_1, \ldots, i_r)(j_0, j_1, i_2, \ldots, i_r) &= (j_0, i_1, i_2, \ldots, i_r)(i_0, j_1, i_2, \ldots i_r) \\ &+ (j_1, i_1, i_2, \ldots, i_r)(j_0, i_0, i_2, \ldots i_r).\end{aligned} \tag{2.19}$$

After the substitution (2.17) these elementary Plücker relations turn into certain determinant identities. For example, choosing $x_{i_0}^{(j)} = \delta_{pj}$, $x_{j_0}^{(j)} = \delta_{qj}$, $q \neq p$, one can recast eq. (2.19) into the form of the Jacobi identity:

$$D[p|p] \cdot D[q|q] - D[p|q] \cdot D[q|p] = D[p, q|p, q] \cdot D\,. \tag{2.20}$$

where D is determinant of a $(r+1)\times(r+1)$-matrix and $D[p_1, p_2|q_1, q_2]$ denotes determinant of the same matrix with $p_{1,2}$-th rows and $q_{1,2}$-th columns removed. Another useful identity contained in eq. (2.19) connects minors $D[l_1, l_2]$ of a $(r+3)\times(r+1)$ rectangular matrix, where the two rows l_1, l_2 are removed:

$$D[l_1, l_2] \cdot D[l_3, l_4] - D[l_1, l_3] \cdot D[l_2, l_4] = D[l_1, l_4] \cdot D[l_2, l_3]\,, \quad l_1 < l_2 < l_3 < l_4\,. \tag{2.21}$$

Identifying terms in eq. (2.19) with terms in HBDE (1.8), one obtains various determinant representations of solutions to HBDE. Two of them follow from the Jacobi identity (2.20):

$$\tau_a(l, m) = \det\left(\tau_1(l+i-a, m-j+a)\right), \qquad i, j = 1, \ldots, a, \qquad \tau_0(l, m) = 1 \tag{2.22}$$

or, in "direct" variables

$$T_s^a(u) = \det\left(T_{s+i-j}^1(u+i+j-a-1)\right), \qquad i,j=1,\ldots,a, \qquad T_s^0(u)=1\,. \tag{2.23}$$

This representation determines an evolution in a from the initial values at $a = 1$. The size of the determinant grows with a. A similar formula exists for the evolution in s:

$$T_s^a(u) = \det\left(T_1^{a+i-j}(u+i+j-s-1)\right), \qquad i,j=1,\ldots,s, \qquad T_0^a(u)=1\,. \tag{2.24}$$

The size of this determinant grows with s. Determinant formulas of this type have been known in the literature on quantum integrable models (see e.g. [22]). They allow one to express $T_s^a(u)$ through $T_1^a(u)$ or $T_s^1(u)$.

A different kind of determinant representation follows from (2.21):

$$\begin{aligned} \tau_a(l,m) &= \det M_{ij}\,, \\ M_{ji} &= \begin{cases} h_i(u+s+a+2j) & \text{if } j=1,\ldots,k-a;\ i=1,\ldots,k \\ \bar{h}_i(u-s+a+2j) & \text{if } j=k-a+1,\ldots,k;\ i=1,\ldots,k \end{cases} \end{aligned} \tag{2.25}$$

where $h_i(x)$ and $\bar{h}_i(x)$ are $2k$ arbitrary functions of one variable. The size of this determinant is equal to k for all $0 \le a \le k$. This determinant formula plays an essential role in what follows.

The determinant representations give a solution to discrete nonlinear equations and expose the essence of the integrability. Let us note that they are simpler and more convenient than their continuous counterparts.

2.3 *Examples of difference and continuous A_1-type equations*

For illustrative purposes we specialize the Hirota equation to the A_1-case and later study it separately. At $k = 2$ eq. (1.7) is

$$T_s(u+1)T_s(u-1) - T_{s+1}(u)T_{s-1}(u) = \phi(u+s)\phi(u-s-2) \tag{2.26}$$

with the condition $T_{-1}(u) = 0$ (here we set $T_s(u) \equiv T_s^1(u)$).

This equation is known as a discrete version of the Liouville equation [20] written in terms of the τ-function. It can be recast to somewhat more universal form in terms of the discrete Liouville field

$$Y_s^1(u) \equiv Y_s(u) = \frac{T_{s+1}(u)T_{s-1}(u)}{\phi(u+s)\phi(u-s-2)} \tag{2.27}$$

(see (2.6)), which hides the function $\phi(u)$ in the r.h.s. of (2.26). The equation becomes

$$Y_s(u-1)Y_s(u+1) = (Y_{s+1}(u)+1)(Y_{s-1}(u)+1)\,. \tag{2.28}$$

(Let us note that the same functional equation but with different analytic properties of the solutions appears in the thermodynamic Bethe Ansatz [30], [31].)

In the continuum limit one should put

$$Y_s(u) = \delta^{-2}\exp(-\varphi(x,t)), \;\; u=\delta^{-1}x, \; s=\delta^{-1}t.$$

An expansion in $\delta \to 0$ then gives the continuous Liouville equation

$$\partial_s^2 \varphi - \partial_u^2 \varphi = 2 \exp(\varphi) . \qquad (2.29)$$

To stress the specifics of the b.c. (2.15) and for further reference let us compare it with the quasiperiodic b.c. Then the A_1-case corresponds to the discrete sine-Gordon (SG) equation [32]:

$$T_s^{a+1}(u) = e^{\alpha} \lambda^{2a} T_s^{a-1}(u-2), \qquad (2.30)$$

where α and λ are parameters. Substituting this condition into (1.7), we get:

$$T_s^1(u+1)T_s^1(u-1) - T_{s+1}^1(u)T_{s-1}^1(u) = e^{\alpha}\lambda^2 T_s^0(u)T_s^0(u-2), \qquad (2.31)$$

$$T_s^0(u+1)T_s^0(u-1) - T_{s+1}^0(u)T_{s-1}^0(u) = e^{-\alpha} T_s^1(u)T_s^1(u+2). \qquad (2.32)$$

Let us introduce two fields $\rho^{s,u}$ and $\varphi^{s,u}$ on the square (s,u) lattice

$$T_s^0(u) = \exp(\rho^{s,u} + \varphi^{s,u}), \qquad (2.33)$$

$$T_s^1(u+1) = \lambda^{1/2} \exp(\rho^{s,u} - \varphi^{s,u}). \qquad (2.34)$$

and substitute them into (2.31), (2.32). Finally, eliminating $\rho^{s,u}$, one gets the discrete SG equation:

$$\sinh(\varphi^{s+1,u} + \varphi^{s-1,u} - \varphi^{s,u+1} - \varphi^{s,u-1}) = \lambda \sinh(\varphi^{s+1,u} + \varphi^{s-1,u} + \varphi^{s,u+1} + \varphi^{s,u-1} + \alpha) . \qquad (2.35)$$

The constant α can be removed by the redefinition $\varphi^{s,u} \to \varphi^{s,u} - \frac{1}{4}\alpha$.

Another useful form of the discrete SG equation appears in variables $X_s^a(u)$ (2.10). Under condition (2.30) one has

$$X_s^{a+1}(u) = X_s^{a-1}(u-2), \qquad \lambda^2 X_s^{a+1}(u+1) X_s^a(u) = 1 , \qquad (2.36)$$

so there is only one independent function

$$X_s^1(u) \equiv x_s(u) = -e^{-\alpha}\lambda^{-1} \exp\big(-2\varphi^{s,u} - 2\varphi^{s,u-2}\big) . \qquad (2.37)$$

The discrete SG equation becomes [32], [33], [34]:

$$x_{s+1}(u)x_{s-1}(u) = \frac{(\lambda + x_s(u+1))(\lambda + x_s(u-1))}{(1+\lambda x_s(u+1))(1+\lambda x_s(u-1))} . \qquad (2.38)$$

In the limit $\lambda \to 0$ eq. (2.38) turns into the discrete Liouville equation (2.28) for $Y_s(u) = -1 - \lambda^{-1} x_s(u)$.

3 Linear problems and Bäcklund transformations

3.1 Zero curvature condition

Consider the square lattice in two light cone variables l and m and a vector function $\psi_a(l,m)$ on this lattice. Let $L_{a,a'}(l,m)$ and $M_{a,a'}(l,m)$ be two shift operators in directions l and m:

$$\sum_{a'} L_{a,a'}(l,m)\psi_{a'}(l+1,m) = \psi_a(l,m),$$

$$\sum_{a"} M_{a,a'}(l,m)\psi_{a'}(l,m+1) = \psi_a(l,m). \qquad (3.1)$$

The zero curvature condition states that the result of subsequent shifts from an initial point to a fixed final point does not depend on the path:

$$L(l,m)\cdot M(l+1,m) = M(l,m)\cdot L(l,m+1). \tag{3.2}$$

HBDE (1.7) possesses [35] a zero-curvature representation by means of the following two-diagonal infinite matrices:

$$\begin{aligned} L_{a,a'} &= \delta_{a,a'-1} + \delta_{a,a'} V_l^a\,, \\ M_{a,a'} &= \delta_{a,a'} + \delta_{a,a'+1} W_m^a\,, \end{aligned} \tag{3.3}$$

where

$$\begin{aligned} V_l^a &= \frac{\tau_a(l+1,m)\tau_{a+1}(l,m)}{\tau_a(l,m)\tau_{a+1}(l+1,m)}\,, \\ W_m^a &= \frac{\tau_{a-1}(l,m+1)\tau_{a+1}(l,m)}{\tau_a(l,m)\tau_a(l,m+1)}\,. \end{aligned} \tag{3.4}$$

More precisely, the compatibility condition of the two linear problems

$$\begin{aligned} \psi_a(l,m) - \psi_{a+1}(l+1,m) &= V_l^a \psi_a(l+1,m)\,, \\ \psi_a(l,m) - \psi_a(l,m+1) &= W_m^a \psi_{a-1}(l,m+1) \end{aligned} \tag{3.5}$$

combined with the b.c. (2.14) yields HBDE (1.8). Introducing an unnormalized "wave function"

$$f_a(l,m) = \psi_a(l,m)\tau_a(l,m)\,, \tag{3.6}$$

we can write the linear problems in the form

$$\begin{aligned} \tau_{a+1}(l+1,m)f_a(l,m) - \tau_{a+1}(l,m)f_a(l+1,m) &= \tau_a(l,m)f_{a+1}(l+1,m)\,, \\ \tau_a(l,m+1)f_a(l,m) - \tau_a(l,m)f_a(l,m+1) &= \tau_{a+1}(l,m)f_{a-1}(l,m+1)\,, \end{aligned} \tag{3.7}$$

or in "direct" variables

$$\begin{aligned} T_{s+1}^{a+1}(u)F^a(s,u) - T_s^{a+1}(u-1)F^a(s+1,u+1) &= T_s^a(u)F^{a+1}(s+1,u)\,, \\ T_{s+1}^a(u-1)F^a(s,u) - T_s^a(u)F^a(s+1,u-1) &= T_s^{a+1}(u-1)F^{a-1}(s+1,u) \end{aligned} \tag{3.8}$$

where $F^a(l+m, l-m-a) \equiv f_a(l,m)$.

An advantage of the light cone coordinates is that they are separated in the linear problems (there are shifts only of l (m) in the first (second) eq. (3.7)).

The wave function and potential possess a redundant gauge freedom:

$$V_l^a \to \frac{\chi(a-l+1)}{\chi(a-l)} V_l^a, \quad W_m^a \to \frac{\chi(a-l)}{\chi(a-l-1)} W_m^a, \quad \psi_a(l,m) \to \chi(a-l+1)\psi_a \tag{3.9}$$

with an arbitrary function χ.

The b.c. (2.4) implies a similar condition for the object of the linear problems

$$F^a(s,u) = 0 \quad \text{as } a<0 \text{ and } a>k-1 \tag{3.10}$$

so that the number of functions F is one less than the number of T's. Then from the second equation of the pair (3.8) at $a = 0$ and from the first one at $a = k-1$ it follows that $F^0(s,u)$ ($F^{k-1}(s,u)$) depends on one cone variable $u+s$ (resp., $u-s$). We introduce a special notation for them:

$$F^0(s,u) = Q_{k-1}(u+s), \;\; F^{k-1}(s,u) = \bar{Q}_{k-1}(u-s). \tag{3.11}$$

Furthermore, it can be shown that the important condition (2.15) relates the functions Q and $\bar{Q}$:

$$\bar{Q}_{k-1}(u) = Q_{k-1}(u-k+1). \tag{3.12}$$

The special form of the functions F^a at the ends of the Dynkin graph ($a = 0, k-1$) reflects the specifics of the "Liouville-type" boundary conditions. This is to be compared with nonlinear equations with the quasiperiodic boundary condition (2.30): in this case all the functions F depend on two variables and obey the quasiperiodic b.c.

3.2 Continuum limit

In the continuum limit $l = -\delta t_+$, $m = -\delta t_-$, $\tau_a \to \delta^{a^2}\tau_a$, $f_a \to \delta^{a^2+a} f_a$, $\delta \to 0$, we recover the auxiliary linear problems for the 2D Toda lattice [36] ($\partial_\pm \equiv \partial/\partial t_\pm$):

$$\begin{aligned} \partial_+\psi_a &= \psi_{a+1} + \partial_+(\log\frac{\tau_{a+1}}{\tau_a})\psi_a, \\ \partial_-\psi_a &= \frac{\tau_{a+1}\tau_{a-1}}{\tau_a^2}\psi_{a-1}, \end{aligned} \tag{3.13}$$

or, in terms of f_a,

$$\begin{aligned} \tau_{a+1}\partial_+ f_a - (\partial_+\tau_{a+1})f_a &= \tau_a f_{a+1}, \\ \tau_a\partial_- f_a - (\partial_-\tau_a)f_a &= \tau_{a+1}f_{a-1}. \end{aligned} \tag{3.14}$$

The compatibility condition of these equations yields the first non-trivial equation of the 2D Toda lattice hierarchy:

$$\partial_+\tau_a\partial_-\tau_a - \tau_a\partial_+\partial_-\tau_a = \tau_{a+1}\tau_{a-1}. \tag{3.15}$$

In terms of

$$\varphi_a(t_+,t_-) = \log\frac{\tau_{a+1}(t_+,t_-)}{\tau_a(t_+,t_-)}$$

it has the familiar form

$$\partial_+\partial_-\varphi_a = e^{\varphi_a-\varphi_{a-1}} - e^{\varphi_{a+1}-\varphi_a}. \tag{3.16}$$

3.3 Duality

The discrete nonlinear equation has a remarkable duality between "potentials" T^a and "wave functions" F^a first noticed in [35]. In the continuum version it is not so transparent. Eqs.(3.8) are symmetric under the interchange of F and T. Then one may treat (3.8) as linear problems for a nonlinear equation on F's. It is not surprising that one again obtains HBDE (1.7):

$$F^a(s,u+1)F^a(s,u-1) - F^a(s+1,u)F^a(s-1,u) = F^{a+1}(s,u)F^{a-1}(s,u)\,. \quad (3.17)$$

Moreover, conditions (3.10)-(3.12) mean that even the b.c. for $F^a(s,u)$ are the same as for $T_s^a(u)$ under a substitution $\phi(u)$ by $Q_{k-1}(u)$. The only change is a reduction of the Dynkin graph: $k \to k-1$. Using this property, one can successively reduce the A_{k-1}-problem up to A_1. Below we use this trick to derive A_{k-1} ("nested") Bethe Ansatz equations.

3.4 Bäcklund flow

To elaborate the chain of these transformations, let us introduce a new variable $t = 0, 1, \ldots, k$ to mark a level of the flow $A_{k-1} \to A_1$ and let $F_{t+1}^a(s,u)$ be a solution to the linear problem at $(k-t)$-th level. In this notation, $F_k^a(s,u) = T_s^a(u)$ and $F_{k-1}^a(s,u) = F^a(s,u)$ is the corresponding wave function. The wave function itself obeys the nonlinear equation (3.17), so $F_{k-2}^a(s,u)$ denotes its wave function and so on. For each level t the function $F_t^a(s,u)$ obeys HBDE of the form (3.17) with the b.c.

$$F_t^a(s,u) = 0 \quad \text{as } a<0 \text{ and } a>t\,. \quad (3.18)$$

As a consequence of (3.18), the first and the last components of the vector $F_t^a(s,u)$ obey the discrete Laplace equation (2.3) and under the condition (3.11) are functions of only one of the light-cone variables ($u+s$ and $u-s$ respectively). We denote them as follows:

$$F_t^0(s,u) \equiv Q_t(u+s)\,, \qquad F_t^t(s,u) \equiv \bar{Q}_t(u-s)\,, \quad (3.19)$$

where it is implied that $Q_k(u) = \phi(u)$. It can be shown that ellipticity requirement (ii) and condition (2.14) impose the relation $\bar{Q}_t(u) = Q_t(u-t)$.

In this notation the linear problems (3.8) at level t,

$$F_{t+1}^{a+1}(s+1,u)F_t^a(s,u) - F_{t+1}^{a+1}(s,u-1)F_t^a(s+1,u+1) = F_{t+1}^a(s,u)F_t^{a+1}(s+1,u)\,, \quad (3.20)$$

$$F_{t+1}^a(s+1,u-1)F_t^a(s,u) - F_{t+1}^a(s,u)F_t^a(s+1,u-1) = F_{t+1}^{a+1}(s,u-1)F_t^{a-1}(s+1,u) \quad (3.21)$$

look as bilinear equations for a functions of 4 variables. However, eq. (3.20) (resp., eq. (3.21)) leaves the hyperplane $u-s+a=\text{const}$ (resp., $u+s+a=\text{const}$) invariant, and actually depends on three variable.

Restricting the variables in eq. (3.20) to the hyperplane $u-s+a=v$ (where v is a constant), by setting

$$\tau_u(t,a) \equiv F_{k-t}^a(u+a-v,u) \quad (3.22)$$

we reduce eq. (3.20) to the form of the same HBDE (1.8) in cone coordinates t and a. The b.c. is

$$\tau_u(t,0) = Q_{k-t}(2u-v), \qquad \tau_u(t,k-t) = \bar{Q}_{k-t}(v+t-k) = \text{const}. \quad (3.23)$$

Similar equations can be obtained from the second linear problem (3.21) by setting

$$\bar{\tau}_u(b,t) = F_{k-t}^{k-t-b}(\bar{v}+b-u, u+t-k) \tag{3.24}$$

($\bar{v}$ is a constant). This function obeys eq. (1.8),

$$\bar{\tau}_u(b+1,t)\bar{\tau}_u(b,t+1) - \bar{\tau}_u(b,t)\bar{\tau}_u(b+1,t+1) = \bar{\tau}_{u+1}(b+1,t)\bar{\tau}_{u-1}(b,t+1)\,, \tag{3.25}$$

where t now plays the role of the light cone coordinate m. The b.c. is

$$\bar{\tau}_u(0,t) = \bar{Q}_{k-t}(2u+t-k-\bar{v}), \qquad \bar{\tau}_u(k-t,t) = Q_{k-t}(v) = \text{const.} \tag{3.26}$$

It is convenient to visualize this array of τ-functions on a diagram; here is an example for the A_3-case ($k=4$):

$$\begin{array}{cccccccc}
0 & 1 & 0 & & & & & \\
0 & Q_1(u+s) & \bar{Q}_1(u-s) & 0 & & & & \\
0 & Q_2(u+s) & F_2^1(s,u) & \bar{Q}_2(u-s) & 0 & & & \\
0 & Q_3(u+s) & F_3^1(s,u) & F_3^2(s,u) & \bar{Q}_3(u-s) & 0 & & \\
0 & \phi(u+s) & T_s^1(u) & T_s^2(u) & T_s^3(u) & \bar{\phi}(u-s) & 0 &
\end{array} \tag{3.27}$$

Functions in each horizontal (constant t) slice satisfy HBDE (3.17), whereas functions on the $u-s+a=$ const slice satisfy HBDE (1.8) with t, a being light cone variables l, m respectively.

A general solution of the bilinear discrete equation (1.7) with the b.c. (2.14) is determined by $2k$ arbitrary functions of one variable $Q_t(u)$ and $\bar{Q}_t(u)$, $t=1,\ldots,k$. The additional requirement (ii) of ellipticity determines these functions through the Bethe Ansatz.

3.5 Nested Bethe Ansatz scheme

Here we elaborate the nested scheme of solving HBDE based on the chain of successive Bäcklund transformations (Sect. 3.4). This is an alternative (and actually the shortest) way to obtain nested Bethe Ansatz equations (3.31). Recall that the function $\tau_u(t,a) = F_{k-t}^a(u+a,u)$ (3.22) (where we put $v=0$ for simplicity) obeys HBDE in light cone variables:

$$\tau_u(t+1,a)\tau_u(t,a+1) - \tau_u(t,a)\tau_u(t+1,a+1) = \tau_{u+1}(t+1,a)\tau_{u-1}(t,a+1)\,. \tag{3.28}$$

Since $\tau_u(t,0) = Q_{k-t}(2u)$, nested Bethe Ansatz equations can be understood as "equations of motions" for zeros of $Q_t(u)$ in discrete time t (level of the Bethe Ansatz). The simplest way to derive them is to consider the auxiliary linear problems for eq. (3.28). Here we present an example of this derivation in the simplest possible form.

Let us assume that $Q_t(u)$ has the form

$$Q_t(u) = e^{\nu_t \eta u} \prod_{j=1}^{M_t} \sigma(\eta(u - u_j^t)) \tag{3.29}$$

(note that we allow the number of roots M_t to depend on t). Since we are interested in dynamics in t at a fixed a, it is sufficient to consider only the first linear equation of the pair (3.7):

$$\tau_{u+1}(t+1,a) f_u(t,a) - \tau_{u+1}(t,a) f_u(t+1,a) = \tau_u(t,a) f_{u+1}(t+1,a)\,. \tag{3.30}$$

An elementary way to derive equations of motion for roots of $\tau_u(t,0)$ is to put u equal to the roots of $f_u(t+1,0)$, $f_u(t,0)$ and $f_{u+1}(t+1,0)$, so that only two terms in (3.30) would survive. Combining relations obtained in this way, one can eliminate f's and obtain the system of equations

$$\frac{Q_{t-1}(u_j^t+2) Q_t(u_j^t-2) Q_{t+1}(u_j^t)}{Q_{t-1}(u_j^t) Q_t(u_j^t+2) Q_{t+1}(u_j^t-2)} = -1\,. \tag{3.31}$$

as the necessary conditions for solutions of the form (3.29) to exist. In the more detailed notation they look as follows:

$$\prod_{k=1}^{M_{t-1}} \frac{\sigma(\eta(u_j^t - u_k^{t-1}+2))}{\sigma(\eta(u_j^t - u_k^{t-1}))} \prod_{k=1}^{M_t} \frac{\sigma(\eta(u_j^t - u_k^t - 2))}{\sigma(\eta(u_j^t - u_k^t + 2))} \prod_{k=1}^{M_{t+1}} \frac{\sigma(\eta(u_j^t - u_k^{t+1}))}{\sigma(\eta(u_j^t - u_k^{t+1} - 2))} = - \exp 2\eta(2\nu_t - \nu_{t+1} - \nu_{t-1})\,. \tag{3.32}$$

With the "boundary conditions"

$$Q_0(u) = 1, \qquad Q_k(u) = \phi(u), \tag{3.33}$$

this system of $M_1 + M_2 + \cdots + M_{k-1}$ equations is equivalent to the nested Bethe Ansatz equations for A_{k-1}-type quantum integrable models with Belavin's elliptic R-matrix. The same equations can be obtained for the right edge of the diagram (3.27) from the second linear equation in (3.7). In Sect. 5 we explicitly identify our Q's with similar objects known from the Bethe Ansatz solution.

Let us remark that the origin of equations (3.32) suggests to consider them as equations of motion for the elliptic Ruijsenaars-Schneider model in discrete time. Taking the continuum limit in t (provided $M_t = M$ does not depend on t), one can check that eqs. (3.32) do yield the equations of motion for the elliptic RS model [17] with M particles. The additional limiting procedure $\eta \to 0$ with finite $\eta u_j = x_j$ yields the well known equations of motion for the elliptic Calogero-Moser system of particles.

However, integrable systems of particles in discrete time seem to have a richer structure than their continuous time counterparts. In particular, the total number of particles in the system may depend on (discrete) time. Such a phenomenon is possible in continuous time models only for singular solutions, when particles can move to infinity or merge to another within a finite period of time.

Remarkably, this appears to be the case for the solutions to eqs. (3.32) corresponding to eigenstates of the quantum model. It is known that the number of excitations M_t at t-th level of the Bethe Ansatz solution does depend on t. In other words, the number of "particles" in the corresponding discrete time RS model is not conserved. Though, the numbers M_t may not be arbitrary.

In the elliptic case degrees of the elliptic polynomials $Q_t(u)$ are equal to $M_t = N/k$ (provided η is incommensurable with the lattice spanned by ω_1, ω_2 and N is divisible by k). This fact follows directly from Bethe equations (3.31). Indeed, the elliptic polynomial form (3.29) implies that if u_j^t is a zero of $Q_t(u)$, i.e. $Q_t(u_j^t) = 0$, then $u_j^t + 2n_1\omega_1 + 2n_2\omega_2$ for all integers n_1, n_2 are its zeros too. Taking into account the well known monodromy properties of the σ-function, one concludes that this is possible if and only if

$$M_{t+1} + M_{t-1} = 2M_t\,, \tag{3.34}$$

which has a unique solution

$$M_t = \frac{N}{k}t \tag{3.35}$$

satisfying b.c. (3.33). This means that the nested scheme for elliptic A_{k-1}-type models is consistent only if N is divisible by k. In trigonometric and rational cases the restrictions on degrees of Q_t's are not so strong.

4 THE A_1-CASE: DISCRETE LIOUVILLE EQUATION

In this section we consider the A_1-case separately. Although in this case the general nested scheme is missing, the construction is more explicit and contains familiar objects from the Bethe Ansatz literature.

4.1 General solution

The basic functional relation is

$$T_s(u+1)T_s(u-1) - T_{s+1}(u)T_{s-1}(u) = \phi(u+s)\bar{\phi}(u-s), \tag{4.1}$$

where we consider a more general b.c. parametrized by the functions ϕ, $\bar{\phi}$ and set $T_s(u) \equiv T_s^1(u)$. The auxiliary linear problems (3.8) acquire the form

$$T_{s+1}(u)Q(u+s) - T_s(u-1)Q(u+s+2) = \phi(u+s)\bar{Q}(u-s-1), \tag{4.2}$$

$$T_{s+1}(u)\bar{Q}(u-s+1) - T_s(u+1)\bar{Q}(u-s-1) = \bar{\phi}(u-s)Q(u+s+2)\,. \tag{4.3}$$

Here we set $Q(u) \equiv Q_1(u)$ and $\phi(u) = Q_2(u)$. Rearranging these equations we obtain

$$\phi(u-2)Q(u+2) + \phi(u)Q(u-2) = A(u)Q(u), \tag{4.4}$$

$$\bar{\phi}(u)\bar{Q}(u+3) + \bar{\phi}(u+2)\bar{Q}(u-1) = \bar{A}(u)\bar{Q}(u+1), \tag{4.5}$$

where we introduce the quantities

$$A(u) = \frac{\phi(u-2)T_{s+1}(u-s) + \phi(u)T_{s-1}(u-s-2)}{T_s(u-s-1)}\,, \tag{4.6}$$

$$\bar{A}(u) = \frac{\bar{\phi}(u+2)T_{s+1}(u+s) + \bar{\phi}(u)T_{s-1}(u+s+2)}{T_s(u+s+1)}. \tag{4.7}$$

Due to consistency condition (4.1) $A(u)$ and $\bar{A}(u)$ are functions of one variable and do not depend on s. The symmetry between u and s allows one to construct similar objects which in turn do not depend on u. Functions $A(u)$ and $\bar{A}(u)$, in the r.h.s. of (4.4), (4.5) are the conservation lows of the s-dynamics.

Running ahead, let us note that the connection between ϕ and $\bar{\phi}$, $\bar{\phi}(u) = \phi(u-2)$, and its consequence $T_{-1}(u) = 0$ (see (2.15)), simplifies eqs. (4.4)-(4.7). Putting $s = 0$ and using the b.c. $T_{-1}(u) = 0$, we find

$$A(u) = \bar{A}(u) = T_1(u). \tag{4.8}$$

Therefore, the following holds

$$T_s(u-1)T_1(u+s) = \phi(u+s-2)T_{s+1}(u) + \phi(u+s)T_{s-1}(u-2), \tag{4.9}$$

$$T_s(u+1)T_1(u-s) = \phi(u-s)T_{s+1}(u) + \phi(u-s-2)T_{s-1}(u+2), \tag{4.10}$$

$$\phi(u-2)Q(u+2) + \phi(u)Q(u-2) = T_1(u)Q(u). \tag{4.11}$$

These equalities are known as fusion relations [6], [37], [38] while eq. (4.11) is Baxter's T-Q-relation [1]. So Baxter's Q function and the T-Q-relation naturally appear in the context of the auxiliary linear problems for HBDE.

A general solution of the discrete Liouville equation (for arbitrary ϕ and $\bar{\phi}$) may be expressed through two independent functions $Q(u)$ and $\bar{Q}(u)$. One may follow the same lines developed for solving the continuous classical Liouville equation (see e.g. [39], [40] and references therein). Let us consider eq. (4.4) (resp., (4.5)) as a second order linear difference equation, where the function $A(u)$ ($\bar{A}(u)$) is determined from the initial data. Let $R(u)$ (resp., $\bar{R}(u)$) be a second (linearly independent) solution of eq. (4.4) (resp., (4.5)) normalized so that the wronskians are

$$W(u) = \begin{vmatrix} R(u) & Q(u) \\ R(u+2) & Q(u+2) \end{vmatrix} = \phi(u), \tag{4.12}$$

$$\bar{W}(u) = \begin{vmatrix} \bar{R}(u) & \bar{Q}(u) \\ \bar{R}(u+2) & \bar{Q}(u+2) \end{vmatrix} = \bar{\phi}(u+1). \tag{4.13}$$

Then the general solution of the eq. (4.1) is given in terms of Q and R:

$$T_s(u) = \begin{vmatrix} Q(u+s+1) & R(u+s+1) \\ \bar{Q}(u-s) & \bar{R}(u-s) \end{vmatrix}, \tag{4.14}$$

This formula is a particular case of the general determinant representation (2.25).

For any given $Q(u)$ and $\bar{Q}(u)$ the second solution $R(u)$ and $\bar{R}(u)$ (defined modulo a linear transformation $R(u) \to R(u) + \alpha Q(u)$) can be explicitly found from the first order recurrence relations (4.12), (4.13), if necessary. Let $Q(u_0)$ and $R(u_0)$ be initial values at $u = u_0$. Then, say, for even $r \geq 0$,

$$R(u_0+r) = Q(u_0+r)\left(-\sum_{j=1}^{r/2} \frac{\phi(u_0+2j-2)}{Q(u_0+2j)Q(u_0+2j-2)} + \frac{R(u_0)}{Q(u_0)}\right) \tag{4.15}$$

and so on for other r's and $\bar{R}(u)$.

Finally, one can express solution to eq. (4.1) through two independent functions $Q(u)$ and $\bar{Q}(u)$:

$$T_s(u+s-1) = Q(u+2s)\bar{Q}(u-1)\left(\frac{T_0(u-1)}{Q(u)\bar{Q}(u-1)} + \sum_{j=1}^{s}\frac{\phi(u+2j-2)}{Q(u+2j)Q(u+2j-2)}\right). \tag{4.16}$$

where $T_0(u)$ can be found from (4.16) by putting $s=0$:

$$-\frac{T_0(u-1)}{Q(u)\bar{Q}(u-1)} + \frac{T_0(u+1)}{Q(u+2)\bar{Q}(u+1)} = \frac{\phi(u)}{Q(u)Q(u+2)} - \frac{\bar{\phi}(u)}{\bar{Q}(u-1)\bar{Q}(u+1)}. \tag{4.17}$$

Note also the following useful representations:

$$A(u) = Q(u+2)R(u-2) - R(u+2)Q(u-2), \tag{4.18}$$

$$\bar{A}(u) = \bar{R}(u+3)\bar{Q}(u-1) - \bar{Q}(u+3)\bar{R}(u-1), \tag{4.19}$$

which are direct corollaries of (4.4), (4.5).

4.2 Equivalent forms of Baxter's equation

The key ingredient of the construction is Baxter's relation (4.11) and its "chiral" versions (4.4), (4.5). For completeness, we gather some other useful forms of them.

Consider first "chiral" linear equations (4.4), (4.5) (thus not implying any specific b.c. in s). Assuming that $T_s(u)$ obeys HBDE (4.1), one can represent eqs. (4.4), (4.5) in the form

$$\begin{vmatrix} T_s(u) & T_{s+1}(u-1) & Q(u+s+1) \\ T_{s+1}(u+1) & T_{s+2}(u) & Q(u+s+3) \\ T_{s+2}(u+2) & T_{s+3}(u+1) & Q(u+s+5) \end{vmatrix} = 0\,, \tag{4.20}$$

$$\begin{vmatrix} T_s(u) & T_{s+1}(u+1) & \bar{Q}(u-s) \\ T_{s+1}(u-1) & T_{s+2}(u) & \bar{Q}(u-s-2) \\ T_{s+2}(u-2) & T_{s+3}(u-1) & \bar{Q}(u-s-4) \end{vmatrix} = 0\,, \tag{4.21}$$

respectively. This representation can be straightforwardly extended to the A_{k-1}-case (see eqs. (5.37), (5.38)).

A factorized form of these difference equations is

$$\left(e^{2\partial_u} - \frac{\phi(u)Q(u-2)}{\phi(u-2)Q(u)}\right)\left(e^{2\partial_u} - \frac{Q(u)}{Q(u-2)}\right) X(u-2) = 0\,, \tag{4.22}$$

$$\left(e^{2\partial_u} - \frac{\bar{\phi}(u+2)\bar{Q}(u-1)}{\bar{\phi}(u)\bar{Q}(u+1)}\right)\left(e^{2\partial_u} - \frac{\bar{Q}(u+1)}{\bar{Q}(u-1)}\right) \bar{X}(u-1) = 0\,. \tag{4.23}$$

Here e^{∂_u} acts as the shift operator, $e^{\partial_u} f(u) = f(u+1)$, and $X(u)$ ($\bar{X}(u)$) stands for any linear combination of $Q(u)$, $R(u)$ ($\bar{Q}(u)$, $\bar{R}(u)$).

Specifying eqs. (4.20), (4.21) to the b.c. $T_{-1}(u) = 0$ (see (4.8)), we see that both of them turn into the equation

$$\sum_{a=0}^{2}(-1)^a T_1^a(u+a-1)X(u+2a-2) = 0 \tag{4.24}$$

that is Baxter's relation (4.11). Furthermore, the difference operator in (4.24) admits a factorization of the form (4.22):

$$\sum_{a=0}^{2}(-1)^a T_1^a(u+a-1)e^{2a\partial_u} = \left(e^{2\partial_u} - \frac{\phi(u)Q(u-2)}{\phi(u-2)Q(u)}\right)\left(e^{2\partial_u} - \frac{Q(u)}{Q(u-2)}\right), \tag{4.25}$$

which is equivalent to the well known formula for $T_1(u)$ in terms of $Q(u)$.

4.3 Double-Bloch solutions to Baxter's equation

In this section we formulate the analytic properties of solutions to Baxter's functional relation (4.4) that are relevant to models on finite lattices.

First let us transform Baxter's relation to a difference equation with *elliptic* (i.e. double-periodic with periods $2\omega_1/\eta$, $2\omega_2/\eta$) coefficients.

The formal substitution

$$\tilde{\Psi}(u) = \frac{Q(u)P(u)}{\phi(u-2)} \tag{4.26}$$

with a (as yet not specified) function $P(u)$ yields

$$\tilde{\Psi}(u+2) + \frac{P(u+2)\phi(u-4)}{P(u-2)\phi(u-2)}\tilde{\Psi}(u-2) = \frac{A(u)P(u+2)}{\phi(u)P(u)}\tilde{\Psi}(u)\,. \tag{4.27}$$

Below we restrict ourselves to the case when the degree N of the elliptic polynomial $\phi(u)$ (1.3) is *even*. Then for any $P(u)$ of the form

$$P(u) = \prod_{j=1}^{N/2} \sigma(\eta(u-p_j)) \tag{4.28}$$

with arbitrary p_j the coefficients in (4.27) are elliptic functions. Indeed, for the coefficient in front of $\tilde{\Psi}(u-2)$ this is obvious. As for the coefficient in the r.h.s. of (4.27), its double-periodicity follows from the "sum rule" (2.8).

Let us represent $\phi(u)$ in the form

$$\phi(u) = \phi_0(u)\phi_1(u)\,, \tag{4.29}$$

where $\phi_0(u)$, $\phi_1(u)$ are elliptic polynomials of degree $N/2$ (of course for $N > 2$ there are many ways to do that). Specifying $P(u)$ as

$$P(u) = \phi_1(u-2)\,, \tag{4.30}$$

we rewrite (4.27) in the form

$$\Psi(u+2)+\frac{\phi_0(u-4)\phi_1(u)}{\phi_0(u-2)\phi_1(u-2)}\Psi(u-2)=\frac{A(u)}{\phi_0(u)\phi_1(u-2)}\Psi(u)\,, \tag{4.31}$$

where

$$\Psi(u)=\frac{Q(u)}{\phi_0(u-2)}\,. \tag{4.32}$$

Now, the coefficients in eq. (4.31) being double-periodic, it is natural to consider its *double-Bloch solutions.* A meromorphic function $f(x)$ is said to be *double-Bloch* if it obeys the following monodromy properties:

$$f(x+2\omega_\alpha)=B_\alpha f(x), \qquad \alpha=1,2. \tag{4.33}$$

The complex numbers B_α are called *Bloch multipliers.* It is easy to see that any double-Bloch function can be represented as a linear combination of elementary ones:

$$f(x)=\sum_{i=1}^{M}c_i\Phi(x-x_i,z)\kappa^{x/\eta}, \tag{4.34}$$

where [16]

$$\Phi(x,z)=\frac{\sigma(z+x+\eta)}{\sigma(z+\eta)\sigma(x)}\left[\frac{\sigma(z-\eta)}{\sigma(z+\eta)}\right]^{x/(2\eta)}. \tag{4.35}$$

and complex parameters z and κ are related by

$$B_\alpha=\kappa^{2\omega_\alpha/\eta}\exp(2\zeta(\omega_\alpha)(z+\eta))\left(\frac{\sigma(z-\eta)}{\sigma(z+\eta)}\right)^{\omega_\alpha/\eta} \tag{4.36}$$

($\zeta(x)=\sigma'(x)/\sigma(x)$ is the Weierstrass ζ-function). Considered as a function of z, $\Phi(x,z)$ is double-periodic:

$$\Phi(x,z+2\omega_\alpha)=\Phi(x,z).$$

For general values of x one can define a single-valued branch of $\Phi(x,z)$ by cutting the elliptic curve between the points $z=\pm\eta$. In the fundamental domain of the lattice defined by $2\omega_\alpha$ the function $\Phi(x,z)$ has a unique pole at the point $x=0$:

$$\Phi(x,z)=\frac{1}{x}+O(1)\,.$$

Coming back to the variable $u=x/\eta$, one can formulate the double-Bloch property of the function $\Psi(u)$ (4.32) in terms of its numerator $Q(u)$. It follows from (4.34) that the general form of $Q(u)$ is

$$Q(u)=Q(u;\nu)=e^{\nu\eta u}\prod_{j=1}^{M}\sigma(\eta(u-u_j))\,, \tag{4.37}$$

where $M=N/2$ and ν determines Bloch multipliers.

For the trigonometric and rational degeneration of eqs. (4.4), (4.31), (4.37) the meaning of ν is quite clear: it plays the role of the "boundary phase" for twisted b.c. in the horizontal (auxiliary) direction. For each ν eq. (4.11) has a solution of the form (4.37). The corresponding value of $T_1(u) = A(u)$ depends on ν as a parameter: $T_1(u) = T_1(u;\nu)$. If there exist $\nu \neq \nu'$ such that $T_1(u;\nu) = T_1(u;\nu')$, one may put $Q(u) = Q(u,\nu)$, $R(u) = Q(u;\nu')$. In the elliptic case the boundary phase in general is not compatible with integrability and so ν should have a different physical sense which is still unclear.

4.4 Bethe equations

It can be shown that for double-Bloch solutions the relation between ϕ and $\bar{\phi}$,

$$\bar{\phi}(u) = \phi(u-2),$$

implies

$$\bar{Q}(u) = Q(u-1), \qquad \bar{R}(u) = R(u-1)\,, \tag{4.38}$$

so that (see (4.14))

$$T_s(u) = \begin{vmatrix} Q(u+s+1) & R(u+s+1) \\ Q(u-s-1) & R(u-s-1) \end{vmatrix}\,. \tag{4.39}$$

It is clear that if $Q(u)$ and $R(u)$ are elliptic polynomials of degree $N/2$ multiplied by an exponential function (as in (4.37)), $T_s(u)$ has the desired general form (2.7).

Under condition (4.38) eq. (4.16) yields the familiar result:

$$T_s(u) = Q(u+s+1)Q(u-s-1)\sum_{j=0}^{s} \frac{\phi(u-s+2j-1)}{Q(u-s+2j+1)Q(u-s+2j-1)}\,. \tag{4.40}$$

This formula was obtained in [37], [38] by direct solution of the fusion recurrence relations (4.9), (4.10).

Let u_j and v_j, $j = 1,\ldots,M$, be zeros of $Q(u)$ and $R(u)$, respectively. Then, evaluating (4.12) at $u = u_j$, $u = u_j - 2$ and $u = v_j$, $u = v_j - 2$ we obtain the relations

$$\phi(u_j) = Q(u_j+2)R(u_j)\,, \qquad \phi(u_j-2) = -Q(u_j-2)R(u_j)\,, \tag{4.41}$$

whence it holds

$$\frac{\phi(u_j)}{\phi(u_j-2)} = -\frac{Q(u_j+2)}{Q(u_j-2)}\,, \tag{4.42}$$

$$\frac{\phi(v_j)}{\phi(v_j-2)} = -\frac{R(v_j+2)}{R(v_j-2)}\,. \tag{4.43}$$

Equations (4.42) are exactly the standard Bethe equations (1.2). We refer to equations (4.43) as *complementary Bethe equations.* It is easy to check that eqs. (4.42) ensure cancellation of poles in (4.40). A more standard way to derive Bethe equations (4.42), (4.43) is to substitute zeros of $Q(u)$ (or $R(u)$) directly into Baxter's relation (4.11). However, the wronskian relation (4.12) is somewhat more informative:

in addition to Bethe equations for u_j, v_j it provides the connection (4.41) between them. In the next section we derive the system of nested Bethe Ansatz equations starting from a proper generalization of eq. (4.12).

In the elliptic case degrees of the elliptic polynomials $Q(u)$, $R(u)$ (for even N) are equal to $M = N/2$ (provided η is incommensurable with the lattice spanned by ω_1, ω_2). This fact follows directly from Bethe equations (4.42), (4.43) by the same argument as in Sect. 3.5.

In trigonometric and rational cases there are no such strong restrictions on degrees M and $\tilde{M}$ of Q and R respectively. This is because a part of their zeros may tend to infinity thus reducing the degree. Whence M and $\tilde{M}$ can be arbitrary integers not exceeding N. However, they must be complementary to each other: $M + \tilde{M} = N$. The traditional choice is $M \le N/2$. In particular, the solution $Q(u) = 1$ $(M = 0)$ corresponds to the simplest reference state ("bare vacuum") of the model.

We already pointed out that the function $Q(u)$ originally introduced by Baxter (see e.g. [1] and references therein) emerged naturally in the context of the auxiliary linear problems. Let us mention that for models with the rational R-matrix this function can be treated as a limiting value of $T_s(u)$ as $s \to \infty$ [6]. Rational degeneration of eqs. (2.7), (4.37) gives

$$T_s(u) = A_s \prod_{j=1}^{N} (u - z_j^{(s)}) , \tag{4.44}$$

$$Q(u) = e^{\nu} \prod_{j=1}^{M} (u - u_j) , \tag{4.45}$$

where

$$A_s = \frac{\sinh(2\nu(s+1))}{\sinh(2\nu)} . \tag{4.46}$$

(The last expression follows from (4.40) by extracting the leading term as $u \to \infty$.) If the "boundary phase" $-i\nu\eta$ is real and $\nu \neq 0$, one has from (4.39):

$$Q(u) = \pm 2 \sinh(2\nu\eta) e^{\nu\eta u} \lim_{s \to \mp\infty} e^{2\nu\eta s} \frac{T_{\mp s-1}(u+s)}{(2s)^{N-M}} . \tag{4.47}$$

For each finite $s \ge 0$ $T_s(u)$ has N zeros but in the limit some of them tend to infinity. The degenerate case $\nu = 0$ needs special analysis since the limits $\nu \to 0$ and $s \to \infty$ do not commute.

Another remark on the rational case is in order. Fusion relations (4.9), (4.10) give "Bethe Ansatz like" equations for zeros of $T_s(u)$ (4.44). Substituting zeros of $T_s(u \pm 1)$ into (4.9), (4.10) and using (4.46) one finds:

$$\frac{\sinh(2\nu(s+2))}{\sinh(2\nu s)} \frac{\phi(z_j^{(s)} + s - 1)}{\phi(z_j^{(s)} + s + 1)} = - \prod_{k=1}^{N} \frac{z_j^{(s)} - z_k^{(s-1)} - 1}{z_j^{(s)} - z_k^{(s+1)} + 1} , \tag{4.48}$$

$$\frac{\sinh(2\nu(s+2))}{\sinh(2\nu s)} \frac{\phi(z_j^{(s)} - s - 1)}{\phi(z_j^{(s)} - s - 3)} = - \prod_{k=1}^{N} \frac{z_j^{(s)} - z_k^{(s-1)} + 1}{z_j^{(s)} - z_k^{(s+1)} - 1} . \tag{4.49}$$

These equations give the discrete dynamics of zeros in s. They are to be compared with dynamics of zeros of rational solutions of classical nonlinear equations [12], [15]. It is an interesting open problem to find elliptic analogues of eqs. (4.47)-(4.49).

5 THE A_{k-1}-CASE: DISCRETE TIME 2D TODA LATTICE

5.1 *General solution*

The family of bilinear equations arising as a result of the Bäcklund flow (Sect. 3.4),

$$F_t^a(s,u+1)F_t^a(s,u-1) - F_t^a(s+1,u)F_t^a(s-1,u) = F_t^{a+1}(s,u)F_t^{a-1}(s,u)\,, \quad (5.1)$$

and the corresponding linear problems,

$$F_{t+1}^{a+1}(s+1,u)F_t^a(s,u) - F_{t+1}^{a+1}(s,u-1)F_t^a(s+1,u+1) = F_{t+1}^a(s,u)F_t^{a+1}(s+1,u)\,, \quad (5.2)$$

$$F_{t+1}^a(s+1,u-1)F_t^a(s,u) - F_{t+1}^a(s,u)F_t^a(s+1,u-1) = F_{t+1}^{a+1}(s,u-1)F_t^{a-1}(s+1,u)\,, \quad (5.3)$$

subject to the b.c.

$$F_t^a(s,u) = 0 \quad \text{as } a<0 \text{ and } a>t \quad (5.4)$$

They may be solved simultaneously by using the determinant representation (2.25). The set of functions $F_t^a(s,u)$ entering these equations as illustrated by the following diagram:

$$\begin{array}{cccccccc} 0 & 1 & 0 & & & & & \\ 0 & F_1^0 & F_1^1 & 0 & & & & \\ 0 & F_2^0 & F_2^1 & F_2^2 & 0 & & & \\ \cdots & \cdots & \cdots & \cdots & \cdots & & & \\ 0 & F_t^0 & F_t^1 & F_t^2 & \cdots & F_t^t & 0 & \end{array} \quad (5.5)$$

(cf. (3.27)). Functions in each horizontal slice satisfy HBDE (5.1). By *level* of eq. (5.1) we understand the number t. Level 0 is introduced for later convenience. At the moment we do not assume any relations between solutions at different levels.

Determinant formula (2.25) gives the solution to these equations for each level t in terms of t arbitrary holomorphic[3] functions $h_t^{(j)}(u+s)$ and t arbitrary antiholomorphic functions $\bar{h}_t^{(j)}(u-s)$. This is illustrated by the diagrams:

$$\begin{array}{cccc} 1 & & & \\ h_1^{(1)} & h_1^{(2)} & & \\ h_2^{(1)} & h_2^{(2)} & h_2^{(3)} & \\ \cdots & \cdots & \cdots & \\ h_t^{(1)} & h_t^{(2)} & \cdots & h_t^{(t+1)} \end{array} \qquad \begin{array}{cccc} 1 & & & \\ \bar{h}_1^{(2)} & \bar{h}_1^{(1)} & & \\ \bar{h}_2^{(3)} & \bar{h}_2^{(2)} & \bar{h}_2^{(1)} & \\ \cdots & \cdots & \cdots & \\ \bar{h}_t^{(t+1)} & \bar{h}_t^{(t)} & \cdots & \bar{h}_t^{(1)} \end{array} \quad (5.6)$$

[3]Here we call *holomorphic* (*antiholomorphic*) a function of $u+s$ (resp., $u-s$).

Then, according to (2.25), the general solution to eq. (5.1) is

$$F^a_{t+1}(s,u) = \chi^a_t(u+s)\bar{\chi}^a_t(u-s) \begin{vmatrix} h_t^{(t+1)}(u+s-a+2) & \cdots & h_t^{(1)}(u+s-a+2) \\ h_t^{(t+1)}(u+s-a+4) & \cdots & h_t^{(1)}(u+s-a+4) \\ \cdots & \cdots & \cdots \\ h_t^{(t+1)}(u+s+a) & \cdots & h_t^{(1)}(u+s+a) \\ \bar{h}_t^{(t+1)}(u-s+a-t) & \cdots & \bar{h}_t^{(1)}(u-s+a-t) \\ \bar{h}_t^{(t+1)}(u-s+a-t+2) & \cdots & \bar{h}_t^{(1)}(u-s+a-t+2) \\ \cdots & \cdots & \cdots \\ \bar{h}_t^{(t+1)}(u-s-a+t) & \cdots & \bar{h}_t^{(1)}(u-s-a+t) \end{vmatrix}, \tag{5.7}$$

where $0 \le a \le t+1$ and the gauge functions $\chi^a_t(u)$, $\bar{\chi}^a_t(u)$ (introduced for normalization) satisfy the following equations:

$$\begin{aligned} \chi^a_t(u+1)\chi^a_t(u-1) &= \chi^{a+1}_t(u)\chi^{a-1}_t(u)\,, \\ \bar{\chi}^a_t(u+1)\bar{\chi}^a_t(u-1) &= \bar{\chi}^{a+1}_t(u)\bar{\chi}^{a-1}_t(u)\,. \end{aligned} \tag{5.8}$$

(cf. (2.5)). The size of the determinant is $t+1$. The first a rows contain functions $h_i^{(j)}$, the remaining $t-a+1$ rows contain $\bar{h}_i^{(j)}$. The arguments of $h_i^{(j)}$, $\bar{h}_i^{(j)}$ increase by 2, going down a column. Note that the determinant in (5.7) (without the prefactors) is a solution itself. At $a=0$ ($a=t+1$) it is an antiholomorphic (holomorphic) function. The required b.c., (3.19) can be satisfied by choosing appropriate gauge functions χ^a_t, $\bar{\chi}^a_t$.

5.2 *Canonical solution*

The general solution (5.7) gives the function $T^a_s(u) \equiv F^a_k(s,u)$ in terms of $2k$ functions of one variable h^i_{k-1} and $\bar{h}^i_{k-1}$. However, we need to represent the solution in terms of another set of $2k$ functions $Q_t(u)$ and $\bar{Q}_t(u)$ by virtue of conditions (5.4) in such a way that eqs. (5.2) (5.3) connecting two adjacent levels are fulfilled. We refer to this specification as the *canonical solution.*

To find it let us notice that at $a=0$ eq. (5.2) consist the holomorphic function $Q_t(u+s)$ and a function F^1. According to eq. (5.7), F^1 is given by the determinant of the matrix with the holomorphic entries $h_t^{(i)}(u+s+1)$ in the first row. Other rows contain antiholomorphic functions only, so $F^1_t(u,s) = \sum_i h_t^{(i)}(u+s+1)\eta_i(u-s)$, where $\eta_i(u-s)$ are corresponding minors of the matrix (5.7) at $a=1$. Substituting this into eq. (5.2) at $a=0$ and separating holomorphic and antiholomorphic functions one gets relations connecting $h_t^{(i)}, h_{t-1}^{(i)}$ and $Q_t(u), Q_{t+1}(u)$. Similar arguments can be applied to eq. (5.3) at another boundary $a = t+1$. As a result one obtains

$$h_t^{(1)}(u+s) = Q_t(u+s)\,, \qquad \bar{h}_t^{(1)}(u-s) = \bar{Q}_t(u-s) \tag{5.9}$$

and

$$Q_{t+1}(u-2)h_{t-1}^{(i)}(u) = \begin{vmatrix} h_t^{(i+1)}(u-2) & Q_t(u-2) \\ h_t^{(i+1)}(u) & Q_t(u) \end{vmatrix}, \tag{5.10}$$

$$\bar{Q}_{t+1}(u+1)\bar{h}_{t-1}^{(i)}(u+1) = \begin{vmatrix} \bar{h}_t^{(i+1)}(u) & \bar{Q}_t(u) \\ \bar{h}_t^{(i+1)}(u+2) & \bar{Q}_t(u+2) \end{vmatrix}, \tag{5.11}$$

where $1 \le i \le t$. Functions χ, $\bar{\chi}$ in front of the determinant (5.7) are then fixed as follows:

$$\begin{aligned} &\chi_t^a(u) = (-1)^{at}\left(\prod_{j=1}^{a-1} Q_{t+1}(u-a+2j)\right)^{-1}, \qquad a \ge 2\,, \\ &\chi_t^0(u) = Q_{t+1}(u)\,, \qquad \chi_t^1(u) = (-1)^t\,, \end{aligned} \tag{5.12}$$

$$\begin{aligned} &\bar{\chi}_t^a(u) = \left(\prod_{j=1}^{t-a} \bar{Q}_{t+1}(u+a-t+2j-1)\right)^{-1}, \qquad a \le t-1\,, \\ &\bar{\chi}_t^t(u) = 1\,, \qquad \bar{\chi}_t^{t+1}(u) = \bar{Q}_{t+1}(u)\,. \end{aligned} \tag{5.13}$$

It is easy to check that they do satisfy eqs. (5.8). The recursive relations (5.10), (5.11) allow one to determine functions $h_t^{(i)}$ and $\bar{h}_t^{(i)}$ starting from a given set of $Q_t(u)$. These formulas generalize wronskian relations (4.12), (4.13) to the A_{k-1}-case.

Let us also note that this construction resembles the Leznov-Saveliev solution [41] to the continuous 2DTL with open boundaries.

5.3 *The Bethe Ansatz and canonical solution*

The canonical solution of the previous section immediately leads to the nested Bethe Ansatz for elliptic solutions.

In this case all functions $h_t^{(i)}$, $\bar{h}_t^{(i)}$ are elliptic polynomials multiplied by an exponential function:

$$h_t^{(i)}(u) = a_t^{(i)} e^{\nu_t^{(i)}\eta u} \prod_{j=0}^{M_t^{(i)}} \sigma(\eta(u-u_j^{t,i}))\,, \tag{5.14}$$

$$\bar{h}_t^{(i)}(u) = \bar{a}_t^{(i)} e^{\bar{\nu}_t^{(i)}\eta u} \prod_{j=0}^{\bar{M}_t^{(i)}} \sigma(\eta(u-\bar{u}_j^{t,i}))\,. \tag{5.15}$$

This implies a number of constraints on their zeros.

The determinant in (5.10) should be divisible by $Q_{t+1}(u-2)$ and $h_{t-1}^{(i)}(u)$, whence

$$\frac{h_t^{(i+1)}(u_j^{t+1})}{h_t^{(i+1)}(u_j^{t+1}+2)} = \frac{Q_t(u_j^{t+1})}{Q_t(u_j^{t+1}+2)}\,, \tag{5.16}$$

$$\frac{h_t^{(i+1)}(u_j^{t-1,i})}{h_t^{(i+1)}(u_j^{t-1,i}-2)} = \frac{Q_t(u_j^{t-1,i})}{Q_t(u_j^{t-1,i}-2)}, \tag{5.17}$$

where $u_j^t \equiv u_j^{t,1}$. Furthermore, it is possible to get a closed system of constraints for the roots of $Q_t(u)$ only. Indeed, choosing $u = u_j^t$, $u = u_j^t + 2$ in (5.10), we get

$$Q_{t+1}(u_j^t - 2)Q_{t-1}(u_j^t) = -Q_t(u_j^t - 2)h_t^{(2)}(u_j^t)\,, \tag{5.18}$$

$$Q_{t+1}(u_j^t)Q_{t-1}(u_j^t + 2) = Q_t(u_j^t + 2)h_t^{(2)}(u_j^t)\,. \tag{5.19}$$

Dividing eq. (5.18) by eq. (5.19) we obtain the system of nested Bethe equations:

$$\frac{Q_{t-1}(u_j^t + 2)Q_t(u_j^t - 2)Q_{t+1}(u_j^t)}{Q_{t-1}(u_j^t)Q_t(u_j^t + 2)Q_{t+1}(u_j^t - 2)} = -1\,, \tag{5.20}$$

which coincides with (3.31) from Sect. 3.5.

Similar relations hold true for the $\bar{h}$-diagram:

$$\frac{\bar{h}_t^{(i+1)}(\bar{u}_j^{t+1}+1)}{\bar{h}_t^{(i+1)}(\bar{u}_j^{t+1}-1)} = \frac{\bar{Q}_t(\bar{u}_j^{t+1}+1)}{\bar{Q}_t(\bar{u}_j^{t+1}-1)}, \tag{5.21}$$

$$\frac{\bar{h}_t^{(i+1)}(\bar{u}_j^{t-1,i}+1)}{\bar{h}_t^{(i+1)}(\bar{u}_j^{t-1,i}-1)} = \frac{\bar{Q}_t(\bar{u}_j^{t-1,i}+1)}{\bar{Q}_t(\bar{u}_j^{t-1,i}-1)}, \tag{5.22}$$

$$\bar{Q}_{t+1}(\bar{u}_j^t + 1)\bar{Q}_{t-1}(\bar{u}_j^t + 1) = \bar{Q}_t(\bar{u}_j^t + 2)\bar{h}_t^{(2)}(\bar{u}_j^t)\,, \tag{5.23}$$

$$\bar{Q}_{t+1}(\bar{u}_j^t - 1)\bar{Q}_{t-1}(\bar{u}_j^t - 1) = -\bar{Q}_t(\bar{u}_j^t - 2)\bar{h}_t^{(2)}(\bar{u}_j^t)\,, \tag{5.24}$$

$$\frac{\bar{Q}_{t-1}(\bar{u}_j^t + 1)\bar{Q}_t(\bar{u}_j^t - 2)\bar{Q}_{t+1}(\bar{u}_j^t + 1)}{\bar{Q}_{t-1}(\bar{u}_j^t - 1)\bar{Q}_t(\bar{u}_j^t + 2)\bar{Q}_{t+1}(\bar{u}_j^t - 1)} = -1\,. \tag{5.25}$$

These conditions are sufficient to ensure that the canonical solution for $T_s^a(u)$ (i.e., for $F_k^a(s,u)$) has the required general form (2.7). To see this, take a generic Q-factor from the product (5.12), $(Q_{t+1}(u-a+2j))^{-1}$. It follows from (5.16) that at its poles the j-th and $j+1$-th rows of the determinant (5.7) become proportional. The same argument repeated for $\bar{Q}$-factors shows that $F_{t+1}^a(s,u)$ has no poles.

Finally, it is straightforward to see from (5.7) that the constraint $\bar{Q}_t(u) = Q_t(u-t)$ leads to condition (2.15) (for $-t \le s \le -1$ two rows of the determinant become equal).

To summarize, the solution goes as follows. First, one should find a solution to Bethe equations (3.31) thus getting a set of elliptic polynomials $Q_t(u)$, $t = 1, \ldots, k-1$, $Q_0(u) = 1$, $Q_k(u) = \phi(u)$ being a given function. To make the chain of equations finite, it is convenient to use the formal convention $Q_{-1}(u) = Q_{k+1}(u) = 0$. Second, one should solve step by step relations (5.10), (5.11) and find the functions $h_t^{(i)}(u)$, $\bar{h}_t^{(i)}(u)$. All these relations are of the same type as the wronskian relation (4.12) in the A_1-case: each of them is a linear inhomogeneous first order difference equation.

5.4 Conservation laws

The solution described in Sects. 5.2 and 5.3 provides compact determinant formulas for eigenvalues of quantum transfer matrices. It also provides determinant representations for conservation laws of the s-dynamics which generalize eqs. (4.6), (4.7) to the A_{k-1}-case. The generalization comes up in the form of eqs. (4.20), (4.21) and (4.24). The conservation laws (i.e., integrals of the s-dynamics) follow from the determinant representation (5.7) of the general solution to HBDE.

Let us consider $(C_k^a+1)\times(C_k^a+1)$-matrices

$$\mathcal{T}^a_{B,B'}(s,u)\equiv T^a_{s+B+B'}(u-s+B-B'),\qquad B,B'=1,\ldots,C_k^a+1\,, \tag{5.26}$$

$$\bar{\mathcal{T}}^a_{B,B'}(s,u)=T^a_{s-B-B'}(u+s+B-B'),\qquad B,B'=1,\ldots,C_k^a+1\,, \tag{5.27}$$

where C_k^a is the binomial coefficient. Let $\mathcal{T}^a[P|R](s,u)$ be minors of the matrix (5.26) with row P and column R removed (similarly for (5.27)).

Theorem 5.1 *Let $T_s^a(u)$ be the general solution to HBDE given by eq. (5.7). Then any ratio of the form*

$$I^{a,R}_{P,P'}(s,u)\equiv\frac{\mathcal{T}^a[P|R](s,u)}{\mathcal{T}^a[P'|R](s,u)} \tag{5.28}$$

does not depend on s. These quantities are integrals of the s-dynamics: $I^{a,R}_{P,P'}(s,u)=I^{a,R}_{P,P'}(u)$. Similarly, minors of the matrix (5.27) give in the same way a complimentary set of conservation laws[4].

A sketch of proof is as follows.

Consider the Laplace expansion of the determinant solution (5.7) with respect to the first a (holomorphic) rows:

$$T_s^a(u)=\sum_{P=1}^{C_k^a}\psi_P^a(u+s)\bar{\psi}_P^a(u-s) \tag{5.29}$$

Here P numbers (in an arbitrary order) sets of indices $(p_1,p_2,\ldots,p_a)$ such that $k\geq p_1>p_2>\ldots>p_a\geq 1$, $\psi_P^a(u+s)$ is minor of the matrix in eq. (5.7) constructed from first a rows and columns $p_1,\ldots,p_a$ (multiplied by $\chi^a_{k-1}(u+s)$), $\bar{\psi}_P^a(u-s)$ is the complimentary minor (multiplied by $\bar{\chi}^a_{k-1}(u-s)$).

Substitute R-th column of the matrix (5.26) by the column vector with components $\psi_P^a(u+2B)$, $B=1,\ldots,C_k^a+1$. The matrix obtained this way (let us call it $(\mathcal{T}^{a;R,P})_{B,B'}$) depends on $R=1,\ldots,C_k^a+1$, $P=1,\ldots,C_k^a$ and $a=1,\ldots,k-1$. The "complementary" matrix $(\bar{\mathcal{T}}^{a;R,P})_{B,B'}$ is defined by the similar substitution of the column vector $\bar{\psi}_P^a(u+2B)$, $B=1,\ldots,C_k^a+1$, into the matrix (5.27).

Lemma 5.1 *Determinants of all the four matrices introduced above vanish:*

$$\det(\mathcal{T}^a)=\det(\bar{\mathcal{T}}^a)=\det(\mathcal{T}^{a;R,P})=\det(\bar{\mathcal{T}}^{a;R,P})=0\,. \tag{5.30}$$

[4]compare with (4.6), (4.7).

The proof follows from the Laplace expansion (5.29). From this representation it is obvious that $C_k^a + 1$ columns of the matrices in (5.30) are linearly dependent. This identity is valid for arbitrary functions $h_t^{(i)}(u+s), \bar{h}_t^{(i)}(u-s)$ in eq. (5.7).

The conservation laws immediately follow from these identities. Indeed, let us rewrite the determinant of the matrix $\mathcal{T}^{a;R,P}$ as a linear combination of entries of the R-th column:

$$\det(\mathcal{T}^{a;R,P}) = \sum_{B'=1}^{C_k^a+1} (-1)^{B'+R} \psi_P^a(u+2B')\mathcal{T}^a[B'|R](s,u) = 0\,. \tag{5.31}$$

Dividing by $\mathcal{T}^a[P'|R](s,u)$, we get, using the notation (5.28):

$$\sum_{B'=1, B'\neq P'}^{C_k^a+1} (-1)^{B'} \psi_P^a(u+2B') I_{B',P'}^{a,R}(s,u) = (-1)^{P'+1}\psi_P^a(u+2P')\,. \tag{5.32}$$

The latter identity is a system of C_k^a linear equations for C_k^a quantities $I_{1,P'}^{a,R}(s,u)$, $I_{2,P'}^{a,R}(s,u)$, ..., $I_{P'-1,P'}^{a,R}(s,u)$, $I_{P'+1,P'}^{a,R}(s,u)$, ..., $I_{C_k^a+1,P'}^{a,R}(s,u)$. In the case of general position wronskians of the functions $\psi_P^a(u)$ is nonzero, whence system (5.32) has a unique solution for $I_{P,P'}^{a,R}(s,u)$. The coefficients of the system do not depend on s. Therefore, $I_{P,P'}^{a,R}(s,u)$ are s-independent too. Similar arguments are applied to minors of the matrix (5.27).

Another form of eq. (5.31) may be obtained by multiplication its l.h.s. by $\bar{\psi}_P^a(u-2s)$ and summation over P. This yields

$$\sum_{B=1}^{C_k^a+1} (-1)^B T_{s+B}^a(u-s+B)\mathcal{T}^a[B|R](s,u) = 0\,, \tag{5.33}$$

which is a difference equation for $T_s^a(u)$ as a function of the "holomorphic" variable $u+s$ with fixed $u-s$.

5.5 Generalized Baxter's relations

Equation (5.31) can be considered as a linear difference equation for a function $\psi^a(u)$ having C_k^a linearly independent solutions $\psi_P^a(u)$. It provides the A_{k-1}-generalization of Baxter's relations (4.4), (4.5). This generalization comes up in the form of eqs. (4.20), (4.21) and (4.24).

The simplest cases are $a=1$ and $a=k-1$. Then there are $k+1$ terms in the sum (5.31). Furthermore, it is easy to see that

$$\psi_i^1(u) = h_{k-1}^{(i)}(u+1), \qquad \bar{\psi}_i^{k-1}(u) = \bar{h}_{k-1}^{(i)}(u)\,. \tag{5.34}$$

Then eq. (5.31) and a similar equation for antiholomorphic parts read:

$$\sum_{j=1}^{k+1} (-1)^j h^{(i)}(u+2j+1)\mathcal{T}^1[j|k+1](s,u) = 0\,, \tag{5.35}$$

$$\sum_{j=1}^{k+1}(-1)^j \bar{h}^{(i)}(u+2j)\bar{\mathcal{T}}^{k-1}[j|k+1](s,u)=0\,, \tag{5.36}$$

where we put $R = k+1$ for simplicity. These formulas may be understood as linear difference equations of order k. Indeed, eq. (5.35) can be rewritten as the following equation for a function $X(u)$:

$$\begin{vmatrix} T_s^1(u) & T_{s+1}^1(u-1) & \cdots & T_{s+k-1}^1(u-k+1) & X(u+s+1) \\ T_{s+1}^1(u+1) & T_{s+2}^1(u) & \cdots & T_{s+k}^1(u-k+2) & X(u+s+3) \\ \cdots & \cdots & \cdots & \cdots & \cdots \\ T_{s+k}^1(u+k) & T_{s+k+1}^1(u+k-1) & \cdots & T_{s+2k-1}^1(u+1) & X(u+s+2k+1) \end{vmatrix} = 0\,, \tag{5.37}$$

This equations has k solutions $h_{k-1}^{(i)}(u)$, $i=1,\ldots,k$. One of them is $Q_{k-1}\equiv h_{k-1}^{(1)}(u)$ (see eq.(5.9)). Similar equation (5.36) for the antiholomorphic parts,

$$\begin{vmatrix} T_s^{k-1}(u) & T_{s-1}^{k-1}(u-1) & \cdots & T_{s-k+1}^{k-1}(u-k+1) & \bar{X}(u-s) \\ T_{s-1}^{k-1}(u+1) & T_{s-2}^{k-1}(u) & \cdots & T_{s-k}^{k-1}(u-k+2) & \bar{X}(u-s+2) \\ \cdots & \cdots & \cdots & \cdots & \cdots \\ T_{s-k}^{k-1}(u+k) & T_{s-k-1}^{k-1}(u+k-1) & \cdots & T_{s-2k+1}^{k-1}(u+1) & \bar{X}(u-s+2k) \end{vmatrix} = 0\,, \tag{5.38}$$

has k solutions $\bar{h}_{k-1}^{(i)}(u)$, $i=1,\ldots,k$. One of them is $\bar{Q}_{k-1}\equiv \bar{h}_{k-1}^{(1)}(u)$.

Difference equations (5.37), (5.38) can be rewritten in the factorized form. This fact follows from a more general statement. Fix an arbitrary level k and set $T_s^a(u) = F_k^a(s,u)$ (as in Sect. 3). Then for each $j=0,1,\ldots,k-1$ it holds:

$$(e^{\partial_s+\partial_u}-R_{j+1}^{(j)}(s,u))(e^{\partial_s+\partial_u}-R_j^{(j)}(s,u))\ldots(e^{\partial_s+\partial_u}-R_1^{(j)}(s,u))F^{k-1-j}(s,u)=0\,, \tag{5.39}$$

$$(e^{\partial_s-\partial_u}-\bar{R}_{j+1}^{(j)}(s,u))(e^{\partial_s-\partial_u}-\bar{R}_j^{(j)}(s,u))\ldots(e^{\partial_s-\partial_u}-\bar{R}_1^{(j)}(s,u))F^{j}(s,u)=0\,, \tag{5.40}$$

where

$$R_i^{(k-1-j)}(s,u)=\frac{T_{s+i-1}^{j}(u+i-1)T_{s+i-2}^{j+i-1}(u-1)T_{s+i}^{j+i}(u)}{T_{s+i-2}^{j}(u+i-2)T_{s+i-1}^{j+i-1}(u)T_{s+i-1}^{j+i}(u-1)}, \tag{5.41}$$

$$\bar{R}_i^{(j)}(s,u)=\frac{T_{s+i-1}^{j+1}(u-l)T_{s+l}^{j-i+1}(u-1)T_{s+i-2}^{j-i+2}(u)}{T_{s+i-2}^{j+1}(u-i+1)T_{s+i-1}^{j-i+1}(u)T_{s+i-1}^{j-i+2}(u-1)} \tag{5.42}$$

The proof is by induction. At $j=0$ eq. (5.39) turns into

$$\left(e^{\partial_s+\partial_u}-\frac{T_{s+1}^k(u)}{T_s^k(u-1)}\right)F^{k-1}(s,u)=0\,.$$

This means that $F^{k-1}(s,u)$ does not depend on $u+s$. Further,

$$F^a(s+1,u) = -\frac{T^a_s(u-1)}{T^{a-1}_s(u)}\left(e^{\partial_s+\partial_u} - \frac{T^a_{s+1}(u)}{T^a_s(u-1)}\right)F^{a-1}(s,u)\,, \tag{5.43}$$

(see (3.8)). The inductive step is then straightforward. The proof of (5.40) is absolutely identical.

Now, putting $j=k-1$ we get the following difference equations in one variable:

$$\begin{aligned}(e^{2\partial_u+\partial_s} - R^{(k-1)}_k(s,u-s))(e^{2\partial_u+\partial_s} - R^{(k-1)}_{k-1}(s,u-s))\\ \dots(e^{2\partial_u+\partial_s} - R^{(k-1)}_1(s,u-s))Q_{k-1}(u) = 0\,,\end{aligned} \tag{5.44}$$

$$\begin{aligned}(e^{-2\partial_u+\partial_s} - \bar{R}^{(k-1)}_k(s,u+s))(e^{-2\partial_u+\partial_s} - \bar{R}^{(k-1)}_{k-1}(s,u+s))\\ \dots(e^{-2\partial_u+\partial_s} - \bar{R}^{(k-1)}_1(s,u+s))\bar{Q}_{k-1}(u) = 0\,.\end{aligned} \tag{5.45}$$

Note that operators $e^{\pm\partial_s}$ act only on the coefficient functions in (5.44), (5.45). These equations provide a version of the discrete Miura transformation of generalized Baxter's operators, which is different from the one discussed in the Ref. [43] (see also below).

Coming back to eq. (5.31) and using relations (5.10), (5.11), one gets after some algebra:

$$\psi^{k-1}_k(u) = h^{(1)}_1(u+k-1) = Q_1(u+k-1)\,, \tag{5.46}$$

$$\bar{\psi}^1_k(u) = \bar{h}^{(1)}_1(u) = \bar{Q}_1(u) \tag{5.47}$$

(the proof is straightforward but too lengthy to present it here).

Then, in complete analogy with eqs. (5.37), (5.38), one obtains from (5.31) the following difference equations:

$$\begin{vmatrix} T^{k-1}_s(u) & T^{k-1}_{s+1}(u-1) & \dots & T^{k-1}_{s+k-1}(u-k+1) & X(u+s+k-1) \\ T^{k-1}_{s+1}(u+1) & T^{k-1}_{s+2}(u) & \dots & T^{k-1}_{s+k}(u-k+2) & X(u+s+k+1) \\ \dots & \dots & \dots & \dots & \dots \\ T^{k-1}_{s+k}(u+k) & T^{k-1}_{s+k+1}(u+k-1) & \dots & T^{k-1}_{s+2k-1}(u+1) & X(u+s+3k-1) \end{vmatrix} = 0\,, \tag{5.48}$$

$$\begin{vmatrix} T^1_s(u) & T^1_{s-1}(u-1) & \dots & T^1_{s-k+1}(u-k+1) & \bar{X}(u-s) \\ T^1_{s-1}(u+1) & T^1_{s-2}(u) & \dots & T^1_{s-k}(u-k+2) & \bar{X}(u-s+2) \\ \dots & \dots & \dots & \dots & \dots \\ T^1_{s-k}(u+k) & T^1_{s-k-1}(u+k-1) & \dots & T^1_{s-2k+1}(u+1) & \bar{X}(u-s+2k) \end{vmatrix} = 0 \tag{5.49}$$

to which $Q_1(u)$ (resp., $\bar{Q}_1(u)$) is a solution. The other $k-1$ linearly independent solutions to eq. (5.48) (resp., (5.49)) are other algebraic complements of the last (first) line of the matrix in eq. (5.7) at $a = k-1$ ($a = 1$) multipiled by χ_{k-1}^{k-1} ($\bar{\chi}_{k-1}^{1}$).

Further specification follows from imposing constraints (3.12) which ensure conditions (2.4) forced by the usual Bethe Ansatz. One can see that under these conditions eqs. (5.48) and (5.49) become the same. Further, substituting a particular value of s, $s = -k$, into, say, eqs. (5.48), (5.37), one gets the following difference equations:

$$\sum_{a=0}^{k}(-1)^a T_1^a(u+a-1)Q_1(u+2a-2) = 0\,, \tag{5.50}$$

$$\sum_{a=0}^{k}(-1)^a \frac{T_1^a(u-a-1)}{\phi(u-2a)\phi(u-2a-2)} Q_{k-1}(u-2a) = 0 \tag{5.51}$$

(we remind the reader that $\phi(u) \equiv Q_k(u)$). The latter equation can be obtained directly from the determinant formula (5.7): notice that under conditions (2.4) the determinants in eq. (5.7) become minors of the matrix $h_{k-1}^{(i)}(u-2k+2j)$, where i numbers columns running from 1 to k, j numbers lines and runs from 0 to k skipping the value $k-a$. Taking care of the prefactors in eq. (5.7) and recalling that $h_{k-1}^{(1)}(u) = Q_{k-1}(u)$, one gets eq. (5.51). These formulas give a generalization of the Baxter equations (4.4), (4.5), (4.11).

At last, we are to identify our Q_t's with Q_t's from the usual nested Bethe Ansatz solution. This is achieved by factorization of the difference operator in (5.50) in terms of $Q_t(u)$:

$$\begin{aligned}\sum_{a=0}^{k}(-1)^a T_1^a(u+a-1)e^{2a\partial_u} &= \left(e^{2\partial_u} - \frac{Q_k(u)Q_{k-1}(u-2)}{Q_k(u-2)Q_{k-1}(u)}\right)\cdots \\ &\times \left(e^{2\partial_u} - \frac{Q_2(u)Q_1(u-2)}{Q_2(u-2)Q_1(u)}\right)\left(e^{2\partial_u} - \frac{Q_1(u)}{Q_1(u-2)}\right).\end{aligned} \tag{5.52}$$

This formula (proved by a straightforward calculation using inductive argument) coincides with the one known in the literature (see e.g. [38], [44]). It yields $T_1^a(u)$ in terms of elliptic polynomials Q_t with roots constrained by the nested Bethe Ansatz equations. They ensure cancellation of poles in the r.h.s. The l.h.s. of eq. (5.52) is known as the generating function for $T_1^a(u)$; $T_s^a(u)$ for $s > 1$ can be found with the help of determinant formula (2.24).

6 Regular elliptic solutions of the HBDE and RS system in discrete time

In this section we study the class of elliptic solutions to HBDE for which the number of zeros M_t of the τ-function does not depend on t. We call them *elliptic solutions of the regular type* since they have a smooth continuum limit. Although it has been argued in the previous section that the situation of interest for the Bethe Ansatz is quite opposite, we find it useful to briefly discuss this class of solutions.

It is convenient to slightly change the notation: $\tau^{l,m}(x) \equiv \tau_u(-m,-l), x \equiv u\eta$. HBDE (1.8) acquires the form

$$\tau^{l+1,m}(x)\tau^{l,m+1}(x) - \tau^{l+1,m+1}(x)\tau^{l,m}(x) = \tau^{l+1,m}(x+\eta)\tau^{l,m+1}(x-\eta). \qquad (6.1)$$

We are interested in solutions that are elliptic polynomials in x,

$$\tau^{l,m}(x) = \prod_{j=1}^{M} \sigma(x - x_j^{l,m}). \qquad (6.2)$$

The main goal of this section is to describe this class of solutions in a systematic way and, in particular, to prove that *all* the elliptic solutions of regular type are finite-gap.

The auxiliary linear problems (3.5) look as follows:

$$\Psi^{l,m+1}(x) = \Psi^{l,m}(x+\eta) + \frac{\tau^{l,m}(x)\tau^{l,m+1}(x+\eta)}{\tau^{l,m+1}(x)\tau^{l,m}(x+\eta)}\Psi^{l,m}(x), \qquad (6.3)$$

$$\Psi^{l+1,m}(x) = \Psi^{l,m}(x) + \frac{\tau^{l,m}(x-\eta)\tau^{l+1,m}(x+\eta)}{\tau^{l+1,m}(x)\tau^{l,m}(x)}\Psi^{l,m}(x-\eta). \qquad (6.4)$$

(The notation is correspondingly changed: $\Psi^{l,m}(u\eta) \equiv \psi_u(-m,-l)$.) The coefficients are elliptic functions of x. Similarly to the case of the Calogero-Moser model and its spin generalizations [14], [15] the dynamics of their poles is determined by the fact that equations (6.3), (6.4) have infinite number of double-Bloch solutions (Sect. 4).

The "gauge transformation" $f(x) \to \tilde{f}(x) = f(x)e^{ax}$ (a is an arbitrary constant) does not change poles of any function and transforms a double-Bloch function into another double-Bloch function. If B_α are Bloch multipliers for f, then the Bloch multipliers for $\tilde{f}$ are $\tilde{B}_1 = B_1 e^{2a\omega_1}$, $\tilde{B}_2 = B_2 e^{2a\omega_2}$, where ω_1, ω_2 are quasiperiods of the σ-function. Two pairs of Bloch multipliers are said to be *equivalent* if they are connected by this relation with some a (or by the equivalent condition that the product $B_1^{\omega_2}B_2^{-\omega_1}$ is the same for both pairs).

Consider first eq. (6.3). Since l enters as a parameter, not a variable, we omit it for simplicity of the notation (e.g. $x_j^{l,m} \to x_j^m$).

Theorem 6.1 *Eq. (6.3) has an infinite number of linearly independent double-Bloch solutions with simple poles at the points x_i^m and equivalent Bloch multipliers if and only if x_i^m satisfy the system of equations*

$$\prod_{j=1}^{M} \frac{\sigma(x_i^m - x_j^{m+1})\sigma(x_i^m - x_j^m - \eta)\sigma(x_i^m - x_j^{m-1} + \eta)}{\sigma(x_i^m - x_j^{m+1} - \eta)\sigma(x_i^m - x_j^m + \eta)\sigma(x_i^m - x_j^{m-1})} = -1. \qquad (6.5)$$

All these solutions can be represented in the form

$$\Psi^m(x) = \sum_{i=1}^{M} c_i(m,z,\kappa)\Phi(x - x_i^m, z)\kappa^{x/\eta} \qquad (6.6)$$

($\Phi(x,z)$ is defined in (4.35)). The set of corresponding pairs (z,κ) are parametrized by points of an algebraic curve defined by the equation of the form

$$R(\kappa, z) = \kappa^M + \sum_{i=1}^{M} r_i(z)\kappa^{M-i} = 0\,. \tag{6.7}$$

Sketch of proof. We omit the detailed proof since it is almost identical to the proof of the corresponding theorem in [16] and only present the part of it which provides the Lax representation for eq. (6.5).

Let us substitute the function $\Psi^m(x)$ of the form (6.6) into eq. (6.3). The cancellation of poles at $x = x_i^m - \eta$ and $x = x_i^{m+1}$ gives the conditions

$$\kappa c_i(m, z, \kappa) + \lambda_i(m) \sum_{j=1}^{M} c_j(m, z, \kappa)\Phi(x_i^m - x_j^m - \eta, z) = 0\,, \tag{6.8}$$

$$c_i(m+1, z, \kappa) = \mu_i(m) \sum_{j=1}^{M} c_j(m, z, \kappa)\Phi(x_i^{m+1} - x_j^m, z)\,, \tag{6.9}$$

where

$$\lambda_i(m) = \frac{\prod_{s=1}^{M} \sigma(x_i^m - x_s^m - \eta)\sigma(x_i^m - x_s^{m+1})}{\prod_{s=1,\neq i}^{M} \sigma(x_i^m - x_s^m) \prod_{s=1}^{M} \sigma(x_i^m - x_s^{m+1} - \eta)}\,, \tag{6.10}$$

$$\mu_i(m) = \frac{\prod_{s=1}^{M} \sigma(x_i^{m+1} - x_s^{m+1} + \eta)\sigma(x_i^{m+1} - x_s^m)}{\prod_{s=1,\neq i}^{M} \sigma(x_i^{m+1} - x_s^{m+1}) \prod_{s=1}^{M} \sigma(x_i^{m+1} - x_s^m + \eta)}\,. \tag{6.11}$$

Introducing a vector $C(m)$ with components $c_i(m, z, \kappa)$ we can rewrite these conditions in the form

$$(\mathcal{L}(m) + \kappa I)C(m) = 0\,, \tag{6.12}$$

$$C(m+1) = \mathcal{M}(m)C(m)\,, \tag{6.13}$$

where I is the unit matrix. Entries of the matrices $\mathcal{L}(m)$ and $\mathcal{M}(m)$ are:

$$\mathcal{L}_{ij}(m) = \lambda_i(m)\Phi(x_i^m - x_j^m - \eta, z), \tag{6.14}$$

$$\mathcal{M}_{ij}(m) = \mu_i(m)\Phi(x_i^{m+1} - x_j^m, z). \tag{6.15}$$

The compatibility condition of (6.12) and (6.13),

$$\mathcal{L}(m+1)\mathcal{M}(m) = \mathcal{M}(m)\mathcal{L}(m) \tag{6.16}$$

is the discrete Lax equation.

By the direct commutation of the matrices $\mathcal{L}$, $\mathcal{M}$ (making use of some non-trivial identities for the function $\Phi(x,z)$ which are omitted) it can be shown that for the matrices $\mathcal{L}$ and $\mathcal{M}$ defined by eqs. (6.14), (6.10) and (6.15), (6.11) respectively, the discrete Lax equation (6.16) holds if and only if the x_i^m satisfy eqs. (6.5). It is worthwhile to remark that in terms of $\lambda_i(m)$, $\mu_i(m)$ equations (6.5) take the form

$$\lambda_i(m+1) = -\mu_i(m), \qquad i = 1, \ldots, M\,. \tag{6.17}$$

Eq. (6.12) implies that

$$R(\kappa, z) \equiv \det(\mathcal{L}(m) + \kappa I) = 0\,. \tag{6.18}$$

The coefficients of $R(\kappa, z)$ do not depend on m due to (6.16). This equation defines an algebraic curve (6.7) realized as a ramified covering of the elliptic curve.

Solutions to eq. (6.5) are implicitly given by the equation

$$\Theta(\vec{U}x_i^{l,m} + \vec{U}_+ l + \vec{U}_- m + \vec{Z}) = 0\,, \tag{6.19}$$

where the Riemann theta-function $\Theta(\vec{X})$ corresponds to the spectral curve (6.7), (6.18), components of the vectors $\vec{U}$, $\vec{U}_+$, $\vec{U}_-$ are periods of certain dipole differentials on the curve, $\vec{Z}$ is an arbitrary vector. Elliptic solutions are characterized by the following property: $2\omega_i\vec{U}$, $i = 1, 2$, belongs to the lattice of periods of holomorphic differentials on the curve. The matrix $\mathcal{L}(m) = \mathcal{L}(l, m)$ is defined by fixing $x_j^{l_0,m_0}$, $x_j^{l_0,m_0+1}$, $i = 1, \ldots, M$. These Cauchy data uniquely define the curve and the vectors $\vec{U}$, $\vec{U}_+$, $\vec{U}_-$ and $\vec{Z}$ in eq. (6.19). The curve and vectors $\vec{U}$, $\vec{U}_+$, $\vec{U}_-$ do not depend on the choice of l_0, m_0. According to eq. (6.19), the vector $\vec{Z}$ depends linearly on this choice and its components are thus angle-type variables.

The same analysis can be repeated for the second linear problem (6.4). Now m enters as a parameter and we set $x^{l,m} \to \hat{x}_i^l$ for simplicity. The theorem is literally the same, the equations of motion for the poles being

$$\prod_{j=1}^{M} \frac{\sigma(\hat{x}_i^l - \hat{x}_j^{l+1} + \eta)\sigma(\hat{x}_i^l - \hat{x}_j^l - \eta)\sigma(\hat{x}_i^l - \hat{x}_j^{l-1})}{\sigma(\hat{x}_i^l - \hat{x}_j^{l+1})\sigma(\hat{x}_i^l - \hat{x}_j^l + \eta)\sigma(\hat{x}_i^l - \hat{x}_j^{l-1} - \eta)} = -1\,. \tag{6.20}$$

The corresponding discrete Lax equation is

$$\hat{\mathcal{L}}(l+1)\hat{\mathcal{M}}(l) = \hat{\mathcal{M}}(l)\hat{\mathcal{L}}(l)\,, \tag{6.21}$$

where[5]

$$\hat{\mathcal{L}}_{ij}(l) = \hat{\lambda}_i(l)\Phi(\hat{x}_i^l - \hat{x}_j^l - \eta, z), \tag{6.22}$$

$$\hat{\mathcal{M}}_{ij}(l) = \hat{\mu}_i(l)\Phi(\hat{x}_i^{l+1} - \hat{x}_j^l - \eta, z), \tag{6.23}$$

and

$$\hat{\lambda}_i(l) = \frac{\prod_{s=1}^M \sigma(\hat{x}_i^l - \hat{x}_s^l - \eta)\sigma(\hat{x}_i^l - \hat{x}_s^{l+1} + \eta)}{\prod_{s=1,\neq i}^M \sigma(\hat{x}_i^l - \hat{x}_s^l)\prod_{s=1}^M \sigma(\hat{x}_i^l - \hat{x}_s^{l+1})}\,, \tag{6.24}$$

$$\hat{\mu}_i(l) = \frac{\prod_{s=1}^M \sigma(\hat{x}_i^{l+1} - \hat{x}_s^{l+1} + \eta)\sigma(\hat{x}_i^{l+1} - \hat{x}_s^l - \eta)}{\prod_{s=1,\neq i}^M \sigma(\hat{x}_i^{l+1} - \hat{x}_s^{l+1})\prod_{s=1}^M \sigma(\hat{x}_i^{l+1} - \hat{x}_s^l)}\,. \tag{6.25}$$

All these formulas can be obtained from (6.5), (6.10)-(6.15) by the formal substitutions $x_i^m \to \hat{x}_i^l$, $x_i^{m\pm1} \to \hat{x}_i^{l\pm1} \mp \eta$. According to the comment after eq. (6.19), the Cauchy data for the l-flow $x_j^{l_0,m_0}$, $x_j^{l_0+1,m_0}$ are uniquely determined by fixing the Cauchy data $x_j^{l_0,m_0}$, $x_j^{l_0,m_0+1}$ for the m-flow and vice versa.

---

[5] A very close version of the discrete L-M pair appeared first in the Ref.[18] as an a priori Ansatz

7 Conclusion and outlook

It turned out that classical and quantum integrable models have a deeper connection than the common assertion that the former are obtained as a "classical limit" of the latter. In this paper we have tried to elaborate perhaps the simplest example of this phenomenon: the fusion rules for quantum transfer matrices coincide with Hirota's bilinear difference equation (HBDE).

We have identified the bilinear fusion relations in Hirota's classical difference equation with particular boundary conditions and elliptic solutions of Hirota equation, with eigenvalues of the quantum transfer matrix. Eigenvalues of the quantum transfer matrix play the role of the τ-function. Positions of zeros of the solution are determined by the Bethe Ansatz equations. The latter have been derived from entirely classical set-up.

We have shown that nested Bethe Ansatz equations can be considered as a natural discrete time analogue of the Ruijsenaars-Schneider system of particles. The discrete time t runs over vertices of the Dynkin graph of A_{k-1}-type and numbers levels of the nested Bethe Ansatz. The continuum limit in t gives the continuous time RS system [17]. This is our motivation to search for classical integrability properties of the nested Bethe Ansatz equations.

In addition we constructed the general solution of the Hirota equation with a certain boundary conditions and obtained new determinant representations for eigenvalues of the quantum transfer matrix. The approach suggested in Sect. 5 resembles the Leznov-Saveliev solution [41] to the 2D Toda lattice with open boundaries. It can be considered as an integrable discretization of the classical W-geometry [42].

We hope that this work gives enough evidence to support the assertion that all spectral characteristics of quantum integrable systems on finite 1D lattices can be obtained starting from a classical discrete soliton equations, not implying a quantization. The Bethe Ansatz technique, which has been thought of as a specific tool of quantum integrability is shown to exist in classical discrete nonlinear integrable equations. The main new lesson is that solving classical discrete soliton equations one recovers a lot of information about a quantum integrable system.

Soliton equations usually have a huge number of solutions with very different properties. To extract the information about a quantum model, one should restrict the class of solutions by imposing certain boundary and analytic conditions. In particular, elliptic solutions to HBDE give spectral properties of quantum models with elliptic R-matrices.

The difference bilinear equation of the same form, though with different analytical requirements, has appeared in quantum integrable systems in another context. Spin-spin correlation functions of the Ising model obey a bilinear difference equation that can be recast into the form of HBDE [46], [47], [48]. More recently, nonlinear equations for correlation functions have been derived for a more general class of quantum integrable models, by virtue of the new approach of Ref. [49].

Thermodynamic Bethe Ansatz equations written in the form of functional relations [30], [31] (see e.g., [45]) appeared to be identical to HBDE with different analytic properties.

All these suggest that HBDE may play the role of a master equation for both classical and quantum integrable systems simultaneously, such that the "equivalence" between quantum systems and discrete classical dynamics might be extended beyond the spectral properties discussed in this paper. In particular, it will be very interesting to identify the quantum group structures and matrix elements of quantum L-0perators and R-matrices with objects of classical hierarchies. We do not doubt that such relation exists.

ACKNOWLEDGEMENTS

We are grateful to O. Babelon, B.Enriquez, L.Faddeev, J.L. Gervais, A. Klumper, V.Korepin, P.Kulish, A.Kuniba, D.Lebedev, A.Mironov, P. Pearce, N.Reshetikhin, R.Seiler, E. Sklyanin, N.Slavnov, T.Takebe and Al. Zamolodchikov for discussions and A. Abanov and J. Talstra for discussions and help. P.W. and O.L. were supported by grant NSF DMR 9509533 and the MRSEC Program of NSF DMR 9400379. The work of A.Z. was supported in part by RFFR grant 95-01-01106, by ISTC grant 015, by INTAS 94 2317 and the grant of the Dutch NWO organization. A.Z. and P.W. thank the Erwin Schrödinger Institute of mathematical physics for hospitality during the semester "Condensed Matter Physics – Dynamics, Geometry and Spectral Theory", where this work was started. P.W. also thanks Ecole Normale Superieure, where a part of this work was done, for hospitality. I.K. and A.Z. are grateful the University of Chicago for the hospitality in November 1995. During these visits they were supported by NSF grant DMR 9509533.

REFERENCES

[1] R.Baxter, *Exactly solved models in statistical mechanics*, Academic Press, 1982.

[2] M.Gaudin, *La fonction d'onde de Bethe*, Masson, 1983.

[3] R.Hirota, *Discrete analogue of a generalized Toda equation*, Journ. Phys. Soc. Japan 50 (1981) 3785-3791.

[4] T.Miwa, *On Hirota's difference equations*, Proc. Japan Acad. 58 (1982) 9-12

[5] L.D.Faddeev and L.A.Takhtadjan, *Quantum inverse scattering method and the XYZ Heisenberg model*, Uspekhi Mat. Nauk 34:5 (1979) 13-63.

[6] P.P.Kulish and N.Yu.Reshetikhin, *On GL_3-invariant solutions of the Yang-Baxter equation and associated quantum systems*, Zap. Nauchn. Sem. LOMI 120 (1982) 92-121 (in russian), Engl. transl.: J. Soviet Math. 34 (1986) 1948-1971.

[7] P.P.Kulish and E.K.Sklyanin, *Quantum spectral transform method. Recent developments*, Lecture Notes in Physics 151 61-119, Springer, 1982.

[8] M.Jimbo, T.Miwa and M.Okado, *An $A_{n-1}^{(1)}$ family of solvable lattice models*, Mod. Phys. Lett. B1 (1987) 73-79.

[9] N.Yu.Reshetikhin, *The functional equation method in the theory of exactly soluble quantum systems*, Sov. Phys. JETP 57 (1983) 691-696.

[10] A.Klümper and P.Pearce, *Conformal weights of RSOS lattice models and their fusion hierarchies*, Physica A183 (1992) 304-350.

[11] A.Kuniba, T.Nakanishi and J.Suzuki, *Functional relations in solvable lattice models, I: Functional relations and representation theory, II: Applications*, Int. Journ. Mod. Phys. A9 (1994) 5215-5312.

[12] H.Airault, H.McKean and J.Moser, *Rational and elliptic solutions of the KdV equation and related many-body problem*, Comm. Pure and Appl. Math. 30 (1977), 95-125.

[13] D.V.Chudnovsky and G.V.Chudnovsky, *Pole expansions of non-linear partial differential equations*, Nuovo Cimento 40B (1977), 339-350.

[14] I.M.Krichever, *Elliptic solutions of Kadomtsev-Petviashvilii equation and integrable systems of particles*, Func. Anal. App 14 (1980), n 4, 282-290.

[15] I.Krichever, O.Babelon, E.Billey and M.Talon, *Spin generalization of the Calogero-Moser system and the Matrix KP equation*, Preprint LPTHE 94/42.

[16] I.M.Krichever and A.V.Zabrodin, *Spin generalization of the Ruijsenaars-Schneider model, non-abelian 2D Toda chain and representations of Sklyanin algebra*, Uspekhi Mat. Nauk, 50:6 (1995), hep-th/9505039.

[17] S.N.M.Ruijsenaars and H.Schneider, *A new class of integrable systems and its relation to solitons*, Ann. Phys. (NY) 170 (1986), 370-405.

[18] F.Nijhof, O.Ragnisco and V.Kuznetsov, *Integrable time-discretization of the Ruijsenaars-Schneider model*, Commun. Math. Phys. 176 (1996) 681-700.

[19] P.P.Kulish, N.Yu.Reshetikhin and E.K.Sklyanin, *Yang-Baxter equation and representation theory: I*, Lett. Math. Phys. 5 (1981) 393-403.

[20] R.Hirota, *Discrete two-dimensional Toda molecule equation*, Journ. Phys. Soc. Japan 56 (1987) 4285-4288.

[21] A.Bilal and J.-L.Gervais, *Extended $c = \infty$ conformal systems from classical Toda field theories*, Nucl. Phys. B314 (1989) 646-686.

[22] V.Bazhanov and N.Reshetikhin, *Restricted solid on solid models connected with simply laced algebras and conformal field theory*, Journ. Phys. A23 (1990) 1477-1492.

[23] Y.Zhou and P.Pearce, *Solution of functional equations of restricted $A_{n-1}^{(1)}$ fused lattice models*, Preprint hep-th/9502067 (1995).

[24] M.Sato, *Soliton equations as dynamical systems on infinite dimensional Grassmann manifolds*, RIMS Kokyuroku 439 (1981) 30-46.

[25] M.Jimbo and T.Miwa, *Solitons and infinite dimensional Lie algebras*, Publ. RIMS, Kyoto Univ. 19 (1983) 943-1001.

[26] G.Segal and G.Wilson, *Loop groups and equations of KdV type*, Publ. IHES 61 (1985) 5-65.

[27] W.V.D.Hodge and D.Pedoe, *Methods of algebraic geometry*, volume I, Cambridge University Press, Cambridge, 1947.

[28] J.W.P.Hirschfeld and J.A.Thas, *General Galois geometries*, Claredon Press, Oxford, 1991.

[29] P.Griffiths and J.Harris, *Principles of algebraic geometry*, A Wiley-Interscience Publication, John Wiley & Sons, 1978.

[30] Al.B.Zamolodchikov, *On the thermodynamic Bethe Ansatz equations for reflectionless ADE scattering theories*, Phys. Lett. B253 (1991) 391-394.

[31] F.Ravanini, A.Valleriani and R.Tateo, *Dynkin TBA's*, Int. Journ. Mod. Phys. A8 (1993) 1707-1727.

[32] R.Hirota, *Nonlinear partial difference equations III; Discrete sine-Gordon equation*, Journ. Phys. Soc. Japan 43 (1977) 2079-2086.

[33] L.D.Faddeev and A.Yu.Volkov, *Quantum inverse scattering method on a space-time lattice*, Teor. Mat. Fiz. 92 (1992) 207-214 (in russian); L.D.Faddeev, *Current-like variables in massive and massless integrable models*, Lectures at E.Fermi Summer School, Varenna 1994, hep-th/9406196.

[34] A.Bobenko, N.Kutz and V.Pinkall, *The discrete quantum pendulum*, Phys. Lett. A177 (1993) 399-404.

[35] S.Saito and N.Saitoh, *Linearization of bilinear difference equations*, Phys. Lett. A120 (1987) 322-326; *Gauge and dual symmetries and linearization of Hirota's bilinear equations*, Journ. Math. Phys. 28 (1987) 1052-1055.

[36] K.Ueno and K.Takasaki, *Toda lattice hierarchy*, Adv. Studies in Pure Math. 4 (1984) 1-95.

[37] A.Kirillov and N.Reshetikhin, *Exact solution of the integrable XXZ Heisenberg model with arbitrary spin*, I, II, Journ. Phys. A20 (1987) 1565-1597.

[38] V.Bazhanov and N.Reshetikhin, *Critical RSOS models and conformal field theory*, Int. Journ. Mod. Phys. A4 (1989) 115-142.

[39] J.-L.Gervais and A.Neveu, *Novel triangle relation and absence of tachyons in Liouville string field theory*, Nucl. Phys. B238 (1984) 125

[40] G.Jorjadze, A.Pogrebkov, M.Polivanov and S.Talalov, *Liouville field theory: IST and Poisson bracket structure*, Journ. Phys. A19 (1986) 121-139.

[41] A.Leznov and M.Saveliev, *Theory of group representations and integration of nonlinear systems* $x_{a,z\bar{z}} = \exp(kx)_a$, Physica 3D (1981) 62-72; *Group-theoretical methods for integartion of nonlinear dynamical systems*, Progress in Physics series 15, Birkhaüser-Verlag, Basel, 1992.

[42] J.-L.Gervais and Y.Matsuo, *W-geometries*, Phys. Lett. B274 (1992) 309-316; *Classical* A_n*-W-geometry*, Comm. Math. Phys. 152 (1993) 317-368.

[43] E.Frenkel and N.Reshetikhin, *Quantum affine algebras and deformations of the Virasoro and W-algebras*, Preprint q-alg/9505025 (1995).

[44] A.Kuniba and J.Suzuki, *Analytic Bethe Ansatz for fundamental representations of Yangians*, hep-th/9406180.

[45] V.Bazhanov, S.Lukyanov and A.Zamolodchikov, *Integrable structure of conformal field theory, quantum KdV theory and thermodynamic Bethe Ansatz*, preprint CLNS 94/1316, RU-94-98, hep-th/9412229.

[46] B.McCoy and T.T.Wu, *Nonlinear partial difference equations for the two-dimensional Ising model*, Phys. Rev. Lett. 45 (1980) 675-678; *Nonlinear partial difference equations for the two-spin correlation function of the two-dimensional Ising model*, Nucl. Phys. B180 (1981) 89-115.

[47] J.H.H.Perk, *Quadratic identities for Ising model correlations*, Phys. Lett. A79 (1980) 3-5.

[48] H.Au-Yang and J.H.H.Perk, *Toda lattice equation and wronskians in the 2d Ising model*, Physica 18D (1986) 365-366.

[49] N.M.Bogoliubov, A.G.Izergin and V.E.Korepin, *Quantum inverse scattering method and correlation functions*, Cambridge University Press, 1993.

CONFORMAL BOOTSTRAP IN LIOUVILLE FIELD THEORY

A. ZAMOLODCHIKOV

DEPARTMENT OF PHYSICS AND ASTRONOMY
RUTGERS UNIVERSITY
P.O.BOX 849, PISCATAWAY, NEW JERSEY 08855-0849, USA
AND
L.D.LANDAU INSTITUTE FOR THEORETICAL PHYSICS
KOSYGINA 2, 117334, MOSCOW, RUSSIA

AL. ZAMOLODCHIKOV

LABORATOIRE DE PHYSIQUE MATHÉMATIQUE
UNIVERSITÉ MONTPELLIER II
PL .E. BATAILLON, 34095 MONTPELLIER, FRANCE

1 INTRODUCTION

Attention to the two-dimensional Liouville Field Theory (LFT) is drawn basically for two reasons. First it was recognized [1] as an effective field theory of the 2d quantum gravity. In particular it is very relevant in the string theory [1–4]. Second, it is an example of non-rational conformal field theory (CFT) which is very likely exactly solvable (e.g., the classical equations of motion are integrable).The interest to LFT was renewed recently with the development of the matrix model approach to 2d gravity [5,6]. It was shown that LFT is able to reproduce some of the predictions of the matrix model approach, in particular the scaling behavior [7–9], the genus one partition functions [10] and some of the integrated correlation functions [11–15]. It is very plausible therefore that LFT describes the same 2d quantum gravity as the matrix models do (at least in the "weak coupling region" $c_L > 25$).

In this lecture we discuss an analytic expression for the three-point function of the exponential fields in LFT on a sphere[1]. From the local point of view these quan-

[1]As we became to know recently, this exact expression has been first proposed by Dorn and Otto

Low Dimensional Applications of Quantum Field Theory
Edited by L. Baulieu *et al.*, Plenum Press, New York, 1997

tities are naturally interpreted as the structure constants in the operator product expansions. In principle a set of the structure constants defines a conformal field theory completely in the sense that all the multipoint correlation functions can be reconstructed from the structure constants by purely "kinematic" procedure. In other words, the set of the structure constants proposed can be considered as a basis of an exact construction of the Liouville CFT. Of course there are still some serious technical problems open (see below). What is more important is that these LFT structure constants are not strictly derived from the LFT dynamics but appear rather as a guess. Even the consistency of the operator product algebra defined by these constants is not obvious and has to be verified. Therefore we perform some consistency tests to support our conjecture. In particular we verify the conformal bootstrap equations which are believed to contain these consistency conditions of the operator product algebra.

The content below is arranged as follows. In sect.2 the LFT on a sphere is introduced and some notations are defined. In sect.3 we describe the exact expression for the LFT three-point function and discuss some of its properties. The Liouville reflection amplitude is also introduced here. Sect.4 contains some calculations in the semiclassical limit.

The exact LFT structure constants together with an effective numerical algorithm for the conformal block permit us to compute numerically the four-point function of the LFT exponential fields. We perform this in sect.5 and verify that the four-point function satisfies the conformal bootstrap equations.

2 Liouville field theory

Local properties of the LFT are derived from the Lagrangian density

$$\mathcal{L} = \frac{1}{4\pi}(\partial_a \phi)^2 + \mu e^{2b\phi} \tag{2.1}$$

where b is the dimensionless Liouville coupling constant and the scale parameter μ is usually called the cosmological constant. Below we often use the complex Euclidean coordinates $z = x_1 + ix_2$; $\bar{z} = x_1 - ix_2$ and denote $\partial = \partial/\partial z$; $\bar{\partial} = \partial/\partial \bar{z}$. The Liouville field $\phi(z, \bar{z})$ is not exactly a scalar but changes under the holomorphic coordinate transformations $z \to w(z)$ as

$$\phi(w, \bar{w}) = \phi(z, \bar{z}) - \frac{Q}{2} \log \left| \frac{dw}{dz} \right|^2 \tag{2.2}$$

where

$$Q = b + 1/b \tag{2.3}$$

The holomorphic Liouville stress tensor

$$\begin{aligned} T(z) &= -(\partial\phi)^2 + Q\partial^2\phi \\ \bar{T}(\bar{z}) &= -(\bar{\partial}\phi)^2 + Q\bar{\partial}^2\phi \end{aligned} \tag{2.4}$$

[16,17] in 1992. We are grateful to Prof. Dorn for bringing our attention to these references.

ensures local conformal invariance [20] of LFT with the Liouville central charge

$$c_L = 1 + 6Q^2 \tag{2.5}$$

To define LFT globally one has to specify the boundary conditions. The LFT on a sphere corresponds to ϕ defined on the whole complex plane with the following asymptotic behavior at $|z| \to \infty$

$$\phi(z, \bar{z}) = -Q \log(z\bar{z}) + O(1) \qquad \text{at } |z| \to \infty \tag{2.6}$$

For specific calculations it is sometimes convenient to set this asymptotic behavior by considering LFT on a large disk Γ of radius $R \to \infty$ and adding a boundary term to the Liouville action

$$A_L = \frac{1}{4\pi} \int_\Gamma \left[(\partial_a \phi)^2 + 4\pi\mu e^{2b\phi} \right] d^2x + \frac{Q}{\pi R} \int_{\partial\Gamma} \phi dl + 2Q^2 \log R \tag{2.7}$$

The last constant term is introduced to make the action finite at $R \to \infty$. This type of boundary condition is conventionally called the background charge $-Q$ at infinity.

The exponential Liouville operators

$$V_\alpha(x) = e^{2\alpha\phi(x)} \tag{2.8}$$

are the spinless primary conformal fields of dimensions

$$\Delta_\alpha = \alpha(Q - \alpha) \tag{2.9}$$

Note that the field $V_{Q-\alpha}$ has the same dimension as (2.8). These two fields are closely related and we shall call $V_{Q-\alpha}$ the reflection image of V_α and vice versa. In particular the perturbation operator V_b in (2.1) have dimension 1 together with its reflection image $V_{1/b}$. The case $\alpha = Q/2$ is degenerate. Here we have two primary fields

$$V_{Q/2}(x) = e^{Q\phi(x)} \tag{2.10}$$

and

$$U_{Q/2}(x) = \frac{1}{2} \frac{\partial}{\partial\alpha} V_\alpha(x)|_{\alpha=Q/2} = \phi(x) e^{Q\phi(x)} \tag{2.11}$$

of dimension $Q^2/4$. The last field (2.11) is called sometimes the puncture operator in LFT [11].

The n-point function of the exponential fields on a sphere is formally defined as a functional integral

$$\mathcal{G}_{\alpha_1,\ldots,\alpha_n}(x_1, \ldots, x_n) = \int V_{\alpha_1}(x_1) \ldots V_{\alpha_n}(x_n) e^{-A_L[\phi]} D\phi \tag{2.12}$$

over the fields $\phi(x)$ with the boundary condition (2.6). Note that in the definition (2.12) we do not divide the right hand side by the zero-point function Z_0.

The scale (μ) dependence of any correlation function (2.12) is predicted by the KPZ scaling [7–9]

$$\mathcal{G}_{\alpha_1,\ldots,\alpha_n}(x_1, \ldots, x_n) = (\pi\mu)^{(Q - \sum \alpha_i)/b} F_{\alpha_1,\ldots,\alpha_n}(x_1, \ldots, x_n) \tag{2.13}$$

with $F_{\alpha_1,\ldots,\alpha_n}(x_1,\ldots,x_n)$ independent on μ. Sometimes it is convenient to consider the n-point functions $\mathcal{G}^{(A)}_{\alpha_1,\ldots,\alpha_n}(x_1,\ldots,x_n)$ with fixed area

$$A = \int e^{2b\phi} d^2x \tag{2.14}$$

These observables are related to (2.12) as

$$\mathcal{G}_{\alpha_1,\ldots,\alpha_n}(x_1,\ldots,x_n) = \int_0^\infty \mathcal{G}^{(A)}_{\alpha_1,\ldots,\alpha_n}(x_1,\ldots,x_n) e^{-\mu A}\frac{dA}{A} \tag{2.15}$$

so that

$$\mathcal{G}^{(A)}_{\alpha_1,\ldots,\alpha_n}(x_1,\ldots,x_n) = \left(\frac{A}{\pi}\right)^{(\sum\alpha_i - Q)/b} \frac{F_{\alpha_1,\ldots,\alpha_n}(x_1,\ldots,x_n)}{\Gamma\left(\left(Q-\sum\alpha_i\right)/b\right)} \tag{2.16}$$

Conformal invariance restricts to some extent the x-dependence of the correlation functions [20]. In particular the three-point function is specified up to an x-independent constant $C(\alpha_1,\alpha_2,\alpha_3)$

$$\mathcal{G}_{\alpha_1,\alpha_2,\alpha_3}(x_1,x_2,x_3) = |x_{12}|^{2\gamma_3} |x_{23}|^{2\gamma_1} |x_{31}|^{2\gamma_2} C(\alpha_1,\alpha_2,\alpha_3) \tag{2.17}$$

where

$$\begin{aligned} \gamma_1 &= \Delta_{\alpha_1} - \Delta_{\alpha_2} - \Delta_{\alpha_3}, \\ \gamma_2 &= \Delta_{\alpha_2} - \Delta_{\alpha_3} - \Delta_{\alpha_1}, \\ \gamma_3 &= \Delta_{\alpha_3} - \Delta_{\alpha_1} - \Delta_{\alpha_2} \end{aligned}$$

(here and below we denote $x_{ij} = x_i - x_j$). The four-point function can be reduced to a function of only one coordinate variable, the projective invariant of the four points $x_1,\ldots,x_4$

$$x = \frac{x_{12}x_{34}}{x_{14}x_{32}} \tag{2.18}$$

as follows

$$\mathcal{G}_{\alpha_1,\ldots,\alpha_4}(x_1,\ldots,x_4) = \prod_{i<j} |x_{ij}|^{2\gamma_{ij}} G_{\alpha_1,\alpha_2,\alpha_3,\alpha_4}(x,\bar{x}) \tag{2.19}$$

where $\gamma_{12} = \gamma_{13} = 0$, $\gamma_{14} = -2\Delta_{\alpha_1}$, $\gamma_{24} = \Delta_{\alpha_1} + \Delta_{\alpha_3} - \Delta_{\alpha_2} - \Delta_{\alpha_4}$, $\gamma_{34} = \Delta_{\alpha_1} + \Delta_{\alpha_2} - \Delta_{\alpha_3} - \Delta_{\alpha_4}$ and $\gamma_{23} = \Delta_{\alpha_4} - \Delta_{\alpha_1} - \Delta_{\alpha_2} - \Delta_{\alpha_3}$. Function $G_{\alpha_1,\alpha_2,\alpha_3,\alpha_4}(x,\bar{x})$ satisfies the following symmetries (sometimes called the crossing symmetry relations) [20]

$$\begin{aligned} G_{\alpha_1,\alpha_2,\alpha_3,\alpha_4}(x,\bar{x}) &= G_{\alpha_1,\alpha_3,\alpha_2,\alpha_4}(1-x,1-\bar{x}) \\ &= |x|^{-4\Delta_{\alpha_1}} G_{\alpha_1,\alpha_4,\alpha_3,\alpha_2}(1/x,1/\bar{x}) \end{aligned} \tag{2.20}$$

In principle one can separate two pairs of operators in the four-point function, say $V_{\alpha_1}(x_1)V_{\alpha_2}(x_2)$ and $V_{\alpha_3}(x_3)V_{\alpha_4}(x_4)$ and represent

$$\mathcal{G}_{\alpha_1,\ldots,\alpha_4}(x_1,\ldots,x_4)$$

as a sum over intermediate physical states [20]. As was established long ago [2,3] the physical LFT space of states $\mathcal{A}$ consists of a continuum variety of primary states corresponding to operators V_α with

$$\alpha = \frac{Q}{2} + iP \tag{2.21}$$

(P is real) and the conformal descendants of these states. The corresponding expression for the four-point function looks as

$$G_{\alpha_1,\alpha_2,\alpha_3,\alpha_4}(x,\bar{x}) = \frac{1}{2}\int_{-\infty}^{\infty} C(\alpha_1,\alpha_2,\frac{Q}{2}+iP) \quad C(\alpha_3,\alpha_4,\tfrac{Q}{2}-iP) \times|\mathcal{F}(\Delta_{\alpha_i},\Delta,x)|^2 dP \tag{2.22}$$

Here $C(\alpha_1,\alpha_2,\alpha_3)$ are the structure constants introduced in eq.(2.17) while the so-called conformal block, $\mathcal{F}(\Delta_{\alpha_i},\Delta,x)$, which sums up all the intermediate descendant states of a given primary one [20]. The conformal block is determined completely by the conformal symmetry of the theory. It depends on the corresponding central charge (2.5), on the dimension

$$\Delta = \frac{Q^2}{4} + P^2 \tag{2.23}$$

of the intermediate primary state and also on the "external" dimensions Δ_{α_i}, $i = 1,\ldots,4$. Unfortunately up to now this function is not known in a closed form. However it is straightforward to evaluate it as a power series in x [20]. Also there are very efficient algorithms for its numerical computation [21].

Of course one can pick up another partition of the four operators into pairs and obtain an expression similar to (2.22) summing up over the intermediate states in the corresponding channel. The resulting four-point function must be the same. In other words the four-point function defined as in eq.(2.22) has to satisfy the crossing symmetry relations (2.20). From this point of view these relations appear as a set of non-trivial restrictions for the structure constants $C(\alpha_1,\alpha_2,\alpha_3)$ which express the associativity of the operator algebra and are known as the conformal bootstrap equations [20]. We discuss more about this point in sect.5. Representations like (2.22) can be written down for higher multipoint functions as well. However they involve the multipoint conformal blocks which are much more complicated objects then the four-point one. Fortunately, the corresponding crossing symmetry relations are not expected to bring up any new restrictions on the structure of the operator algebra [20].

To finish this section let us discuss the classical limit $b \to 0$ of LFT. Here it is more convenient to use the field

$$\varphi = 2b\phi \tag{2.24}$$

which becomes a classical Liouville field. Its dynamics is governed by the classical action

$$S_{\text{Liouv}}[\varphi] = b^2 A_L[\phi] \qquad \text{at} \quad b \to 0 \tag{2.25}$$

where explicitly

$$S_{\text{Liouv}}[\varphi] = \frac{1}{8\pi}\int_{\Gamma}\left[\frac{1}{2}(\partial_a\varphi)^2 + 8\pi\mu b^2 e^{\varphi}\right] d^2x + \varphi_\infty + 2\log R \tag{2.26}$$

and

$$\varphi_\infty = \frac{1}{2\pi R}\int_{\partial\Gamma}\varphi dl \tag{2.27}$$

The field $\varphi(x)$ satisfies the classical Liouville equation

$$\partial\bar{\partial}\varphi = 2\pi\mu b^2 e^{\varphi} \tag{2.28}$$

and locally describes a surface of constant negative curvature $-8\pi\mu b^2$.

The leading (exponential) asymptotic of the n-point function (2.12) in the classical limit

$$\mathcal{G}_{\alpha_1,\ldots,\alpha_n}(x_1,\ldots,x_n) \sim \exp\left(-\frac{1}{b^2}S^{(\mathrm{cl})}\right) \tag{2.29}$$

is governed by the classical Liouville action on an appropriate solution to the Liouville equation. Note that the insertion of any $V_{\alpha_i}(x_i)$ in (2.12) affects the classical field dynamics only if the corresponding α_i is "heavy", i.e. if $\alpha_i = \eta_i/b$ is of order $O(b^{-1})$. Technically one has to distinguish two possibilities. If $\sum_i \eta_i > 1$ a classical solution with negative curvature exists and in (2.29)

$$S^{(\mathrm{cl})}_{\eta_1,\ldots,\eta_n}(x_1,\ldots,x_n) = S_{\mathrm{Liouv}}\left[\varphi_{\eta_1,\ldots,\eta_n}(x|x_1,\ldots,x_n)\right] \tag{2.30}$$

where $\varphi_{\eta_1,\ldots,\eta_n}(x|x_1,\ldots,x_n)$ is a solution to (2.28) with the following boundary conditions

$$\begin{aligned} \varphi(z,\bar{z}) &= -2\log|z|^2 + O(1) && \text{at } \ |z| \to \infty \\ \varphi(z,\bar{z}) &= -2\eta_i \log|z-x_i|^2 + O(1) && \text{at } \ z \to x_i \end{aligned} \tag{2.31}$$

This field configuration is singular at $z \to x_i$. Therefore it is better to cut out small disks of radius ϵ_i around each point x_i and define the regularized Liouville action on the remaining part Γ of the complex plane

$$\begin{aligned} S_{\mathrm{Liouv}}[\varphi] \;=\; & \frac{1}{8\pi}\int_{\Gamma}\left[\frac{1}{2}(\partial_a\varphi)^2 + 8\pi\mu b^2 e^{\varphi}\right]d^2x + \varphi_{\infty} \\ & +2\log R - \sum_i\left(\eta_i\varphi_i + 2\eta_i^2\log\epsilon_i\right) \end{aligned} \tag{2.32}$$

Here φ_∞ is as in eq.(2.27) while the boundary terms with

$$\varphi_i = \frac{1}{2\pi\epsilon_i}\int_{\partial\Gamma_i}\varphi dl \tag{2.33}$$

are added for each small circle to ensure the behavior (2.31) near x_i. Also we include some field independent terms such that S_{Liouv} is finite and independent on ϵ_i at $\epsilon_i \to 0$.

At $\sum\eta_i < 1$ there is no solution to (2.28) and (2.31) with negative curvature. In this case it is relevant to consider the n-point function (2.17) of fixed area. The leading classical behavior is again governed by the classical action

$$\mathcal{G}^{(A)}_{\alpha_1,\ldots,\alpha_n}(x_1,\ldots,x_n) \sim \exp\left(-\frac{1}{b^2}S^{(\mathrm{cl})}_A\right) \tag{2.34}$$

but now one has to impose the constant area constraint

$$A = \int_{\Gamma} e^{\varphi} d^2x \tag{2.35}$$

and solve the positive curvature Liouville equation

$$\partial\bar{\partial}\varphi = \frac{2\pi}{A}\left(\sum \eta_i - 1\right)e^{\varphi} \tag{2.36}$$

with the same boundary conditions (2.31). The corresponding constant area Liouville action is again (2.32) but without the cosmological term $\mu b^2 A$ so that

$$S_{\text{Liouv}}[\varphi] = S_A^{(\text{cl})}[\varphi] + \mu b^2 \int_\Gamma e^{\varphi} d^2x \tag{2.37}$$

Note that the same fixed area calculation can be performed in the first case $\sum \eta_i > 1$ too. The integral over A in eq.(2.15) is now dominated by the saddle point $A = (\sum \eta_i - 1)/(\mu b^2)$ which corresponds to the classical solution of eq.(2.28).

3 Three-point function

Naively one can try to expand the N-point function (2.12) into a perturbative series in the cosmological constant μ

$$\mathcal{G}_{\alpha_1,\ldots,\alpha_N}(x_1,\ldots,x_N) = \sum_{n=0}^{\infty} \mathcal{G}^{(n)}_{\alpha_1,\ldots,\alpha_N}(x_1,\ldots,x_N) \tag{3.1}$$

where

$$\mathcal{G}^{(n)}_{\alpha_1,\ldots,\alpha_N} = \frac{(-\mu)^n}{n!}\int \langle V_{\alpha_1}(x_1)\ldots V_{\alpha_N}(x_N)V_b(u_1)\ldots V_b(u_n)\rangle\, d^2u_1\ldots d^2u_n \tag{3.2}$$

and $\langle\ldots\rangle$ denotes the functional integral over a free field (i.e., (2.12) at $\mu = 0$)

$$\begin{aligned}\langle V_{\alpha_1}(x_1)\ldots V_{\alpha_N}(x_N)\rangle &= \int \prod_{i=1}^{N} e^{2\alpha_i\phi(x_i)} \exp\left(-\frac{1}{4\pi}\int(\partial_a\phi)^2 d^2x\right) D\phi \\ &= \prod_{i>j} |x_i - x_j|^{-4\alpha_i\alpha_j}\end{aligned} \tag{3.3}$$

However it is well known that in LFT expression (3.1) in general does not work. The point is that the free field functional integral (3.3) for the n-th term (3.2) matches the spherical boundary condition (2.6) only if

$$\sum_{i=1}^{N} \alpha_i = Q - nb \tag{3.4}$$

Following refs.[12–14] we shell interpret eq.(3.4) as a kind of "on-mass-shell" condition. Namely, $\mathcal{G}_{\alpha_1,\ldots,\alpha_N}(x_1,\ldots,x_N)$ exhibits a pole in the variable $\alpha = \sum \alpha_i$ every time eq.(3.4) is satisfied for $n = 0, 1, 2, \ldots$, the residue being specified by the corresponding perturbative integral

$$\operatorname*{res}_{\sum\alpha_i = Q-nb} \mathcal{G}_{\alpha_1,\ldots,\alpha_N}(x_1,\ldots,x_N) = \mathcal{G}^{(n)}_{\alpha_1,\ldots,\alpha_N}(x_1,\ldots,x_N)\Big|_{\sum\alpha_i = Q-nb} \tag{3.5}$$

It seems unlikely that the on-mass-shell condition (3.5) alone is enough to determine the N-point function. However the situation is simplified in the three-point case $N = 3$. The x-dependence of the on-mass-shell integrals turns out to be the same as that of the conformal three-point function (2.17)

$$\mathcal{G}^{(n)}_{\alpha_1,\alpha_2,\alpha_3}(x_1,x_2,x_3)\Big|_{\sum\alpha_i=Q-nb} = |x_{12}|^{2\gamma_3}\,|x_{23}|^{2\gamma_1}\,|x_{31}|^{2\gamma_2}\,I_n(\alpha_1,\alpha_2,\alpha_3) \tag{3.6}$$

with the same γ_1, γ_2 and γ_3 as in eq.(2.17), while $I_n(\alpha_1,\alpha_2,\alpha_3)$ has been carried out explicitly in ref.[23]

$$I_n(\alpha_1,\alpha_2,\alpha_3) = \frac{\left(-\pi\mu/\gamma(-b^2)\right)^n \prod_{j=1}^{n}\gamma(-jb^2)}{\prod_{k=0}^{n-1}\left[\gamma(2\alpha_1 b+kb^2)\gamma(2\alpha_2 b+kb^2)\gamma(2\alpha_3 b+kb^2)\right]} \tag{3.7}$$

Here and below the standard notation

$$\gamma(x) = \Gamma(x)/\Gamma(1-x) \tag{3.8}$$

is used. The on-mass-shell condition now reads

$$\underset{\sum\alpha_i=Q-nb}{\mathrm{res}}\; C(\alpha_1,\alpha_2,\alpha_3) = I_n(\alpha_1,\alpha_2,\alpha_3) \tag{3.9}$$

At this step we need to introduce some special function $k(b,x)$. Below we consider b as a parameter and suppress it in the notation. In the strip $0 < \mathrm{Re}x < Q$ function $k(x)$ is defined by the integral representation

$$\log k(x) = \int_0^\infty \frac{dt}{t}\left[\left(\frac{Q}{2}-x\right)^2 e^{-2t} - \frac{\sinh^2\left(\frac{Q}{2}-x\right)t}{\sinh bt \sinh\frac{t}{b}}\right] \tag{3.10}$$

From the definition it is clear that

$$\begin{aligned} k(x) &= k(Q-x) \\ k(Q/2) &= 1 \end{aligned} \tag{3.11}$$

and that $k(x)$ is "self-dual", *i.e.* remains unchanged if $b \to 1/b$. Also the following functional relations are derived

$$\begin{aligned} k(x+b) &= \gamma(bx)b^{1-2bx}k(x) \\ k(x+1/b) &= \gamma(x/b)b^{2x/b-1}k(x) \end{aligned} \tag{3.12}$$

Using (3.10) and (3.12) it is easy to verify that $k(x)$ is an entire function of x with zeroes located at $x = -m/b - nb$ and $x = (m+1)/b + (n+1)b$, where m and n run over all non-negative integers. Below we also use the notation

$$k_0 = dk(x)/dx|_{x=0} \tag{3.13}$$

With the relations (3.12) it is straightforward to verify that

$$C(\alpha_1,\alpha_2,\alpha_3) = \left[\pi\mu\gamma(b^2)b^{2-2b^2}\right]^{(Q-\sum\alpha_i)/b} \times \frac{k_0 k(2\alpha_1)k(2\alpha_2)k(2\alpha_3)}{k(\alpha_1+\alpha_2+\alpha_3-Q)k(\alpha_1+\alpha_2-\alpha_3)k(\alpha_2+\alpha_3-\alpha_1)k(\alpha_3+\alpha_1-\alpha_2)} \tag{3.14}$$

satisfies the on-mass-shell condition (3.9). We propose this expression as the exact three-point function in LFT.

As a function of $\alpha = \sum_{i=1}^{3} \alpha_i$ expression (3.14) has more poles then predicted by eq.(3.9). They appear at $\alpha = Q - m/b - nb$ and at $\alpha = 2Q + m/b + nb$ for any pair of non-negative integers m and n. It seems suggestive to note that the corresponding residues are related to the more general multiple integrals, also evaluated explicitly in ref.[23]. Namely

$$\operatorname*{res}_{\sum \alpha_i = Q - m/b - nb} \mathcal{G}_{\alpha_1,\alpha_2,\alpha_3}(x_1,x_2,x_3) = \frac{(-\tilde{\mu})^m(-\mu)^n}{m!n!} \times \int \left\langle \prod_{i=1}^{3} V_{\alpha_i}(x_i) \prod_{j=1}^{m} V_{1/b}(u_j) \prod_{j=1}^{n} V_b(v_j) \right\rangle d^2u_1 \ldots d^2u_m d^2v_1 \ldots d^2v_n \quad (3.15)$$

where the "dual" cosmological constant $\tilde{\mu}$ is related to μ as

$$\pi\tilde{\mu}\gamma(1/b^2) = \left(\pi\mu\gamma(b^2)\right)^{1/b^2} \quad (3.16)$$

The whole expression (3.14) is self-dual in the sense that it is invariant under the substitution $b \to 1/b$, $\mu \to \tilde{\mu}$.

The three-point function (3.14) reveals some interesting features. In particular the reflection $\alpha \to Q - \alpha$ of each of the three operators introduces the Liouville reflection amplitude $S(P)$

$$C(Q-\alpha_1,\alpha_2,\alpha_3) = C(\alpha_1,\alpha_2,\alpha_3)S(i\alpha_1 - iQ/2) \quad (3.17)$$

which reads explicitly

$$S(P) = -\left(\pi\mu\gamma(b^2)\right)^{-2iP/b} \frac{\Gamma(1+2iP/b)\Gamma(1+2iPb)}{\Gamma(1-2iP/b)\Gamma(1-2iPb)} \quad (3.18)$$

We shall also use the notation

$$G(\alpha) = S(-i\alpha + iQ/2) = \frac{\left(\pi\mu\gamma(b^2)\right)^{(Q-2\alpha)/b}}{b^2} \frac{\gamma(2\alpha b - b^2)}{\gamma\left(2 - 2\alpha/b + 1/b^2\right)} \quad (3.19)$$

associating $G(\alpha)$ with the two-point function of operators $V_\alpha(x_1)V_\alpha(x_2)$ in LFT.

There are several different classical limits of the three-point function (3.14) dependent on how the parameters α_i behave as $b \to 0$. We consider here only the case when all the three operators are "heavy", i.e. $\alpha_i = \eta_i/b$ at $b \to 0$ with η_i fixed. The leading asymptotic is

$$C(\alpha_1,\alpha_2,\alpha_3) \sim \exp\left(\frac{1}{b^2} S^{(\text{cl})}(\eta_1,\eta_2,\eta_3)\right) \quad (3.20)$$

where

$$\begin{aligned} S^{(\text{cl})}(\eta_1,\eta_2,\eta_3) = \; & (\textstyle\sum \eta_i - 1)\log(\pi\mu b^2) \\ & + F(\eta_1+\eta_2+\eta_3-1) + F(\eta_1+\eta_2-\eta_3) \\ & + F(\eta_2+\eta_3-\eta_1) + F(\eta_3+\eta_1-\eta_2) \\ & - F(0) - F(2\eta_1) - F(2\eta_2) - F(2\eta_3) \end{aligned} \quad (3.21)$$

and we have denoted

$$F(\eta)=\int_{1/2}^{\eta}\log\gamma(x)dx \tag{3.22}$$

Note that $S^{(\mathrm{cl})}(\eta_1,\eta_2,\eta_3)$ vanishes if $\sum_{i=1}^{3}\eta_i=1$.

4 Classical limit

The asymptotic (3.20), (3.21) has to be compared with the classical action (2.32) evaluated at the solution $\varphi_{\eta_1,\eta_2,\eta_3}(z|x_1,x_2,x_3)$ of the classical Liouville equation (2.28) with three singular points x_1, x_2 and x_3. The boundary conditions (2.31) at these points are automatically imposed by the boundary terms in (2.32). In the three-point case the solution $\varphi_{\eta_1,\eta_2,\eta_3}(z|x_1,x_2,x_3)$ can be found explicitly in terms of hypergeometric functions

$$\varphi_{\eta_1,\eta_2,\eta_3}(z|x_1,x_2,x_3)=-2\log\left[a_1\psi_1(z)\psi_1(\bar z)+a_2\psi_2(z)\psi_2(\bar z)\right] \tag{4.1}$$

Here

$$\begin{aligned}\psi_1(z) &= (z-x_1)^{\eta_1}(z-x_2)^{1-\eta_1-\eta_3}(z-x_3)^{\eta_3}\\ &\quad\times {}_2F_1(\eta_1+\eta_3-\eta_2,\eta_1+\eta_2+\eta_3-1,2\eta_1,x)\\ \psi_2(z) &= (z-x_1)^{1-\eta_1}(z-x_2)^{\eta_1+\eta_3-1}(z-x_3)^{1-\eta_3}\\ &\quad\times {}_2F_1(1+\eta_2-\eta_1-\eta_3,2-\eta_1-\eta_2-\eta_3,2-2\eta_1,x)\end{aligned} \tag{4.2}$$

and we have denoted

$$x=\frac{(z-x_1)x_{32}}{(z-x_2)x_{31}} \tag{4.3}$$

while the constants a_1 and a_2 read explicitly

$$\begin{aligned}a_1^2 &= \frac{\pi\mu b^2}{|x_{13}|^{4\eta_3+4\eta_1-2}\,|x_{12}|^{2-4\eta_2}\,|x_{23}|^{2-4\eta_1}}\times\\ &\quad\frac{\gamma(\eta_1+\eta_2+\eta_3-1)\gamma(\eta_1+\eta_3-\eta_2)\gamma(\eta_1+\eta_2-\eta_3)}{\gamma^2(2\eta_1)\gamma(\eta_2+\eta_3-\eta_1)}\end{aligned} \tag{4.4}$$

and

$$a_2=-\frac{\pi\mu b^2}{|x_{13}|^2\,(1-2\eta_1)^2a_1} \tag{4.5}$$

Now we have to evaluate the integral (2.32). Calculation of the Liouville action can be simplified as follows. Since $\varphi_{\eta_1,\dots,\eta_n}(x|x_1,\dots,x_n)$ is an extremum of (2.32) we have

$$\frac{\partial}{\partial\eta_i}S_{\mathrm{Liouv}}\left[\varphi_{\eta_1,\dots,\eta_n}(x|x_1,\dots,x_n)\right]=-\varphi_i-4\eta_i\log\epsilon_i \tag{4.6}$$

where φ_i is defined in eq.(2.33). Near the singular point x_i the solution behaves as

$$\varphi_{\eta_1,\dots,\eta_n}(z|x_1,\dots,x_n)=-2\eta_i\log|z-x_i|^2+X_i+O\left(|z-x_i|^{2-4\eta_i}\right) \tag{4.7}$$

(we suppose here that all $\eta_i<1/2$). Therefore in the limit $\epsilon_i\to 0$

$$\frac{\partial}{\partial\eta_i}S_{\mathrm{Liouv}}=-X_i \tag{4.8}$$

This equation implies that the form

$$dS_{\text{Liouv}} = -\sum_{i=1}^{n} X_i d\eta_i \tag{4.9}$$

can be integrated, defining S_{Liouv} up to a constant independent on η_i. The integration ambiguity can be fixed if one notes that the Liouville action (2.32) is explicitly evaluated if $\sum_{i=1}^{n} \eta_i = 1$

$$S_{\text{Liouv}}|_{\sum \eta_i = 1} = \sum_{i<j} 2\eta_i\eta_j \log|x_i - x_j|^2 \tag{4.10}$$

In the three-point case the coefficients X_i are found from the explicit solution (4.1–5). E.g.

$$\begin{aligned} X_1 \;=\; & -(1-2\eta_1)\log\left|\frac{x_{12}x_{13}}{x_{23}}\right|^2 - \log\pi\mu b^2 - \\ & \log\frac{\gamma(\eta_1+\eta_2+\eta_3-1)\gamma(\eta_1+\eta_2-\eta_3)\gamma(\eta_1+\eta_3-\eta_2)}{\gamma^2(2\eta_1)\gamma(\eta_2+\eta_3-\eta_1)} \end{aligned} \tag{4.11}$$

while X_2 and X_3 are obtained from (4.11) by obvious permutations of η_i and x_i. Integrating the form (4.9) we find

$$\begin{aligned} S_{\text{Liouv}} = \;& (\textstyle\sum \eta_1 - 1)\log\pi\mu b^2 + (\delta_1+\delta_2-\delta_3)\log|x_{12}|^2 \\ & +(\delta_2+\delta_3-\delta_1)\log|x_{23}|^2 + (\delta_3+\delta_1-\delta_2)\log|x_{13}|^2 \\ & +F(\eta_1+\eta_2+\eta_3-1) + F(\eta_1+\eta_2-\eta_3) \\ & +F(\eta_2+\eta_3-\eta_1) + F(\eta_3+\eta_1-\eta_2) \\ & -F(2\eta_1) - F(2\eta_2) - F(2\eta_3) - F_0(x_i) \end{aligned} \tag{4.12}$$

where $\delta_i = \eta_i(1-\eta_i)$, function $F(\eta)$ is defined by eq.(3.22) and $F_0(x_i)$ is the integration constant independent on η_i. Comparing with (4.10) we find $F_0(x_i) = F(0)$ in complete agreement with the asymptotic (3.21) of the proposed exact three-point function.

5 Conformal bootstrap

In this section we study numerically the conformal bootstrap equations (2.20) using the representation (2.22) of the four-point function in LFT. The structure constants $C(\alpha_1, \alpha_2, \alpha_3)$ are described explicitly in sect.3. The conformal block $\mathcal{F}(\Delta_{\alpha_i}, \Delta, x)$ which also enters eq.(2.22) is not known in general in a closed form. In refs.[21] the following convenient representation has been derived

$$\begin{aligned} \mathcal{F}(\Delta_{\alpha_i}, \Delta, x) \;=\; & (16q)^{P^2} x^{Q^2/4-\Delta_{\alpha_1}-\Delta_{\alpha_2}}(1-x)^{Q^2/4-\Delta_{\alpha_1}-\Delta_{\alpha_3}} \times \\ & [\theta_3(q)]^{3Q^2-4(\Delta_{\alpha_1}+\Delta_{\alpha_2}+\Delta_{\alpha_3}+\Delta_{\alpha_4})}\, H(\lambda_i^2, \Delta|q) \end{aligned} \tag{5.1}$$

Here

$$\theta_3(q) = \sum_{n=-\infty}^{\infty} q^{n^2} \tag{5.2}$$

and

$$q = e^{i\pi\tau} \tag{5.3}$$

is related to x by the equation

$$\tau = i\frac{K(1-x)}{K(x)} \tag{5.4}$$

where

$$K(x) = \frac{1}{2}\int_0^1 \frac{dt}{\sqrt{t(1-t)(1-xt)}} \tag{5.5}$$

Function $H(\lambda_i^2, \Delta|q)$ is better parameterized in terms of the variables

$$\lambda_i^2 = \frac{Q^2}{4} - \Delta_{\alpha_i} = \left(\frac{Q}{2} - \alpha_i\right)^2 \tag{5.6}$$

instead of the external dimensions Δ_{α_i}. It satisfies the following recurrence relation

$$H(\lambda_i^2, \Delta|q) = 1 + \sum_{m,n} \frac{q^{mn} R_{m,n}(\lambda_i)}{\Delta - \Delta_{m,n}} H(\lambda_i^2, \Delta_{m,n} + mn|q) \tag{5.7}$$

Here the sum is over all pairs (m,n) of positive integers and

$$\Delta_{m,n} = \frac{Q^2}{4} - \frac{(m/b + nb)^2}{4} \tag{5.8}$$

are the dimensions of degenerate representations of the Virasoro algebra with the central charge (2.5). With the notation

$$\lambda_{m,n} = (m/b + nb)/2 \tag{5.9}$$

the multipliers $R_{m,n}(\lambda_i)$ read explicitly

$$R_{m,n}(\lambda_i) = 2 \times \frac{\prod_{r,s}(\lambda_1+\lambda_2-\lambda_{r,s})(\lambda_1-\lambda_2-\lambda_{r,s})(\lambda_3+\lambda_4-\lambda_{r,s})(\lambda_3-\lambda_4-\lambda_{r,s})}{\prod'_{k,l}\lambda_{k,l}} \tag{5.10}$$

The products in (5.10) are over the following sets of integers (r,s) and (k,l)

$$\begin{aligned} r &= -m+1, -m+3, \ldots, m-3, m-1 \\ s &= -n+1, -n+3, \ldots, n-3, n-1 \end{aligned} \tag{5.11}$$

and

$$\begin{aligned} k &= -m+1, -m+2, \ldots, m-1, m \\ l &= -n+1, -n+2, \ldots, n-1, n \end{aligned} \tag{5.12}$$

while the prime sign near the last product symbol $\prod'_{k,l}$ means that the two pairs $(k,l) = (0,0)$ and (m,n) are missing. From the numerical point of view the expression (5.7) is almost analytic in the sense that it permits us to compute $H(\lambda_i^2, \Delta|q)$ very fast and precisely.

In view of eq.(5.1) we find it more convenient to write down the four-point function (2.19) as

$$G_{\alpha_1,\alpha_2,\alpha_3,\alpha_4}(x,\bar{x}) = (x\bar{x})^{Q^2/4-\Delta_{\alpha_1}-\Delta_{\alpha_2}}[(1-x)(1-\bar{x})]^{Q^2/4-\Delta_{\alpha_1}-\Delta_{\alpha_3}} \times[\theta_3(q)\theta_3(\bar{q})]^{3Q^2-4\sum_i\Delta_{\alpha_i}} g_{\alpha_1,\alpha_2,\alpha_3,\alpha_4}(x,\bar{x}) \tag{5.13}$$

and study the crossing properties of the reduced function

$$g_{\alpha_1,\alpha_2,\alpha_3,\alpha_4}(\tau,\bar{\tau}) = \tfrac{1}{2}\int C(\alpha_1,\alpha_2,\tfrac{Q}{2}+iP)C(\alpha_3,\alpha_4,\tfrac{Q}{2}-iP)\left|(16q)^{P^2}H(\lambda_i^2,\tfrac{Q^2}{4}+P^2|q)\right|^2 dP \tag{5.14}$$

From eqs.(2.20) it follows that

$$\begin{aligned} g_{\alpha_1,\alpha_2,\alpha_3,\alpha_4}(\tau,\bar{\tau}) &= g_{\alpha_1,\alpha_2,\alpha_4,\alpha_3}(\tau+1,\bar{\tau}+1) \\ g_{\alpha_1,\alpha_2,\alpha_3,\alpha_4}(\tau,\bar{\tau}) &= |\tau|^{3Q^2-4\sum_i\Delta_{\alpha_i}} g_{\alpha_1,\alpha_3,\alpha_2,\alpha_4}(-1/\tau,-1/\bar{\tau}) \end{aligned} \tag{5.15}$$

The first of these equations is identically satisfied by (5.14) due to the following property of the function $H(\lambda_i^2,\Delta|q)$

$$H(\lambda_1^2,\lambda_2^2,\lambda_3^2,\lambda_4^2,\Delta|q) = H(\lambda_1^2,\lambda_2^2,\lambda_4^2,\lambda_3^2,\Delta|-q) \tag{5.16}$$

which is easily derived from the relation (5.7) and eq.(5.10). The second equation in (5.15) still remains a non-trivial condition for the structure constants which is believed to contain the whole information about the associativity of the operator algebra.

For the numerical calculations we have chosen the correlation function of four puncture operators (2.11)

$$g(\tau,\bar{\tau}) = \frac{1}{16}\left.\frac{\partial^4 g_{\alpha_1,\alpha_2,\alpha_3,\alpha_4}(\tau,\bar{\tau})}{\partial\alpha_1\partial\alpha_2\partial\alpha_3\partial\alpha_4}\right|_{\alpha_1=\alpha_2=\alpha_3=\alpha_4=Q/2} \tag{5.17}$$

After separating some irrelevant overall factors

$$g(\tau,\bar{\tau}) = \frac{k_0^8}{\pi^2}\left[\pi\mu\gamma(b^2)b^{2-2b^2}\right]^{-Q/b}\left(\frac{\tau-\bar{\tau}}{2i}\right)^{-Q^2/2} f(\tau,\bar{\tau}) \tag{5.18}$$

we have to verify the relation

$$f(\tau,\bar{\tau}) = f(-1/\tau,-1/\bar{\tau}) \tag{5.19}$$

for the function

$$f(\tau,\bar{\tau}) = \frac{1}{2}\left(\frac{\tau-\bar{\tau}}{2i}\right)^{Q^2/2}\int r(P)\left|(16q)^{P^2}H(0,Q^2/4+P^2|q)\right|^2 dP \tag{5.20}$$

Here

$$\begin{aligned} r(P) = \frac{\pi^2 k(2iP)k(-2iP)}{k_0^2 k^8(Q/2+iP)} &= \sinh\frac{2\pi P}{b}\sinh(2\pi bP) \\ &\times\exp\left[-8\int_0^\infty\frac{dt}{t}\frac{\sin^2 Pt\left(1-e^{-Qt}\cos^2 Pt\right)}{\sinh(bt)\sinh(t/b)}\right] \end{aligned} \tag{5.21}$$

We have computed $f(\tau,\bar{\tau})$ for several values of b and found that eq.(6.19) is satisfied up to high numerical accuracy. For example in figs.1 and 2 functions $f(\tau,\bar{\tau})$ and $f(-1/\tau,-1/\bar{\tau})$ are compared for purely imaginary $\tau = it$ at $b = 0.8$ and $b = (1+i)/\sqrt{2}$. This last complex value corresponds to $c_L = 13$.

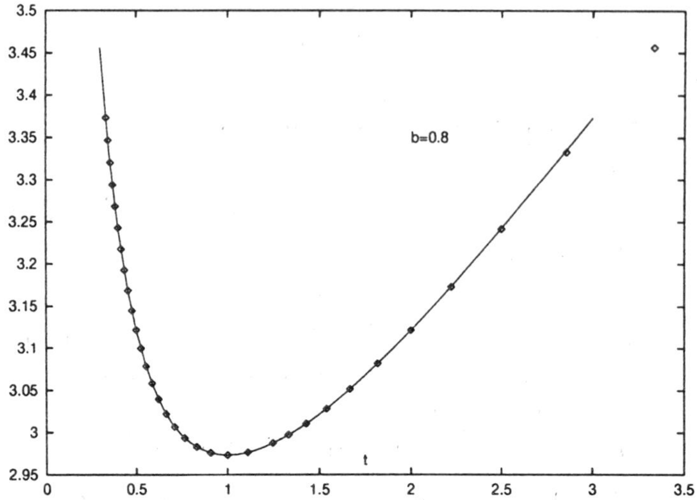

Figure 1. Reduced four-point function $f(\tau, \bar{\tau})$ of eq.(7.20) (continuous line) at $b = 0.8$ and real $t = -i\tau$. It is compared with $f(-1/\tau, -1/\bar{\tau})$ (points) to verify the crossing symmetry relation (7.19).

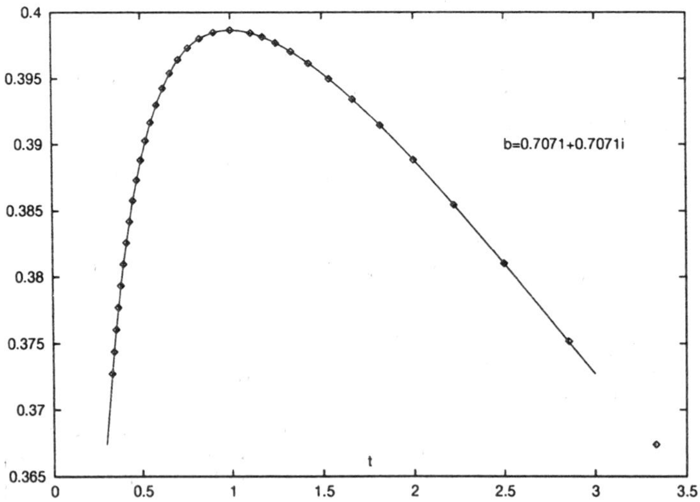

Figure 2. The same as in fig.1 but at $b = (1+i)/\sqrt{2}$ corresponding to $c_L = 13$.

6 Conclusions

The main subject of this lecture is the explicit expression (3.14) for the three-point function together with the supporting evidences in Sects.4,5. Although (3.14) is a conjecture we find the evidences convincing enough to take it as the starting point in addressing some important questions in 2d quantum gravity.

The structure constants (3.14) allow one to construct (in principle) the multipoint correlation functions of LFT through the decompositions similar to (2.22). This gives access to the multipoint correlation functions of quantum gravity (coupled to a matter theory) at fixed conformal moduli. This is in contrast with the integrated correlation functions considered in [11-15]. The moduli dependence of the correlation functions is expected to add some insight on the nature of physical states in 2d quantum gravity.

One can also try to analyze the physics of non minimal CFT or non conformal matter QFT coupled to quantum gravity.

The most interesting question is exactly what is it that happens to 2d quantum gravity when the central charge c_M of matter theory ($c_M = 26 - c_L$) exceeds 1. We hope that the structure constants (3.14) could form a convenient frame to approach this still rather mysterious domain.

And of course it is important to extend the above analysis to incorporate SUSY.

References

[1] A. Polyakov. *Phys. Lett.*, B103 (1981) 207.

[2] T. Curtright and C. Thorn. *Phys. Rev. Lett.*, 48 (1982) 1309; E. Braaten, T. Curtright and C. Thorn. *Phys. Lett.*, B118 (1982) 115; *Ann. Phys.*, 147 (1983) 365.

[3] J.-L. Gervais and A. Neveu. *Nucl. Phys.*, B238 (1984) 125; B238 (1984) 396; B257[FS14] (1985) 59.

[4] E. D'Hoker and R. Jackiw. *Phys. Rev.*, D26 (1982) 3517.

[5] V. Kazakov. *Phys. Lett.*, 150 (1985) 282; F.David. *Nucl. Phys.*, B257 (1985) 45; V. Kazakov, I. Kostov and A. Migdal. *Phys. Lett.*, 157 (1985) 295.

[6] E. Brézin and V. Kazakov. *Phys. Lett.*, B236 (1990) 144; M. Douglas and S. Shenker. *Nucl. Phys.*, B335 (1990) 635; D. Gross and A. Migdal. *Phys. Rev. Lett.*, 64 (1990) 127.

[7] V. Knizhnik, A. Polyakov and A. Zamolodchikov. *Mod. Phys. Lett.*, A3 (1988) 819.

[8] F. David. *Mod. Phys. Lett.*, A3 (1988) 1651.

[9] J. Distler and H. Kawai. *Nucl. Phys.*, B321 (1989) 509.

[10] M. Bershadsky and I. Klebanov. *Phys. Rev. Lett.*65(1990)3088.

[11] J. Polchinski. Remarks on Liouville Field theory, in "Strings 90", R. Arnowitt et al, eds, World Scientific, 1991; *Nucl. Phys.* B357 (1991) 241.

[12] M. Goulian and M. Li. *Phys. Rev. Lett.*, 66 (1991) 2051.

[13] A. Polyakov. *Mod. Phys. Lett.* A6 (1991) 635.

[14] P. Di Francesco, D. Kutasov. *Phys. Lett.* B261 (1991) 385.

[15] Vl. Dotsenko. *Mod. Phys. Lett.* A6 (1991) 3601.

[16] H. Dorn and H.-J. Otto. *Phys. Lett.*, B291 (1992) 39.

[17] H. Dorn and H.-J. Otto. *Nucl. Phys.*, B429 (1994) 375.

[18] Vl.S. Dotsenko, V.A. Fateev. *Phys. Lett.* 154B (1985) 291.

[19] V. Fateev and A. Zamolodchikov.*Sov. Phys. JETP* 82(1985)215; Sov.J.*Nucl. Phys.* 43 (1986)657.

[20] A. Belavin, A. Polyakov and A. Zamolodchikov. *Nucl. Phys.* B241 (1984) 333.

[21] Al. Zamolodchikov. *Comm. Math. Phys.*, 96 (1984) 419; *Theor. Math. Phys.*, 73 (1987) 1088.

[22] N. Seiberg. Notes on Quantum Liouville Theory and Quantum Gravity, in *Random Surfaces and Quantum Gravity*, ed. O. Alvarez, E. Marinari, P. Windey, Plenum Press, 1990.

[23] Vl. Dotsenko and V. Fateev. *Nucl. Phys.*, B251 (1985) 691.

THERMODYNAMIC BETHE ANSATZ FOR EXCITED STATES

A. Zamolodchikov

Department of Physics and Astronomy
Rutgers University
P.O.Box 849, Piscataway, New Jersey 08855-0849, USA
and
L.D.Landau Institute for Theoretical Physics
Kosygina 2, 117334, Moscow, Russia

LECTURE I

1.1. Introduction

In these lectures I am going to discuss Integrable Quantum Field Theories in two dimensions. Integrable QFT possess infinite series of commuting local Integrals of Motion (IM). Relatively simple but very important Integrable QFT are Conformal Field Theories (CFT). More general non-conformal Integrable QFT are obtained by perturbing CFT with suitably chosen relevant operators [1]. These Integrable Perturbed CFT (PCFT) are typically massive QFT and their exact on-shell solutions can be given in terms of Factorizable S-matrices[1,2]. The off-shell information although encoded in the S-matrix is much more difficult to extract. Very interesting off-shell characteristic of QFT is its finite-size energy spectrum, *e.g.* the spectrum of its Hamiltonian in the situation when the spatial coordinate is compactified on a circle of circumference R. Important progress towards calculating this spectrum has been made lately with the help of Thermodynamic Bethe Ansatz (TBA) technique [3]. This approach allows one to obtain the ground-state energy $E_0(R)$ of the finite-size system provided the on-shell solution is known, the problem being reduced to solving non-linear integral equations (TBA equations).

The ideas behind traditional derivation of the TBA equations can not be directly generalized to incorporate excited states, and therefore the power of the approach based on TBA was limited to the ground-state energy (or ground-state energies in

Low Dimensional Applications of Quantum Field Theory
Edited by L. Baulieu *et al.*, Plenum Press, New York, 1997

"twisted" sectors). However it turns out that basic quantities appearing in TBA ("Y-functions") admit interpretation in terms of the ground-sate eigenvalues of certain operators ("**T**-operators") which act in the space of states of the finite-size system [4]. Remarkable properties of these operators make it possible to derive the nonlinear integral equations which generalize TBA equations to excited states [5].

In these lectures I will give a short review of the above circle of ideas. Throughout the lectures I will use simple model of Integrable QFT, so called Lee-Yang QFT[6,3,15], to illustrate various general results. In the first lecture I will discuss integrable perturbations of CFT, Factorizable S-matrices and TBA. In the second lecture the construction and basic properties of the **T**-operators will be described. This construction will lead us to the generalization of the TBA technique to excited states. All new results presented in the second lecture are obtained in collaboration with V.Bazhanov and S.Lukyanov [4,5,7].

1.2. Local IM in CFT

Conformal symmetry in CFT is generated by the "left" and "right" chiral components $T(z)$ and $\bar{T}(\bar{z})$ of its (traceless) energy-momentum tensor [8]; here z and $\bar{z}$ are complex coordinates

$$z = x + iy, \qquad \bar{z} = x - iy. \tag{1.2.1}$$

The operators

$$L_n = \frac{c}{24}\,\delta_{n,0} + \frac{R}{2\pi}\int_0^R \frac{dx}{2\pi}\,T(x+iy)\,e^{i\frac{2\pi}{R}nx} \tag{1.2.2}$$

satisfy the Virasoro algebra vir

$$[L_n, L_m] = (n-m)\,L_{n+m} + \frac{c}{12}\,(n^3 - n)\,\delta_{n+m,0} \tag{1.2.3}$$

where c denotes the parameter known as the Virasoro central charge which takes specific value in each particular CFT. The "right" Virasoro generators $\bar{L}_n$ are defined by similar integrals in terms of the field $\bar{T}(x - iy)$; they form the "right" Virasoro algebra $\overline{vir}$ identical to (1.2.3). In writing (1.2.2) we have assumed that CFT is defined in the geometry of Euclidean cylinder, with the spatial coordinate compactified on a circle of the circumference R; correspondingly, the fields T and $\bar{T}$ satisfy the periodicity condition

$$T(z+R) = T(z), \qquad \bar{T}(\bar{z}+R) = \bar{T}(\bar{z}). \tag{1.2.4}$$

The space of states of a the CFT decomposes into a sum

$$\mathcal{H}_{CFT} = \oplus_a \mathcal{H}_a \tag{1.2.5}$$

where

$$\mathcal{H}_a = \left(\mathcal{V}_{\Delta_a} \otimes \bar{\mathcal{V}}_{\Delta_a}\right) \tag{1.2.6}$$

is an irreducible highest weight representation of $vir \oplus \overline{vir}$ with the highest weight (Δ_a, Δ_a). The space $\mathcal{V}_\Delta$ is spanned by the vectors

$$L_{-n_1}L_{-n_2}\ldots L_{n_N}\mid\Delta\rangle, \quad 0 \le n_1 \le n_2 \le \ldots \le n_N, \tag{1.2.7}$$

where $| \Delta \rangle$ is the highest-weight state which satisfies

$$L_n \mid \Delta\rangle = 0 \quad \text{for} \quad n > 0; \qquad L_0 \mid \Delta\rangle = \Delta \mid \Delta\rangle. \tag{1.2.8}$$

The space $\bar{\mathcal{V}}_\Delta$ is defined similarly in terms of the "right" Virasoro generators $\bar{L}_n$. Note that the space $\mathcal{V}_\Delta$ (and of course $\bar{\mathcal{V}}_\Delta$ as well) admits the following decomposition

$$\mathcal{V}_\Delta = \oplus_l \mathcal{V}_\Delta^{(l)}, \tag{1.2.9}$$

it terms of the finite-dimensional "level subspaces" $\mathcal{V}_\Delta^{(l)}$,

$$L_0 \, \mathcal{V}_\Delta^{(l)} = (\Delta + l)\, \mathcal{V}_\Delta^{(l)}. \tag{1.2.10}$$

The sum in (1.2.5) includes certain set of representations which is specific for each given CFT [8]. In writing (1.2.5) we ignored for simplicity the possibility of so called non-diagonal CFT [8].

Any CFT possesses infinitely many commuting local Integrals of Motion (IM) [9,10]. These are the operators acting in (1.2.5) which can be written as the integrals

$$\mathbf{I}_{2k-1} = \int_0^R \frac{dx}{2\pi} \, T_{2k}(x+iy), \tag{1.2.11}$$

where $T_{2k}(z)$, $k = 1, 2, \ldots$ are certain local fields built from $T(z)$ and its derivatives, for example

$$\begin{aligned} &T_2(z) = T(z), \quad T_4(z) =: T^2 : (z), \\ &T_6(z) =: T^3 : (z) + \frac{c+2}{12} : (\partial_z T)^2 : (z), \ \ldots \ , \\ &\qquad T_{2k}(z) =: T^k : (z) + \text{terms with derivatives}. \end{aligned} \tag{1.2.12}$$

Here : : denote appropriately regularized operator products, see [1] for details. The densities (1.2.12) can be chosen such a way that all the integrals (1.2.11) commute [9,10,11][1]

$$[\mathbf{I}_{2k-1}, \mathbf{I}_{2l-1}] = 0. \tag{1.2.13}$$

Of course there are also "right" local IM

$$\bar{\mathbf{I}}_{2k-1} = \int_0^R \frac{dx}{2\pi} \, \bar{T}_{2k}(x-iy), \tag{1.2.14}$$

built similar way in terms of $\bar{T}(\bar{z})$, which commute among themselves and also commute with all $\mathbf{I}_{2k-1}$.

Simultaneous diagonalization of the above operators $\mathbf{I}_{2k-1}$ in the space $\mathcal{V}_\Delta$ is an interesting mathematical problem. Note that $\mathbf{I}_1 = (2\pi/R)\,(L_0 - c/24)$ and so all the operators $\mathbf{I}_{2k-1}$ invariantly act in each of the level subspaces $\mathcal{V}_\Delta^{(l)}$ where their diagonalization reduces to a finite-dimensional algebraic problem. However this problem rapidly becomes very complex for large l and so straightforward approach does not provide the full solution. We will say more about this problem in the second Lecture.

[1] In fact there is more then one different choice of the such commuting set of local IM. The one we discuss here is suitable for perturbing the theory with the operator $\Phi_{(1,3)}$ (see below). Another interesting commuting set is discussed in [1] (see also [2]).

1.3. Integrable Perturbations of CFT

Given a CFT, one can construct a family of more general non-conformal QFT by perturbing the CFT by relevant operators Φ_A with conformal dimensions $\Delta_A < 1$ [1,2]. These QFT are defined by the formal action

$$\mathcal{A}_{PCFT} = \mathcal{A}_{CFT} + \sum_A \hat{\mu}_A \int \Phi_A(z,\bar{z})\, d^2z \ , \tag{1.3.1}$$

where $\mathcal{A}_{CFT}$ is the action of CFT and $\hat{\mu}_A$ are perturbation parameters which carry the dimensions

$$\hat{\mu}_A \sim [length]^{2(\Delta_A - 1)}. \tag{1.3.2}$$

The condition $\Delta_A < 1$ guarantees that in the short-distance limit the perturbation in (1.3.1) can be neglected and so the short distance behavior of this QFT is controlled by the original CFT. From the Renormalization Group (RG) point of view the CFT corresponds to certain fixed point of the RG and the QFT (action) span so called unstable manifold associated with this fixed point [12].

In many cases one can find among the QFT (1.3.1) particular perturbations in which all or some part of the local IM (1.2.11) of CFT remain commutative IM when the perturbation is added [1]. The best known example is $c < 1$ CFT perturbed with the degenerate field $\Phi_{(1,3)}$, the primary field which satisfies the third-level null-vector equations [8]

$$[(\Delta+1)(\Delta+2)\, L_{-3} - 2\,(\Delta+1)\, L_{-1}L_{-2} + L_{-3}]\Phi_{(1,3)} = 0,$$

$$[(\Delta+1)(\Delta+2)\, \bar{L}_{-3} - 2\,(\Delta+1)\, \bar{L}_{-1}\bar{L}_{-2} + \bar{L}_{-3}]\Phi_{(1,3)} = 0. \tag{1.3.3}$$

where $\Delta = \Delta_{(1,3)}$ is the conformal dimension of $\Phi_{(1,3)}$. Although in the perturbed theory

$$\mathcal{A}_{PCFT} = \mathcal{A}_{CFT} + \hat{\mu} \int \Phi_{(1,3)}(z,\bar{z})\, d^2z \tag{1.3.4}$$

the fields T_{2k} are no longer holomorphic fields, *i.e.* they depend both on z and $\bar{z}$, they satisfy the continuity equations

$$\partial_{\bar{z}} T_{2k}(z,\bar{z}) = \hat{\mu}\, \partial_z \Theta_{2k-2}(z,\bar{z}), \tag{1.3.5}$$

where $\Theta_{2k-2}(z,\bar{z})$ are certain local fields. Similarly,

$$\partial_z \bar{T}_{2k}(\bar{z},z) = \hat{\mu}\, \partial_{\bar{z}} \bar{\Theta}_{2k-2}(\bar{z},z), \tag{1.3.6}$$

It follows that the integrals

$$\mathbf{I}_{2k-1} = \int_0^R (T_{2k}(x+iy, x-iy) + \hat{\mu}\, \Theta(x+iy, x-iy))\, dx$$

$$\bar{\mathbf{I}}_{2k-1} = \int_0^R (\bar{T}_{2k}(x-iy, x+iy) + \hat{\mu}\, \bar{\Theta}(x-iy, x+iy))\, dx \tag{1.3.7}$$

in fact do not depend on the "Euclidean time" y, *i.e.* they are Integrals of Motion (IM). In writing (1.3.7) we have assumed again that the perturbed theory (1.3.4) is defined on the cylinder with the spatial coordinate x compactified on the circle of circumference R, in particular the field $\Phi_{(1,3)}$ in (1.3.4) is periodic, $\Phi_{(1,3)}(z+R,\bar{z}+R)=\Phi_{(1,3)}(z,\bar{z})$. It is possible to show that the operators (1.3.7) commute among themselves,

$$[\mathbf{I}_{2k-1},\mathbf{I}_{2l-1}]=[\mathbf{I}_{2k-1},\bar{\mathbf{I}}_{2l-1}]=[\bar{\mathbf{I}}_{2k-1},\bar{\mathbf{I}}_{2l-1}]=0. \tag{1.3.8}$$

Note that the field $\hat{\mu}\,\Theta_0=\hat{\mu}\,\bar{\Theta}_0$ coincides with the trace T^{α}_{α} of the energy-momentum tensor $T_{\alpha\beta}$ of the QFT (1.3.4), and the IM $\mathbf{I}_1$ and $\bar{\mathbf{I}}_1$ are related to the Hamiltonian $\mathbf{H}$ and the spatial momentum $\mathbf{P}$ of this QFT,

$$\mathbf{H}=\mathbf{I}_1+\bar{\mathbf{I}}_1, \qquad \mathbf{P}=\mathbf{I}_1-\bar{\mathbf{I}}_1. \tag{1.3.9}$$

Throughout these lectures I will illustrate main ideas and methods using relatively simple example of Integrable QFT - so called Lee-Yang QFT. This QFT is obtained by perturbing one of the simplest "minimal CFT" (see [8]), the model $\mathcal{M}_{2/5}$.

The CFT $\mathcal{M}_{2/5}$ has the central charge

$$c(\mathcal{M}_{2/5})=-22/5 \tag{1.3.10}$$

and its space of states (1.2.5) contains only two irreducible representations

$$\mathcal{H}_{\mathcal{M}_{2/5}}=\mathcal{H}_0\oplus\mathcal{H}_1\ , \tag{1.3.11}$$

with the highest weights

$$\Delta_0=-1/5\ , \qquad \Delta_1=0. \tag{1.3.12}$$

This CFT is non-unitary, *i.e.* the natural scalar product in the space (1.3.11) is not positive. As was shown by Cardy [13] the CFT $\mathcal{M}_{2/5}$ describes the criticality associated with Lee-Yang edge singularity [14] and therefore it is often referred to as the Lee-Yang CFT.

The Lee-Yang QFT is described by the action (1.3.4), i.e.

$$\mathcal{A}_{LY}=\mathcal{A}_{\mathcal{M}_{2/5}}+\mu\int\Phi(z,\bar{z})\,d^2z\ , \tag{1.3.13}$$

where $\Phi=\Phi_{(1,3)}$ is the primary field with the conformal dimension $\Delta=-1/5$ and $\hat{\mu}\sim[length]^{-\frac{12}{5}}$; if the field Φ is normalized so that

$$\langle\Phi(z_1,\bar{z}_1)\Phi(z_2,\bar{z}_2)\rangle\simeq-|z_1-z_2|^{\frac{4}{5}} \quad \text{as} \quad |z_1-z_2|\to 0 \tag{1.3.14}$$

the parameter $\hat{\mu}$ should be taken real [6].

As the perturbation in (1.3.13) is relevant (*i.e.* the mass dimension of the parameter $\hat{\mu}$ is positive) the perturbation theory in $\hat{\mu}$ generated by the action (1.3.13) is superrenormalizable; in fact this perturbation theory does not have any short-distance divergences at all. Therefore in the finite-size system with $R<\infty$ where possible infrared divergences are also well contained there are all reasons to assume that the space of states $\mathcal{H}_{LY}$ of the perturbed theory (1.3.13) coincides with the space of states of the CFT $\mathcal{M}_{2/5}$,

$$\mathcal{H}_{LY}\sim\mathcal{H}_{\mathcal{M}_{2/5}}. \tag{1.3.15}$$

Of course for $\hat{\mu} \neq 0$ the highest-weight states $| -\frac{1}{5}\rangle$ and $| 0\rangle$ in (1.3.11) are no longer stationary states of the perturbed Hamiltonian (1.3.9); the exact eigenstates of the perturbed Hamiltonian (1.3.9) are embedded into (1.3.11) in complicated way and the associated eigenvalues $E_n(R)$ are very complicated functions of R. I will show how one can calculate the ground-state energy eigenvalue $E_0(R)$ using so called Thermodynamic Bethe Ansatz (TBA) approach by the end of this Lecture; the general method for calculating the excited-state energies $E_n(R)$ will be developed in the next one.

The QFT (1.3.13) possesses infinite number of commuting IM (1.3.7), *i.e.* it is integrable. In fact an important remark is in order here. The representations entering (1.3.11) are "strongly degenerate" ones (*i.e.* the Virasoro moduli with the highest weights (1.3.12) exhibit greater number of null vectors then the generic ones [CFT]), as the result some of the local densities (1.2.12) identically vanish. It is possible to show that only those IM $\mathbf{I}_{2k-1}$ and $\bar{\mathbf{I}}_{2k-1}$ are nonzero which have the spins $2k-1 \neq 0 \,(\mathrm{mod}\ 3)$, *i.e.* $2k-1 = 1, 5, 7, 11, \ldots$ [6].

1.4. Integrable QFT and Factorizable S-matrices

In general case a massive QFT in infinite space-time can be interpreted as a scattering theory of relativistic particles [16]. A massive Integrable QFT in two-dimensional infinite space-time relates to so called Factorizable Scattering Theory (FST)[17]. To make connection with our previous discussion in Sect.1.1 and 1.2 let us consider an Integrable QFT defined, say, by the action (1.3.4), in *infinite volume*, i.e at $R = \infty$. Making continuation to pure imaginary values of the coordinate y,

$$y \to it\,, \tag{1.4.1}$$

one obtains the Lorenz invariant QFT in infinite two-dimensional Minkowski space-time. The space of states $\mathcal{H}_\infty$ of such QFT is the Fock space of multiparticle states, the particle content (i.e the number of different species of particles and their masses) being specific for given QFT. Let us assume that there are N species of particles $A_j\,,\ j = 1,\,2,\,\ldots,\,N$, with the masses m_j. The multiparticle states can be classified as asymptotic *in*- or *out*- states, which we denote as

$$| A_{j_1}(\theta_1) A_{j_2}(\theta_2) \cdots A_{j_n}(\theta_n)\rangle_{\substack{in \\ out}} \tag{1.4.2}$$

where $\theta_1, \theta_2, \ldots, \theta_n$ are the asymptotic (at $t \to \mp\infty$) *rapidities* of the particles which relate to their asymptotic momenta as

$$p_j^0 = m_j \cosh\theta_j; \qquad p_j^1 = m_j \sinh\theta_j, \tag{1.4.3}$$

where p^0 denote the time and p^1 the space components of the two-momenta. According to *asymptotic completeness* [16] the *in*-states (1.4.2) form a basis in $\mathcal{H}_\infty$; the same is true for the *out*-states (1.4.2) and the relation between these two bases is provided by the S-matrix.

In Integrable QFT the S-matrix is subject to very strong restrictions. The operators $\mathbf{I}_{2k-1}$ and $\bar{\mathbf{I}}_{2k-1}$ defined by the integrals similar to (1.3.7), but with the integration extending from $-\infty$ to ∞, act in $\mathcal{H}_\infty$. It is easy to show that one-particle states must be the eigenstates of these operators,

$$\mathbf{I}_{2k-1} \mid A_j(\theta)\rangle = M_j^{(2k-1)} \, e^{\,(2k-1)\theta} \mid A_j(\theta)\rangle$$

$$\bar{\mathbf{I}}_{2k-1} \mid A_j(\theta)\rangle = M_j^{(2k-1)} \, e^{-(2k-1)\theta} \mid A_j(\theta)\rangle \tag{1.4.4}$$

where $M_j^{(2k-1)}$ are constants which carry the dimensions $[mass]^{2k-1}$, in particular $M_j^{(1)} = m_j$. The θ dependence of the eigenvalues in (1.4.4) follow from the commutation relations

$$[\mathbf{M}, \mathbf{I}_{2k-1}] = (2k-1)\,\mathbf{I}_{2k-1}\,, \quad [\mathbf{M}, \bar{\mathbf{I}}_{2k-1}] = -(2k-1)\,\bar{\mathbf{I}}_{2k-1}\,, \tag{1.4.5}$$

where $\mathbf{M}$ is the Lorenz boost generator. As for the multiparticle states (1.4.2), the structure (1.3.7) of the local IM and the very idea of the asymptotic states as the states where the particles are infinitely separated in the asymptotic past or future suggest that

$$\mathbf{I}_{2k-1} \mid A_{j_1}(\theta_1)A_{j_2}(\theta_2)\cdots A_{j_n}(\theta_n)\rangle_{\substack{in\\out}} = \tag{1.4.6}$$

$$(M_{j_1}^{(2k-1)}\, e^{(2k-1)\theta_1} + \cdots + M_{j_n}^{(2k-1)}\, e^{(2k-1)\theta_n}) \mid A_{j_1}(\theta_1)A_{j_2}(\theta_2)\cdots A_{j_n}(\theta_n)\rangle_{\substack{in\\out}}\,,$$

and similarly for $\bar{\mathbf{I}}_{2k-1}$, with $e^{\,(2k-1)\theta}$ replaced by $e^{-(2k-1)\theta}$. Being the Integrals of Motion the operators $\mathbf{I}_{2k-1}$ and $\bar{\mathbf{I}}_{2k-1}$ must commute with the S-matrix and therefore (1.4.6) provide very restrictive conditions upon possible rapidities of the outgoing particles associated with given *in*-state. More careful analysis of the situation [18,17,19] shows that when enough local IM of the form (1.3.7) is present QFT the associated scattering theory possesses the following remarkable properties

i. The number of particles and the set of their individual rapidities do not change in the process of the scattering, *i.e.* $n_{out} = n_{in}$ and $\{\theta_1, \ldots, \theta_n\}_{out} = \{\theta_1, \ldots, \theta_n\}_{in}$.

ii. The n-particle S-matrix factorizes into the product of $n(n-1)/2$ two-particle S-matrices, in full agreement with the idea of n-particle scattering process as the series of successive pairwise collisions.

The scattering theory with the above properties is known as *Factorizable.* Clearly the two-particle S-matrix is the central object in Factorizable Scattering Theory (FST). Its elements $S_{i_1 i_2}^{j_1 j_2}$ are defined as the coefficients in the expansion

$$\mid A_{i_1}(\theta_1)A_{i_2}(\theta_2)\rangle_{in} = \sum_{j_1 j_2} S_{i_1 i_2}^{j_1 j_2}(\theta_1 - \theta_2) \mid A_{j_1}(\theta_1)A_{j_2}(\theta_2)\rangle_{out}. \tag{1.4.7}$$

The fact that they depend only on the difference $\theta_1 - \theta_2$ of the rapidities follow from the Lorenz invariance. In FST the two-particle S-matrix must obey powerful restrictions coming from the general requirements of S-matrix theory such as unitarity, analyticity and crossing symmetry conditions [16] as well as specific "factorizability conditions"[17] which guarantee the consistency of the above factorizable structure of the full S-matrix. I am not going to discuss here any details of the Factorizable S-matrix theory; one can find them in original papers and reviews (see *e.g.* [2,17]).

Instead I will describe particular example, the factorizable S-matrix associated with the Lee-Yang QFT.

The FST corresponding to the QFT (1.3.13) was found by Mussardo and Cardy [6]. The theory contains one sort of scalar particles A. As follows from simple dimensional analysis the mass m of the particle A relates to the parameter $\hat{\mu}$ in (1.3.13) as

$$m = \rho\, \mu^{\frac{5}{12}}, \tag{1.4.8}$$

where ρ is certain number. The two-particle S-matrix has only one element $S(\theta)$ which describes the process $AA \to AA$, i.e.

$$\mid A(\theta_1)A(\theta_2)\rangle_{in} = S(\theta_1 - \theta_2) \mid A(\theta_1)A(\theta_2)\rangle_{out}. \tag{1.4.9}$$

It follows from the standard analyticity conditions of S-matrix theory that $S(\theta)$ is meromorphic function in the complex θ plane, real at $\Re e\theta = 0$.

The cross-channel of the scattering $AA \to AA$ describes the same process and therefore the crossing-symmetry condition for this S-matrix can be written as

$$S(\theta) = S(i\pi - \theta). \tag{1.4.10}$$

As was mentioned above the Lee-Yang QFT is non-unitary in the sense that the natural scalar product in $\mathcal{H}_\infty$ is not positive defined. Nonetheless its S-matrix satisfy the "unitarity" condition $S\,S^\dagger = 1$, where dagger denotes the conjugation with respect to that non-positive scalar product. This "unitarity" condition can be expressed as

$$S(\theta)S(-\theta) = 1 \tag{1.4.11}$$

It follows from (1.4.10) and (1.4.11) that $S(\theta)$ is periodic function with the imaginary period $2\pi i$.

The domain $0 < \text{Imm}\theta < \pi$ is called *physical strip*. The only singularities of the amplitude $S(\theta)$ admitted into the physical strip are poles at the axis $\Re e\theta = 0$ which are interpreted as the stable particles of the same QFT - the "bound states" of two particles A. In the Lee-Yang QFT there are no symmetries which could forbid appearance of the particle A itself as such bound state. Therefore let us assume, following Mussardo and Cardy [6], that the amplitude $S(\theta)$ indeed has the pole corresponding to the particle A. Simple calculation shows that this pole must be located at $\theta = 2\pi i/3$. Then according to (1.4.10) the amplitude $S(\theta)$ must have another pole at $\theta = i\pi/3$ corresponding to the "bound state" A in the "cross-channel".

The multiparticle elements of the S-matrix must also exhibit similar poles due to the fact that the particle A could appear as the bound state of any pair of the particles involved in the scattering process. Therefore the $n-1$-particle element can be related to the residue of the n-particle amplitudes taken at one of the above poles. As in FST the multiparticle S-matrix is the product of two-particle S-matrices this "bootstrap condition" reduces to the simple relation for the two-particle S-matrix itself,

$$S(\theta + i\pi/3)S(\theta - i\pi/3) = S(\theta). \tag{1.4.12}$$

The "minimal" S-matrix which satisfy all the above relations (1.4.10) , (1.4.11) and (1.4.12) has the form

$$S(\theta) = \frac{\sinh\theta + i\sin(\pi/3)}{\sinh\theta - i\sin(\pi/3)} \, ; \tag{1.4.13}$$

it was proposed in [6] as the exact two-particle S-matrix associated with the QFT (1.3.13). As the full S-matrix is expressed in a simple way through the amplitude (1.4.13) it gives the complete on-shell solution to the Lee-Yang QFT.

One remark concerning the local IM (1.3.7) in Lee-Yang QFT can be added. According to (1.4.6) the operators $\mathbf{I}_{2k-1}$ act on the two-particle state as

$$\mathbf{I}_{2k-1} \mid A(\theta_1)A(\theta_2)\rangle_{in} = M^{(2k-1)}\left(e^{(2k-1)\theta_1} + e^{(2k-1)\theta_2}\right) \mid A(\theta_1)A(\theta_2)\rangle_{in}. \tag{1.4.14}$$

Let us analytically continue this eigenvalue to the complex values of the rapidities and evaluate it at $\theta_1 - \theta_2 = 2\pi i/3$ where the S-matrix in (1.4.9) exhibits the bound-state pole. We obtain

$$M^{(2k-1)}\left(e^{(2k-1)(\theta_0 + i\pi/3)} + e^{(2k-1)(\theta_0 - i\pi/3)}\right) = 2\,M^{(2k-1)}\,e^{(2k-1)\theta_0}\,\cos\left(\frac{2k-1}{3}\pi\right).$$

where $\theta_0 = \frac{1}{2}(\theta_1 + \theta_2)$. As near the pole this two-particle state is dominated by the "bound state" $\mid A(\theta_0)\rangle$ the above expression must coincide with the eigenvalue $M^{(2k-1)}\,e^{(2k-1)\theta_0}$ of $\mathbf{I}_{2k-1}$ on the one-particle state $\mid A(\theta_0)\rangle$. Clearly, this requires

$$M^{(2k-1)} = 0 \quad \text{for} \quad 2k - 1 = 0 \ (\mathrm{mod}\, 3). \tag{1.4.15}$$

Indeed, as was mentioned in Sect.1.3 in the Lee-Yang QFT the IM $\mathbf{I}_{2k-1}$ with $2k-1 = 0 \ (\mathrm{mod}\, 3)$ do not exist.

1.5. Thermodynamic Bethe Ansatz

In principle, all physical information about a QFT is contained in its S-matrix. However it is only the large-scale behavior of the QFT which is immediately available through its particle spectrum and scattering amplitudes. The information about short-distance properties of the QFT is hidden and in most cases very difficult to extract. Nevertheless, I am going to show here how one can calculate the ground-state energy $E_0(R)$ of an Integrable QFT in finite-size geometry (with the spatial coordinate compactified on a circle of the circumference R) starting from its factorizable S-matrix. I will use the approach known as Thermodynamic Bethe Ansatz (TBA); its origin is traced back to the pioneering works of Yang and Yang [20], in local relativistic QFT it was first introduced in [3].

Let us consider, as we did in Sect.1.2 and 1.3, the Euclidean cylinder with the compact coordinate x, $x \sim x + R$, and the coordinate y along the cylinder. Obviously, in this geometry there are two natural Hamiltonian pictures to choose between. First, one can interpret y as the (Euclidean) time and associate the space of states to the "equal-time" section y = const which is the circle of circumference R; this choice was actually implied throughout our discussion in Sect.1.2 and 1.3 and the spaces

(1.2.5), (1.3.11) and (1.3.15) correspond in fact to this Hamiltonian picture. The "time" evolution of the states in this picture is described by the Hamiltonian

$$\mathbf{H}_R = \frac{1}{2\pi} \int_0^R T_{yy}(x, y)\, dx \tag{1.5.1}$$

which is the same as (1.3.9) in Sect.1.3; here $T_{\alpha\beta}$ is the energy-momentum tensor[2]. If $E_0(R)$ is the lowest eigenvalue of (1.5.1) the partition function of the QFT on very long but finite cylinder (of the length L) is

$$Z = e^{-L\, E_0(R)} \tag{1.5.2}$$

Alternatively, one could take the compact coordinate x for the Euclidean time (this means considering the QFT at a finite temperature $T = R^{-1}$) so that the space of states would be associated with the infinite (or very long) line $x =$ const; this is the space $\mathcal{H}_\infty$ which we discussed in Sect.1.4. the associated Hamiltonian is

$$\mathbf{H}_\infty = \frac{1}{2\pi} \int_{-\infty}^{\infty} T_{xx}(x, y)\, dy\,, \tag{1.5.3}$$

and the partition function (1.5.2) can be calculated in this Hamiltonian picture as

$$Z = \mathrm{tr}_{\mathcal{H}_\infty} \left[e^{-R\mathbf{H}_\infty} \right]. \tag{1.5.4}$$

As was explained in Sect.1.4 the space $\mathcal{H}_\infty$ is spanned by the multiparticle states (1.4.2). Therefore using the knowledge about the S-matrix one can calculate the trace in (1.5.4) and thus find through (1.5.2) the ground-state energy $E_0(R)$ of the finite-size system. Let us see how it is done. In the following consideration I again will have in mind the simple example of Lee-Yang QFT, in particular I will assume that the FST under consideration contains only one sort of scalar particles A.

Our goal is to calculate the energy density of states in the space $\mathcal{H}_\infty$. Clearly, for truly infinite system this density is infinite and so to avoid working with infinite quantities let us consider first large but finite system, with the coordinate y also compactified on a circle of large circumference L (this was already assumed in writing (1.5.2)) and then take the limit $L \to \infty$. If L is much larger then $R_c = m^{-1}$ the states still can be classified as the multiparticle states (1.4.2) although their rapidities are no longer arbitrary. The requirement of periodicity of the particle's "wave functions[3]" leads to quantization of these rapidities. Indeed, consider the state of n particles A with the rapidities $\theta_1,\ \theta_2,\ \ldots,\ \theta_n$. To compare the wave function of, say, i-th particle at some point y to its value at $y + L$ one has to bring this particle around the large circle of the length L. In doing that one has to take into account the phase factors $S(\theta_i - \theta_j)$ the particle's wave function acquires in scatterings with the other $n-1$ particles; here $S(\theta)$ is the two-particle S-matrix (1.4.9). Therefore the required periodicity condition takes the form

$$e^{mL\sinh\theta_i} \prod_{j\neq i} S(\theta_i - \theta_j) = 1, \tag{1.5.5}$$

[2]Our normalization of $T_{\alpha\beta}$ corresponds to the definition $\delta\mathcal{A} = (1/2\pi)\int \sqrt{g}\, \delta g_{\alpha\beta}\, T^{\alpha\beta}$.

[3]Of course in QFT in view of virtual creation and annihilation of particles the notion of particle wave function applies only in the situations when all the particles are well separated in space. Note that in the arguments presented here one always can restrict attention to this domain.

where the first factor is just the wave factor $e^{ip_i L}$ with $p_i = m \sinh\theta_i$. The equations (1.5.5), which must hold for every $i = 1, 2, \ldots, n$, are known as the Bethe Ansatz equations. It must be stressed that in relativistic QFT the equations (1.5.5) hold only asymptotically at $L \to \infty$; it can be shown that the effects of virtual particle creation neglected in the above picture are exponentially small in $m L$. And of course the above considerations leading to (1.5.5) apply only to Integrable QFT as in non-integrable cases for instance the number of particles does not conserve even on-shell.

The system of equations (1.5.5) is just what we need in order to calculate the partition function (1.5.4). In the limit $L \to \infty$ the number of particles in any thermal state grows as L, and the spectrum of their rapidities becomes very dense with the typical rapidity interval between two adjacent levels $\theta_i - \theta_{i+1} \sim 1/mL \to 0$. Therefore one can introduce the density of particles in the rapidity space, $\rho_p(\theta)$, so that the number of particles Δn having their rapidities in the interval $\Delta\theta$ is

$$\Delta n = L\,\rho_p(\theta)\,\Delta\theta. \tag{1.5.6}$$

Let us recall now that the actual values that the rapidities θ_i can take are determined by the Bethe-Ansatz equations (1.5.5). The later can be written as

$$m\,L\,\sinh\theta_i + \sum_{j\neq i}\delta(\theta_i - \theta_j) = 2\pi\,n_i\ , \tag{1.5.7}$$

where $\delta(\theta) = -i\,\log S(\theta)$ is the scattering phase and n_i are integers. In the limit $L \to \infty$ the sum in (1.5.8) can be replaced by the integral, and one obtains the equation

$$m\,\sinh\theta_i + \int_{-\infty}^{\infty}\delta(\theta_i - \theta')\,\rho_p(\theta')\,d\theta' = 2\pi\,n_i/L. \tag{1.5.8}$$

With appropriate choice of the integers n_i in the r.h.s this equation returns the particle rapidities θ_i which constitute the density ρ_p. However one can consider the equation (1.5.8) with arbitrary integers n_i in the r.h.s. , quite independent of the actual particle rapidities. In this case the Eq.(1.5.8) is interpreted as the equation for the "rapidity levels", *i.e.* possible rapidity positions the particles could assume given the overall particle density $\rho_p(\theta)$. Now we can introduce the "density of levels" $\rho_s(\theta)$ so that $L\,\rho_s(\theta)\,\Delta\theta$ is the number of the levels in the rapidity interval $\Delta\theta$, and write the Eq.(1.5.8) as

$$\rho_s(\theta) = \frac{m}{2\pi}\cosh\theta + \int_{-\infty}^{\infty}\Phi(\theta - \theta')\,\rho_p(\theta')\,d\theta'\ , \tag{1.5.9}$$

where

$$\Phi(\theta) = \frac{1}{2\pi i}\frac{d}{d\theta}S(\theta). \tag{1.5.10}$$

For $L \to \infty$ there is large number of different "microscopic" states corresponding to any given particle density $\rho_p(\theta)$; this number grows exponentially with L,

$$N_{states}\,[\rho_p] \sim e^{L\,S[\rho_p]}\ , \tag{1.5.11}$$

where $S[\rho_p]$ is specific (per unit length) entropy of the "macroscopic" state characterized by the density ρ_p. This entropy can be expressed through the densities ρ_p and ρ_s as [20,3]

$$S[\rho_p] = \int_{-\infty}^{\infty} [\rho_s(\theta)\,\log\rho_s(\theta) - \rho_p(\theta)\,\log\rho_p(\theta) -$$

$$(\rho_s(\theta) - \rho_p(\theta))\,\log\left(\rho_s(\theta) - \rho_p(\theta)\right)]\,d\theta\ , \tag{1.5.12}$$

where we have assumed that $S(0) = -1$ (as it is the case for the lee-Yang S-matrix (1.4.13)) and so the particles effectively have Fermi statistics [3]. The density ρ_s entering (1.5.12) is related to ρ_p by the Eq. (1.5.9).

The energy of the state can be expressed through the rapidity density of the particles as

$$E[\rho_p] = m\,L \int_{-\infty}^{\infty} \cosh\theta\,\rho_p(\theta)\,d\theta + L\,\varepsilon_0\ , \tag{1.5.13}$$

where ε_0 is the specific vacuum energy. This equation together with the expression (1.5.12) for the entropy S in (1.5.11) contain just enough data to calculate the energy density of states and whence the partition function (1.5.5). The easiest way to do that is to find actual density ρ_p^{eq} corresponding to the thermal equilibrium state at temperature R^{-1} which minimizes the (specific) free energy

$$R\,F[\rho_p] = R\,E[\rho_p] - S[\rho_p]. \tag{1.5.14}$$

With little algebra one finds the extremum condition

$$\epsilon(\theta) = m\,R\,\cosh\theta + \int_{-\infty}^{\infty} \Phi(\theta-\theta')\,\log\left(1+e^{-\epsilon(\theta')}\right)d\theta\ , \tag{1.5.15}$$

where $\epsilon(\theta)$ is related to $\rho_p(\theta)$ as

$$e^{-\epsilon(\theta)} = \frac{\rho_p(\theta)}{\rho_s(\theta)-\rho_p(\theta)}\ , \tag{1.5.16}$$

and ρ_s is given by (1.5.15). The desired finite-size ground-state energy in (1.5.2) is directly related to the extremal value of the free energy (1.5.14). Simple calculation gives

$$E_0(R) = \frac{m}{2\pi}\int_{-\infty}^{\infty} \cosh\theta\,\log\left(1+e^{-\epsilon(\theta')}\right)d\theta + R\,\varepsilon_0. \tag{1.5.17}$$

The nonlinear integral equation (1.5.15) together with the expression (1.5.17) are known as the TBA equations [20,3].

For the Lee-Yang QFT the kernel $\Phi(\theta)$ in (1.5.15) can be derived from (1.4.13),

$$\Phi(\theta) = \frac{\sqrt{3}}{\pi}\,\frac{\sinh(2\theta)}{\sinh(3\theta)}\ , \tag{1.5.18}$$

The TBA equations with this kernel was studied in [3] where it is shown in particular that the ground-state energy $E_0(R)$ determined by these equations behaves as

$$E_0(R) \simeq -\frac{\pi}{15R}\ , \tag{1.5.19}$$

in complete agreement with (1.3.10).

A number of important properties of the function $\epsilon(\theta)$ can be extracted from the equation (1.5.15) and explicit form of the kernel (1.5.18) [3,21]. First, it is clear from the form of (1.5.15) that the first term in its r.h.s always dominate at $\theta \to \pm\infty$ and therefore the large-θ asymptotic of $\epsilon(\theta)$ is

$$\epsilon(\theta) \simeq m\,R\,\exp|\theta| \qquad \text{for} \qquad |\theta| \to \infty. \tag{1.5.20}$$

Next, it can be shown using (1.5.18) that the equation (1.5.15) defines the function $\epsilon(\theta)$ which is analytic in the strip $-\pi/3 < \mathrm{Imm}\theta < \pi/3$ in the complex θ-plane. And finally, it turns out that the integral operator $\hat{\Phi}$ defined by the kernel (1.5.18) is the inverse to the following finite-difference operator

$$\hat{\Phi} = \hat{D}^{-1}; \quad (\hat{D}\,f)(\theta) = f(\theta + i\pi/3) + f(\theta - i\pi/3) - f(\theta). \tag{1.5.21}$$

It is also easy to check that the "free term" $mR\cosh\theta$ in (1.5.15) is annihilated by this operator $\hat{D}$ and therefore the function

$$Y(\theta) = e^{\epsilon(\theta)} \tag{1.5.22}$$

satisfies the functional equation

$$Y(\theta + i\pi/3)\,Y(\theta - i\pi/3) = 1 + Y(\theta). \tag{1.5.23}$$

Appearance of the functional equations similar to (1.5.23) is typical feature of the TBA systems associated with the integrable perturbations of rational CFT [21]. These functional equations turn out to be very useful in analyzing analytic properties of the solutions to the TBA equations. In particular, it follows from (1.5.23) that the function (1.5.22) satisfy the periodicity condition

$$Y(\theta + 5i\pi/3) = Y(\theta). \tag{1.5.24}$$

In the next Lecture I will show that the function (1.5.22) coincides with the ground-state eigenvalue of certain operator $\mathbf{Y}(\theta)$ which acts in the space (1.3.11). The relations (1.5.23) and (1.5.24) turn out to be simple consequences of general properties of this operator. Moreover, using the properties of the operator $\mathbf{Y}(\theta)$ one can find the equations which generalize the TBA system (1.5.15) and (1.5.17) to excited states.

LECTURE II

2.1. KdV Theory

As we already mentioned in Sect.1.1 any CFT possesses infinitely many commuting IM, *i.e.* the operators $\mathbf{I}_{2k-1}$ having the structure (1.2.11) which act in the highest weight Virasoro moduli $\mathcal{V}_\Delta$ and satisfy there the equations (1.2.13). Simultaneous diagonalization of these operators in the space $\mathcal{V}_\Delta$ appears to be very interesting mathematical problem. As was remarked in [9,10,22] this problem can be viewed as the quantum version of famous Korteweg-de Vries (KdV) problem as it reduces to the classical KdV theory in the "classical limit" $c \to -\infty$.

The classical KdV equation

$$u_\tau = u_{zzz} + 6\, u\, u_z, \tag{2.1.1}$$

where $u = u(z,\tau)$ and the subscripts denote the derivatives, is known to be completely integrable Hamiltonian equation [23]. One possible Hamiltonian form (the "second Hamiltonian structure") of this equation is provided by the Poisson bracket

$$\{u(z), u(z')\} = 2\,(u(z) + u(z'))\,\delta_z(z-z') + \delta_{zzz}(z-z') \tag{2.1.2}$$

and the Hamiltonian functional

$$H = -\frac{1}{2}\int_0^R u^2(z)\, dz\ , \tag{2.1.3}$$

so that the equation (2.1.1) is written as

$$u_\tau(z) = \{H, u(z)\}. \tag{2.1.4}$$

We have assumed here that the field u satisfies the periodicity condition $u(z+R) = u(z)$, *i.e.* we consider the so-called periodic KdV problem [23]. Therefore the function $u(z)$ can be written as the Fourier series

$$u(z) = \frac{1}{2} - \sum_{-\infty}^{\infty} l_n\, e^{-i\frac{2\pi}{R}nz} \tag{2.1.5}$$

and the Poisson bracket (2.1.2) reduce to

$$\{l_n, l_m\} = (n-m)\, l_{n+m} + \frac{1}{2}(n^3 - n)\,\delta_{n+m,0}. \tag{2.1.6}$$

The KdV system possesses infinitely many Integrals of Motion (IM)

$$\begin{aligned}
\mathbf{I}_1^{(cl)} &= \textstyle\int_0^R \frac{dz}{2\pi}\ u(z),\\
\mathbf{I}_3^{(cl)} &= \textstyle\int_0^R \frac{dz}{2\pi}\ u^2(z),\\
\mathbf{I}_5^{(cl)} &= \textstyle\int_0^R \frac{dz}{2\pi}\ [u^3(z) - \frac{1}{2}\, u_z^2(z)],\\
&\dots\\
\mathbf{I}_{2k-1}^{(cl)} &= \textstyle\int_0^R \frac{dz}{2\pi}\ [u^k(z) + \text{terms with derivatives}],
\end{aligned} \tag{2.1.7}$$

which are in involution with respect to the Poisson bracket (2.1.2),

$$\{\mathbf{I}^{(cl)}_{2k-1}, \mathbf{I}^{(cl)}_{2l-1}\} = 0.$$

Obviously, the Hamiltonian (2.1.4) is simply proportional to $\mathbf{I}^{(cl)}_1$.

It is clear that the above IM (2.1.4) can be obtained from CFT IM (1.2.11) in the "classical limit" $c \to -\infty$. Indeed, if one regards $\hbar = -6/c$ as the Planck's constant, in this limit the commutators can be replaced by the Poisson brackets and under the substitution

$$\frac{ic}{6}[\ ,\ ] \to \{\ ,\ \}\ , \qquad -\frac{6}{c}\,T(z) \to u(z), \quad \frac{6}{c}\,L_n \to l_n \tag{2.1.8}$$

the algebra (1.2.3) becomes the Poisson bracket algebra (2.1.6) while the densities (1.2.12) assume the classical form which appear in (2.1.7), so that $\mathbf{I}_{2k-1} \to (-\frac{c}{6})^k \mathbf{I}^{(cl)}_{2k-1}$.

It is well known [24] that the time evolution according to the KdV equation, as well as each of the "higher KdV flows"

$$\frac{1}{2\pi}\, u_{\tau_{2k-1}} = \{\mathbf{I}^{(cl)}_{2k-1}, u\}\ , \tag{2.1.9}$$

generates the isospectral deformation of the second order differential operator

$$L = \frac{d^2}{dz^2} + u(z) - \lambda^2. \tag{2.1.10}$$

In particular, if we define the 2×2 Monodromy Matrix $\mathbf{m}(\lambda)$, which belongs to the group $SL(2)$, as

$$(\psi_1(z+R), \psi_2(z+R)) = (\psi_1(z), \psi_2(z))\ \mathbf{m}(\lambda)\ , \tag{2.1.11}$$

where $\psi_1(z), \psi_2(z)$ are two linearly independent solutions to the equation $L\psi = 0$, then the eigenvalues of $\mathbf{m}(\lambda)$ are involutive (with respect to the Poisson structure (2.1.2)) IM of the KdV flows (2.1.9), and the trace

$$\mathbf{t}(\lambda) = \mathrm{tr}\,[\mathbf{m}(\lambda)] \tag{2.1.12}$$

is the generating function for the local IM (2.1.7) as it expands in the asymptotic series

$$\log\,[\mathbf{t}(\lambda)] \simeq R\,\lambda - \sum_{n=1}^{\infty} c_n\ \mathbf{I}^{(cl)}_{2n-1}\ \lambda^{1-2n}\ , \tag{2.1.13}$$

where $c_1 = \pi, \quad c_n = 2\pi\, 2^{-n}\,(2n-3)!!/n!\ ,\ n > 1$.

It is more convenient for our purposes to start with the first order differential operator

$$\mathcal{L} = \frac{d}{dz} - \varphi_z(z)\ h - \lambda\ (e+f)\ , \tag{2.1.14}$$

where $\varphi(z)$ is the canonical field variable with the Poisson bracket

$$\{\varphi(z), \varphi(z')\} = \frac{1}{2}\,\epsilon\,(\frac{z-z'}{R})\ , \tag{2.1.15}$$

$$\epsilon(z) = 2n+1 \quad \text{for} \quad n < u < n+1 \; ; \quad n \in \mathbf{Z} \; , \tag{2.1.16}$$

which is related to $u(z)$ by Miura transform [25]

$$u(z) = -\varphi_z^2(z) - \varphi_{zz}(z) \; , \tag{2.1.17}$$

and e, f, h are the generators of the Lie algebra $sl(2)$,

$$[h,e] = 2e, \quad [h,f] = -2f, \quad [e,f] = 2h. \tag{2.1.18}$$

In general, the classical field $\varphi(z)$ has to be taken quasiperiodic

$$\varphi(z+R) = \varphi(z) + 2\pi i p \tag{2.1.19}$$

to guarantee the periodicity of $u(z)$.

In order to define the monodromy matrix for the operator (2.1.14) one has to pick some matrix representation for the $sl(2)$ algebra (2.1.18). Let $\pi_j[e]$, $\pi_j[f]$, $\pi_j[h]$ be $(2j+1)\times(2j+1)$ matrices, $j = 0,\ 1/2,\ 1,\ 3/2,\ \ldots$, representing (2.1.18) , such that $\pi_j[h] = \mathrm{diag}(2j, 2j-2, \ldots, -2j)$. Then, for given j, the solution to the equation $\mathcal{L}\Psi(z) = 0$ is

$$\Psi(z) = \pi_j \left[e^{\varphi(z)\, h}\, \mathcal{P} \exp\left(\lambda \int_0^z dw\ (e^{-2\phi(w)}\, e + e^{2\phi(w)}\, f) \right) \right] \Psi_0 \; , \tag{2.1.20}$$

where the symbol $\mathcal{P}$ denotes the "path ordered" exponential and Ψ_0 is arbitrary vector in $\mathbf{C}^{2j+1}$. The associated monodromy matrix has the form

$$\mathbf{m}_j(\lambda) = \pi_j \left[e^{2\pi i p\, h} \mathcal{P} \exp\left(\lambda \int_0^R dz (e^{-2\phi(z)}\, e + e^{2\phi(z)}\, f) \right) \right]. \tag{2.1.21}$$

It can be shown that the matrices

$$\mathbf{l}_j(\lambda) = \pi_j[e^{-\pi i p h}]\ \mathbf{m}_j(\lambda). \tag{2.1.22}$$

satisfy the "r-matrix" Poisson bracket algebra [26]

$$\{\mathbf{l}_j(\lambda) \otimes \mathbf{l}_{j'}(\lambda')\} = [\mathbf{r}_{jj'}(\lambda/\lambda'), \mathbf{l}_j(\lambda) \otimes \mathbf{l}_{j'}(\lambda')] \; , \tag{2.1.23}$$

where $\mathbf{r}_{jj'}(x) = \pi_j \otimes \pi_{j'}[\ \mathbf{r}\ ](x)$ is the "classical r-matrix"

$$\mathbf{r}(x) = \frac{1}{2}\, \frac{x + x^{-1}}{x - x^{-1}}\ h \otimes h + \frac{2}{x - x^{-1}}\ (e \otimes f + f \otimes e). \tag{2.1.24}$$

It follows immediately that the quantities

$$\mathbf{t}_j(\lambda) = \mathrm{tr}\,[\mathbf{m}_j(\lambda)] \tag{2.1.25}$$

are in involution with respect to the Poisson bracket (2.1.2),

$$\{\mathbf{t}_j(\lambda), \mathbf{t}_{j'}(\lambda')\} = 0. \tag{2.1.26}$$

In particular, $\mathbf{t}_{\frac{1}{2}}(\lambda)$ coincides with (2.1.12).

Of course the functionals $\mathbf{t}_j(\lambda)$ can not be all independent. It is possible to show that they satisfy the relation

$$[\mathbf{t}_j(\lambda)]^2 = \mathbf{1} + \mathbf{t}_{j+\frac{1}{2}}(\lambda)\mathbf{t}_{j-\frac{1}{2}}(\lambda) \ , \tag{2.1.27}$$

which allows one to express all of them all of them through any one, say $\mathbf{t}(\lambda) \equiv \mathbf{t}_{\frac{1}{2}}(\lambda)$.

After this brief review of known results in classical KdV theory we can return to the quantum case to construct the quantum operators $\mathbf{T}_j(\lambda)$ corresponding to the above functionals $\mathbf{t}_j(\lambda)$ and study their remarkable properties.

2.2. Operators $\mathbf{T}_j$ in CFT

Quantum version of the Miura transform (2.1.17) is known as the Feigin-Fuchs "free-field realization" of the Virasoro algebra *vir* [27]. Introduce the "left" chiral bose field

$$\phi(z) = \beta\left(\, Q + P\frac{2\pi z}{R} - i \sum_{n\neq 0} \frac{A_{-n}}{n} e^{i\frac{2\pi}{R}nz} \, \right) \tag{2.2.1}$$

where the operators Q, P and A_n satisfy the canonical commutation relations

$$[Q,P] = \frac{1}{2}\, i, \qquad [A_n, A_m] = \frac{n}{2}\, \delta_{n+m,0} \tag{2.2.2}$$

and β is a parameter which will be defined below. Also, let $\mathcal{F}_p$ be the Fock space, i.e the space generated by the action of A_n with $n < 0$ on the "vacuum" state $|p>$ which satisfies the equations

$$P|p>= p|p> \ , \qquad A_n|p>= 0 \quad \text{for} \quad n > 0. \tag{2.2.3}$$

Note that (2.2.1) implies that the field $\phi(z)$ is quasiperiodic, $\phi(z+R) = \phi(z) + 2\pi P$. As is well known the energy-momentum tensor

$$T(z) = \beta^{-2} \ :(\partial_z\phi)^2: +i\,(1-\beta^{-2})\,\partial_z^2\phi - \frac{1}{24} \tag{2.2.4}$$

generates the Virasoro algebra (1.2.3) with the central charge

$$c = 1 - 6(\beta - \beta^{-1})^2 \tag{2.2.5}$$

which therefore is represented in the Fock space $\mathcal{F}_p$. We will always assume that $c < 1$ and hence that β is real. For generic values of β and p the space $\mathcal{F}_p$ is isomorphic to the irreducible highest weight module $\mathcal{V}_\Delta$ with

$$\Delta = \Delta(p) \equiv (\frac{p}{\beta})^2 + \frac{c-1}{24}; \tag{2.2.6}$$

while for particular values of these parameters, when the "null-vectors" appear (so called "degenerate representations"), the irreducible representation $\mathcal{V}_\Delta$ can be obtained from $\mathcal{F}_p$ by factoring out the corresponding invariant subspaces. In fact, in what follows we always will imply that all such invariant subspaces in $\mathcal{F}_p$ are factored out, so that $\mathcal{F}_p$ will actually denote not the original Fock space defined above but the "Virasoro irreducible Fock space", $\mathcal{F}_p = \mathcal{F}_p^{full} \,/\, (invariant\ subspaces)$; with this convention the space $\mathcal{F}_p$ is always isomorphic to $\mathcal{V}_{\Delta(p)}$.

Note that in the classical limit $c \to -\infty$ the Feigin-Fuchs field $\phi(z)$ relates to the Miura field $\varphi(z)$ in Sect.2.1 as

$$\phi(z) \to -i\,\varphi(z) \quad \text{as} \quad \beta^2 \to 0 \tag{2.2.7}$$

Let $E,\ F$ and H be canonical generating elements of the quantum algebra $U_q(sl(2))$,

$$[H, E] = 2E, \qquad [H, F] = -2F, \qquad [E, F] = \frac{q^H - q^{-H}}{q - q^{-1}}, \tag{2.2.8}$$

with

$$q = e^{i\pi\beta^2}, \tag{2.2.9}$$

Like in Sect.2.1 we denote $\pi_j\ ,\ \ j = 0,\ 1/2,\ 1,\ 3/2,\ \ldots$ the $2j+1$ dimensional representation of the above quantum algebra (2.2.8), so that $\pi_j(E)$, $\pi_j(F)$ and $\pi_j(H)$ are matrices satisfying (2.2.8), with $\pi_j(H) = \mathrm{diag}(2j, 2j-2, \ldots, -2j)$. Let us define the following $(2j+1) \times (2j+1)$ matrix

$$\mathbf{L}_j(\lambda) = \pi_j \left[e^{i\pi P H}\, \mathcal{P} \exp\left(\lambda \int_0^R (\, : e^{-2i\phi} : (z) q^{\frac{H}{2}} E + : e^{2i\phi} : (z) q^{-\frac{H}{2}} F)\, dz \right) \right], \tag{2.2.10}$$

where the symbol $\mathcal{P}$ denotes the "path ordering" of both the operator and matrix products. Its elements are operators which act in the space

$$\hat{\mathcal{F}}_p = \oplus_{n=-\infty}^{\infty} \mathcal{F}_{p+n\beta^2}. \tag{2.2.11}$$

Note that the exponential fields $: e^{\pm 2i\phi} : (z)$ in (2.2.10) have anomalous dimensions and therefore in quantum case the parameter λ carries the dimension

$$\lambda \sim [length]^{\beta^2 - 1}. \tag{2.2.12}$$

Strictly speaking the expression (2.2.10) applies directly only to the cases when $\beta^2 < 1/2$ (*i.e.* for $c < -2$) for otherwise straightforward treatment of the path-ordered integral in (2.2.10) leads to divergences; in fact, for $\beta^2 > 1/2$ the integral in (2.2.10) can be defined through appropriate analytic continuation in β^2 [7].

The operator matrices $\mathbf{L}_j(\lambda)$ are designed to satisfy the full quantum Yang-Baxter relation

$$\mathbf{R}_{jj'}(\lambda/\lambda')\, (\mathbf{L}_j(\lambda) \otimes \mathbf{1})\, (\mathbf{1} \otimes \mathbf{L}_{j'}(\lambda')) = (\mathbf{1} \otimes \mathbf{L}_{j'}(\lambda'))\, (\mathbf{L}_j(\lambda) \otimes \mathbf{1})\, \mathbf{R}_{jj'}(\lambda/\lambda')\ , \tag{2.2.13}$$

where $\mathbf{R}_{j\,j'}(x)$ is the R-matrix acting in $2j+1$ and $2j'+1$ dimensional representations of $U_q(sl(2))$. As a simple consequence of (2.2.13) the operators

$$\mathbf{T}_j(\lambda) = \mathrm{tr}_{\pi_j}[e^{i\pi P H}\, \mathbf{L}_j(\lambda)] \tag{2.2.14}$$

form a commuting family, *i.e.* they satisfy

$$[\mathbf{T}_j(\lambda), \mathbf{T}_{j'}(\lambda')] = 0\,. \tag{2.2.15}$$

Note that although the off-diagonal elements of the matrices (2.2.10) act between different components of (2.2.11) the operators (2.2.14) invariantly act in the spaces $\mathcal{F}_p$,

$$\mathbf{T}_j(\lambda): \qquad \mathcal{F}_p \quad \rightarrow \quad \mathcal{F}_p. \tag{2.2.16}$$

It is possible to show using (2.2.10) and the properties of $U_q(sl(2))$ R-matrix that the operators $\mathbf{T}_j(\lambda)$ satisfy the following "fusion relations"

$$\mathbf{T}_j(q^{\frac{1}{2}}\lambda)\mathbf{T}_j(q^{-\frac{1}{2}}\lambda) = \mathbf{I} + \mathbf{T}_{j-\frac{1}{2}}(\lambda)\mathbf{T}_{j+\frac{1}{2}}(\lambda). \tag{2.2.17}$$

which allow one to express any operator $\mathbf{T}_j(\lambda)$ with $j > 1/2$ through the "fundamental" one $\mathbf{T}(\lambda) \equiv \mathbf{T}_{1/2}(\lambda)$. Note that these relations are very similar to the functional relation obeyed by the commuting transfer-matrices of the integrable XXZ model [28,29,30]. In fact this is well expected as our $\mathbf{T}$- operators appear to be the continuous field theory versions of the lattice transfer-matrices.

Obviously the matrices (2.2.10) and the operators (2.2.14) generalize the monodromy matrices (2.1.22) and the generating functions $\mathbf{t}_j(\lambda)$ of the KdV theory to the quantum case $\beta^2 > 0$. The fusion relation (2.2.17) generalize the algebraic relation (2.1.27).

It can be shown [4] that like in the classical case the operator $\mathbf{T}(\lambda)$ is entire function of λ^2 (the same is true for all the operators $\mathbf{T}_j(\lambda)$). Its power series expansion in λ^2 follows directly from the definitions (2.2.10), (2.2.14);

$$\mathbf{T}(\lambda) = 2\ \cos(2\pi\mathbf{P}) + \sum_{k=1}^{\infty} t^k \mathbf{G}_k\ , \tag{2.2.18}$$

where $\mathbf{G}_k$ are basic "nonlocal IM"

$$\mathbf{G}_k = q^n \int_{\mathcal{D}_R(x_1,\ldots,x_{2k})} \Big(e^{2i\pi P} : e^{-2i\phi(x_1)} :: e^{+2i\phi(x_2)} : \cdots : e^{+2i\phi(x_{2k})} : +$$

$$e^{-2i\pi P} : e^{+2i\phi(x_1)} :: e^{-2i\phi(x_2)} :: e^{+2i\phi(x_3)} :: e^{-2i\phi(x_4)} : \cdots : e^{-2i\phi(x_{2k})} : \Big). \tag{2.2.19}$$

Here the integration is performed over the domain $D_R(x_1, \ldots, x_{2k}) = \{R > x_1 > x_2 > \cdots > x_{2k}\}$. In the opposite limit $\lambda^2 \to \infty$ the operator $\mathbf{T}(\lambda)$ admits asymptotic expansion [4]

$$\log \mathbf{T}(\lambda)\ \simeq\ (\kappa\, R/2\pi)\ \lambda^{(1+\xi)}\ \mathbf{I} - \sum_{n=1}^{\infty} C_n\ \lambda^{(1-2n)(1+\xi)}\ \mathbf{I}_{2n-1} \tag{2.2.20}$$

in powers of $\lambda^{-(1+\xi)}$ where

$$\xi = \frac{\beta^2}{1-\beta^2}, \tag{2.2.21}$$

with the coefficients proportional to the local IM (1.2.11). The numerical factors κ and C_n can be obtained in closed form [4,7],

$$\kappa = \frac{2\sqrt{\pi}\ \Gamma(\frac{1}{2} - \frac{\xi}{2})}{\Gamma(1 - \frac{\xi}{2})} \left(\Gamma(1-\beta^2)\right)^{1+\xi}\ ;$$

$$C_n = \frac{\sqrt{\pi}}{n!}\,(\beta^2)^n\,\frac{\Gamma((n-\frac{1}{2})(1+\xi))}{\Gamma(1+(n-\frac{1}{2})\xi)}\,\Big(\Gamma(1-\beta^2)\Big)^{-(2n-1)(1+\xi)}. \tag{2.2.22}$$

Note that (2.2.20) generalizes the classical expansion (2.1.13), in particular it is easy to check that $\kappa \to 2\pi$ and $C_n \to \beta^{2n}\, c_n$ as $\beta^2 \to 0$. Note also that in the quantum case the expansion (2.2.20) involves fractional powers of λ; this is related to the fact that the exponential fields in (2.2.10) have anomalous dimensions.

The above expansions suggest that the eigenvalues $T(\lambda)$ of the operator $\mathbf{T}(\lambda)$ unite all the information about the eigenvalues of both nonlocal IM (2.2.19) and local IM (1.2.11) in a very interesting manner. Consider for example the vacuum eigenvalues

$$\mathbf{T}(\lambda)\mid p\rangle = T(t,p)\mid p\rangle\ ; \qquad t \equiv \lambda^2\Big(\frac{R}{2\pi}\Big)^{2-2\beta^2}. \tag{2.2.23}$$

According to (2.2.18) and (2.2.20) this function expands both in convergent series around $t=0$ and in asymptotic series in the vicinity of $t=\infty$,

$$T(p,t) = 2\cos(2\pi p) + \sum_{k=1}^{\infty} t^k G_k^{vac}(p) \;\simeq$$

$$\exp\Big((\kappa R/2\pi)\, t^{\frac{1+\xi}{2}} - \sum_{n=1}^{\infty} C_n\, t^{(1/2-n)(1+\xi)}\, I_{2n-1}^{vac}(p)\Big), \tag{2.2.24}$$

where $G_k^{vac}(p)$ and $I_{2n-1}^{vac}(p)$ are the vacuum eigenvalues of the nonlocal IM (2.2.19) and the local IM (1.2.11), respectively, calculated at $R=2\pi$; the last expression in (2.2.24) is in fact an asymptotic series with zero radius of convergence. The eigenvalues $G_k^{vac}(p)$ are given by the integrals

$$\begin{aligned} G_k^{vac}(p) &= \textstyle\int_0^{2\pi} du_1 \int_0^{u_1} dv_1 \int_0^{v_1} du_2 \int_0^{u_2} dv_2 \cdots \int_0^{v_{k-1}} du_k \int_0^{u_k} dv_n \\ &\textstyle\prod_{j>i}^k \Big[\Big(2\,\sin\big(\frac{u_i-u_j}{2}\big)\Big)^{2\beta^2}\Big(2\,\sin\big(\frac{v_i-v_j}{2}\big)\Big)^{2\beta^2}\Big]\prod_{j\geq i}^k\Big(2\,\sin\big(\frac{u_i-v_j}{2}\big)\Big)^{-2\beta^2} \\ &\textstyle\prod_{j>i}^k\Big(2\,\sin\big(\frac{v_i-u_j}{2}\big)\Big)^{-2\beta^2}\ 2\,\cos\Big(2p\,\big(\pi+\sum_{i=1}^k(v_i-u_i)\big)\Big). \end{aligned} \tag{2.2.25}$$

while the eigenvalues $I_{2n-1}(p)$ can be calculated (in principle) from the explicit formulas for the associated local densities $T_{2k}(z)$ in (1.2.11), for example

$$\begin{aligned} I_1^{vac}(p) &= \Delta(p) - \tfrac{c}{24}\,, \\ I_3^{vac}(p) &= \Delta^2(p) - \tfrac{c+2}{12}\,\Delta(p) + \tfrac{c\,(5c+22)}{2880}\,, \\ I_5^{vac}(p) &= \Delta^3(p) - \tfrac{c+4}{8}\,\Delta^2(p) + \tfrac{(c+2)\,(3c+20)}{576}\,\Delta(p) - \tfrac{c\,(3c+14)\,(7c+68)}{290304}\,, \end{aligned} \tag{2.2.26}$$

where $\Delta(p)$ is the same as in (2.2.6) (many more eigenvalues $I_{2k-1}^{vac}(p)$ are presented in [5]). Unfortunately for general k neither the integrals (2.2.25) can be calculated explicitly nor the densities $T_{2k}(z)$ are known in closed form and so one needs less straightforward approach to calculate the eigenvalues of $\mathbf{T}_j(\lambda)$.

Let us return to the fusion relations (2.2.17). For generic values of c these relations just allow one to express the higher $\mathbf{T}$ operators in terms of the lower ones. However it is well known in the lattice theory that at particular values of the parameters, when the XXZ system can be reduced to the $RSOS$ model [31] , the functional relations truncate to become a finite system of functional equations [32, 29]. Of course, similar phenomenon happens in our continuous theory. For example, very simple truncation occurs at

$$c = 1 - 3\,\frac{(2n+1)^2}{2n+3}, \quad \xi = \frac{2}{2n+1}; \qquad n = 1, 2, 3, \ldots \tag{2.2.27}$$

For given n in (2.2.27) consider the finite sum of the "irreducible Fock spaces"

$$\oplus_{k=0}^{n} \mathcal{F}_{p_k}\ ; \ \ p_k = \frac{2k+1}{2\,(2n+3)}. \tag{2.2.28}$$

This space coincides with the space of chiral states of the "minimal CFT" $\mathcal{M}_{2/2n+3}$ [8]. It is possible to show that the operators $\mathbf{T}_j(\lambda)$ with $j = 0,\ 1/2,\ 1,\ \ldots,\ n+1/2$ invariantly act in the space (2.2.28) and, being restricted to this space, these operators satisfy the symmetry relation [4]

$$\mathbf{T}_{n+\frac{1}{2}-j}(\lambda) = \mathbf{T}_j(\lambda); \quad j = 0,\ 1/2,\ \ldots\ ,\ n+1/2\ , \tag{2.2.29}$$

in particular $\mathbf{T}_{n+\frac{1}{2}}(\lambda) = \mathbf{I}$. Clearly, with this additional relation the fusion relation (2.2.17) closes within n first operators $\mathbf{T}_{\frac{1}{2}}(\lambda),\ \mathbf{T}_1(\lambda),\ \ldots\ \mathbf{T}_{\frac{n}{2}}(\lambda)$. So, the eigenvalues of these $\mathbf{T}$-operators in the above "minimal CFT" satisfy closed functional equations. More complicated cases of truncation are discussed in [5].

The simplest of the above "minimal CFT" is the Lee-Yang CFT $\mathcal{M}_{2/5}$ already discussed in the first lecture. In this case the space of chiral states (2.2.28) consists of only two components

$$\mathcal{F}_{p_0} \oplus \mathcal{F}_{p_1}\ , \qquad p_0 = 1/10, \quad p_1 = 3/10. \tag{2.2.30}$$

It follows from (2.2.17) and (2.2.29) that any eigenvalue $T(t)$ of the operator $\mathbf{T}(\lambda) \equiv \mathbf{T}_{\frac{1}{2}}(\lambda)$ in the space (2.2.30) satisfies the functional equation

$$T(q^{\frac{1}{4}}\,t)\,T(q^{-\frac{1}{4}}\,t) = 1 + T(t) \tag{2.2.31}$$

where $t = \lambda^2\,(R/2\pi)^{2-2\beta^2}$.

2.3. Operators $\mathbf{T}_j(\mu|\lambda)$ in the $\Phi_{1,3}$ Perturbed Theory

The operators $\mathbf{T}_j$ constructed above are natural objects in $c < 1$ CFT in the sense that they generate the local IM $\mathbf{I}_{2k-1}$ of this CFT. As was explained in Sect.1.3 one obtains more general Integrable QFT (1.3.4) by perturbing the CFT with operator $\Phi_{(1,3)}$. It turns out that very simple modifications of the operators $\mathbf{T}_j(\lambda)$ apply to this perturbed CFT.

As in Sect.2.2 it is convenient for our purposes to use the Feigin-Fuchs free-field realization for the original CFT. The space of states (1.2.5) can be realized as

$$\mathcal{H}_{CFT} = \oplus_p \mathcal{H}_p; \qquad \mathcal{H}_p \equiv \mathcal{F}_p \otimes \bar{\mathcal{F}}_{-p}\ , \tag{2.3.1}$$

where $\mathcal{F}_p$ is the "irreducible Fock space" defined in Sect.2.2 and $\bar{\mathcal{F}}_{-p}$ is similar space generated by the "right-moving" chiral Bose field

$$\bar{\phi}(\bar{z}) = \beta\,(\bar{Q} - \frac{2\pi\bar{z}}{R}\bar{P} - i\sum_{n\neq 0}\frac{\bar{a}_{-n}}{n}\,e^{-i\frac{2\pi}{R}n\bar{z}}). \tag{2.3.2}$$

Each of the spaces $\mathcal{H}_p$ forms an irreducible representation of $vir \oplus \overline{vir}$ with the highest weight $(\Delta(p), \Delta(p))$ and the highest weight vector

$$| \Delta(p)\rangle = | p\rangle \otimes \overline{| -p\rangle} \,, \tag{2.3.3}$$

where $| p\rangle$ and $\overline{| -p\rangle}$ are the vacuum vectors in $\mathcal{F}_p$ and $\bar{\mathcal{F}}_{-p}$, respectively. The sum in (2.3.1) is taken over certain set of admitted values of the parameter p which is specific for given CFT; it determines its Virasoro representation content. In different theories this set can be discrete or continuous.

One remark is in order here. The perturbed QFT (1.3.4) can be constructed only if the original CFT meets certain criteria. For instance, as in CFT the space of local fields is isomorphic to the space of states (2.3.1), the sum in (2.3.1) must contain the term with $2p = 1 - 3\,\beta^2$ corresponding to the field $\Phi_{1,3}$ (and its conformal descendants). Moreover, according to the fusion rules which follow from the null-vector equations (1.3.3) the operator $\Phi_{1,3}$ acts as

$$\Phi_{1,3} \;:\quad \mathcal{H}_p \to \mathcal{H}_{p-\beta^2} \oplus \mathcal{H}_p \oplus \mathcal{H}_{p+\beta^2} \,, \tag{2.3.4}$$

and therefore to admit the action of this operator the space (2.3.1) must contain the whole sum

$$\hat{\mathcal{H}}_p = \oplus_n \, \mathcal{H}_{p+n\beta^2} \tag{2.3.5}$$

together with each $\mathcal{H}_p$ included therein. For particular values of p (corresponding to the "completely degenerate" representations of vir) this sum truncates and contains in fact only finitely many terms (see [8]). In any case we will assume that the space (2.3.1) has the structure

$$\mathcal{H}_{CFT} = \oplus_a \, \hat{\mathcal{H}}_{p_a} . \tag{2.3.6}$$

Obvious example of CFT with this structure is "minimal CFT" $\mathcal{M}_{r/r'}$ with $\beta^2 = r/r'$ [8] in which case p in (2.3.1) runs over finitely many values $2\,p_{l,k} = l - k\,\beta^2$ ($l = 1, 2, \ldots, r-1$; $k = 1, 2, \ldots, r'-1$). Another example is $c = 1$ CFT of uncompactified free Bose field where p in (2.3.1) spans continuous range of values.

In Sect.2.2 we have defined the operator valued matrices $\mathbf{L}_j(\lambda)$ for the "left" chiral sector of the CFT; these matrices satisfy the Yang-Baxter relation (2.2.13). Similar way one can define the "right" chiral counterparts of these matrices,

$$\bar{\mathbf{L}}_j(\lambda) = \pi_j \left[\mathcal{P} \exp\left(\lambda \int_0^R (: e^{-2i\bar{\phi}(\bar{z})} : q^{\frac{H}{2}} E + : e^{2i\bar{\phi}(\bar{z})} : q^{-\frac{H}{2}} F) \, d\bar{z} \right) \, e^{-i\pi \bar{P} H} \right] , \tag{2.3.7}$$

where $\bar{\phi}(\bar{z})$ is the right chiral Bose field (2.3.2), E, F and H denote the same elements of $U_q(sl(2))$ as in (2.2.10) and like in (2.2.10) $\mathcal{P}$ indicates the path ordering of both operators and matrices along the integration path. The elements of the matrices (2.3.7) act in the space $\hat{\bar{\mathcal{F}}}_p$ defined similarly to (2.2.11) in terms of the "right" Fock spaces $\bar{\mathcal{F}}_p$. It is possible to check that the operator matrices (2.3.7) also satisfy the Yang-Baxter relation in the form

$$(\bar{\mathbf{L}}_j(\lambda) \otimes \mathbf{1}) \, (\mathbf{1} \otimes \bar{\mathbf{L}}_{j'}(\lambda')) \, \mathbf{R}_{jj'}(\lambda/\lambda') \; = \mathbf{R}_{jj'}(\lambda/\lambda') \, (\mathbf{1} \otimes \bar{\mathbf{L}}_{j'}(\lambda')) \, (\bar{\mathbf{L}}_j(\lambda) \otimes \mathbf{1}) \tag{2.3.8}$$

where $R_{jj'}(x)$ are exactly the same R-matrices as in (2.2.13). Note that the R-matrices enter (2.3.8) different way as compared to (2.2.13), the difference being traced down to the sign difference in the commutation relations

$$[\phi(x), \phi(x')] = -[\bar{\phi}(x), \bar{\phi}(x')] = \frac{\pi\beta^2}{2i}\, \epsilon\Big(\frac{x-x'}{R}\Big)\,, \tag{2.3.9}$$

where $\epsilon(x)$ is the function defined in (2.1.16).

Now we can introduce the following operators

$$\mathbf{T}_j(\mu|\lambda) = tr_{\pi_j}\Big[\mathbf{L}_j(\lambda)\,\bar{\mathbf{L}}_j(\mu/\lambda)\Big]\,, \tag{2.3.10}$$

where μ is a parameter. It is easy to check that (2.3.10) defines the operator which acts in the space (2.3.5), i.e.

$$\mathbf{T}_j(\mu|\lambda)\ : \qquad \hat{\mathcal{H}}_p \to \hat{\mathcal{H}}_p. \tag{2.3.11}$$

It follows from (2.2.13), (2.3.8) and the fact that the R-matrices satisfy the "unitarity condition"

$$R_{jj'}(x)\ R_{jj'}(x^{-1}) = 1\,, \tag{2.3.12}$$

that these operators form the commuting family, i.e.

$$[\mathbf{T}_j(\mu|\lambda), \mathbf{T}_{j'}(\mu|\lambda')] = 0 \tag{2.3.13}$$

for all values of j, j'. The following simple properties of the operators (2.3.10) are immediately established

1. For $\mu = 0$ the operator $\mathbf{T}_j$ reduces to the operator $\mathbf{T}_j$ defined in Sect.2.2,

$$\mathbf{T}_j(0|\lambda) = \mathbf{T}_j(\lambda) \otimes \bar{\mathbf{I}}\,, \tag{2.3.14}$$

i.e. it acts as $\mathbf{T}_j(\lambda)$ in the "left" spaces $\mathcal{F}_p$ and as the identity operator $\bar{\mathbf{I}}$ in the "right" spaces $\bar{\mathcal{F}}_p$ in (2.3.1)[4]. Similarly,

$$\mathbf{T}_j(\mu|\mu/\lambda)\big|_{\mu=0} = \mathbf{I} \otimes \bar{\mathbf{T}}_j(\lambda)\,, \tag{2.3.15}$$

where $\mathbf{I}$ is the identity in the "left" spaces $\mathcal{F}_p$ and $\bar{\mathbf{T}}_j(\lambda) = tr_{\pi_j}[\,\bar{\mathbf{L}}_j(\lambda)\, e^{-i\pi\bar{P}H_j}\,]$ act in the spaces $\bar{\mathcal{F}}_p$ as the "right" counterparts of the operators (2.3.10).

2. Like $\mathbf{T}_j(\lambda)$ the operators $\mathbf{T}_j(\mu|\lambda)$ are in fact the single-valued functions of λ^2.

3. As follows directly from the definition (2.3.10) the operators $\mathbf{T}_j(\mu|\lambda)$ inherit their large λ asymptotic from those of the operators $\mathbf{T}_j(\lambda)$,

$$\log \mathbf{T}_j(\mu|\lambda) \sim \kappa_j\, R/(2\pi)\ \lambda^{1+\xi}, \qquad \lambda \to \infty\,, \tag{2.3.16}$$

where

$$\kappa_j = \frac{\sin(\pi j\xi)}{\sin(\pi\xi/2)}\ \kappa \tag{2.3.17}$$

---

[4] Obviously, the zero mode operator $\bar{P}$, when acting on the space (2.3.1), can be identified with $-P$.

and κ is given by (2.2.22). Similarly, the asymptotic of these operators at $\lambda \to 0$ is controlled by the second factor in (2.3.10) and so

$$\log \mathbf{T}_j(\mu|\lambda) \sim \kappa_j\, R/(2\pi)\ (\mu/\lambda)^{1+\xi}, \qquad \lambda \to 0. \tag{2.3.18}$$

In particular, it follows from (2.3.18) that $\mathbf{T}_j(\mu|\lambda)$ has an essential singularity at $\lambda^2 = 0$.

It is clear that for generic value of the parameter μ the operators $\mathbf{T}_j(\mu|\lambda)$ do not commute with the CFT local IM (1.2.11). Instead, it is natural to expect that the operators $\mathbf{T}_j(\mu|\lambda)$ play the same role in perturbed theory (1.3.4) as $\mathbf{T}_j(\lambda)$ did in the CFT, namely, the asymptotic expansions of $\mathbf{T}_j(\mu|\lambda)$ near the essential singularities at $\lambda^2 = 0$ and $\lambda^2 = \infty$ generate the local IM (1.3.7) of the perturbed theory (1.3.4). Note that the spaces (2.3.5) of the perturbed theory admit the action of the perturbed local IM (1.3.7). The parameter μ in (2.3.10) is expected to be related to the coupling parameter $\hat{\mu}$ in (1.3.4).

By analogy with the operators $\mathbf{T}_j(\lambda)$ of CFT it seems natural to expect also that

4. The operators $\mathbf{T}_j(\mu|\lambda)$ are single-valued functions of λ^2, regular everywhere in the complex λ^2 plane except at $\lambda^2 = 0$ and $\lambda^2 = \infty$ where they have essential singularities.

5. The asymptotic expansions of the operators $\mathbf{T}_j(\mu|\lambda)$ near $\lambda^2 = \infty$ and $\lambda^2 = 0$ are controlled by the perturbed local IM (1.3.7); in particular, for $\mathbf{T}(\mu|\lambda) \equiv \mathbf{T}_{1/2}(\mu|\lambda)$ one has

$$\log \mathbf{T}(\mu|\lambda) = \kappa\, R/(2\pi)\ \lambda^{1+\xi} - \sum_{n=1}^{\infty} C_n\ \lambda^{(1-2n)(1+\xi)}\ \mathbf{I}_{2n-1}\ , \tag{2.3.19a}$$

$$\log \mathbf{T}(\mu|\lambda) = \kappa\, R/(2\pi)\ (\mu/\lambda)^{1+\xi} - \sum_{n=1}^{\infty} C_n\ (\mu/\lambda)^{(1-2n)(1+\xi)}\ \bar{\mathbf{I}}_{2n-1}\ , \tag{2.3.19b}$$

where κ and C_k are exactly the same constants as in (2.2.22) and the parameter μ is related to $\hat{\mu}$ in (1.3.4), $\mu = \zeta\, \mu^2$ with some numerical coefficient ζ (exact value of this parameter is given in [5]). In particular, the "transfer-matrices" $\mathbf{T}_j$ commute with all local IM (1.3.7)

$$[\mathbf{T}_j(\mu|\lambda), \mathbf{I}_{2n-1}] = [\mathbf{T}_j(\mu|\lambda), \bar{\mathbf{I}}_{2n-1}] = 0. \tag{2.3.20}$$

Although I do not yet know how to prove the statements 4 and 5 above, they can be supported by studies of several examples. In what follows I'll assume them valid.

It is possible to show that the operators (2.3.10) satisfy exactly the same "fusion relations" as the operators $\mathbf{T}_j$, namely

$$\mathbf{T}_j(\mu|q^{\frac{1}{2}}\lambda)\mathbf{T}_j(\mu|q^{-\frac{1}{2}}\lambda) = \mathbf{I} + \mathbf{T}_{j+1/2}(\mu|\lambda)\mathbf{T}_{j-1/2}(\mu|\lambda). \tag{2.3.21}$$

Also, if q is a root of unity the fusion relations (2.3.21) truncate exactly the same way as it happens in the corresponding unperturbed CFT. For example, in the Lee-Yang QFT the operator $\mathbf{T}(\nu|\lambda)$ acting in the space (1.3.11) satisfy the relation

$$\mathbf{T}(\mu|q^{\frac{1}{2}}\,\lambda)\,\mathbf{T}(\mu|q^{-\frac{1}{2}}\,\lambda) = \mathbf{I} + \mathbf{T}(\mu|\lambda). \tag{2.3.22}$$

2.4. Excited State Energies in Lee-Yang QFT

In this Section I will study the eigenvalues of $\mathbf{T}(\mu|\lambda)$ in Lee-Yang QFT. It is convenient to introduce the rapidity variable

$$\theta = \frac{5}{6}\log(\lambda^2/\mu) \tag{2.4.1}$$

and denote

$$\mathbf{Y}(r|\theta) = \mathbf{T}(\mu|\lambda)\ , \tag{2.4.2}$$

where I also decided to trade μ for the dimensionless parameter

$$r = \kappa\,(R/\pi)\,\mu^{\frac{5}{6}} \tag{2.4.3}$$

The operator $\mathbf{T}(\mu|\lambda)$ is single-valued function of λ^2 and consequently the operator (2.4.2) is periodic

$$\mathbf{Y}(r|\theta + 5i\pi/3) = \mathbf{Y}(r|\theta). \tag{2.4.4}$$

In these new variables the expansion (2.3.19a) can be rewritten as

$$\log \mathbf{Y}(r|\theta) \simeq (r/2)\,e^{\theta} - \sum_{n=1}^{\infty} C_n\,\mu^{\frac{5}{6}(1-2n)}\,e^{(1-2n)\theta}\,\mathbf{I}_{2n-1} \quad \text{for} \quad \theta \to +\infty\,, \tag{2.4.5}$$

and (2.3.19b) gives similar expansion for $\theta \to -\infty$. The equation (2.3.22) takes the form

$$\mathbf{Y}(r|\theta + i\pi/3)\,\mathbf{Y}(r|\theta - i\pi/3) = \mathbf{I} + \mathbf{Y}(r|\theta). \tag{2.4.6}$$

Let $Y(\theta)$ be any eigenvalue of the operator $\mathbf{Y}(r|\theta)$ in the space (1.3.11). It follows from (2.4.5) and (2.4.6) that this function has the large θ asymptotics

$$\log Y(r|\theta) \simeq (r/2)\,e^{\pm\theta} \quad \text{as} \quad \theta \to \pm\infty\,, \tag{2.4.7}$$

and satisfies the functional relation

$$Y(\theta + i\pi/3)\,Y(\theta - i\pi/3) = 1 + Y(\theta). \tag{2.4.8}$$

Note that this functional equation and the asymptotic condition coincide with the Eq.(1.5.23) and (1.5.20) found in TBA approach. Here we arrived at (2.4.7) and (2.4.8) from quite different starting point, completely bypassing the on-shell solution of the Lee-Yang QFT. In fact, I will show later that the particle content and the S-matrix of this QFT can be recovered from these equations.

The functional equation identical to (2.4.8) was studied in relation to certain solvable lattice models by Klumper and Pearce [33], who developed the method of transforming the functional equations to nonlinear integral equations, the knowledge about analytic properties of the solutions being key ingredient in this approach. Although in field theory the analytic properties of the functions $Y(\theta)$ are somewhat different from those considered in [33], the method of Klumper and Pearce can be easily adopted to our case. The derivation in [33] is based on the following

LEMMA. If a function $f(\theta)$ is regular and bounded in the strip $\mathrm{Imm}\,\theta \in (-\pi/3, \pi/3)$ and satisfies the relation

$$(\hat{D}\, f)(\theta) = f(\theta + i\pi/3) + f(\theta - i\pi/3) - f(\theta) = g(\theta) \tag{2.4.9}$$

where $\hat{D}$ is the finite-difference operator (1.5.21), and $g(\theta)$ is any function then

$$f(\theta) = \int_{-\infty}^{\infty} d\theta' \; \Phi(\theta - \theta')\, g(\theta') \;, \tag{2.4.10}$$

where $\Phi(\theta)$ is given by (1.5.10).

This lemma specifies the class of functions on which the operator $\hat{D}$ is invertible.

If $Y(\theta)$ would be free of zeroes in the strip $\mathrm{Imm}\,\theta \in (-\pi/3, \pi/3)$ then the function $f(\theta) = \log Y(\theta) - r\,\cosh(\theta)$ would satisfy the conditions of this lemma with $g(\theta) = \log\left(1 + Y^{-1}(\theta)\right)$, and integral equation would follow immediately. However a typical eigenvalue $Y(\theta)$ does have zeroes in the above strip and so we need to take care of them.

It follows from the definition (2.2.14) that the functions $Y(\theta)$ must be real at real θ. Therefore zeroes of $Y(\theta)$ are either real or appear as complex-conjugated pairs. Let us ignore for simplicity the possibility of real zeroes and assume that $Y(\theta)$ has N complex-conjugated pairs of zeroes $\theta_a\,, \theta_a^*$, $a = 0,\ 1,\ \ldots,\ N-1$ located in the strip $\mathrm{Imm}\,\theta \in (-\pi/3, \pi/3)$,

$$\begin{matrix} \theta_a = \vartheta_a + i\eta_a \\ \theta_a^* = \vartheta_a - i\eta_a \end{matrix} \;; \qquad 0 < \eta_a < \pi/3. \tag{2.4.11}$$

Introduce the function

$$\sigma(\theta, \eta) = \frac{\cosh(\theta) - \cos(\eta)}{\cosh(\theta) + \cos(\eta)}\, \frac{\cosh(\theta) - \sin(\pi/6 - \eta)}{\cosh(\theta) + \sin(\pi/6 - \eta)} \tag{2.4.12}$$

As the function of θ (2.4.12) has zeroes at $\theta = \pm i\,\eta$ and $\theta = \pm i\,(\pi/3 + \eta)$, and it satisfies the relation

$$\sigma(\theta + i\pi/3, \eta)\, \sigma(\theta - i\pi/3, \eta) = \sigma(\theta, \eta). \tag{2.4.13}$$

Note that $\sigma(\theta, \eta)$ can be expressed through the S-matrix (1.4.13),

$$\sigma(\theta, \eta) = S(\theta - i\pi/3 + i\eta)\, S(\theta - 2\pi i/3 - i\eta). \tag{2.4.14}$$

Let us write $Y(\theta)$ as

$$Y(\theta) = e^{r\,\cosh\theta} \prod_{a=0}^{N-1} \sigma(\theta - \vartheta_a,\, \eta_a)\, X(\theta). \tag{2.4.15}$$

In view of (2.4.8) and (2.4.13) the function $X(\theta)$ satisfies the equation

$$\frac{X(\theta + i\pi/3)\, X(\theta - i\pi/3)}{X(\theta)} = \left(1 + Y^{-1}(\theta)\right). \tag{2.4.16}$$

Note that because the σ-factors in (2.4.15) absorb all zeroes of $Y(\theta)$ in the strip $\mathrm{Im}m\ \theta \in (-\pi/3, \pi/3)$ and the exponential prefactor saturates the asymptotic (2.4.7) the function $\log X(\theta)$ is regular and bounded in this strip. Applying the above lemma we obtain the integral equation

$$\epsilon(\theta) = r\ \cosh\theta + \sum_{a=0}^{N-1} \log\left(\sigma(\theta-\vartheta_a, \eta_a)\right) + \int_{-\infty}^{\infty} d\theta'\ \Phi(\theta-\theta')\ \log\left(1+e^{-\epsilon(\theta')}\right), \quad (2.4.17)$$

where $\epsilon(\theta) = \log\ Y(\theta)$.

Given a set of real parameters ϑ_a, η_a the equation (2.4.17) uniquely determines the function $\epsilon(\theta)$ and whence $Y(\theta)$. Of course these parameters can not be taken arbitrary. With generic values of these parameters the function $Y(\theta)$ defined through the solution to (2.4.17) would have poles in the domain $\mathrm{Im}m\,\theta \in (\pi/3, 2\pi) \bmod 5\pi/3$. Indeed, if, for instance, $Y(\vartheta_a + i\eta_a) = 0$ then according to (2.4.8) the function $Y(\theta)$ must exhibit a pole at $\theta = \vartheta_a + i\eta_a - 2\pi i/3$, unless, of course, $Y(\vartheta_a + i\eta_a - i\pi/3) = -1$. Therefore the condition that $Y(\theta)$ is entire function leads to the equations

$$Y(\vartheta_a + i\eta_a - i\pi/3) = Y(\vartheta_a - i\eta_a + i\pi/3) = -1, \quad a = 1, \dots\ N-1. \quad (2.4.18)$$

which determine the parameters ϑ_a and η_a in (2.4.15).

The integral equation (2.4.17) together with the transcendental equations (2.4.18) enables one to determine (in principle) any eigenvalue $Y(\theta)$ of the operator $\mathbf{T}(\lambda)$ in the space (1.3.11).

Ones a solution $\epsilon(\theta)$, $\{\vartheta_a\,,\ \eta_a\}$ to the coupled equations (2.4.17), (2.4.18) is found the corresponding eigenvalues I_{2n-1} of the local IM $\mathbf{I}_{2k-1}$ can be obtained from the expansion (2.4.5). Note that both the terms $\log\sigma$ and the kernel Φ in the r.h.s. of (2.4.17) can be expanded for $\Re e\theta > 0$ in powers of $e^{-\theta}$. Substituting these expansions into (2.4.17) and comparing it with (2.4.5) one can find the desired eigenvalues. In particular for the energy $E(R) = I_1 + \bar{I}_1$ one obtains

$$E(R) = \frac{\kappa}{\pi}\,\mu^{\frac{5}{6}}\Big(-\frac{r}{4\sqrt{3}} + 2\sum_{a=0}^{N-1}\cos(\pi/6+\eta_a)\,\cosh\vartheta_a - \int_{-\infty}^{\infty}\cosh\theta\,\log\left(1+e^{-\epsilon(\theta)}\right)\frac{d\theta}{2\pi}\Big). \quad (2.4.19)$$

Just like the TBA equations (1.5.15) the Eq.(2.4.17), (2.4.18) for excited states can be studied numerically. I am not going to discuss this analysis here, see [5]. Instead I will show how the mass spectrum and factorizable S-matrix of the Lee-Yang QFT can be recovered from these equations in the limit $R \to \infty$. Simple estimate shows that for $r >> 1$ the first term in the r.h.s. of Eq.(2.4.17) dominates everywhere in the strip $\mathrm{Im}m\,\theta \in (-\pi/3 + \varepsilon\,,\ \pi/3 - \varepsilon)$ except in the small disks of the size ε around the zeroes $\vartheta_a \pm i\,\eta_a$, where ε is some number exponentially small in r. In this domain the first term in the r.h.s of (2.4.8) can be neglected and whence the function $Y(\theta)$ can be written as

$$Y(\theta) \simeq \prod_a S(\theta - \vartheta_a - i\pi/2)\exp(r\cosh\theta)\ ,$$

$$r >> 1; \qquad |\theta - \vartheta_a| > \varepsilon; \qquad \mathrm{Imm}\,\theta \in (-\pi/3 + \varepsilon,\ \pi/3 - \varepsilon)\ , \tag{2.4.20}$$

where $S(\theta)$ is the function (1.4.13) and ϑ_a are some real numbers. In writing (2.4.20) we have taken into account the asymptotic conditions (2.4.7) and the fact that $Y(\theta)$ must be real at real θ[5]. The form (2.4.20) shows in particular that for r sufficiently large the zeroes (2.4.11) have the form

$$\theta_a = \vartheta_a \pm i\,(\pi/6 - x_a)\ , \tag{2.4.21}$$

where x_a are exponentially small in r. The equation (2.4.18) then takes the form

$$\exp(i\,r\,\sinh\vartheta_a)\prod_{b\neq a} S(\vartheta_a - \vartheta_b) = 1, \quad a = 0,\ 1, \ldots,\ N-1\ , \tag{2.4.22}$$

where we have taken into account that $S(0) = -1$. This equations shows that for $r >> 1$ the eigenstate associated with the eigenvalue (2.4.20) can be understood as N-particle state, the parameters ϑ_a being interpreted as the rapidities of the particles, so that the equations (2.4.22) have the meaning of the Bethe Ansatz equations (1.5.5) for N scalar particles with the S-matrix (1.4.13). We should stress here that in the field theory the Bethe Ansatz equations of the form (2.4.22) make sense only in the limit $r \to \infty$; for finite r these equations acquire corrections due to the vacuum polarization. From (2.4.22) and (2.4.3) we read out the mass m of the particle A,

$$m = \frac{\kappa}{\pi}\,\mu^{\frac{5}{6}}. \tag{2.4.23}$$

Finally, we can use (2.4.19) to calculate large R asymptotics of the eigenvalues $E_N(R)$ corresponding to this N-particle state,

$$E_N(R) = m\,(-\frac{mR}{4\sqrt{3}} + \sum_a \cosh\vartheta_a) + O(\exp(-\frac{\sqrt{3}\,m}{2}R))\,. \tag{2.4.24}$$

The first term in the brackets is the bulk vacuum energy (which already appeared in (1.5.13)) and we find [3]

$$\varepsilon_0 = \frac{m^2}{4\sqrt{3}}\,. \tag{2.4.25}$$

References

1. A.B.Zamolodchikov. Adv. Stud. in Pure Math. 19 (1989) 641.

2. G.Mussardo. Phys. Rep. 218 (1992) 215.

3. Al.B.Zamolodchikov. Nucl. Phys. B342 (1990) 695.

[5]The factors of the form $S(\theta + i\pi/6)\,S(\theta + 5i\pi/6)$ also satisfy (2.4.8) with the first term in the r.h.s. in this equation omitted, and they agree with the real analyticity condition. More detailed analysis shows however that these factors would be incompatible with (2.4.18).

4. V.V.Bazhanov, S.L.Lukyanov, A.B.Zamolodchikov. Commun. Math. Phys. 177 (1996) 381.

5. V.V.Bazhanov, S.L.Lukyanov, A.B.Zamolodchikov. Preprint CLNS 96/1416, LPM 96/24 (1996).

6. G.Mussardo, J.Cardy. Phys. Lett. B225 (1989) 275.

7. V.Bazhanov, S.Lukyanov, A.Zamolodchikov. Preprint CLNS 96/1405, LPTENS 96/18, hep-th #9604044

8. C.Itzykson, H.Saleur, J.B.Zuber (eds.). Conformal Invariance and Applications to Statistical Mechanics, World Scientific 1988

9. R.Sassaki, I.Yamanaka. Adv. Stud. in Pure Math. 16 (1988) 271.

10. T.Eguchi, S.K.Yang. Phys. Lett. B224 (1989) 373.

11. B.Feigin, E.Frenkel. Proceeding of C.I.M.E. Summer School on "Integrable systems and Quantum groups", hep-th/9310022

12. K.G.Wilson, J.Kogut. Phys.Rep. 12C (1974) 75.

13. J.L.Cardy. Phys. Rev. Lett. 54 (1985) 1354.

14. M.E.Fisher. Phys.Rev.Letters. 40 (1978) 1610.

15. Al.B.Zamolodchikov. Nucl. Phys. B348 (1991) 619.

16. J.D.Bjorken, S.D.Drell. Relativistic Quantum Fields. McGraw Hill.

17. A.B.Zamolodchikov, Al.B.Zamolodchikov. Ann.Phys. 120 (1979) 253.

18. L.D.Faddeev, V.K.Korepin. Phys.Rep. 420 (1978) 1.

19. D.Iagolnitzer. Saclay Preprint DPh-T/77-130 (1977).

20. C.N.Yang, C.P.Yang. J. Math. Phys. 10 (1969) 1115.

21. Al.Zamolodchikov. Phys. Lett. B253 (1991) 391.

22. B.A.Kupershmidt, P.Mathieu. Phys. Lett. B227 (1989) 245.

23. V.E.Zakharov, L.D.Faddeev. Funct. Anal. 4 (1971) 18.

24. P.D.Lax. Comm. Pure Appl. Math. 21 (1968) 467.

25. R.M.Miura. Phys. Rev. Lett. 19 (1968) 1202.

26. L.D.Faddeev, L.A.Takhtajan. Hamiltonian Method in the Theory of Solitons. Springer, New York 1987

27. B.L.Feigin, D.B.Fuchs. Representations of the Virasoro algebra. Lect. Notes in Math. 1060. Springer. Berlin, Heidelberg, New York 1984.

28. A.N.Kirilov, N.Yu.Reshetikhin. J. Phys. A20 (1987) 1565.

29. V.V.Bazhanov, N.Yu.Reshetikhin. Int. J. Mod. Phys. A4 (1989) 115.

30. A.Klümper, P.A.Pearce. J. Phys., A183 (1992) 304.

31. G.Andrews, R.Baxter and J.Forrester. J. Stat. Phys. 35 (1984) 193.

32. R.J.Baxter, P.A.Pearce. J. Phys. A15 (1982) 897.

33. A.Klümper, P.A.Pearce. J. Stat. Phys. 64 (1991) 13.

List of Speakers

Alvarez Orlando Department of Physics, University of Miami, P.O. Box 248046, Coral Gables, FL 33124 USA, alvarez@phyvax.ir.miami.edu

Bernard Denis Service de physique théorique, CEN-Saclay, 91191 Gif-sur-Yvette, France, dbernard@amoco.saclay.cea.fr

Bilal Adel Laboratoire de physique théorique, École normale supérieure, 24, rue Lhomond, 75231 Paris Cedex 05, France, bilal@physique.ens.fr

David François Service de physique théorique, CEN-Saclay, 91191 Gif-sur-Yvette, France, david@amoco.saclay.cea.fr

Di Francesco Philippe Service de physique théorique, CEN-Saclay, 91191 Gif-sur-Yvette, France, philippe@amoco.saclay.cea.fr

Dijkgraaf Robbert University of Amsterdam, Faculty of Maths & Computer Science, Plantage Muidergracht 24, 1018 TV Amsterdam, The Netherlands, rhd@fwi.uva.nl

Douglas Michael Rutgers University, Department of Physics and Astronomy, P.O. Box 849, Piscataway, NJ 08855-0849 USA, mrd@physics.rutgers.edu

Eguchi Tohru Department of Physics, Faculty of Science, University of Tokyo, Bunkyo-ku, Tokyo, Japan 113, eguchi@danjuro.phys.s.u-tokyo.ac.jp

Gervais Jean-Loup Laboratoire de physique théorique, École normale supérieure, 24, rue Lhomond, 75231 Paris Cedex 05, France, gervais@physics.ens.fr

Gross David Princeton University, Joseph Henry Lab., Princeton, NJ 08544, USA, gross@puhep1.princeton.edu

Intriligator Ken Rutgers University, Department of Physics and Astronomy, P.O. Box 849, Piscataway, NJ 08855-0849 USA, keni@physics.rutgers.edu

Kostov Ivan Service de physique théorique, CEN-Saclay, 91191 Gif-sur-Yvette, France, kostov@amoco.saclay.cea.fr

Laughlin Robert Stanford University, Department of Physics, Stanford, CA 94305 USA, rbl@large.stanford.edu

Matsuo Yutaka Uji Research Institute, Yukawa Institute for Theoretical Physics, Kyoto University, Gokajo, Uji, Kyoto, Japan, yutaka@yukawa.kyoto-u.ac.jp

Migdal Alexandre Princeton University, Joseph Henry Laboratory, Princeton, NJ 08544 USA, migdal@puhep1.princeton.edu

Mukhi Sunil Theoretical Physics Group, Tata Institute of Fundamental Research, Homi Bhabha Road, Bombay 400 005, India, mukhi@theory.tifr.res.in

Mussardo Giuseppe SISSA, Strada Costiera 12, 34014 Trieste, Italy, mussardo@tsmi19.sissa.it

Polyakov Alexander Princeton University, Joseph Henry Laboratory, Princeton, NJ 08544 USA, polyakov@puhep1.princeton.edu

Rabinovici Eliezer Racah Institute of Physics, The Hebrew University, Jerusalem, 94904, Israel, eliezer@vms.huji.ac.il

Saleur Hubert Department of Physics & Astronomy, University of Southern California, Los Angeles, CA 90089-0484 USA, saleur@diderot.usc.edu

Staudacher Matthias Laboratoire de physique théorique, École normale supérieure, 24, rue Lhomond, 75231 Paris Cedex 05, France, matthias@physique.ens.fr

Verlinde Erik CERN, Theory Division, 1211 Geneva 23, Switzerland, verlinde@nxth04.cern.ch

Verlinde Herman University of Amsterdam, Inst. for Theoretical Physics, Valckenierstraat 65, 1018 XE Amsterdam, The Netherlands, verlinde@phys.uva.nl

Wiegmann Paul James Franck Institute, The University of Chicago, 5640 South Ellis avenue, Chicago, IL 60637-1433 USA, wiegmann@rainbow.uchicago.edu

Zamolodchikov Alexander Rutgers University, Department of Physics and Astronomy, P.O. Box 849, Piscataway, NJ 08855-0849 USA, sashaz@physics.rutgers.edu

Zamolodchikov Alexei Laboratoire de physique mathématique, Université de Montpellier II, Place E. Bataillon, 34095 Montpellier Cedex 05, France, zamolod@lpm.univ-montp2.fr

Zuber Jean-Bernard Service de physique théorique, CEN-Saclay, 91191 Gif-sur-Yvette, France, zuber@amoco.saclay.cea.fr

List of Participants

Ahn Changrim Dept. of Physics, EWHA Womans Univ., 120-750 Seoul, Korea, ahn@ewhahp3.ewha.ac.kr

Antonov Alexander Landau Institute For Theoretical, GSP-1, 117940 Moscow, V-334 Russia, antonov@itp.ac.ru

Attal Romain LPTHE - Université Pierre et Marie Curie, Boîte 126, 4 place Jussieu, 75252 Paris Cedex 05, France, attal@lpthe.jussieu.fr

Barbarin François ENSLAPP, Chemin de Bellevue, BP 110, 74941 Annecy le Vieux, France, barbarin@lapp.in2p3.fr

Biet Franck Laboratoire de physique théorique, 19 rue du Solarium, 33175 Gradignan Cedex, France, biet@frcpn11.in2p3.fr

Broda Boguslaw Department of Theoretical Physics, Univiversity of Lodz, Pomorska 149/153, 90236 Lodz, Poland, bobroda@mvii.uni.lodz.pl

Buffenoir Eric Ecole polytechnique, CPT, 91128 Palaiseau, France, buffenoir@orphee.polytechnique.fr

Comellas Jordi Facultat de Fisica, Dept. d'Estructura i Constituents de la Matèria, Diagonal 647, 08028 Barcelone, Spain, comellas@sophia.ecm.ub.es

Coste Antoine Institut Henri Poincaré, Centre Emile Borel, 11 rue Pierre et Marie Curie, 75005 Paris, France, coste@ihp.jussieu.fr

De Mirleau Olivier University of Amsterdam, Faculty of Maths & Computer Science, Plantage Muidergracht 24, 1018 TV Amsterdam, The Netherlands, mirleau@fwi.uva.nl

Dick Rainer Institute for Advanced Study, School of Natural Science, Princeton, NJ 08540 USA, dick@guinness.ias.edu

Egger Reinhold Fakultät Physik, Univ. Freiburg, Hermann-Herder Str. 3, 79104 Freiburg, Germany, egger@tfp.physics.uni-freiburg.de

Fabre Cyrille Ecole polytechnique, CPT, 91128 Palaiseau, France, fabre@orphee.polytechnique.fr

Fring Andreas Freie Universität Berlin, Fachbereich Physik, Arnimalle 14, 14195 Berlin, Germany, fring@omega.physik.fu-berlin.de

Furtlehner Cyril I.P.N., Division de physique théorique, 91406 Orsay, France, furtlehner@ipncls.in2p3.fr

Gallot Laurent ENSLAPP, 46, allée d'Italie, 69007 Lyon, France, lgallot@enslapp.ens-lyon.fr

Gandenberger Georg Cambridge University, DAMTP, Cambridge CB3 9EW, UK, gmg1002@damtp.cam.ac.uk

Garajeu Daniela CNRS Luminy, Centre de physique théorique, 13288 Marseille Cedex 9, France, garajeu@cpt.univ-mrs.fr

Giusto Stefano Dipartimento di Fisica, Via Dodecaneso, 16146 Genova, Italy, giusto@vaxgea.ge.infn.it

Govindarajan Suresh Tata Institute of Fundamental Research, Theoretical Physics Group, 400 005 Bombay, India, suresh@theory.tifr.res.in

Greiter Martin CERN, Theory Division, 1211 Geneve 23, Switzerland, greiter@surya11.cern.ch

Gurarie Victor Physics Department, Princeton University, Jadwin Hall, Princeton, NJ 08544-1000 USA, gurarie@puhep1.princeton.edu

Gutperle Michael CERN, Theory Division, 1211 Geneve 23, Switzerland, gutperle@nxth04.cern.ch

Hacquebord Feike University of Amsterdam, Institute for Theoretical Physics, Vackenterstraat 65, 1018 XE, Amsterdam, The Netherlands, hacquebo@phys.uva.nl

Hauer Tamás Research Institute for Particle and Nuclear Physics, P.O. Box 49, 1525 Budapest 114, Hungary, hauer@ludens.elte.hu

Horava Petr Princeton University, Joseph Henry Laboratory, Princeton, NJ 08544-1000, USA, horava@puhep1.princeton.edu

Houart Laurent LPTHE - Université Pierre et Marie Curie, Boîte 126, 4 place Jussieu, 75252 Paris Cedex 05, France, lhouart@lpthe.jussieu.fr

Kadeishvili Alexey Landau Institute for Theoretical Physics, GSP-1, 117940 Moscow, V-334 Russia, kadeishv@itp.ac.ru

Kedem Rinat Research Institute for Mathematical Sciences, Kyoto University, Kyoto 606, Japan, rinat@kurims.kyoto-u.ac.jp

Kristjansen Charlotte Service de physique théorique, CEN-Saclay, 91191 Gif-sur-Yvette, France, kristja@amoco.saclay.cea.fr

Lashkevich Michael Landau Institute for Theoretical Physics, GSP-1, 117940 Moscow, V-334 Russia, lashkevi@landau.ac.ru

Lozano Yolanda Princeton University, Joseph Henry Laboratory, Princeton, NJ 08544-1000, USA, yolanda@puhep1.princeton.edu

Maassarani Ziad Service de physique théorique, CEN-Saclay, 91191 Gif-sur-Yvette, France, ziad@amoco.saclay.cea.fr

Mohammedi Noureddine Southampton University, Department of Physics, Highfield Southampton, SO17 1BJ, UK, nouri@hep.phys.sotom.ac.uk

Nagura Masaru University of Tokyo, Department of Physics, Bunkyo-ku, Tokyo 113, Japan, nagura@danjuro.phys.s.u-tokyo.ac.jp

Nassar Tarek Laboratoire de physique mathématique, Université de Montpellier II, Place E. Bataillon, 34095 Montpellier Cedex 05, France, nassar@lpm.univ-montp2.fr

Polyubin Igor Landau Institute For Theoretical Physics, GSP-1, 117940 Moscow, V-334 Russia, polyub@itp.ac.ru

Pujol Pierre LPTHE - Université Pierre et Marie Curie, Boîte 126, 4 place Jussieu, 75252 Paris Cedex 05, France, pujol@lpthe.jussieu.fr

Schaden Martin Physics Department, New York University, 4 Washington Place, New York, NY 10003 USA, schaden@mafalda.physics.nyu.edu

Schmidhuber Christof Physics Department, Princeton University, Jadwin Hall, Princeton, NJ 08544-1000 USA, schmidhu@puhep1.princeton.edu

Skorik Sergei Physics Department, University of Southern California, Los Angeles, CA 90089-0484 USA, sergei@physics.usc.edu

Talon Michel LPTHE - Université Pierre et Marie Curie, Boîte 126, 4 place Jussieu, 75252 Paris Cedex 05, France, talon@lpthe.jussieu.fr

Toppan Francesco Università di Padova, Dipt. di Fisica, Via Marzolo 8, 35131 Padova, Italy, toppan@padova.infn.it

Vergnano Luca Università di Firenze, Dipt. di Fisica, Largo E. Fermi 2, 50125 Florence, Italy, vergnano@fi.infn.it

Villain-Guillot Simon Université de Cergy Pontoise, Groupe de physique statistique, BP 8428, 95806 Cergy-Pontoise, France, villain@paris.u-cergy.fr

Vinet Luc Centre de recherches mathématiques, Université de Montréal, CP 6128, Montréal, QC H3C3J7, Canada, vinet@crm.umontreal.ca

Vinnakota Sreedhar Department of Theoretical Physics, Uppsala Univ., Box 803, Uppsala 75108, Sweden, sreedhar@rhea.teorfys.uu.se

Watabiki Yoshiyuki Niels Bohr Institute, Blegdamsvej 17, 2100 Copenhagen, Denmark, watabiki@nbivax.nbi.dk

Watts Paul CNRS Luminy, Centre de physique théorique, 13288 Marseille Cedex 9, France, watts@cpt.univ-mrs.fr

Wegrzyn Pawel Department of Field Theory, Institute of Physics, Reymonta 4, Jagellonian University, 30-009 Cracow, Poland, wegrzyn@ztc386a.if.uj.edu.pl

Wexler Mark LPTHE - Université Pierre et Marie Curie, Boîte 126, 4 place Jussieu, 75252 Paris Cedex 05, France, wexler@lpthe.jussieu.fr

Wiese Kay Service de physique théorique, CEN-Saclay, 91191 Gif-sur-Yvette, France, wiese@amoco.saclay.cea.fr

Wynter Tom Laboratoire de physique théorique, École normale supérieure, 24, rue Lhomond, 75231 Paris Cedex 05, France, wynter@physique.ens.fr

Zinn-Justin Paul Laboratoire de physique théorique, École normale supérieure, 24, rue Lhomond, 75231 Paris Cedex 05, France, pzinn@clipper.ens.fr

INDEX